Lecture Notes in Business Information Processing 436

More information about this series at https://link.springer.com/bookseries/7911

Andrea Marrella · Barbara Weber (Eds.)

Business Process Management Workshops

BPM 2021 International Workshops
Rome, Italy, September 6–10, 2021
Revised Selected Papers

Springer

Editors
Andrea Marrella
Sapienza University of Rome
Rome, Italy

Barbara Weber
University of St. Gallen
St. Gallen, Switzerland

ISSN 1865-1348 ISSN 1865-1356 (electronic)
Lecture Notes in Business Information Processing
ISBN 978-3-030-94342-4 ISBN 978-3-030-94343-1 (eBook)
https://doi.org/10.1007/978-3-030-94343-1

This Springer imprint is published by the registered company Springer Nature Switzerland AG
The registered company address is: Gewerbestrasse 11, 6330 Cham, Switzerland

Preface

The International Conference on Business Process Management (BPM) was established about 19 years ago as the conference where people from academia and industry meet and discuss the latest developments in the area of business process management. In 2021, the conference was organized in Rome, Italy. This year's BPM also featured eight different workshops, selected after a thorough evaluation phase, which provided a forum for novel research ideas.

Workshops were held on September 6, 2021, right before the main conference. While the main conference was meant to present finished research, the workshops were meant to discuss research that is still in progress. Each of the workshops focused on particular aspects of business process management. These proceedings present the work that was discussed during the workshops.

In total, the BPM 2021 workshops were able to attract 60 submissions. 31 full papers were accepted for publication, leading to an overall acceptance rate of 51.6%. The resulting workshop programs were complemented by many keynote talks from renowned experts. Two of them have been included in these proceedings as invited papers. BPM 2021 featured the following workshops:

- 5th International Workshop on Artificial Intelligence for Business Process Management (AI4BPM 2021)
- 17th International Workshop on Business Process Intelligence (BPI 2021)
- 9th International Workshop on DEClarative, DECision and Hybrid approaches to processes (DEC2H 2021)
- 14th International Workshop on Social and Human Aspects of Business Process Management (BPMS2 2021)
- 1st International Workshop on BPM Governance for and Beyond Digital Transformation (BPMGOV 2021)
- The Fourth Workshop on Security and Privacy-enhanced Business Process Management (SPBP 2021)
- 5th International Workshop on Business Processes Meet the Internet-of-Things (BP-Meet-IoT 2021)
- 1st International Workshop on Business Process Management and RoutineDynamics (BPM&RD 2021)

As in previous years, the BPM workshop proceedings are post-proceedings. We are confident that this process of selection, presentation, and revision has led to a collection of high-quality papers. We hope that the reader will enjoy this volume and find some inspiration for future work.

We would like to thank the workshop organizers, the reviewers, the authors, and, of course, all the people from the BPM community that helped make the BPM 2021 workshops a success. We particularly thank the general chair, Massimo Mecella, for organizing such an outstanding conference, despite the challenges that came with the COVID-19 pandemic.

November 2021

Andrea Marrella
Barbara Weber

Organization

Workshop Chairs

Andrea Marrella	Sapienza University of Rome, Italy
Barbara Weber	University of St. Gallen, Switzerland

Program Committee

Rafael Accorsi	Accenture Applied Intelligence, Switzerland
Lars Ackermann	University of Bayreuth, Germany
Annalisa Appice	University of Bari Aldo Moro, Italy
Aleksandre Asatiani	University of Gothenburg, Sweden
Adriano Augusto	University of Melbourne, Australia
Ahmed Awad	University of Tartu, Estonia
Saimir Bala	WU Vienna, Austria
Matteo Baldoni	University of Turin, Italy
Markus Becker	University of Southern Denmark, Denmark
Abdelmalek Benzekri	University Paul Sabatier Toulouse III, France
Nicholas Berente	University of Notre Dame, USA
Ralph Bergmann	University of Trier, Germany
Agnieszka Bitkowska	Warsaw University of Technology, Poland
Maria Booth	Stockholm School of Economics, Sweden
Jan Bosch	Chalmers University of Technology, Sweden
Marco Brambilla	Polytechnic University of Milan, Italy
Lars Brehm	Munich University of Applied Science, Germany
Achim D. Brucker	University of Exeter, UK
Edyta Brzychczy	AGH University of Science and Technology, Poland
Andrea Burattin	Technical University of Denmark, Denmark
Alessio Cecconi	Vienna University of Economics and Business, Austria
Sylvain Cherrier	University of Marne-la-Vallée, France
Federico Chesani	University of Bologna, Italy
Marco Comuzzi	Ulsan National Institute of Science and Technology, South Korea
Francesco Corcoglioniti	Free University of Bozen-Bolzano, Italy
Carl Corea	University of Koblenz-Landau, Germany
João Costa Seco	NOVA University Lisbon, Portugal
Massimiliano de Leoni	University of Padua, Italy
Riccardo De Masellis	Uppsala University, Sweden
Silvia Inês Dallavalle de Pádua	University of Sao Paulo, Brazil
Johannes De Smedt	KU Leuven, Belgium
Benoît Depaire	Hasselt University, Belgium

Claudio Di Ciccio	Sapienza University of Rome, Italy
Chiara Di Francescomarino	Bruno Kessler Center in Information and Communication Technology, Italy
Rik Eshuis	Eindhoven University of Technology, The Netherlands
Joerg Evermann	Memorial University of Newfoundland, Canada
Stephan Fahrenkrog-Petersen	Humboldt University of Berlin, Germany
Peter Fettke	German Research Center for Artificial Intelligence, Germany
Francesco Folino	National Research Council of Italy, Italy
Vladislav Fomin	Vilnius University, Lithuania
Virginia Franqueira	University of Kent, UK
Daniela Garbin Praničević	University of Split, Croatia
Luciano García-Bañuelos	Monterrey Institute of Technology and Higher Education, Mexico
Robert Golan	DBmind Technologies, USA
Gianluigi Greco	University of Calabria, Italy
Norbert Gronau	University of Potsdam, Germany
Xunhua Guo	Tsinghua University, China
Holger Günzel	Munich University of Applied Sciences, Germany
Emna Hachicha Belghith	University of Caen Normandy, France
Michael Henke	TU Dortmund, Germany
Tomislav Hernaus	University of Zagreb, Croatia
Julius Holderer	University of Freiburg, Germany
Rick Hull	New York University, USA
Mubashar Iqbal	University of Tartu, Estonia
Amin Jalali	Stockholm University, Sweden
Christian Janiesch	TU Dortmund, Germany
Gert Janssenswillen	Hasselt University, Belgium
Jan Jürjens	Fraunhofer Institute for Software and Systems Engineering ISST and University of Koblenz-Landau, Germany
Anna Kalenkova	University of Melbourne, Australia
Christos Kalloniatis	University of the Aegean, Greece
Udo Kannengiesser	Johannes Kepler University Linz, Austria
Dimitris Karagiannis	University of Vienna, Austria
Dimka Karastoyanova	University of Groningen, The Netherlands
Barbara Keller	Munich University of Applied Sciences, Germany
Kathrin Kirchner	Technical University of Denmark, Denmark
Ralf Klamma	RWTH Aachen University, Germany
Krzysztof Kluza	AGH University of Science and Technology, Poland
Stephan Kuehnel	Martin Luther University Halle-Wittenberg, Germany
Sander J.J. Leemans	Queensland University of Technology, Australia
Henrik Leopold	Kühne Logistics University, Germany
Francesco Leotta	Sapienza University of Rome, Italy
Michael Leyer	University of Rostock, Germany

Milanović-Glavan Ljubica	University of Zagreb, Croatia
Zakaria Maamar	Zayed University, United Arab Emirates
Fabrizio Maria Maggi	Free University of Bozen-Bolzano, Italy
Christian Mahringer	University of Stuttgart, Germany
Lukas Malina	Brno University of Technology, Czech Republic
Felix Mannhardt	Eindhoven University of Technology, The Netherlands
Ivan Matić	University of Split, Croatia
Martin Matzner	Friedrich-Alexander-Universität Erlangen-Nürnberg, Germany
Paola Mello	University of Bologna, Italy
Giovanni Meroni	Polytechnic University of Milan, Italy
Roberto Micalizio	University of Turin, Italy
Hamid Motahari	IBM Almaden Research Center, USA
Jorge Munoz-Gama	Pontifical Catholic University of Chile, Chile
Michael Möhring	Munich University of Applied Sciences, Germany
Andrzej Niesler	Wroclaw University of Economics and Business, Poland
Alex Norta	Tallinn University of Technology, Estonia
Selmin Nurcan	University of Paris 1 Panthéon-Sorbonne, France
Andreas Oberweis	Karlsruhe Institute of Technology, Germany
Oscar Pastor Lopez	Polytechnic University of Valencia, Spain
Constantinos Patsakis	University of Piraeus, Greece
Jens Myrup Pedersen	Aalborg University, Denmark
Marco Pegoraro	RWTH Aachen University, Germany
Vicente Pelechano	Polytechnic University of Valencia, Spain
Brian Pentland	Michigan State University, USA
Günther Pernul	University of Regensburg, Germany
Gregor Polančič	University of Maribor, Slovenia
Artem Polyvyanyy	University of Melbourne, Australia
Luigi Pontieri	National Research Council of Italy, Italy
Natalia Potoczek	Polish Academy of Sciences, Poland
Gerald Quirchmayr	University of Vienna, Austria
Mohammad Ehson Rangiha	City, University of London, UK
Jan Recker	University of Hamburg, Germany
Manfred Reichert	University of Ulm, Germany
Andrey Rivkin	Free University of Bozen-Bolzano, Italy
Williams Rizzi	Bruno Kessler Center in Information and Communication Technology, Italy
Christoph Rosenkranz	University of Cologne, Germany
Gustavo Rossi	Universidad Nacional de La Plata, Argentina
Uwe Roth	Luxembourg Institute of Science and Technology, Luxembourg
Flavia Santoro	Rio de Janeiro State University, Brazil
Thomas Schaberreiter	CS-Aware Corporation, Austria
Stefan Schauer	AIT Austrian Institute of Technology GmbH, Austria
Rainer Schmidt	Munich University of Applied Sciences, Germany

Stefan Schönig	University of Regensburg, Germany
Stefan Seidel	University of Liechtenstein, Liechtenstein
Miguel-Angel Sicilia	University of Alcala, Spain
Tijs Slaats	University of Copenhagen, Denmark
Pnina Soffer	University of Haifa, Israel
Dragana Stojanović	University of Belgrade, Serbia
Mark Strembeck	Vienna University of Economics and Business, Austria
Heiner Stuckenschmidt	University of Mannheim, Germany
Jianwen Su	University of California, Santa Barbara, USA
Dalia Suša Vugec	University of Zagreb, Croatia
Niek Tax	Booking.com, The Netherlands
Irene Teinemaa	Booking.com, The Netherlands
Johannes Tenschert	Friedrich-Alexander-Universität Erlangen-Nürnberg, Germany
Daniele Theseider Dupré	University of Eastern Piedmont, Italy
Lucinéia Heloisa Thom	Federal University of Rio Grande do Sul, Brazil
Martina Tomicic Furjan	University of Zagreb, Croatia
Inge van de Weerd	Utrecht University, The Netherlands
Han van der Aa	University of Mannheim, Germany
Will van der Aalst	RWTH Aachen University, Germany
Boudewijn Van Dongen	Eindhoven University of Technology, The Netherlands
Sebastiaan J. van Zelst	Fraunhofer Institute for Applied Information Technology and RWTH Aachen University, Germany
Seppe Vanden Broucke	KU Leuven, Belgium
Irene Vanderfeesten	Open University of the Netherlands, The Netherlands
Eric Verbeek	Eindhoven University of Technology, The Netherlands
Hagen Voelzer	IBM Zurich Research Lab, Switzerland
Matthias Weidlich	Humboldt University of Berlin, Germany
Hans Weigand	Tilburg University, The Netherlands
Mathias Weske	University of Potsdam, Germany
Moe Wynn	Queensland University of Technology, Australia
Xiwei Xu	Commonwealth Scientific and Industrial Research Organisation, Australia
Youngjin Yoo	Case Western Reserve University, USA
Alfred Zimmermann	Reutlingen University, Germany

Contents

14th International Workshop on Social and Human Aspects of Business Process Management (BPMS2 2021)

9th International Workshop on DEClarative, DECision and Hybrid Approaches to Processes (DEC2H 2021)

5th International Workshop on Business Processes Meet the Internet-of-Things (BP-Meet-IoT 2021)

The Fourth Workshop on Security and Privacy-enhanced Business Process Management (SPBP 2021)

5th International Workshop on Artificial Intelligence for Business Process Management (AI4BPM 2021)

5th International Workshop on Artificial Intelligence for Business Process Management (AI4BPM 2021)

As the field of artificial intelligence (AI) continues to grow, novel methodologies and techniques continue to emerge and are being applied across numerous areas. In the past few years, there has been a strong interest from both industry and academia in applying AI techniques in the area of business process management (BPM). Indeed, the application of AI is impacting additional areas where process management perspectives become relevant, including industrial engineering, IoT, and healthcare. The use of AI in BPM has been discussed as the next disruptive technology that will touch almost all business process activities performed by humans. In some cases, AI will dramatically simplify human interaction with processes, while in other cases it will enable full automation of tasks that have traditionally required manual contributions. We believe that over time, AI may lead to entirely new paradigms for business process management in all of its aspects: modeling, analysis, automation, and monitoring. For example, instead of BPM models centered either on process or on case management, we anticipate models that are based fundamentally on goal achievement. Moreover, these models will fully enable continuous improvement and adaptation based on experiential learning with little to no human intervention after the learning phase has been completed.

The goal of this workshop is to establish a forum for researchers and professionals interested in understanding, envisioning, and discussing the challenges and opportunities of moving from current, largely programmatic approaches for BPM to emerging forms of AI-enabled BPM. Given the recent adoption in the BPM context of explainable AI techniques, this year, a keynote speech by Fettke and Mehdiyev on "Explainable AI for Predictive Process Monitoring" enriched the program of the workshop. In addition, the workshop attracted eight international submissions on different topics including recommendations for business process modeling, predictive process monitoring, and process monitoring through generative adversarial networks (GANs). All submissions were reviewed by at least three Program Committee members or their sub-reviewers and eventually four papers were accepted. We believe that the accepted papers provide a novel mix of conceptual and technical contributions that are of interest for the AI4BPM community.

Sola et al. propose an approach for the provision of activity recommendations on the way models under development can be expanded by leveraging knowledge graphs. In their work, they investigate different approaches to apply embedding-based and rule-based knowledge graph completion methods to support this task and evaluate these methods in an experimental study. Weytjens and De Weerdt identify and demonstrate the importance of several issues affecting the pre-processing of event logs to be used for predictive process monitoring. They propose a number of standard pre-processing steps to derive unbiased benchmark datasets in a principled way, thus creating representative test sets without data leakage. In his work, Weinzerl investigates the usage of graph neural networks (GNNs) - and in particular of gated graph sequence neural networks - for the next activity prediction task. GNNs can directly compute graph-oriented data inputs, generally structured into nodes and edges, thus allowing for the explicit modeling of event relationships. Lastly, Hoffman et al. present the ProGAN framework for implementing three (Monitor, Analyze and Plan) of the four phases of the MAPE-K control loop. ProGAN leverages a deep learning architecture that builds upon generative adversarial networks. The discriminator is used for monitoring the process in its environment by using sensor data and for detecting deviations with respect to the desired process state, while the generator is used afterwards for analyzing the detected deviations and its symptoms. These contributions demonstrate the latest advances in the AI4BPM research area.

October 2021

Chiara Di Francescomarino
Fabrizio Maria Maggi
Arik Senderovich
Emilio Sulis

Organization

Organizing Committee

Chiara Di Francescomarino	Fondazione Bruno Kessler, Italy
Fabrizio Maria Maggi	Free University of Bozen-Bolzano, Italy
Arik Senderovich	University of Toronto, Canada
Emilio Sulis	University of Turin, Italy

Program Committee

Han van der Aa	University of Mannheim, Germany
Annalisa Appice	University of Bari, Italy
Matteo Baldoni	University of Turin, Italy
Ralph Bergmann	University of Trier, Germany
Federico Chesani	University of Bologna, Italy
Marco Comuzzi	Ulsan Institute, South Korea
Francesco Corcoglioniti	Free University of Bozen-Bolzano, Italy
Riccardo De Masellis	Uppsala University, Sweden
Claudio Di Ciccio	Sapienza University of Rome, Italy
Joerg Evermann	Memorial University of Newfoundland, Canada
Stephan Fahrenkrog-Petersen	Humboldt University of Berlin, Germany
Francesco Folino	CNR, Italy
Emna Hachicha Belghith	University of Namur, Belgium
Rick Hull	New York University, USA
Krzysztof Kluza	AGH University of Science and Technology, Poland
Henrik Leopold	Kühne Logistics University, Germany
Francesco Leotta	Sapienza University of Rome, Italy
Paola Mello	University of Bologna, Italy
Roberto Micalizio	University of Turin, Italy
Hamid R. Motahari Nezhad	IBM Almaden Research Center, USA
Artem Polyvyanyy	University of Melbourne, Australia
Luigi Pontieri	CNR, Italy
Andrey Rivkin	Free University of Bozen-Bolzano, Italy
Williams Rizzi	Fondazione Bruno Kessler, Italy
Tijs Slaats	University of Copenhagen, Denmark
Heiner Stuckenschmidt	University of Mannheim, Germany
Niek Tax	Booking.com, The Netherlands
Irene Teinemaa	Booking.com, The Netherlands
Daniele Theseider Dupré	University of Eastern Piedmont, Italy
Hagen Voelzer	IBM Zurich Research Lab, Switzerland
Matthias Weidlich	Humboldt University of Berlin, Germany

On the Use of Knowledge Graph Completion Methods for Activity Recommendation in Business Process Modeling

Diana Sola[1,2(✉)], Christian Meilicke[2], Han van der Aa[2], and Heiner Stuckenschmidt[2]

[1] Intelligent Robotic Process Automation, SAP SE, Walldorf, Germany
[2] Data and Web Science Group, University of Mannheim, Mannheim, Germany
{diana,christian,han,heiner}@informatik.uni-mannheim.de

Abstract. Business process modeling is essential for organisations. However, it is a time-consuming task that requires expert knowledge. In particular, this is the case when modeling domain-specific processes, which often involves the consistent use of technical terminology. Process modelers can be supported through the provision of recommendations on how the model under development can be expanded. Activity recommendation is one such support approach, in which suitable activities to be inserted at a user-defined position are recommended. Recently, it has been suggested to treat activity recommendation as a knowledge graph completion task and to apply methods from this discipline. In this paper, we investigate different approaches to apply embedding- and rule-based knowledge graph completion methods out of the box and evaluate them in an experimental study. Additionally, we compare them to two methods that have specifically been designed for activity recommendation.

Keywords: Activity recommendation · Knowledge graph completion · Rule learning · Embeddings

1 Introduction

Business processes are an integral part of every organisation. They refer to sets of activities, which are performed to achieve an outcome that is of interest to an organization or its customers. Capturing information on such processes, *process models* are present in all phases of the business process management lifecycle [5]. Despite the relevance of process models for the documentation, analysis and improvement of business processes, creating them is a time-consuming and error-prone task that requires substantial expertise [7,8]. Modeling business processes is even more challenging in the case of domain-specific processes, which often require the consistent use of specialized and technical vocabulary.

It is possible to support process modelers in their modeling task by providing recommendations on how the process model that they are developing can be

A. Marrella and B. Weber (Eds.): BPM 2021 Workshops, LNBIP 436, pp. 5–17, 2022.
https://doi.org/10.1007/978-3-030-94343-1_1

expanded [6]. One possibility for such a support is *activity recommendation*. An activity recommendation system suggests suitable activities to extend a given business process model under development at a user-defined position. A repository of already available process models can be used as a basis for this task. Activity recommendation systems should be context-aware, i.e., they should take the current content and state of a process model into account. Figure 1 shows context-aware recommendations for a process model under development. The user has just inserted an unlabeled activity at the model's bottom. The activity-recommendation task is to suggest suitable activities for this position, i.e., to find an appropriate label for the so far unlabeled activity node. Since the process that has been modeled up to this point contains activities that are commonly associated with *order-to-cash* processes, a context-aware recommender system provides recommendations that correspond to activities that occur at a comparable position in similar processes (found in a given model repository), such as *Create sales order*, *Analyze purchase order*, and *Update sales order*.

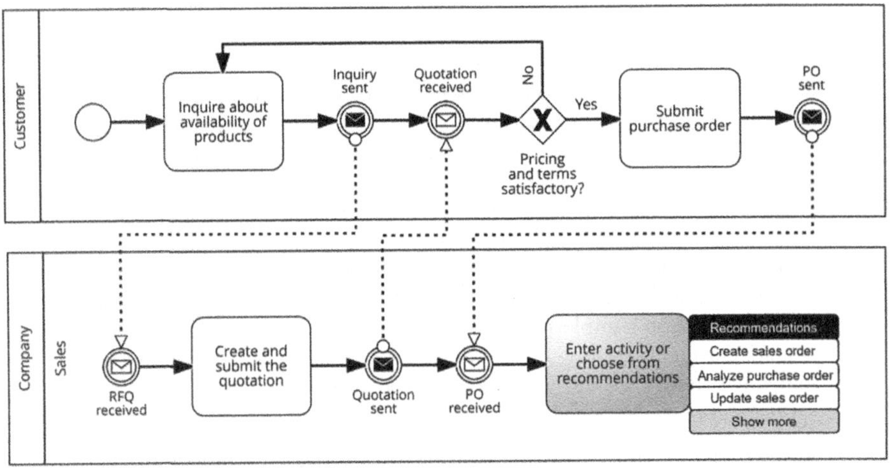

Fig. 1. A business process model under development

The activity-recommendation problem can be framed in terms of a *knowledge graph completion* task [21]. A knowledge graph captures data on a specific domain in the form of entities and their inter-relations. In our setting, we propose to turn a process model repository and a process model under development into a large knowledge graph, which thus captures knowledge on the relations between activities in the models, e.g., activities that commonly follow each other. Then, the recommendation of an appropriate activity is interpreted as a completion task in the graph, to which so-called knowledge graph completion approaches can be applied. This completion task has attracted lots of attention in the last decade with the majority of the approaches being based on the idea to embed the knowledge graph into a low dimensional space [22]. More recently, rule-based approaches have shown to be competitive alternatives [12].

In this paper, we thus investigate how methods that have originally been developed for knowledge graph completion can be applied for activity recommendation in business process modeling and compare the performance of different approaches in an experimental study. In an error analysis, we found some weaknesses of the methods that can be fixed through problem-specific post-processing.

While our main focus lies on the application of knowledge graph completion techniques to the field of activity recommendation for business process modeling, the results are also interesting from a general point of view. Most papers that propose knowledge graph completion techniques evaluate these methods on the same evaluation datasets that have been used in the community for many years. Within this paper, we instead apply existing methods to a different dataset and an evaluation scenario that reflects a real-world downstream task and investigate if the methods are flexible enough to adapt to our task.

2 Related Work

Besides activity recommendation approaches that make use of graph mining techniques [3,4,11], our work primarily relates to embedding- and rule-based methods for activity recommendation.

Wang et al. developed the embedding-based recommendation approach RLRecommender [21], which embeds activities and relations between them into a continuous low-dimensional space. However, as we already proved in prior work [19], the performance of RLRecommender is comparably low, since the recommendations for an unlabeled activity are based on one related activity in the process model only. Therefore, the method generates different recommendations depending on the chosen related activity that is used to determine the recommendations for the unlabeled activity. In other words, RLRecommender lacks of an aggregation method which combines all possible recommendations given the process model under development in one recommendation list.

In prior work [19], we presented our rule-based activity recommendation approach. First, the method learns rules that describe regularities in the use of labels in the given process repository. For this purpose, we have defined various rule types that capture different co-occurrence and structural patterns. Then, it applies the learned rules to the model under development making use of the full given context. In an extensive experimental study the rule-based method outperformed a variety of other approaches [9,10,21].

Beyond the scope of activity recommendation, our work relates to knowledge graph embedding (KGE) models and rule-based systems that can be used for knowledge graph completion. Two prominent KGE models are TransE [1] and DistMult [23]. While TransE models relations between entities in the knowledge graph as translations operating on the embeddings of the entities, DistMult uses a simple bilinear interaction between entities and relations. An alternative to the KGE models is given by AnyBURL [12], which is a rule learner designed for knowledge graph completion and has been shown to perform on the same level as current state of the art KGE models [15].

3 Activity Recommendation Through Knowledge Graph Completion

In this section, we formally define the activity-recommendation problem and discuss different approaches for applying knowledge graph completion methods.

3.1 Problem Definition

Our work is independent from a particular modeling notation. Therefore, we define the activity-recommendation problem based on the notion of business process graphs. Such a graph represents an abstracted version of a business process model in a particular modeling notation, such as a Petri net or BPMN model. Specifically, we employ the following definition:

Definition 1 (Business process graph). *Let $\mathcal{L}$ be a set of labels. A business process graph is a tuple (N, E, λ), where N is a set of nodes, $E \subseteq N \times N$ is a set of directed edges and $\lambda : N \rightarrow \mathcal{L}$ is a function that maps a node to a label.*

Given a process model, the set of nodes N of a business process graph corresponds to a subset of the nodes of the model, e.g., the (non-silent) transitions of a Petri net, whereas the set of edges E reflects the directly-follows relation between these nodes, defined by the model's execution semantics. The mapping function λ follows straightforwardly from the process model as well.

Given a process model under development, the activity-recommendation problem is concerned with recommending suitable activities to extend the model at a user-defined position. The position of the activity that has to be recommended is given by the activity that was last added to the model. Therefore, the activity-recommendation problem breaks down to finding a suitable label for the last added, and so far unlabeled, activity node $\tilde{n}$.

Definition 2 (Activity-recommendation problem). *Let $\mathcal{B}$ be a set of business process graphs and $\mathcal{L}_{\mathcal{B}}$ the set of activity labels that are used in $\mathcal{B}$. Let $B = (N, E, \lambda)$ be a given business process graph under development, where each node $n \in N$ except one node $\tilde{n}$ is labeled, i.e., $\lambda(n)$ is given for all $n \in N \backslash \{\tilde{n}\}$. The activity-recommendation problem is to find a suitable label $\lambda(\tilde{n}) \in \mathcal{L}_{\mathcal{B}}$ for $\tilde{n}$.*

In the next section, we discuss different approaches to construct a knowledge graph from given process models such that we can apply knowledge graph embedding models and rule-based methods.

3.2 Knowledge Graph Construction

A knowledge graph is a collection of triples *(head entity, relation, tail entity)*. While the entities are represented as nodes in the graph, the relation between two entities is given as a labeled edge. To apply methods that have originally been developed for knowledge graph completion to the activity-recommendation problem, we need to describe each business process graph $B \in \mathcal{B}$ of the given repository and the process model under development in terms of such triples. The partial knowledge graph resulting from a graph $B = (N, E, \lambda)$ can, for example, comprise the nodes N and the activity labels $\mathcal{L}$ as entities. We developed three different approaches to translate a business process graph into a set of triples, which vary in the used sets of entities and relations.

The simplest approach uses the relations *hasLabel* and *followedBy*. These two relations are the basis for all approaches, as they capture the core information of a business process graph as knowledge graph, i.e., a triple $(n, \textit{hasLabel}, \lambda(n))$ expresses that a node n has the label $\lambda(n)$, whereas an edge (m, n) becomes the triple $(m, \textit{followedBy}, n)$. The other two approaches are alternative extensions which, in addition to the structural patterns captured by *followedBy*, consider co-occurrence patterns by using the relations *inSameProcess* and *inProcess*, respectively. This results in the following three translation approaches:

1. For each node $n \in N$ we add a triple $(n, \textit{hasLabel}, \lambda(n))$. Furthermore, we add for each edge $(m, n) \in E$ a triple $(m, \textit{followedBy}, n)$.
2a. This approach extends the first one by additionally using the relation *inSameProcess*: In addition to the first approach, we add for each pair of nodes $m \neq n \in N$ the triples $(m, \textit{inSameProcess}, n)$ and $(n, \textit{inSameProcess}, m)$ to link all nodes that belong to the same process.
2b. This approach is another extension of the first one and an alternative to 2a, where the additional relation is given by *inProcess*: Let $\mathcal{P}$ be a set of process identifiers and π be a function that maps each given business process graph B to its unique identifier $\pi(B) \in \mathcal{P}$. In addition to the triples of the first approach, we add for each node $n \in N$ a triple $(n, \textit{inProcess}, \pi(B))$.

The different translation approaches are closely related. In particular, in 2b the two triples $\textit{inProcess}(m, p)$ and $\textit{inProcess}(n, p)$ imply $\textit{inSameProcess}(m, n)$ in 2a and vice versa. However, there is a slight difference. In 2a two activities m and n are not in the same process, if the triple $\textit{inSameProcess}(m, n)$ does not exist. In 2b they are not in the same process, because we have $\textit{inProcess}(m, p)$ and $\textit{inProcess}(n, p')$ with $p \neq p'$. Detecting that two activities m and n are in the same process with approach 1 is a bit more complicated, as it requires to identify a sequence of *followedBy* triples (the direction does not matter) which establishes a path that connects m and n. This path might be relatively long. When working with translation approach 1 it is thus more complicated to make use of the implicit information that two activities are or are not in the same process model. As we will see later, the different approaches have their merits depending on the choice of the applied knowledge graph completion method.

For the construction of a knowledge graph from the business process graphs of the given repository and the process model under development, each of the business process graphs is translated to a partial knowledge graph by one of the above approaches. The partial knowledge graphs of the different processes are connected by the activity labels that they share, while an activity node always belongs to exactly one process. If, for example, two processes both contain an activity *Register*, then the two respective activity nodes are both linked to the node that represents the label *Register* in the graph.

Given an obtained knowledge graph, we are interested in predictions of the completion task $(\tilde{n}, hasLabel, ?)$, where $\tilde{n}$ denotes the unlabeled node in the process model under development for which we want to suggest a suitable label. For this task, existing completion approaches can be employed, as shown next.

4 Experimental Study

In this section, we report on the design and results of our study in which we investigate the performance of different approaches for applying existing knowledge graph completion methods that have not specifically been designed for activity recommendation. Additionally, we compare the approaches to the embedding-based activity recommendation technique RLRecommender [21] and to the rule-based method from our earlier work [19].

4.1 Dataset

For the experiments we employ the model collection of the Business Process Management Academic Initiative [13]. Unlike in our previous work [18,19], we only use the last revisions of the BPMN 2.0 models in the collection. In contrast to using all revisions, we thus represent the possible case that the recommendation methods sometimes only have few or even none domain-specific reference models for the suggestion of activities available. Moreover, the runtimes of the evaluated KGE models remain in a reasonable range of maximum 48 h for the hyperparameter search given a particular translation approach[1]. Out of all last revision BPMN 2.0 models, we use those with 3 to 50 activities and English labels. This choice results in a dataset consisting of 3,688 process models. On average, the processes involve 14.3 activities while the standard deviation equals 8.3.

4.2 Evaluation Setup

We separate the dataset into train, validation and test splits. More precisely, we train each method on 80% of the process models while we use 10% of the models for validation and 10% for evaluation.

Evaluation Procedure. We want to evaluate the approaches on realistic recommendation tasks. Therefore, we use an evaluation procedure in which we simulate the current status of a process model under development from a given business process graph by specifying the amount of information that is available for the recommendation. The basic idea is to remove some of the nodes and all edges connected to these nodes from a given business process graph while treating the remaining graph as the intermediate result of a modeling process. The

[1] Note that we performed the experiments dividedly on two computers: Intel® Xeon® CPU E5-2640 v3@40x2.40 GHz and Intel® Xeon® Silver 4114 CPU@40x2.20 GHz.

employed evaluation procedure is based on breadth-first search. In this procedure, we randomly select one activity, which is neither source nor sink node, as the one to be predicted. During the evaluation we therefore filter out processes which do not have a chain of at least three different activities when executed. Then we hide the label of the chosen activity and determine the shortest path s from a source node to the activity. After that we hide all other activities that are not on a path of length s starting from a source node while the remaining activities and edges between them serve as a context for the recommendation task. For the experiments, we create one recommendation task for every process model in the validation and test split.

Evaluated Methods. We combine each of the three translation approaches defined in Sect. 3.2 with three existing knowledge graph completion methods. For this, we employ the TransE [1] and DistMult [23] KGE models (also referred to as *embedding-based methods*), as well as the rule learner AnyBURL [12][2]. For the application of the KGE models we use the PyTorch-based framework libKGE [2]. While these approaches are rather old, they have proven to yield good results when trained in an appropriate way [16]. Additionally, we include the results of our previously proposed rule-based activity recommendation method [19] and of the embedding-based technique RLRecommender by Wang et al. [21], which have both been developed with a special focus on activity recommendation.

Dataset Statistics. Table 1 shows the statistics of the dataset for the embedding-based methods, i.e., the number of entities and relations as well as the number of triples in the training, validation, and test sets for each translation approach. While the translation approaches 1 and 2a yield 70,876 entities, which is the sum of the total number of activity nodes (48,677) and activity labels (22,199), approach 2b also considers processes as entities, which adds the total number of processes in the evaluation (3,672) to the sum.

Table 1. Overview of the dataset statistics for the embedding-based methods

Translation approach	Entities	Relations	Training triples	Validation triples	Test triples
1	70,876	2	105,150	358	362
2a	70,876	3	955,492	358	362
2b	74,548	3	153,827	358	362

The large difference in the number of triples between the training set and the validation and test sets has two reasons. First, we are only interested in one special prediction task, which is the tail prediction of triples $(\tilde{n}, \mathit{hasLabel}, \lambda(\tilde{n}))$, where $\tilde{n}$ denotes the node for which we want to recommend an activity label

[2] We also tested other popular KGE models (ComplEx, ConvE) but they yielded comparatively poor results that we do not report here.

$\lambda(\tilde{n})$. Therefore, the validation and test sets comprise for each process model in the validation or test split only one triple $(\tilde{n}, \textit{hasLabel}, \lambda(\tilde{n}))$, whereas the training set contains all triples of the process models in the training split.

Second, we want to consider the context of the process model under development, i.e., the activities and relations so far included in the process model. Since, for embedding-based approaches, the entity that appears in the completion task needs to be part of the learned embeddings, this means that the contexts of the process models in the validation and test splits must be included in the training set. As such, the training set additionally contains all context triples of the validation and test process models. This is not necessary for rule-based approaches such as AnyBURL, since they naturally allow for the consideration of the context, which thus does not have to be included in the training set.

Hyperparameters. For the KGE models, we basically employed the large hyperparameter space presented by Ruffinelli et al. [16]. However, we additionally considered the embedding sizes 32 and 64. While our previously proposed rule-based activity recommendation method [19] has no hyperparameters, we have set the following hyperparameters for AnyBURL: We increased the length of cyclic rules to 5 and the length of acyclic rules to 2 (a higher parameter is not supported by AnyBURL)[3]. Aside from these changes we used AnyBURL in its default setting. Also note that rule-based approaches usually do not fine-tune their hyperparameters against the validation set. Thus, both AnyBURL and the rule-based activity recommendation method from our previous work do not make any use of the validation set.

Evaluation Metrics. We use two different metrics for quantifying the relevance of provided recommendations. First, we report on the *hit rate* Hits@10, which captures the fraction of *hits* in the top 10 recommendations, i.e., the fraction of cases where the activity label that was actually used in a process model is among the ten recommendations which are according to the employed method the most likely. Second, we report on the Mean Reciprocal Rank (MRR), which additionally takes the position of a hit in the recommendation list into account. The reciprocal rank of a recommendation list has the value 0 if the actually chosen activity is not in the provided list and $1/p$ otherwise, where p denotes the position of the hit in the list. For the MRR we take the mean of the reciprocal ranks of all generated recommendation lists. Note that we consider a recommendation list of length 10 to compute MRR, which is a close approximation of the MRR that is based on the full ranking. However, this is more realistic, as the list of recommendations shown to the user has to be limited as well.

4.3 Results

The results of our experiments are shown in Table 2. With Hits@10 numbers that are at least 10 % worse than the rule-based activity recommendation approach and low MRR numbers, no combination of tested methods and translation

[3] These parameter settings are specified by `MAX_LENGTH_CYCLIC = 5` and `MAX_LENGTH_ACYCLIC = 2`.

Table 2. Experimental results

Translation approach	Method	Hits@10	MRR
1	TransE	34.0%	8.4%
	DistMult	35.9%	15.0%
	AnyBURL	36,1%	14,8%
2a	TransE	20.7%	4.2%
	DistMult	34.3%	19.1%
	AnyBURL	37.2%	15.2%
2b	TransE	33.4%	8.0%
	DistMult	36.5%	15.6%
	AnyBURL	35.8%	14.5%
RLRecommender [21]		35.1%	23.8%
Rule-based approach [19]		47.5%	41.4%

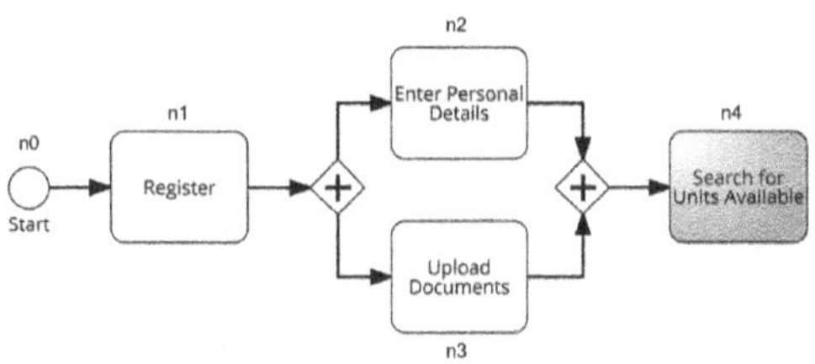

1. Register
2. Upload Documents
3. Enter Personal Details
4. Enter Details of Unit for Rent
5. Upload Pictures
6. n4
7. Select Units for Inspectation
8. Search For Units Available
9. Start
10. n3

Fig. 2. An example of a process model in the validation set where the completion task (n_4, *hasLabel*, ?) has to be solved and the top-10 recommendations of TransE in combination with translation approach 1

approaches works well. The two specialized methods, in particular the rule-based method, that are shown in the last two lines of the table work significantly better.

Examination of the Predictions. To spot the reason for these unexpected results, we looked at some concrete cases. Figure 2 shows a process model in the validation set, where the activity *Search for Units Available* has been randomly selected as the one to be predicted. The use of translation approach 1 combined with TransE yields the top 10 recommendations shown on the right.

The correct activity *Search for Units Available* is at position eight of the recommendation list, which means that the MRR would be $1/8 = 0.125$, if this was the only completion task of the whole evaluation. Other items of the recommendation list include some nodes of the given process model, i.e., n3 and n4, as well as activities like *Register* or *Enter Personal Details* that have already been used in the process model. Clearly, the recommendation of nodes is not useful since we are interested in the prediction of activities. Also, it is likely that activities that have already been inserted into the process model are not added a second time. Therefore, we decided to do a post-processing in which we

Table 3. Experimental results with post-processing

Translation approach	Method	Hits@10	MRR
1	TransE	41.7% (+ 7.7%)	24.9% (+ 21.1%)
	DistMult	38.7% (+ 2.8%)	29.8% (+ 14.8%)
	AnyBURL	36.9% (+ 0.1%)	24.4% (+ 9.6%)
2a	TransE	37.0% (+ 16.3%)	20.8% (+ 16.6%)
	DistMult	37.3% (+ 3.0%)	29.2% (+ 10.1%)
	AnyBURL	38.6% (+ 1.4%)	24.2% (+ 9.0%)
2b	TransE	42.5% (+ 9.1%)	29.3% (+ 21.2%)
	DistMult	39.8% (+ 3.3%)	31.4% (+ 15.8%)
	AnyBURL	36.4% (+ 0.5%)	24.4% (+ 9.8%)
RLRecommender [21]		35.1%	23.8%
Rule-based approach [19]		47.5%	41.4%

filter out other recommendations than labels, i.e., nodes and processes, as well as activities that are already present in the given process model. In the example of Fig. 2, this means that the correct prediction moves to position four of the recommendation list which is an MRR improvement from 0.125 to 0.25.

Results with Post-processing. The results of the experiments with post-processing are shown in Table 3. The percentages in brackets indicate the improvement of the associated Hits@10 or MRR numbers in comparison to the results without post-processing.

The KGE model TransE profits most from the post-processing. However, the TransE results strongly depend on the translation approach. This effect is smaller when using DistMult or AnyBURL. For the embedding-based methods TransE and DistMult, translation approach 2b is the best, which shows that the additional explicit information given by the triples with the relation *inProcess* is useful. In contrast, approach 2a works comparably poor in combination with the embedding-based methods. One reason for this could be that in approach 2a the nodes are strongly interconnected via the relation *inSameProcess*. Thus, the interconnection of the nodes via *inSameProcess* is more prominent than via the relation *followedBy*. This can be disadvantageous since the co-occurrence patterns depicted by *inSameProcess* are often less relevant than the structural patterns captured by *followedBy*, which avoid recommending activities that have high co-occurrence statistics, but are not relevant at the current model position.

Unlike the embedding-based methods, AnyBURL achieves the best results when using the translation approach 2a. While this approach only needs one triple $(m, inSameProcess, n)$ to express that two nodes m and n are in a process p, approach 2b needs the two triples $(m, inProcess, p)$ and $(n, inProcess, p)$. This has a direct impact on the regularities that can be captured by AnyBURL. A rule as $hasLabel(X, register) \leftarrow inSameProcess(X, Y), hasLabel(Y, upload\ documents)$

is within the supported language bias while the equivalent rule will have a body length of three based on translation approach 2b and is thus out of scope.

If we now compare the results of the standard knowledge graph completion methods with post-processing to the results of the specialized methods RLRecommender and the rule-based activity recommendation approach, we observe that the gap between standard and specialized methods has become less significant after post-processing. As described in Sect. 2, RLRecommender [21] is based on a rather specific approach to use embeddings in which only one related activity in the process model is used for the recommendation of an activity, thus its results are slightly worse compared to the results of our approaches for using TransE and DistMult. This holds in particular for translation approach 2b where both TransE and DistMult are better in Hits@10 and MRR.

The rule-based method [19], which has specifically been designed for activity recommendation, is still at least 5 % better in Hits@10 and 10 % better in MRR. These significant differences illustrate that a standard knowledge graph completion method cannot compete with an approach which has specifically been designed for activity recommendation in business process modeling. This holds even though we tried out different translation approaches and developed a problem-specific filtering as post-processing step to increase the quality of the results.

It seems that the specific regularities which are important for making a good activity recommendation seem to influence the resulting embedding space only to a limited degree. These specifics are reflected in the types of rules supported by the rule-based recommendation method [19] that are also more expressive compared to the general rule types supported by AnyBURL. We can conclude that standard methods for knowledge graph completion are not flexible enough to adapt to the given problem resulting in a relatively low prediction quality.

5 Conclusion and Future Work

In this paper, we presented different approaches to use knowledge graph completion methods for activity recommendation in business process modeling. A problem-specific filtering as post-processing step improved the quality of the predictions. However, the rule-based activity recommendation method still worked better than the application of various standard knowledge graph completion methods, which revealed their lack of flexibility to adapt to the given problem.

In the future, alternative techniques working on (RDF) knowledge graphs, e.g., Graph Convolutional Networks [17] or RDF2Vec [14], could be tested. Until now we have only considered co-occurrence and structural patterns, which suffers from the sparseness of the knowledge graphs. We would also like to take textual patterns into account in future, for example, by applying KG-Bert [24].

Another future direction is related to the current state of the process model under development that should be taken into account as a context for the recommendation of an activity. When putting the context into the training set, as done for the embedding-based methods, the KGE models have to be retrained after every activity that has been added to the model under development. This is very impractical for the application of such methods for real-time recommendations, as the training takes too much time. In future work, we would like to explore ways of avoiding the need for complete retraining when the process model under development has been extended, as it has, for example, been done by Song et al. [20] for evolving knowledge graphs and translation-based embeddings.

References

1. Bordes, A., Usunier, N., García-Durán, A., Weston, J., Yakhnenko, O.: Translating embeddings for modeling multi-relational data. In: NIPS, pp. 2787–2795 (2013)
2. Broscheit, S., Ruffinelli, D., Kochsiek, A., Betz, P., Gemulla, R.: LibKGE - a knowledge graph embedding library for reproducible research. In: EMNLP: System Demonstrations, pp. 165–174 (2020)
3. Cao, B., Yin, J., Deng, S., Wang, D., Wu, Z.: Graph-based workflow recommendation: on improving business process modeling. In: CIKM, pp. 1527–1531. ACM (2012)
4. Deng, S., et al.: A recommendation system to facilitate business process modeling. IEEE Trans. Cybern. **47**(6), 1380–1394 (2017)
5. Dumas, M., La Rosa, M., Mendling, J., Reijers, H.A.: Fundamentals of Business Process Management. Springer, Berlin (2013). https://doi.org/10.1007/978-3-642-33143-5
6. Fellmann, M., Zarvic, N., Metzger, D., Koschmider, A.: Requirements catalog for business process modeling recommender systems. In: WI, pp. 393–407 (2015)
7. Frederiks, P.J., Van der Weide, T.P.: Information modeling: the process and the required competencies of its participants. DKE **58**(1), 4–20 (2006)
8. Friedrich, F., Mendling, J., Puhlmann, F.: Process model generation from natural language text. In: Mouratidis, H., Rolland, C. (eds.) CAiSE 2011. LNCS, vol. 6741, pp. 482–496. Springer, Heidelberg (2011). https://doi.org/10.1007/978-3-642-21640-4_36
9. Jannach, D., Fischer, S.: Recommendation-based modeling support for data mining processes. In: RecSys, pp. 337–340 (2014)
10. Jannach, D., Jugovac, M., Lerche, L.: Supporting the design of machine learning workflows with a recommendation system. ACM TiiS **6**(1), 1–35 (2016)
11. Li, Y., et al.: An efficient recommendation method for improving business process modeling. IEEE Trans. Industr. Inf. **10**(1), 502–513 (2014)
12. Meilicke, C., Chekol, M.W., Ruffinelli, D., Stuckenschmidt, H.: Anytime bottom-up rule learning for knowledge graph completion. In: IJCAI, pp. 3137–3143. AAAI Press (2019)
13. Model collection of the BPM Academic Initiative. http://bpmai.org/
14. Ristoski, P., Paulheim, H.: RDF2Vec: RDF graph embeddings for data mining. In: Groth, P., et al. (eds.) ISWC 2016. LNCS, vol. 9981, pp. 498–514. Springer, Cham (2016). https://doi.org/10.1007/978-3-319-46523-4_30
15. Rossi, A., Barbosa, D., Firmani, D., Matinata, A., Merialdo, P.: Knowledge graph embedding for link prediction: a comparative analysis. ACM Trans. Knowl. Discov. Data (TKDD) **15**(2), 1–49 (2021)
16. Ruffinelli, D., Broscheit, S., Gemulla, R.: You CAN teach an old dog new tricks! On training knowledge graph embeddings. In: ICLR. OpenReview.net (2020)

17. Schlichtkrull, M., Kipf, T.N., Bloem, P., van den Berg, R., Titov, I., Welling, M.: Modeling relational data with graph convolutional networks. In: Gangemi, A., et al. (eds.) ESWC 2018. LNCS, vol. 10843, pp. 593–607. Springer, Cham (2018). https://doi.org/10.1007/978-3-319-93417-4_38
18. Sola, D.: Towards a rule-based recommendation approach for business process modeling. In: Hacid, H., et al. (eds.) ICSOC 2020. LNCS, vol. 12632, pp. 25–31. Springer, Cham (2021). https://doi.org/10.1007/978-3-030-76352-7_4
19. Sola, D., Meilicke, C., van der Aa, H., Stuckenschmidt, H.: A rule-based recommendation approach for business process modeling. In: La Rosa, M., Sadiq, S., Teniente, E. (eds.) CAiSE 2021. LNCS, vol. 12751, pp. 328–343. Springer, Cham (2021). https://doi.org/10.1007/978-3-030-79382-1_20
20. Song, H.J., Park, S.B.: Enriching translation-based knowledge graph embeddings through continual learning. IEEE Access **6**, 60489–60497 (2018)
21. Wang, H., Wen, L., Lin, L., Wang, J.: RLRecommender: a representation-learning-based recommendation method for business process modeling. In: Pahl, C., Vukovic, M., Yin, J., Yu, Q. (eds.) ICSOC 2018. LNCS, vol. 11236, pp. 478–486. Springer, Cham (2018). https://doi.org/10.1007/978-3-030-03596-9_34
22. Wang, Q., Mao, Z., Wang, B., Guo, L.: Knowledge graph embedding: a survey of approaches and applications. IEEE Trans. Knowl. Data Eng. **29**(12), 2724–2743 (2017)
23. Yang, B., tau Yih, W., He, X., Gao, J., Deng, L.: Embedding entities and relations for learning and inference in knowledge bases. In: ICLR (Poster) (2015)
24. Yao, L., Mao, C., Luo, Y.: KG-BERT: BERT for knowledge graph completion. CoRR abs/1909.03193 (2019)

Creating Unbiased Public Benchmark Datasets with Data Leakage Prevention for Predictive Process Monitoring

Hans Weytjens(✉) and Jochen De Weerdt

Research Centre for Information Systems Engineering (LIRIS), KU Leuven, Leuven, Belgium
{hans.weytjens,jochen.deweerdt}@kuleuven.be

Abstract. Advances in AI, and especially machine learning, are increasingly drawing research interest and efforts towards predictive process monitoring, the subfield of process mining (PM) that concerns predicting next events, process outcomes and remaining execution times. Unfortunately, researchers use a variety of datasets and ways to split them into training and test sets. The documentation of these preprocessing steps is not always complete. Consequently, research results are hard or even impossible to reproduce and to compare between papers. At times, the use of non-public domain knowledge further hampers the fair competition of ideas. Often the training and test sets are not completely separated, a data leakage problem particular to predictive process monitoring. Moreover, test sets usually suffer from bias in terms of both the mix of case durations and the number of running cases. These obstacles pose a challenge to the field's progress. The contribution of this paper is to identify and demonstrate the importance of these obstacles and to propose preprocessing steps to arrive at unbiased benchmark datasets in a principled way, thus creating representative test sets without data leakage with the aim of levelling the playing field, promoting open science and contributing to more rapid progress in predictive process monitoring.

Keywords: Predictive process monitoring · Remaining time prediction · Bias · Benchmarking · Reproducibility · Datasets · Preprocessing

1 Introduction

Process mining analyzes event data logged by information systems with the goal of process discovery, process conformance checking and process enhancement. Predictive process monitoring is an important sub-field of process mining and concerns predicting next events, process outcomes and remaining execution times. The field is enjoying increased interest and is progressing thanks to opportunities offered by developments in AI and machine learning and the availability of data. Consequently, an increasing number of papers concerning predictive process monitoring are being published. Studying them, however, we identified three major obstacles to further progress in the field.

A. Marrella and B. Weber (Eds.): BPM 2021 Workshops, LNBIP 436, pp. 18–29, 2022.
https://doi.org/10.1007/978-3-030-94343-1_2

The first obstacle involves the use of different datasets that complicates or even impedes the comparison of results between papers. Even though many researchers use publicly available datasets, differences in their pruning and training/test set splits lead to significantly diverging results. Somewhat unfairly, some of these preprocessing decisions may have been guided by domain knowledge unavailable to all researchers. The second obstacle concerns the proper splitting of training and test sets. When using standard temporal splitting or cross-validation techniques, a number of cases will have prefixes in both the training and test set which is a form of data leakage and affects prediction performance. Most papers do not remedy this issue. For correct science, however, predictive process monitoring problems require a specific data split technique. The third obstacle relates to two forms of bias that are often ignored: when crudely determining start and end times for training and test sets, the number and average length of running cases at any given time may be greatly impacted at the chronological beginning and ending of these training and test sets.

To overcome these three obstacles, we believe predictive process monitoring needs benchmark datasets with predefined targets, training and test sets. With such datasets available, researchers could process the training sets at will while having to test their results on the given test sets. Benchmark datasets have always played an important role in major machine learning areas such as natural language[1], reinforcement learning[2], image recognition, etc. They allow the research community to compare different proposed methods and spur the development of new state-of-the-art ideas, often in the form of competitions. Well-known examples in the field of image recognition include –sorted by increasing size and complexity– MNIST[3] (60,000 black-white handwritten digit images), CIFAR-10[4] (60,000 color images in 10 classes) and ImageNet[5] (14 million images in 20,000 classes), progressively used as the field matured.

The contribution of this paper is twofold. First, we want to expose the three obstacles mentioned above by investigating nine of the most commonly used public process mining datasets. Second, based on this analysis, we derive a set of proposals on how to design benchmark datasets for predictive process monitoring. These will ensure fairness, enable reproducibility and facilitate progress in the field. The code to implement our proposals is publicly available[6].

The next section refers to influential process outcome prediction research and relates it to the three obstacles. We then detail our approach in Sect. 3 and describe our experimental setup in Sect. 4. We derive preprocessing steps that convert public datasets into unbiased benchmark datasets in Sect. 5. Using a simple convolutional neural network (CNN), Sect. 6 exposes the impact of dataset design decisions before concluding in Sect. 7.

---

[1] https://gluebenchmark.com/.
[2] https://arxiv.org/abs/2004.07219.
[3] http://yann.lecun.com/exdb/mnist/.
[4] https://www.cs.toronto.edu/~kriz/cifar.html.
[5] https://image-net.org/.
[6] https://github.com/hansweytjens/predictive-process-monitoring-benchmarks.

2 Related Work

When exploring some of the most influential original papers and surveys on outcome [1,2], next event [5,6,8] or remaining time [3,5,7] prediction, the first obstacle of poor comparability becomes immediately apparent. Even where researchers work with the same datasets, direct comparisons of their results prove impossible. For example, the classes for the BPIC_2011 and Traffic datasets are defined differently by Teinemaa et al. [1] and Kratsch et al. [2]. In next event prediction, Tax et al. [5] consider only events of type "complete" whilst Evermann et al. [6] create new features by concatenating the "activity" and "resource" features. In the remaining time prediction problem, Tax et al. [5] and Verenich et al. [3] use different ratios (67%/33% and 80%/20% respectively) when chronologically splitting training and test sets, whereas Polato et al. [7] use fivefold cross-validation. In their literature overview study, Neu et al. [4] acknowledge the latter problem and propose to "use a unified evaluation approach for further publications." Other than Teinemaa et al. [1] largely solving and Verenich et al. [3] recognizing (but also accepting) the data leakage problem when naively splitting training and test sets, we found no mention of the second obstacle in these papers. To the best of our knowledge, no research diagnoses the third obstacle of bias.

3 Approach

To counter the first obstacle's lack of comparability, we will formulate a standardized preprocessing approach to construct benchmark datasets in Sect. 5. Along the steps in this preprocessing procedure, we will observe and remedy both the second obstacle of data leakage between test and training sets and the third obstacle of bias as they appear. We will use nine datasets to analyze the problems and illustrate the preprocessing steps and their effects on prediction. Before doing so, we first define the four main guiding principles that steer this work:

- **A principled approach:** increases transparency and enables reproducibility to other datasets / problems,
- **Proximity to real-life:** creates datasets that resemble the original underlying business problems as much as possible,
- **Enabling of good science:** (only) where required, we deviate from the first two principles for the benefit of science,
- **Domain agnosticism:** as domain knowledge is often sparse and unevenly distributed amongst researchers, we refrain from using it altogether.

We demonstrate how different preprocessing impacts prediction results. It is rather unconventional to illustrate the importance of the problem towards the end of a paper; however, in our case, it offers the benefit of having clearly-defined preprocessing steps that each generate different variants of the original datasets on which we ran a simple convolutional network to predict remaining times.

4 Experimental Setup

4.1 Preliminaries

In predictive process monitoring, datasets are event logs describing processes, often called *cases*. These cases consist of *events*. A number of *features* describe these cases and events. Every event is also associated with a *target* feature. A *prefix* is an ongoing case, with the *prefix length* its number of completed events. The learning problem is to train an algorithm on a training dataset containing events, described by their features and organized in prefixes labelled with targets, with the goal of predicting the targets of unseen prefixes.

We distinguish two phases in the preprocessing process. We will restrict the use of the term *preprocessing* to refer to the first phase steps that are only performed once and lead to the construction of unique benchmark training and test sets. The second phase involves further steps, such as feature selection and standardisation taken by individual researchers to prepare these fixed training and test sets to train and test their models. This latter phase is not subject of this paper and belongs to the area where researchers and ideas compete.

4.2 Datasets

We used five publicly available datasets from the 4TU.ResearchData repository. BPIC_2012[7] contains logs of a loan application process at a Dutch bank. BPIC_2015[8] is a collection of building permit applications in five Dutch municipalities, which we concatenated into one dataset. BPIC_2017[9] is a richer, cleaner and larger set of logs of a Dutch bank loan application process. BPIC_2019[10], while comparable in size, has much shorter cases and concerns a purchase order handling process. BPIC_2020[11] is a collection of five smaller datasets related to travel administration at a university. The five subsets are records of processes covering domestic declaration documents (Domestic Declarations), international declaration documents (Intl. Declarations), pre-paid travel costs and requests for payment (Payments), travel permits (Permits) and expense claims (Travel Costs). Our target for all of these datasets was defined as the fractional number of days until case completion. Table 1 provides some key statistics of these datasets.

4.3 Models

To illustrate the relevance of the preprocessing choices to construct datasets at the hand of the remaining time prediction problem, we calculated the mean absolute error (MAE) of a CNN model on various preprocessing versions of the

[7] https://data.4tu.nl/articles/dataset/BPI_Challenge_2012/12689204.
[8] https://data.4tu.nl/collections/BPI_Challenge_2015/5065424.
[9] https://data.4tu.nl/articles/dataset/BPI_Challenge_2017/12696884.
[10] https://data.4tu.nl/articles/dataset/BPI_Challenge_2019/12715853.
[11] https://data.4tu.nl/collections/BPI_Challenge_2020/5065541.

Table 1. Statistics of the used datasets*.

Dataset	Nr. cases	Nr. events	Median nr. events	Avg. nr. events	Median nr. days	Mean nr. days	Max nr. days	Dataset nr. days
BPIC_2012	12,183	228,873	9	18.8	0.5	7.8	137.2	152
BPIC_2015	5,641	262,328	45	46.5	68.4	100.8	1,512	1,617
BPIC_2017	31,497	1,201,390	35	38.1	19.1	21.9	286.1	397
BPIC_2019	251,465	1,587,810	5	6.3	64.0	69.5	380.0	830
BPIC_2020:								
Domestic. Declarations	6,087	56,359	5	9.3	14.3	101.7	735.2	773
Intl. Declarations	1,440	71,735	13	49.8	416.1	382.9	781.7	1,010
Payments	6,593	35,046	5	5.3	7.9	11.1	1,238	711
Permits	7,063	86,560	11	12.3	71.7	87.1	502.4	1121
Travel Costs	2,097	18,238	8	8.7	24.0	36.7	325.0	749

*Outliers already removed as per Subsect. 5.2.

datasets. For each dataset, we included one categorical ("activity") and one range ("elapsed": time since case start) feature. We used embedding to feed the categorical feature in the models and reduced its dimensionality to the square root. Two convolutional layers followed the embedding layer, each with 40 kernels of size three and a max-pooling operation with a window of size three. The stride size for both the kernels and pooling was one. Two dense layers of sizes 100 and ten and with a ReLu activation completed the architecture. We reduced overfitting with early stopping using a validation set of the 20% latest-starting cases in the training sets and a patience of 30. The sequence length was ten, the batch size 2048.

5 Constructing Benchmark Datasets

This section applies to remaining time prediction and classification problems. There are some important differences with next event predictions that will be covered in Subsect. 5.8.

5.1 Target Definition

Remaining time targets are defined as the remaining time between an event's time stamp and the case's last time stamp. For classification tasks, however, determining targets requires weakening the domain agnosticism principle and recourse to an authority for a final decision.

5.2 Elimination of Chronological Outliers and Duplicates

The BPIC_2015 dataset contains cases starting in the period 2009–2015. As is shown in the top row in Fig. 1, the cases in the year 2009 appear aberrant as they are few in number and last much longer than cases in the following years. When zooming in on cases starting before 2011-01-31st (bottom row), we see that the log starts behaving "normally" as of October 2010. In the following, we only included cases starting after Oct 1st, 2010. We found similar outliers in BPIC_2019.

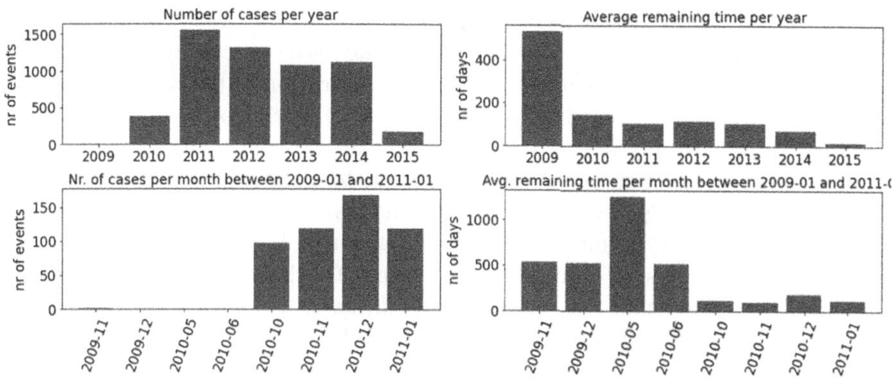

Fig. 1. BPIC_2015 contains remote, faulty outliers at the beginning of the dataset.

Similar patterns arose for the case's ending times at the end of all datasets involved. To remove those outliers, we imposed new end dates. An overview can be found in Table 2.

Table 2. Start and end dates before and after eliminating outliers.

Dataset	Original dataset		After removing outliers	
	First event	Last event	Start in/after	End before/in
BPIC_2012	2011-10	2012-03	-	2012-02
BPIC_2015	2009-11	2015-07	2010-10	–
BPIC_2017	2016-01	2017-02	–	2017-01
BPIC_2019	1948-01	2020-04	2018-01	2019-02
BPIC_2020:			–	
Domestic Declarations	2017-01	2019-06	–	2019-02
Intl. Declarations	2016-10	2020-05	–	2019-07
Payments	2017-01	2019-08	–	2018-12
Permits	2016-10	2021-08	–	2019-10
Travel Costs	2017-01	2019-02	–	2019-01

A dataset may also contain faulty outliers. These are not detectable without domain knowledge and therefore we did not attempt to remove them. The elimination of duplicates requires no further discussion. None were found in our datasets. For clarity, all tables and graphs (outside of this subsection) in this paper, including the preceding Table 1, rely on datasets without chronological outliers.

5.3 Debiasing the End of the Dataset

In order to know their targets, a dataset should only contain completed cases. Without domain knowledge, however, we cannot be 100% certain which cases are completed. We therefore drop the potentially incomplete cases that start in the yellow zone (both light and dark maroon) in the artificial example of Fig. 2. To be on the safe side, the zone's width equals the duration of the dataset's longest case. This introduces two forms of bias towards the end of the dataset (third obstacle). First, the black-grey cases that end in the yellow zone have a longer average duration than the grey cases ending before the yellow zone. Moreover, the number of running cases in the yellow zone (black prefixes) no longer reflects the underlying reality, which is the second form of bias. It prohibits the use of inter-case variables to compute for example the load of a resource dealing with the cases. To debias the ending of the dataset, we must reject the (black) *prefixes* ending in the yellow zone. *Cases* for which we removed the longer (black) prefixes, still remain in the dataset with their shorter (grey) prefixes whose targets are known.

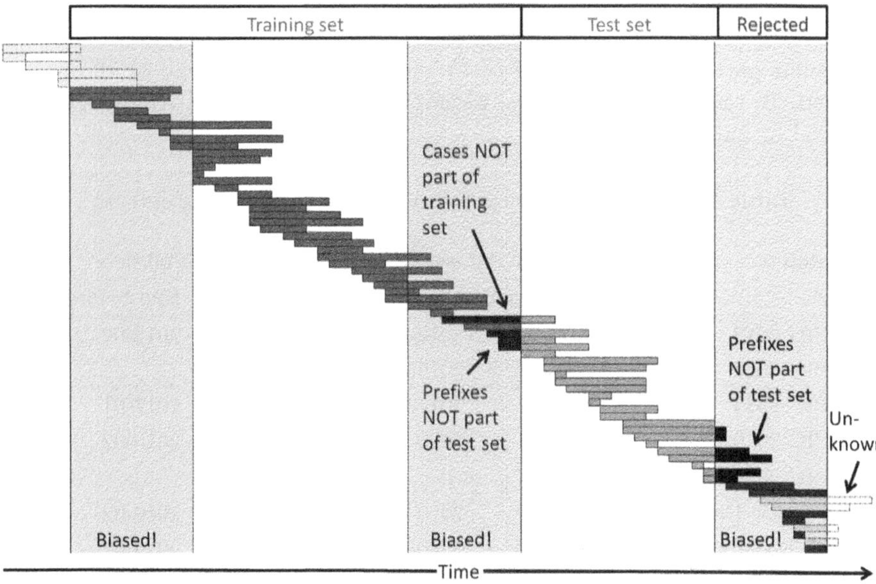

Fig. 2. An artificial dataset with cases ranked by starting time. (Color figure online)

Note that one (possibly faulty) long case in the dataset could suffice to severely reduce the training set size or even annihilate it when debiasing the dataset's end. This problem will be tackled below in Subsect. 5.7.

5.4 Choice of Test Set

Some authors (e.g. [6,7]) use cross-validation for predictions (not just whilst training) to make their models robust against concept drift. However, some of that concept drift can be attributed to the end-of-dataset bias we described and remedied above. Moreover, concept drift happens in real-life situations too. It may be even learnable to some extent. As the strict separation of training and test sets (second obstacle) is also more difficult to implement in a cross-validation setting, we recommend using a classic temporal training/test split for all of these reasons. Larger test sets are more likely to be representative for the whole dataset, but will reduce the size of the corresponding training set and carry a higher risk of real concept drift. As a compromise, we propose that test sets consist of the last 20% of the datasets' cases chronologically.

5.5 Temporal Splitting

To isolate a test set, many authors rank cases by the timestamp of their first event and determine a separation time, with the x% last cases starting after that time going to the test set and the remainder constituting the training set. From a real-life perspective, the logical flaw is that the outcome (remaining time or class) of incomplete cases in the training set (red/grey in Fig. 2) is not known at that separation time. Moreover, some cases from the training set and test will be running concurrently and influencing each other. This is the second obstacle mentioned in the introduction. To counter it, we recommend to always apply "strict temporal splitting" by only retaining in the training set those cases that are completed before the separation time. These are colored in green in Fig. 2. As a result, the training set will contain two biased zones (light blue), at its beginning and end. It is left to researchers to debias these zones as they see fit.

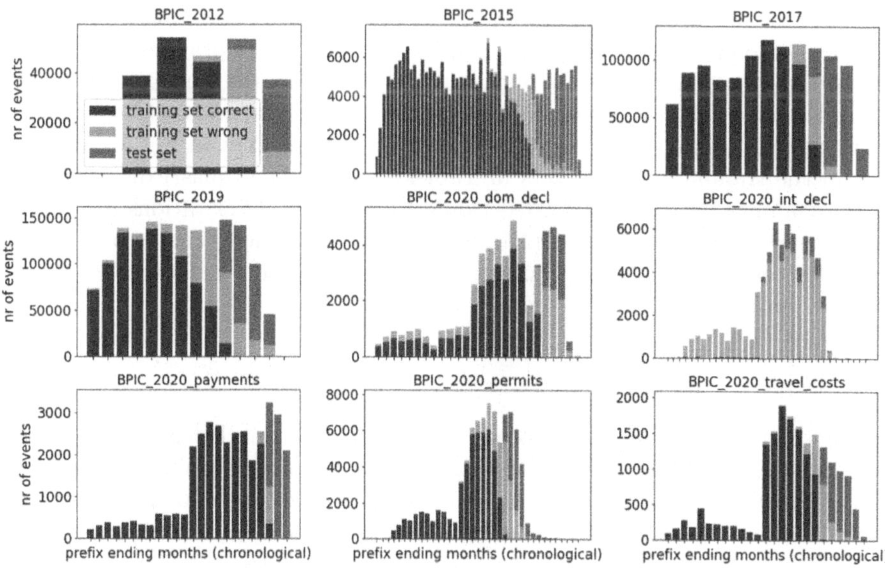

Fig. 3. Grey bars represent test set events (20%). Green bars are events in training set cases ending before the separation time (strict temporal splitting), red ones belong to cases ending after that separation time (regular temporal splitting). As debiasing was not done to retain sufficient samples, the bias at the datasets' ends is clearly visible. (Color figure online)

A compelling reason to make this imperative is not present. Depending on the duration of the cases in relation to the time span covered by the dataset, the difference between strict and regular temporal splitting can be rather significant, as demonstrated in Fig. 3.

5.6 Debiasing the Beginning of the Test Set

We now face both obstacle three biases at the beginning of the test set. Of the (red-grey) cases that start before the separation time but end after it, we also include in the test set those (grey) *prefixes* that end after the separation time. This restores balance both in terms of the case lengths and the number of running cases. It also better resembles real-life settings where a maximum of available data at any given time would be considered.

5.7 Removal of Long Cases

Applying our third guideline, we suggest discarding extremely long cases from the dataset in the interest of the dataset's survival when removing biases at the end of the dataset as described in Subsect. 5.3. This clearly changes the nature of the original dataset, but we believe that leaving the biases in would have a more negative impact. Figure 4 plots the size of the training and test sets (in number of events) constructed as described above for different maximum case durations. Keeping the longest cases would drastically, often fatally shrink the datasets as can be seen on the right side of the graphs. Some degree of pragmatism is called for: we propose removing up to, but no more than, 5% of the longest-lasting cases to find the case duration within the window that corresponds with the largest training set (in number of cases) as shown by the red lines in Fig. 4. We, thus, obtain the training and test sets in Tables 3 and 4. The (quasi) disappearance of the training sets for BPIC_2020 Domestic and Intl. Declarations disqualifies them for serious research. Of course, steps 5.3, 5.4, 5.5 and 5.6 have to be carried out after removing the longest cases. We only discussed the removal last, as it assumes an understanding of these same steps 5.3, 5.4, 5.5 and 5.6.

5.8 Next Event Prediction

As the targets in the next event prediction problem do not depend on the knowledge of cases' final events, Subsects. 5.3, 5.4, 5.5, 5.6 and 5.7 no longer apply: we can work with prefixes rather than completed cases. After removing the chronological outliers, it suffices to separate training and test sets by chronologically sorting all remaining events in the log and determining the 80%/20% separation time. All prefixes (not cases) ending before that separation time belong in the training set, and those ending after it constitute the test set. Viewed on Fig. 2, this means we retain all cases in the yellow zone and keep the red prefixes in the training set.

6 Model Predictions

With the example of the remaining time prediction problem, Fig. 5 visualizes the impact of the preprocessing decisions on predictions. For every dataset, we ran seven simulations. The first simulation concerned a base case with a train/test split of the first 90% and last 10% starting cases. In the following six simulations, we successively introduced the measures from Sect. 5. The maximal durations used in the last simulation correspond to the red vertical lines in Fig. 4. The resulting quality of the predictions from the CNN model, expressed as MAE, is represented by the black lines (left vertical axis). The obtained sizes of training (green bars) and test set (grey bars) are calibrated to the right vertical axis.

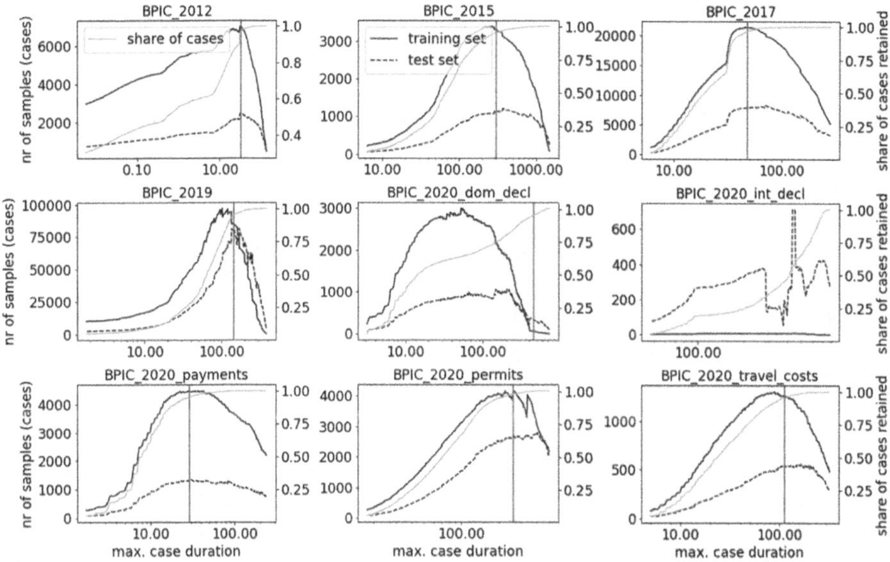

Fig. 4. Size of training and test sets in function of maximal case lengths. Debiasing and strict temporal splitting applied. Vertical red lines show our proposals. (Color figure online)

Table 3. Proposed benchmark datasets* (test set 20%): timing.

Dataset	Original datasets*				Proposed datasets			
	Max. duration cases	Dataset nr. days	Dataset start	Dataset end	Max. duration cases	Dataset nr. days	Test set start	Test (= Data-) set end
BPIC_2012	137.2	152	2011-10	2012-02	32.3	119.7	2012-01-05	2012-01-28
BPIC_2015	1,512	1,617	2010-10	2015-03	302.8	1,314.2	2013-06-10	2014-05-10
BPIC_2017	286.1	397	2016-01	2017-01	47.8	348.5	2016-10-10	2016-12-14
BPIC_2019	380.0	830	2018-01	2019-02	143.3	253.6	2018-07-19	2018-09-11
BPIC_2020:								
Payments	1,238	711	2017-01	2018-12	28.9	682.2	2018-09-14	2018-11-22
Permits	502.4	1,121	2016-10	2019-10	258.8	806.2	2018-09-07	2018-12-20
Travel Costs	325.0	749	2017-01	2019-01	114.3	634.0	2018-07-06	2018-10-06

*Outliers removed as per Subsect. 5.2.

Table 4. Proposed benchmark datasets* (test set 20%). For the test sets, the total number of cases are listed first, followed by cases at the beginning of the test set missing some of their shorter prefixes (red/grey in Fig. 2), cases at the end of the test set missing their longer prefixes (grey/black in Fig. 2) and cases with all prefixes present in the test sets (grey in Fig. 2). The last column is the total number of cases in the dataset, but with the average of both kinds of incomplete test set cases rather than their sum.

Dataset	Number of cases							
	Original dataset	Training set	Test set				Total dataset	
			All	Missing short prefixes	Missing long prefixes	Complete	All	Full case equivalent
BPIC_2012	12,183	7.019	2,468	570	621	1,340	9,487	8,955
BPIC_2015	5,641	3,311	1,116	229	221	666	4,427	4,202
BPIC_2017	31,497	21,404	7,902	2,040	1,754	4,108	29,306	27,409
BPIC_2019	251,465	84,233	86,905	52,676	46,423	5,327	171,138	139,110
BPIC_2020:								
Payments	6,593	4,494	1,293	135	171	987	5,787	5,634
Permits	7,063	4,168	2,679	1,309	831	667	6,847	5,905
Travel Costs	2,097	1,265	533	173	100	260	1,798	1,662

*Outliers removed as per Subsect. 5.2.

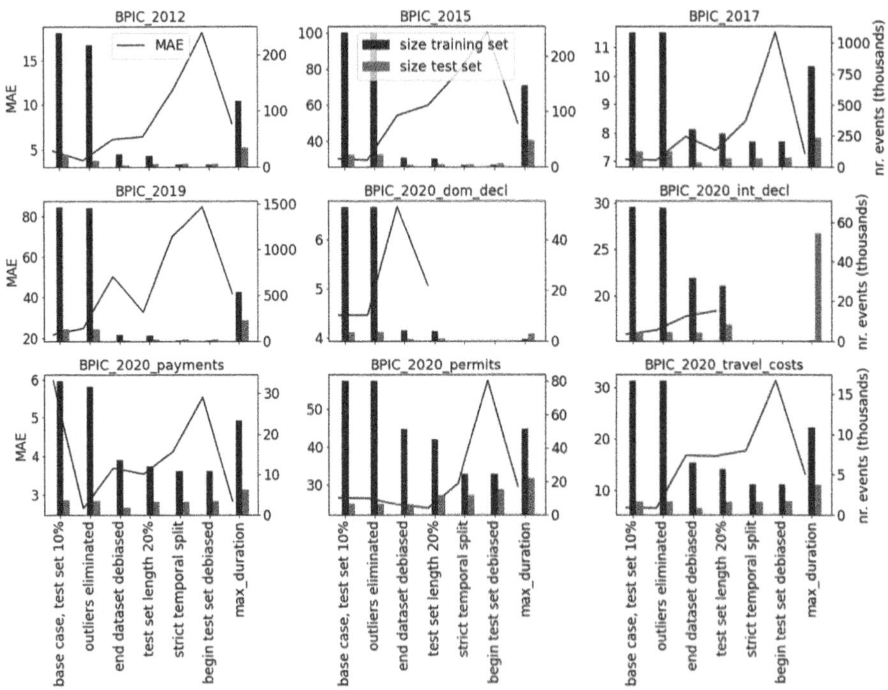

Fig. 5. Prediction results (MAE, lower is better) vary significantly in function of the chosen training and test data sets, calling for an agreement on proper benchmarks.

7 Conclusion and Future Work

In this paper, we demonstrated the impact of dataset preprocessing choices and pointed out the problem of overlapping training and test sets and at two sources of bias in test sets. Left unaddressed, these three obstacles risk slowing down progress in the field of predictive process monitoring. We argued for the use of unbiased benchmark datasets with strict temporal splitting of training and test sets. Using the example of nine BPIC datasets and the remaining time prediction problem, we established preprocessing steps to construct such benchmark datasets. The scripts for our examples are publicly available (see footnote 6). It is our hope that researchers will use these benchmarks and thus facilitate a more rapid advancement of predictive process monitoring as the comparability and quality (no data leakage, no bias) of their results improves. In the future, perhaps recommendations for metrics and statistical analysis of results could be formulated to further enhance comparability and interpretability. We also hope that other publicly available datasets will be constructed using our proposed transformation steps. The scientific marketplace will then decide which of these datasets will become predictive process monitoring's next MNIST, CIFAR-10 or ImageNet.

References

1. Teinemaa, I., Dumas, M., La Rosa, M., Maggi, F.M.: Outcome-oriented predictive process monitoring: review and benchmark. ACM Trans. Knowl. Discov. Data (TKDD) **13**(2), Article No. 17 (2019)
2. Kratsch, W., Manderscheid, J., Roeglinger, M., Seyfried, J.: Machine learning in business process monitoring: a comparison of deep learning and classical approaches used for outcome prediction. Bus. Inf. Syst. Eng. **63**, 261–276 (2020). https://doi.org/10.1007/s12599-020-00645-0
3. Verenich, I., Dumas, M., La Rosa, M., Maggi, F.M., Teinemaa, I.: Survey and cross-benchmark comparison of remaining time prediction methods in business process monitoring. ACM Trans. Intell. Syst. Technol. (TIST) **10**(4), 1–34 (2019)
4. Neu D. A., Lahann J., Fettke P.: A systematic literature review on state-of-the-art deep learning methods for process prediction. Artif. Intell. Rev. (2021). https://doi.org/10.1007/s10462-021-09960-8
5. Tax, N., Verenich, I., La Rosa, M., Dumas, M.: Predictive business process monitoring with LSTM neural networks. In: Dubois, E., Pohl, K. (eds.) CAiSE 2017. LNCS, vol. 10253, pp. 477–492. Springer, Cham (2017). https://doi.org/10.1007/978-3-319-59536-8_30
6. Evermann, J., Rehse, J.-R., Fettke, P.: Predicting process behaviour using deep learning. Decis. Support Syst. **100**, 129–140 (2017)
7. Polato, M., Sperduti, A., Burattin, A, de Leoni, M.: Data-aware remaining time prediction of business process Instances. In: Proceedings of the International Joint Conference on Neural Networks, pp. 816–823 (2014)
8. Camargo, M., Dumas, M., González-Rojas, O.: Learning accurate LSTM models of business processes. In: Hildebrandt, T., van Dongen, B.F., Röglinger, M., Mendling, J. (eds.) BPM 2019. LNCS, vol. 11675, pp. 286–302. Springer, Cham (2019). https://doi.org/10.1007/978-3-030-26619-6_19

Exploring Gated Graph Sequence Neural Networks for Predicting Next Process Activities

Sven Weinzierl(✉)

Institute of Information Systems, Friedrich-Alexander-Universität Erlangen-Nürnberg, Fürther Straße 248, 90429 Nürnberg, Germany
sven.weinzierl@fau.de

Abstract. A current trend in predictive business process monitoring is to construct predictive models using deep neural networks (DNNs), especially long short-term memory neural networks, convolutional neural networks, or multilayer perceptron neural networks. While these DNN types typically require data defined on the Euclidean space (e.g., grids), graph neural networks (GNNs), a relatively new type of DNNs, can compute data defined on the non-Euclidean space (e.g., graphs). Because GNNs can directly compute graph-oriented data inputs, generally structured into nodes and edges, they can explicitly model event relationships. This paper investigates gated graph sequence neural networks (GGNNs) for the next activity prediction. First results with two real-life event logs show that GGNNs can outperform traditional DNNs regarding predictive quality, especially if nodes of an input graph are assumed as events, and the graph's adjacency matrix only describes event relationships of a process instance prefix.

Keywords: Predictive business process monitoring · Graph neural networks · Deep learning · Machine learning

1 Introduction

The highly volatile and uncertain digital economy increases the pressure on organizations to proactively manage their business processes [16]. Consequently, predictive business process monitoring (PBPM) is gaining momentum in business process management (BPM) [2]. It provides a set of techniques to predict properties of operational business processes such as future process behavior (e.g., next activities) or process outcomes (e.g., a process performance indicator). Predictions from these techniques enable process stakeholders to make decisions that can improve the efficiency of operational business processes. However, this assumes correct predictions, which cannot yet be fully achieved by existing PBPM techniques, especially in real business processes [11,18,24].

Most recent PBPM techniques construct predictive models from historical event log data using machine learning (ML) algorithms. An ML algorithm automatically discovers structures (i.e., patterns) in data and captures those within a predictive model [1]. In the context of ML, deep learning (DL) has "turned out to be very good at discovering the intricate structures in high-dimensional data and is therefore applicable to many domains" [7, p. 436]. This also applies to PBPM, where DL, especially deep

A. Marrella and B. Weber (Eds.): BPM 2021 Workshops, LNBIP 436, pp. 30–42, 2022.
https://doi.org/10.1007/978-3-030-94343-1_3

neural networks (DNNs), have shown to outperform techniques relying on traditional ML algorithms (e.g., regulated probabilistic automata [12] or extreme gradient boosting [23]).

A current trend in PBPM is to use DNNs such as long short-term memory neural networks (LSTM-NNs), convolutional neural networks (CNNs), or multi-layer perceptron neural networks (MLP-NNs) [17]. However, these DNN types typically require data defined on the Euclidean space (e.g., grids) for model training and application [25]. Therefore, a business process instance is represented by a feature vector or matrix, and features of these structures describe event relationships. Consequently, DNNs of current PBPM techniques are not explicitly aware of event relationships, i.e., they do not explicitly model them. Naturally, this makes it difficult for current DNN-based PBPM techniques to increase the predictive quality further because important information (i.e., domain knowledge) of process instances may not be considered. Consequently, they cannot identify some crucial intricate structures in the process data.

A relatively new group of DNNs are graph neural networks (GNNs) [19] that can compute data defined on the non-Euclidean space (e.g., graphs) directly [25]. Here, the structure of input graphs can be matched directly onto the topology of the GNNs, and direct inferences can be made between the network nodes and nodes of the input graphs. In PBPM, first works [10,15,20,22] have proven GNNs to be useful because the control-flow of process instances can intuitively be represented as graphs, and relationships between process activities can explicitly be modeled. However, these works present either an approach designed for the process outcome prediction or apply graph convolutional neural networks (GCNs) [5] for the next activity prediction. GCNs belong to another type of GNNs computing input graphs from a spatial perspective.

This paper investigates gated graph sequence neural networks (GGNNs) [9,19] for the next activity prediction. This type of GNNs is designed for sequential graphs and integrates a gated recurrent unit (GRU) [3] that explicitly considers the temporal aspect of sequences. Since the high expressiveness of GGNNs allows us to represent the input graphs in different ways, we investigate three forms of representing these.

The remainder of this paper is structured as follows: Sect. 2 presents preliminaries and related work on PBPM techniques using GNNs and reveals the research gap of investigating GGNNs for the next activity prediction. Section 3 describes the three investigated representation forms of input graphs and presents the developed GGNN architecture. While Sect. 5 discusses the results and limitations of this paper, Sect. 6 concludes it with a summary and points out future research.

2 Background

2.1 Preliminaries

PBPM techniques receive as input event logs. An event log is structured into traces, and in turn, a trace is structured into events.

Definition 1 (Event, Trace, Event Log). *An event is a tuple* $e = (c, a, t, d_1, \ldots, d_n)$*, where* c *is the process instance or case id,* a *is the activity,* t *is the timestamp, and* d_n *is the* n^{th} *context feature assigned to the event* e*. A trace is a non-empty sequence*

$\sigma = \langle e_1, \ldots, e_{|\sigma|} \rangle$ of events such that $\forall i, j \in \{1, \ldots, |\sigma|\}$ $e_i.c = e_j.c$, with $i > j$, where . denotes the process instance id c of the event e_i or e_j. A trace referring to a process instance can also be represented as a sequence of vectors $\langle \mathbf{x}_1, \ldots, \mathbf{x}_{|\sigma|} \rangle$, where $\mathbf{x} \in \mathbb{R}^m$ is a vector with size m. A vector can store all event information or a part of it (e.g., information belonging to the event's activity and its nth context feature). An event log $\mathcal{L}$ is a set of traces $\{\sigma_1, \ldots, \sigma_{|\mathcal{L}|}\}$.

The next activities are predicted based on prefixes of a trace. Labels are the next activities and represent the learning target.

Definition 2 (Prefix, Label). *A prefix is a non-empty sub-sequence of a trace $\sigma = \langle \mathbf{x}_1, \ldots, \mathbf{x}_k, \ldots, \mathbf{x}_{|\sigma|} \rangle$ with a length k. It is defined as $f_{pre}^{(k)}(\sigma) = \langle \mathbf{x}_1, \ldots, \mathbf{x}_k \rangle$, with $0 < k < |\sigma|$. For instance, possible prefixes for $\sigma_1 = \langle \mathbf{x}_1, \mathbf{x}_2, \mathbf{x}_3 \rangle$ are $\langle \mathbf{x}_1 \rangle$ or $\langle \mathbf{x}_1, \mathbf{x}_2 \rangle$. A label is an annotation for a prefix (i.e., the next activity) of a trace $\sigma = \langle \mathbf{x}_1, \ldots, \mathbf{x}_k, \ldots, \mathbf{x}_{|\sigma|} \rangle$ with a length k. It is defined as $f_l^{(k)}(\sigma) = \mathbf{a}_{k+1}$, with $0 < k < |\sigma|$, where $\mathbf{a}_{k+1}$ includes features describing the activity of the next event $\mathbf{x}_{k+1}$. For instance, possible labels for $\sigma_1 = \langle \mathbf{x}_1, \mathbf{x}_2, \mathbf{x}_3 \rangle$ are $\mathbf{a}_2$ or $\mathbf{a}_3$ only storing information of the next events' activities.*[1]

2.2 Related Work

GNNs have been used in four PBPM works so far. Philipp et al. [15] developed DNNs with two graph convolutional [5] layers for predicting real-valued process outcomes based on finished process instances. The graph layers receive an adjacency matrix that captures activity relationships of the entire event log. A graph node is assumed as an activity. It is described by a feature representing the sum of how often the activity was performed.

Harl et al. [10] developed DNNs with a gated graph [9] layer for binary process outcome prediction based on finished process instances. The authors constructed an adjacency matrix per process instance that captures the activity relationships of the respective process instance's activities. Even though they considered a node as an activity, such as Philipp et al. [15], they described it by the one-hot encoding of the activity and model edges between nodes to describe their relation type. Finally, they extracted relevance values per node from *softmax* attention to explain the GNN models' predictions. Stierle et al. [20] presented an extension of Harl et al. [10]'s approach. They aggregated the relevance values across graphs to create a process model in which each activity is colored depending on its aggregated relevance value.

Venugopal et al. [22] developed DNNs with two graph convolutional layers [5] for the next activity and next timestamp prediction based on process instance prefixes. Like Philipp et al. [15], they constructed adjacency matrices based on the entire event log. Additionally, they investigated different approaches to normalize the matrices' values. A node is assumed as an activity and described by temporal features constructed from the control-flow data of the event logs.

Against this background, we are the first to investigate the use of GGNNs for the next activity prediction that can learn sequence representations from graph data.

[1] Note: The functions $f_{pre}(\cdot)$ and $f_l(\cdot)$ can be applied to process instances σ with a log-oriented or vector-oriented representation.

3 Gated Graph Sequence NNs for Next Activity Prediction

3.1 Three Representation Forms of Process Instance Graphs

GNNs, such as GGNNs, require inputs with a graph-oriented representation. Therefore, we transform prefixes created from process instances of the event log $\mathscr{L}$ into process instance graphs. However, before we present the three investigated representation forms of process instance graphs, we define the terms *graph, node*, and *edge*.

Definition 3 (Graph, Node, Edge). *A graph g is a two-element tuple (V,E), where V is the set of nodes and E is the set of edges. A node $v \in V$ is represented by a single value, and a set of node features can be assigned to it. An edge $\mathring{e}$ is a tuple $(v,v') \in V \times V$, and a set of edge features can be assigned to it.*

GGNNs, as used here, cannot directly compute forward and backward propagation based on (raw) graphs. Therefore, we transform graphs into *instance graphs.*

Definition 4 (Instance Graph, Adjacency Matrix, Node Matrix, and Edge Matrix). *An instance graph ψ is a three-element tuple $(\mathbf{A}, \mathbf{V},$ and $\mathbf{E})$. $\mathbf{A}$ is an adjacency matrix storing which nodes of the graph are connected by an edge and lies in $\mathbb{R}^{|V|\times|V|}$, where $|V|$ is the number of nodes of the graph. $\mathbf{V}$ is a node matrix storing features that describe the graph's nodes and lies in $\mathbb{R}^{|V|\times q}$, where q is the number of node features. $\mathbf{E}$ is an edge matrix storing features that describe the edges of the graph and lies in $\mathbb{R}^{|V|\times p}$, where p refers to the number of edge features, i.e., the source node, the target node, and features describing the edge.*[2]

A process instance graph is an instance graph that represents a prefix of a process instance or an entire process instance. Because nodes and edges of a process instance graph can be interpreted in different ways, there are different forms of how a process instance graph can represent a prefix or an entire process instance. In this paper, we investigate three conceptually different representation forms of process instance graphs. The first representation form ($RF1$) assumes a node as an event, and the edges are prefix-based. The second representation form ($RF2$) assumes a node as an activity, and the edges are prefix-based. The third representation form ($RF3$) assumes a node as an activity, and edges are event-log-based. In the following, we present the three representation forms based on the exemplary prefix $f_{pre}^{(4)}(\sigma_1)$[3]:

$$\begin{aligned} f_{pre}^{(4)}(\sigma_1) = \langle &(1,\text{A},\text{2021-01-01T00:10:00},\text{Le}),\\ &(1,\text{B},\text{2021-01-02T00:10:00},\text{Hu}),\\ &(1,\text{B},\text{2021-01-03T00:10:00},\text{Hu}),\\ &(1,\text{C},\text{2021-01-04T00:10:00},\text{Le})\rangle. \end{aligned} \tag{1}$$

[2] Note: $\mathbf{E}$ can also be described as a third-order tensor E, that lies in $\mathbb{R}^{|V|\times|V|\times p}$. In this case, E's three dimensions refer to the source node, target node, and edge features.

[3] Note: In Eq. (1), the last element per event is a context feature indicating to the employee who performed the event's activity.

RF1 - A Node is an Event and Edges are Prefix-Based: The first representation form of a process instance graph assumes (time-ordered) events of a prefix as graph nodes and considers edges between the prefix's events. The process instance graph ψ_ε^{RF1} for the prefix $f_{pre}^{(4)}(\sigma_1) = \varepsilon$ is shown in Eq. (2).[4]

$$\psi_\varepsilon^{RF1} = \left(\begin{bmatrix} 0.0 & 1.0 & 0.0 & 0.0 \\ 0.0 & 0.0 & 1.0 & 0.0 \\ 0.0 & 0.0 & 0.0 & 1.0 \\ 0.0 & 0.0 & 0.0 & 0.0 \end{bmatrix}, \begin{bmatrix} A & 0.0 & 0.0 & 0.42 & 5.0 & Le \\ B & 1.0 & 1.0 & 0.42 & 6.0 & Hu \\ B & 1.0 & 2.0 & 0.42 & 7.0 & Hu \\ C & 1.0 & 3.0 & 0.42 & 1.0 & Le \end{bmatrix}, \begin{bmatrix} 1 & 2 & Forward \\ 2 & 3 & Repeat \\ 3 & 4 & Forward \end{bmatrix} \right) \quad (2)$$

The first matrix of the process instance graph ψ_ε^{RF1} is the adjacency matrix.[5] It stores the connection between events. For example, the first vector $[0.0, 1.0, 0.0, 0.0]$ indicates that an edge goes from the first event to the second event of the prefix. The second matrix is the node matrix. It stores all information for a node. In this matrix, the first feature is the activity, the next four are temporal features calculated based on the events' timestamps, and the last feature is the resource feature. We add the four temporal features to the node features proposed by Tax et al. [21]; that are, (1) the time since the previous event in the process instance, (2) the time since the process instance started, (3) the time since midnight, and (4) the day of the week for the event. Additionally, we transform the values of the first three temporal features from seconds into days to allow smooth learning of the internal model parameters; otherwise, values of these features can be very high, and therefore, negatively affect the learning of the internal model parameters. The third matrix is the edge matrix. It stores the id of the source event, the id of the target node, and the type of edge defined by the source and target node. Regarding edge types, we differ between (1) *Repeat* (activity of a target event is equal to an activity of a source event), (2) *Backward* (activity of a target event has been observed in a previous event of the current prefix), and (3) *Forward* (activity of a target event has not been observed in previous events of the current prefix).

RF2 - A Node is an Activity and Edges are Prefix-Based: The second representation form of a process instance graph assumes activities of a prefix as graph nodes and considers edges between the performed activities of the prefix. The process instance graph ψ_ε^{RF2} for the prefix $f_{pre}^{(4)}(\sigma_1)$ is shown in Eq. (3).

$$\psi_\varepsilon^{RF2} = \left(\begin{bmatrix} 0.0 & 1.0 & 0.0 \\ 0.0 & 1.0 & 1.0 \\ 0.0 & 0.0 & 0.0 \end{bmatrix}, \begin{bmatrix} A & 1.0 & 0.0 & 0.0 & 0.42 & 5.0 & Le \\ B & 2.0 & 1.0 & 1.0 & 0.42 & 6.0 & Hu \\ C & 1.0 & 1.0 & 3.0 & 0.42 & 1.0 & Le \end{bmatrix}, \begin{bmatrix} A & B & Forward \\ B & B & Repeat \\ B & C & Forward \end{bmatrix} \right) \quad (3)$$

Like in ψ_ε^{RF1}, the first matrix of the process instance graph ψ_ε^{RF2} is the adjacency matrix. In contrast to ψ_ε^{RF1}, recurring activities cannot be marked because a single node

[4] Note: For better readability, we show the raw values of the categorical features (i.e., activity, resource, and edge type) for the three representation forms and not the features' one-hot encoded values.

[5] Note: We normalize the adjacency matrix for each prefix of each representation form by $\mathbf{A} \leftarrow \mathbf{D}^{-1/2}\mathbf{A}\mathbf{D}^{-1/2}$.

represents an activity. Consequently, this adjacency matrix is smaller than the adjacency matrix of ψ_{ε}^{RF1}. Further, as Philipp et al. [15] suggested, we store the number of how often an activity has been performed in the second column of this representation form's node matrix. The other features are equal to the first representation form. However, since activities are nodes and can be performed more than once in a sequence, we follow the work of Venugopal et al. [22] and only store the feature values of an activity's last occurrence. The edge matrix stores the source activity, the target activity, and the edge type. Concerning edges, we differ again between the types (1) *Repeat*, (2) *Backward*, and (3) *Forward*.

RF3 - A Node is an Activity and Edges are Event-Log-Based: The third representation form of a process instance graph assumes activities as graph nodes and considers edges between activities of the entire event log's process instances. The process instance graph ψ_{ε}^{RF3} for the prefix $f_{pre}^{(4)}(\sigma_1)$ is shown in Eq. (4).

$$\psi_{\varepsilon}^{RF3} = \left(\begin{bmatrix} 594.0 & 515.0 & 0.0 & 355.0 & 0.0 & 0.0 \\ 4.0 & 960.0 & 209.0 & 803.0 & 0.0 & 0.0 \\ 0.0 & 1.0 & 922.0 & 751.0 & 0.0 & 8.0 \\ 0.0 & 0.0 & 0.0 & 0.0 & 0.0 & 0.0 \\ 515.0 & 0.0 & 0.0 & 235.0 & 7.0 & 0.0 \\ 2.0 & 0.0 & 33.0 & 57.0 & 32 & 0.0 \end{bmatrix}, \begin{bmatrix} A \ldots Le \\ B \ldots Hu \\ C \ldots Le \end{bmatrix}, \begin{bmatrix} A\ B\ Forward \\ B\ B\ \ Repeat \\ B\ C\ Forward \end{bmatrix} \right) \tag{4}$$

The first matrix is the adjacency matrix, representing the edges between activities of the entire event log. To determine the edges, we create a directly-follows graph (DFG) in its native variant, counting the number of directly follows occurrences in all process instances of the event log. In contrast to the previous two representation forms, the adjacency matrix is not prefix-based but event-log-based. Therefore, each process instance graph has assigned the same adjacency matrix, representing the DFG created based on the entire event log. The node and edge matrix are equally constructed like in the second representation form.

3.2 Learning Gated Graph Sequence Neural Network Models

A GGNN model's architecture can be structured into two phases, a message-passing and a readout phase [4] (see Fig. 1).

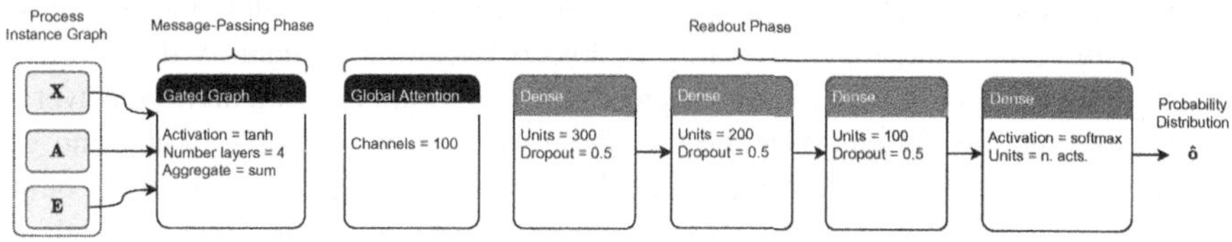

Fig. 1. Architecture of the gated graph sequence neural network model.

The message-passing phase receives as input a set of process instance graphs $\{(\mathbf{A}_1, \mathbf{V}_1, \mathbf{E}_1), (\mathbf{A}_{\ldots}, \mathbf{V}_{\ldots}, \mathbf{E}_{\ldots}), (\mathbf{A}_n, \mathbf{V}_n, \mathbf{E}_n)\}$, where n is the number of process instance graphs. This phase is realized by a gated graph layer, calculating abstract node representations $\mathbf{h}_v^{(t+1)}$ for each node v of a process instance graph in two steps. First, it calculates for each node v messages $\mathbf{m}_v^{(t+1)}$, expressing interactions between nodes, through the message function $f_{msg}^{(t)}(\cdot)$, as shown in Eq. (5).

$$\mathbf{m}_v^{(t+1)} = \sum_{v' \in f_{nbr}(v)} f_{msg}^{(t)}(\mathbf{h}_v^{(t)}, \mathbf{h}_{v'}^{(t)}, \mathbf{e}_{(v,v')}) = \sum_{v' \in f_{nbr}(v)} \mathbf{A}_{e_{(v,v')}} \mathbf{h}_{v'}^{(t)}. \tag{5}$$

In Eq. (5), the function $f_{nbr(\cdot)}$ returns for a node v its neighboring nodes v'. For each node v, the interactions to neighboring nodes are aggregated by calculating the sum of them. Consequently, the messages for each node v are $\mathbf{m}_v^{(t+1)}$. Second, this phase calculates the new node representation $\mathbf{h}_v^{(t+1)}$ based on the messages $\mathbf{m}_v^{(t+1)}$ and the previous node representation $\mathbf{h}_v^{(t)}$ by applying the node update function $f_{gru}(\cdot)$, as shown in Eq. (6).

$$\mathbf{h}_v^{(t+1)} = f_{gru}(\mathbf{h}_v^{(t)}, \mathbf{m}_v^{(t+1)}). \tag{6}$$

In Eq. (6), the function $f_{gru(\cdot)}$ is a GRU [3], with a *tanh* activation. The number of (sub-)layers within the gated graph layer is four and refers to the number of iterations with the GRU. The number of GRU iterations and the number of time steps (i.e., the sequence length) determine the number of updates. In each update, the node representations for each node of a process instance graph are updated roughly simultaneously.

The readout phase receives as input the final node representations $\mathbf{h}_v^{(T)}$ from the message-passing phase. It maps these to a probability distribution vector $\hat{\mathbf{o}}$ through the readout function $f_{rea}(\cdot)$, as shown in Eq. (7). The vector $\hat{\mathbf{o}}$ stores the probabilities of the next activities for a process instance graph. The readout function $f_{rea}(\cdot)$ is realized by a global attention layer [9], followed by four dense layers. The first three dense layers include 300, 200, and 100 neurons, and the *tanh* activation function is applied. A dropout with a ratio of 0.5 is applied to the outgoing connections of these layers to avoid overfitting. The last dense layer's number of units refers to the number of activities, and a *softmax* activation function is applied to calculate the vector $\hat{\mathbf{o}}$.

$$\hat{\mathbf{o}} = f_{rea}(\{\mathbf{h}_v^{(T)} | v \in G\}). \tag{7}$$

We update the internal parameters of the GGNN models batch-wise and over 100 epochs. Per epoch, the process instance graphs are partitioned into batches, including 32 process instance graphs. For each process instance graph of a batch, the *categorical cross-entropy* (loss function) calculates the loss between its probability distribution $\hat{\mathbf{o}}$ and the assigned ground label $\mathbf{y}$. Subsequently, the cost function calculates the average of the loss function outputs over the batch's process instance graphs. Then, the gradient descent algorithm *ADAM* calculates the derivation of the cost function's output and updates the internal parameters of the model. After completing the last epoch, the internal model parameters are adjusted.

4 Evaluation

4.1 Event Logs

In our evaluation, we used the following two real-life event logs (see Table 1 for the event logs' characteristics):

bpi2012w[6] originates from a Dutch Financial Institute. It includes three sub-processes: A (states of application), O (states of the offer belonging to the application), and W (states of work items belonging to the application). We only considered the sub-process W because work items are preformed by humans. As a context feature, we included the *resource* feature for the evaluation.

sepsis[7] stores sepsis cases. One case represents patient-related activities in a hospital. For instance, receiving treatments or taking measurements. We included the context feature *org_group* for the evaluation.

Table 1. Characteristics of event logs used.

Event log	# Instances	# Inst. variants	# Events	# Act.	# Events per Inst.*	# Context attr.'s values
bpi2012w	9,658	2,263	72,413	6	[1, 74, 7]	60
sepsis	1,050	846	15,214	16	[3, 185, 14]	26

*[min, max, mean]

4.2 Procedure

Event Logs: We only considered traces of the event logs with a size greater than two, and we created prefixes from the event log's traces with a minimum size of two for model training and application. This is necessary because prefixes with a smaller size do not provide sufficient data for predicting the next activities [21]. Further, we added an artificial *End* event after each trace. This is a typical pre-processing step in the next activity prediction because most event logs do not include such an event, and we also want to predict the end of traces [17].

Validation Strategy: We performed a stratified ten-fold cross-validation with random shuffling to keep the bias and variance of the models low [6]. We used the last 20% of each training set as the validation set. We applied early stopping to the validation set after ten epochs to avoid overfitting. For a fair comparison, we performed this validation strategy for the GGNN and the baseline models.

ML Metrics: We calculated ML metrics in two ways. First, we calculated the weighted *Precision*, *Recall*, and *F1-Score* per iteration of the ten-fold cross-validation for both event logs. Then, we calculated the average and the standard deviation over the ten values per metric. A definition of these ML metrics can be found in other PBPM works, such as Mehdiyev et al. [12]. Second, we calculated the averaged *Recall* and *Precision* over the ten folds of the cross-validation prefix-wise for the bpi2012w event log. We

[6] https://data.4tu.nl/articles/dataset/BPI_Challenge_2012/12689204/1.

[7] https://data.4tu.nl/articles/dataset/Sepsis_Cases_-_Event_Log/12707639.

start the range of prefixes with size two, increment the prefix sizes by step size one, and end the range with size 15. This prefix range allows us to assess how well the DNN models perform in the course of process instances.

Baselines: We benchmarked the GNN models with an LSTM-NN, an MLP-NN, and a CNN model. For the baselines, we describe an event of a prefix by a one-hot encoded activity, four temporal features (see Sect. 3.1), and a one-hot encoded resource value. The LSTM-NN model included one hidden LSTM layer with an internal cell element size of 100. We applied a dropout with a ratio of 0.3 to this layer's inputs to avoid overfitting. The last layer of the LSTM-NN model and the other baselines is equal to the GGNN models' final layer, calculating a probability distribution over the next activities. The multi-layer perceptron neural network (MLP-NN) included three hidden dense layers with 100 neurons. After each hidden layer, a dropout with a ratio of 0.5 is applied to the layer's outgoing connections to avoid overfitting. The convolutional neural network (CNN) included six hidden layers. The first five layers were one-dimensional convolutional layers with *max pooling*. For each convolutional layer, we set the number of filters to 128, the kernel size to 10, the padding to *same*, the strides to 1, and the activation function to *relu*. The last hidden layer was a dense layer with 100 neurons and a *relu* activation. Like for the GGNN models, we applied *ADAM* with standard values and *categorical cross-entropy* loss to optimize the baseline models' internal parameters.

4.3 Results

$GGNN_{RF1}$ achieves the highest *Precision*, *Recall*, and *F1-Score* for both event logs (see Table 2). For the bpi2012w event log, $GGNN_{RF2}$ outperforms $GGNN_{RF3}$ and all baselines regarding each metric, except the *Precision* of *MLP-NN*. $GGNN_{RF3}$ also achieves a higher *Recall*, and *F1-Score* than the baselines for bpi2012w, and its *Precision* is lower than for *MLP-NN*. For the sepsis event log, $GGNN_{RF2}$ has a higher *Precision* than $GGNN_{RF3}$. While $GGNN_{RF2}$ outperforms $GGNN_{RF3}$, *MLP-NN*, and *CNN* regarding *Recall*, it achieves a higher *F1-Score* than *CNN* and $GGNN_{RF3}$. For sepsis, $GGNN_{RF3}$ achieves the lowest predictive quality regarding all metrics.

Looking at the *Precision* and *Recall* values (see Fig. 2), $GGNN_{RF1}$ has the highest values for most of the prefix sizes. In contrast, *Precision* and *Recall* values for $GGNN_{RF2}$ and $GGNN_{RF3}$ are only higher than the baselines for certain prefix sizes.

5 Discussion

Findings: We derived two main findings from our results. First, *RF*1, where each event is represented as a graph node, works better for GGNNs than the other two representation forms, assuming activities as graph nodes (see Table 2 and Fig. 2). This effect can be attributed to the fact that *RF*1 stores all event information, even in the case of recurrent activities in process instances. Second, GGNNs with all representation forms perform better than the baselines for bpi2012w as for sepsis (see Table 2). We suppose that GGNNs can better learn from process instance graphs with a homogeneous structure than with a diffuse structure. Consequently, GGNNs, as used in this paper, may

Table 2. Predictive quality of the GGNN and baseline models (average over ten folds).

Event log	DL model	Precision (w)	Recall (w)	F1-Score (w)
bpi2012w	$GGNN_{RF1}$	**0.7941** (±0.0056)	**0.7834** (±0.0046)	**0.7817** (±0.0048)
	$GGNN_{RF2}$	0.7742 (±0.0044)	0.7768 (±0.0030)	0.7646 (±0.0044)
	$GGNN_{RF3}$	0.7724 (±0.0110)	0.7724 (±0.0050)	0.7572 (±0.0063)
	$LSTM\text{-}NN$	0.7610 (±0.0065)	0.7707 (±0.0052)	0.7559 (±0.0066)
	$MLP\text{-}NN$	0.7763 (±0.0076)	0.7649 (±0.0091)	0.7553 (±0.0164)
	CNN	0.7496 (±0.0043)	0.7579 (±0.0058)	0.7505 (±0.0035)
sepsis	$GGNN_{RF1}$	**0.6141** (±0.0113)	**0.6192** (±0.0099)	**0.6050** (±0.0100)
	$GGNN_{RF2}$	0.5851 (±0.0193)	0.5956 (±0.0134)	0.5816 (±0.0130)
	$GGNN_{RF3}$	0.4614 (±0.0305)	0.5018 (±0.0154)	0.4591 (±0.0273)
	$LSTM\text{-}NN$	0.6101 (±0.0105)	0.6137 (±0.0068)	0.6040 (±0.0072)
	$MLP\text{-}NN$	0.5892 (±0.0134)	0.5913 (±0.0088)	0.5819 (±0.0088)
	CNN	0.5881 (±0.0094)	0.5864 (±0.0082)	0.5800 (±0.0071)

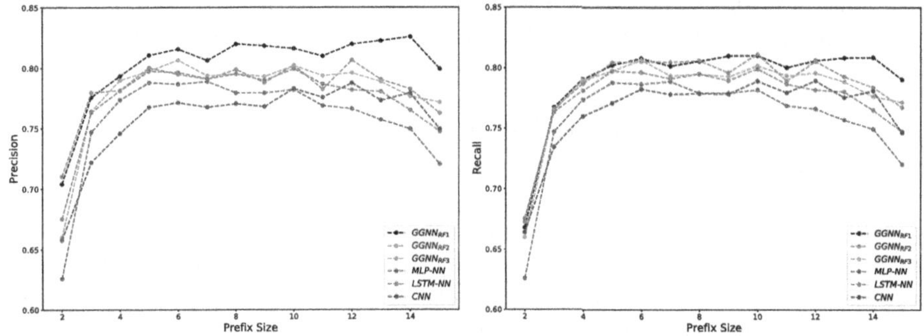

Fig. 2. *Precision* and *Recall* of the GGNN and Baseline models per prefix size for bpi2012w (Average over Ten Folds).

perform insufficiently for the sepsis event log because it includes a higher number of different activities and a lower instance per variant ratio than the bpi2012w event log.

The Problem of Recurrence with GNNs: While $RF1$ assumes a node as event, $RF2$ and $RF3$ consider a node as activity. Naturally, an activity can be performed more than once within a process instance. However, a node is typically described by a static feature vector, and it does not exist an individual feature vector for each repetition of an activity. To address this problem, we follow the suggestion by Venugopal et al. [22] and only store the node features of the last occurrence of an activity. To address this problem, a sliding window approach may help to exploit more of the historical information of a node that represents an activity. For example, with a window of size three, information on the last three activity occurrences can be stored in a node vector. Another approach to address this problem can be creating an embedding vector per feature and calculating the average over the vectors of the same feature. For example, such embedding vectors can be created using *word2vec* [13].

Limitations: Even though the results are promising, our work has two limitations. First, we selected standard values for most hyper-parameters of the DNN models. We suppose that the DNN model's predictive quality can be further improved if their hyper-parameters are selected for each event log by performing a hyper-parameter search method such as tree-structured parzen estimators. Second, the third representation form's adjacency matrix reflects a DFG created based on the entire event log. Since a DFG is also created based on prefixes from the test set, information from the test set is used in the training phase of these models. However, we decided to create the DFGs based on the entire event log to avoid a representation bias between a DFG created based on the training set and prefixes from the test set. A promising approach to address this bias can be trace alignment [8].

6 Conclusion

This paper investigated GGNNs for the prediction of the next activities in operational business processes. Motivated by the high expressiveness of GGNNs, we explored three conceptually different representation forms of process instance graphs. Our results show that the first representation form, assuming events as nodes and using a prefix-based adjacency matrix, works best, and GGNNs with this representation form achieve a higher *Precision*, *Recall*, and *F1-Score* than the baselines for both event logs used.

In future research, we plan to develop new GNN architectures for PBPM, such as spatial, temporal GCNs, combining graph convolutions with LSTM cells or GRUs, or graph-based transformer networks. It would also be interesting to investigate unsupervised graph-based approaches, such as *graph2vec* [14], to learn expressive representations of event log data that can be used as input for subsequent models, addressing a specific prediction task like the next activity prediction. Another avenue for future research is to investigate different discovery algorithms for process model creation. Since the created process models can be used as the underlying structure for process instance graphs, DL-based PBPM techniques can be developed that are process-aware [11]. Finally, we interpreted the next activity prediction as a graph-based classification task in this paper. However, since GNNs can also be designed for node-based or edge-based classification tasks, researchers can poof whether and how these classification approaches can be applied in BPM.

References

1. Bishop, C.M.: Pattern Recognition and Machine Learning. Springer, Heidelberg (2006)
2. Breuker, D., Matzner, M., Delfmann, P., Becker, J.: Comprehensible predictive models for business processes. MIS Q. **40**(4), 1009–1034 (2016)
3. Cho, K., van Merriënboer, B., et al.: Learning phrase representations using RNN encoder-decoder for statistical machine translation. In: Proceedings of the Conference on Empirical Methods in Natural Language Processing (EMNLP), pp. 1724–1734 (2014)

4. Gilmer, J., Schoenholz, S.S., Riley, P.F., Vinyals, O., Dahl, G.E.: Neural message passing for quantum chemistry. In: Proceedings of the 34th International Conference on Machine Learning (ICML), pp. 1263–1272. PMLR (2017)
5. Kipf, T.N., Welling, M.: Semi-supervised classification with graph convolutional networks. In: Proceedings of the 4th International Conference on Learning Representations (ICLR) (2016)
6. Kohavi, R., et al.: A study of cross-validation and bootstrap for accuracy estimation and model selection. In: Proceedings of the 14th International Joint Conference on Artificial Intelligence (IJCAI), pp. 1137–1145 (1995)
7. Wick, C.: Deep learning. Informatik-Spektrum **40**(1), 103–107 (2016). https://doi.org/10.1007/s00287-016-1013-2
8. Leemans, S.J., Poppe, E., Wynn, M.T.: Directly follows-based process mining: exploration & a case study. In: Proceedings of the 1st International Conference on Process Mining (ICPM), pp. 25–32. IEEE (2019)
9. Li, Y., Tarlow, D., et al.: Gated graph sequence neural networks. In: Proceedings of the 4th International Conference on Learning Representations (ICLR) (2016)
10. Harl, M., Weinzierl, S., Stierle, M., Matzner, M.: Explainable predictive business process monitoring using gated graph neural networks. J. Decis. Syst. (2020)
11. Márquez-Chamorro, A.E., Resinas, M., Ruiz-Cortés, A.: Predictive monitoring of business processes: a survey. IEEE Trans. Serv. Comput. **11**(6) (2017)
12. Mehdiyev, N., Evermann, J., Fettke, P.: A novel business process prediction model using a deep learning method. Bus. Inf. Syst. Eng. **62**(2), 143–157 (2020)
13. Mikolov, T., Sutskever, I., Chen, K., Corrado, G., Dean, J.: Distributed representations of words and phrases and their compositionality. In: Proceedings of the 26th Advances in Neural Information Processing Systems (NIPS) (2013)
14. Narayanan, A., Chandramohan, M., Venkatesan, R., Chen, L., Liu, Y., Jaiswal, S.: graph2vec: learning distributed representations of graphs. In: Proceedings of the 16th International Workshop on Mining and Learning Graphs (2017)
15. Philipp, P., Georgi, R.X.M., Beyerer, J., Robert, S.: Analysis of control flow graphs using graph convolutional neural networks. In: Proceedings of the 6th International Conference on Soft Computing & Machine Intelligence (ISCMI), pp. 73–77. IEEE (2019)
16. Poll, R., Polyvyanyy, A., Rosemann, M., Röglinger, M., Rupprecht, L.: Process forecasting: towards proactive business process management. In: Weske, M., Montali, M., Weber, I., vom Brocke, J. (eds.) BPM 2018. LNCS, vol. 11080, pp. 496–512. Springer, Cham (2018). https://doi.org/10.1007/978-3-319-98648-7_29
17. Rama-Maneiro, E., Vidal, J.C., Lama, M.: Deep Learning for Predictive Business Process Monitoring: Review and Benchmark. arXiv preprint arXiv:2009.13251 (2020)
18. Rizzi, W., Di Francescomarino, C., Maggi, F.M.: Explainability in predictive process monitoring: when understanding helps improving. In: Fahland, D., Ghidini, C., Becker, J., Dumas, M. (eds.) BPM 2020. LNBIP, vol. 392, pp. 141–158. Springer, Cham (2020). https://doi.org/10.1007/978-3-030-58638-6_9
19. Scarselli, F., Gori, M., Tsoi, A.C., Hagenbuchner, M., Monfardini, G.: The graph neural network model. IEEE Trans. Neural Netw. **20**(1), 61–80 (2008)
20. Stierle, M., Weinzierl, S., Harl, M., Matzner, M.: A technique for determining relevance scores of process activities using graph-based neural networks. Decis. Support Syst. **144**, 113511 (2021)
21. Tax, N., Verenich, I., La Rosa, M., Dumas, M.: Predictive business process monitoring with LSTM neural networks. In: Dubois, E., Pohl, K. (eds.) CAiSE 2017. LNCS, vol. 10253, pp. 477–492. Springer, Cham (2017). https://doi.org/10.1007/978-3-319-59536-8_30

22. Venugopal, I., Töllich, J., Fairbank, M., Scherp, A.: A comparison of deep-learning methods for analyzing and predicting business processes. In: Proceedings of the International Joint Conference on Neural Networks (IJCNN) (2021)
23. Verenich, I., Dumas, M., La Rosa, M., Nguyen, H.: Predicting process performance: a white-box approach based on process models. J. Softw.: Evol. Process **31**(6), e2170 (2019)
24. Weinzierl, S., Revoredo, K.C., Matzner, M.: Predictive business process monitoring with context information from documents. In: Proceedings of the 27th European Conference on Information Systems (ECIS), pp. 1–10. AISeL (2019)
25. Zhou, J., Cui, G., et al.: Graph neural networks: a review of methods and applications. AI Open **1**, 57–81 (2020)

ProGAN: Toward a Framework for Process Monitoring and Flexibility by Change via Generative Adversarial Networks

Maximilian Hoffmann[1,2(✉)], Lukas Malburg[1,2], and Ralph Bergmann[1,2]

[1] Artificial Intelligence and Intelligent Information Systems, University of Trier, 54296 Trier, Germany
{hoffmannm,malburgl,bergmann}@uni-trier.de

[2] German Research Center for Artificial Intelligence (DFKI) Branch University of Trier, Behringstraße 21, 54296 Trier, Germany
{maximilian.hoffmann,lukas.malburg,ralph.bergmann}@dfki.de
http://www.wi2.uni-trier.de

Abstract. Monitoring the state of currently running processes and reacting to deviations during runtime is a key challenge in *Business Process Management (BPM)*. The MAPE-K control loop describes four phases for approaching this challenge: Monitor, Analyze, Plan, Execute. In this paper, we present the ProGAN framework, an idea of an approach for implementing the monitor, analyze, and plan phases of MAPE-K. For this purpose, we leverage a deep learning architecture that builds upon *Generative Adversarial Networks (GANs)*: The discriminator is used for monitoring the process in its environment by using sensor data and for detecting deviations w.r.t. the desired process state (monitor phase). The generator is used afterwards for analyzing the detected deviation and its symptoms as well as for adapting the current process to resolve the deviation and to restore the desired state. Both components are trained together by utilizing each other's feedback in a self-supervised way. We demonstrate the application of our approach for an exemplary scenario in the manufacturing domain.

Keywords: Business process prediction · Generative Adversarial Networks · Flexibility by change · Process adaptation

1 Introduction

Business Process Management (BPM) [3] is a research field of high interest with different areas of usage, e.g., in companies, in public administration, or in smart environments [23] such as smart homes or smart factories (e.g., [11,13,14,21]). Especially environments that are characterized by high context sensitivity and

M. Hoffmann and L. Malburg—These authors contributed equally to the work.

A. Marrella and B. Weber (Eds.): BPM 2021 Workshops, LNBIP 436, pp. 43–55, 2022.
https://doi.org/10.1007/978-3-030-94343-1_4

huge amounts of real-time data can benefit from the advantages of BPM solutions [7]. One such key advantage is the ability to react to changes in the environment flexibly [11] to satisfy process performance indicators or to improve the processes. In order to check whether the actual state of an executed process in the environment is accordant with its desired state w.r.t. its process model, MAPE-K (*Monitor*, *Analyze*, *Plan*, and *Execute* phases as well as the knowledge *K* [6]) control loops can be implemented [22,23]. Several approaches that can be applied for the monitoring and analyze phases of the MAPE-K control loop utilize deep learning techniques and, primarily, monitor processes based on predictions of the next event or the outcome of a process (e.g., [10,18–20,24,25]). Other approaches based on deep learning (e.g., [15]) or based on other AI techniques (e.g., [14,23,28]) can be used for the plan phase by adapting the processes to the current environmental context by inserting, deleting, or modifying the activities of a process. In this paper, we propose to link the monitor phase with the analyze and plan phases to resolve deviations between the actual and desired state during process execution in order to reduce the manual effort for the user. For this purpose, we present a new framework, called ProGAN, for applying *Generative Adversarial Networks (GANs)* [5] to monitor the current environment by sensors and to detect deviations during process execution as well as to further analyze and plan process adaptations to enable (momentary) flexibility by change [26] for resolving at runtime. GANs, or deep learning approaches in general, are a suitable approach to tackle the described problem due to their learning capabilities that reduce the effort of knowledge acquisition compared to other AI approaches. We also want to leverage the learning capabilities by closely linking the monitor, analyze, and plan phases with the components of the GAN architecture. The aim of this joint integration is to improve the performance of discriminator and generator with the reciprocal feedback of each other. The paper is structured as follows: Sect. 2 shows foundations of our semantic graph representation, GANs, and MAPE-K control loops and, additionally, presents related work. Further, Sect. 3 presents our concept for using GANs for process prediction (monitor phase) and flexibility by change (analyze and plan phases). Section 4 presents a specific explanation of our proposed framework in the context of a real-world exemplary scenario from the manufacturing domain. Finally, Sect. 5 concludes the paper and shows areas of future work.

2 Foundations and Related Work

The foundations comprise an introduction to our process representation format, i.e., semantically annotated graphs (see Sect. 2.1), Generative Adversarial Networks (GANs; see Sect. 2.2), and MAPE-K control loops (see Sect. 2.3). We also cover related work from the fields of (predictive) process monitoring and adaptation, papers with and without a deep learning background (see Sect. 2.4).

2.1 Semantic Graph Process Representation

In our work, we use semantically annotated graphs named *NEST* graphs introduced by Bergmann and Gil [2] to represent processes. In contrast to other process representations such as Petri nets or BPMN 2.0, semantic graphs allow to model complex semantic information for its nodes and edges as well as to represent state changes during the production by using explicitly modeled data nodes, e.g., for intermediate products. Nevertheless, the presented concept is independent of the underlying process representation and, in fact, we use the de facto standard workflow modeling language BPMN to model the processes and afterwards convert them into NEST graphs.

NEST graphs consists of four elements: *N*odes, *E*dges, *S*emantic descriptions, and *T*ypes. Figure 1 illustrates a data-driven sheet metal manufacturing process represented as a *NEST* graph with semantic annotations. One type of nodes are task nodes that represent manufacturing operations in the process (e.g., `Burn`) and another type are data nodes that specify the material that is processed or produced during the execution of a task (e.g., `Sheet Metal`). The set of edge types consists of control-flow edges that define the execution order in a process (e.g., the task `Unload from Warehouse` is executed before the `Pick Up from Warehouse and Transport to Oven` task). Another edge type are data-flow edges that specify the flow of the material in the manufacturing process, for example, the `Burn` task processes the unprocessed steel slab and produces a rolled piece of sheet metal of a certain size and thickness. Semantic descriptions can be used to attach specific information to nodes and edges. They are based on a semantic metadata language (e.g., ontologies [9]) that expresses domain-independent and domain-dependent knowledge. In the *NEST* graph illustrated in Fig. 1, semantic descriptions are used to express properties of the tasks (e.g., the status of the resource that executes the manufacturing operation or the concrete parameter settings) or of the data nodes (e.g., the position of the material or the concrete properties such as the size). By using *NEST* graphs, it is possible to express processes as semantically annotated graphs.

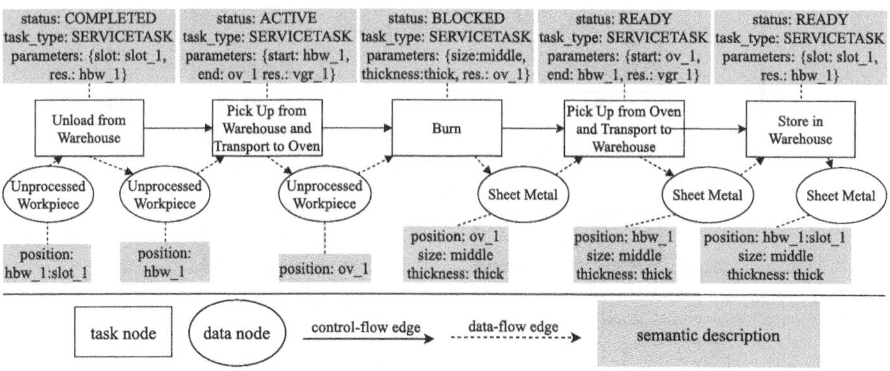

Fig. 1. Exemplary semantic graph representing a sheet metal manufacturing process.

2.2 Generative Adversarial Networks

Generative Adversarial Networks (GANs) introduced by Goodfellow et al. [5] leverage an adversarial training procedure of two deep learning models, i.e., a *Generator* G and a *Discriminator* D. Training both models recreates a two-player game situation where D tries to correctly distinguish generated samples of G from original training examples. At the same time, G tries to generate samples from random input noise that are classified as training examples by D. The classification output of D is usually a binary value (sample comes from the training examples or not). This training strategy has the advantage that both parts are trained in a self-supervised way, leading to a generator G that generates samples matching the training data distribution and a discriminator D that learns to detect samples from the training data distribution.

2.3 MAPE-K Control Loops

MAPE-K control loops [6] provide a blueprint for self-managing information systems (see Fig. 2). The entity of interest is the managed element such as, in the example, a process (cf. [22,23]).

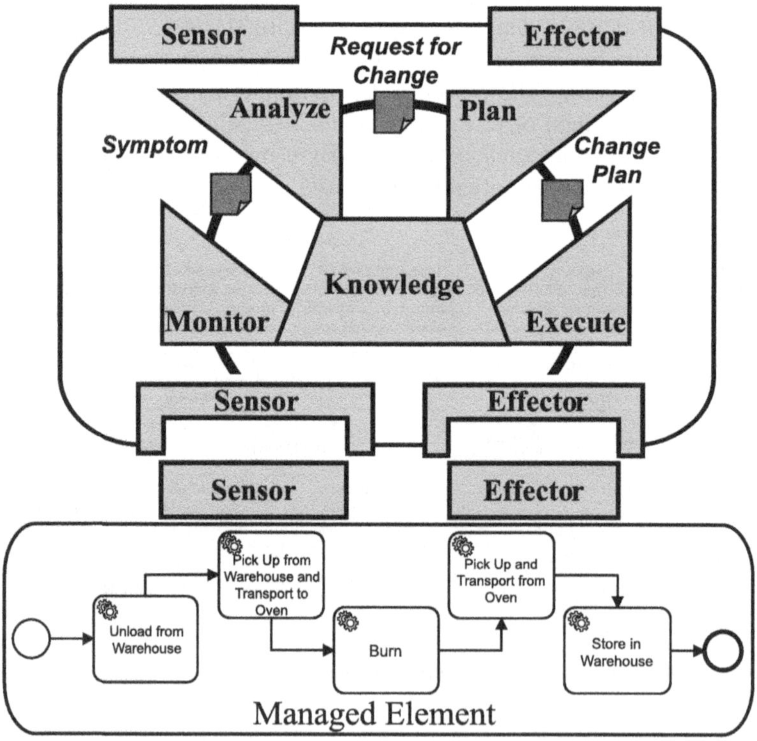

Fig. 2. MAPE-K control loop for processes (Based on: [6] and [23])

The MAPE-K cycle consists of four phases that are designed as a continuous loop:

1. *Monitor*: The environment in which the managed element is executed is continuously monitored by using sensor data. In case deviations are detected in this data (*Symptoms*), the data is further analyzed.
2. *Analyze*: In the analyze phase, the current situation and the symptoms are checked in more detail. Based on this, a *Change Request* is created.
3. *Plan*: Based on the *Change Request*, the plan phase creates a *Change Plan* with corresponding actions. Applying these actions to the managed element leads to the process no longer deviating from the desired state.
4. *Execute*: Eventually, the *Change Plan* is executed in the environment in which the managed element operates.

Knowledge is shared between the methods of all phases. Relevant knowledge for autonomic systems can be topology information, historical logs, metrics, symptoms, and policies.

2.4 Related Work

Related work for our paper covers approaches based on deep learning (e.g., [4,10,15,18,19,24,27]) and based on other techniques (e.g., [14,23,28]) for business process monitoring and adaptation of processes. Rama-Maneiro et al. [19] provide a survey of different deep learning techniques, benchmarked in predictive process monitoring tasks on multiple process logs. See their work for a systematic overview from which we want to highlight some approaches in the following: Tax et al. [24] and Evermann et al. [4] apply *Long-Short Term Memory Neural Networks (LSTMs)* to perform business process monitoring, in particular next activity prediction based on event logs. In [10], Kratsch et al. compare deep learning and classical machine learning approaches in the context of business process monitoring by predicting process outcomes on publicly available event logs. Poll et al. [18] present their framework for proactive process forecasting, what can be seen as a paradigm of proactively predicting future process behavior rather than reactively responding to happened events. In [15], Metzger et al. approach the task of process monitoring with a novel online reinforcement learning setup where they learn when to trigger alerts based on process performance indicators. Weinzierl et al. [27] present an approach of prescriptive business process monitoring where process predictions of currently executed processes are examined regarding *Key Performance Indicators (KPIs)*. The presented approach utilizes a learning method that can be used for recommending beneficial actions that are optimized regarding certain KPIs. A new type of approach in context of predictive process monitoring is presented by Taymouri et al. [25]: They use *Generative Adversarial Networks (GANs)* and show that the respective adversarial learning paradigm outperforms other approaches based on non-adversarial neural network architectures. In their work, the trained generator can be used for process predictions, e.g., predicting the next label given a process trace.

In [14], Marrella et al. present the *SmartPM* system that detects deviations during process execution. Their system is motivated by knowledge-intensive processes such as in emergency management scenarios with structured processes and ad-hoc exceptions. To resolve deviations during process execution, SmartPM uses automated planning techniques that reduce the gap between the physical reality and the expected reality. Similar to this is the *PROtEUS* system presented by Seiger et al. [22,23] that enables self-healing processes in smart homes by compensating occurred mismatches in the environment by searching for other resources that can perform the required activity. Similar to our proposed approach, they also apply MAPE-K control loops to cyber-physical processes in order to check the synchronization and consistency of them in smart environments. Wieland et al. [28] present an approach for using situation-aware adaptive workflows in the manufacturing domain. In addition to the standard workflow model, situational workflow fragments are manually constructed that define which actions should be performed in certain real world contexts to adapt the currently running process.

Compared to our self-supervised learning approach, the presented ones require sophisticated domain knowledge such as for automated planning (e.g., [14]) or need manually modeled compensations (e.g., [28]) to resolve deviations in the managed environment. In addition, the single phases of the MAPE-K loop are considered separately, whereas we combine the monitor with the analyze and plan phases in one approach to enable self-supervised learning.

3 Implementing MAPE-K Control Loops with GANs

This section contains the definition of our ProGAN framework that uses GANs in an implementation of the MAPE-K control loop. In this section, we first elaborate on the generic GAN architecture including its training procedure and relevant data. Given this foundation, we explain how this procedure is integrated in ProGAN and what characterizes the data in our GAN setup (see Sect. 3.1). Afterwards, we discuss design and application of the trained generator and discriminator w.r.t. the MAPE-K control loop (see Sect. 3.2).

3.1 Training Procedure and Data

We use a similar training procedure as in the generic setup (see Sect. 2.2) and apply it to data from the context of business processes. Figure 3 visualizes the GAN's training procedure and components with data being depicted as cylinders and data-flow as arrows. We also redefine the purpose of the discriminator that is used to monitor running processes as managed elements in the MAPE-K loop as well as the generator that is used to analyze the observed deviation and to generate adapted processes that are in line with the desired state, i.e., resolve the deviation (see Sect. 3.2). A central aspect in our framework is the concept of deviation, which is important for the training procedure, all involved data, and the relation to MAPE-K control loops. We do not technically define this concept but rather harness the capabilities of GANs to automatically learn to recognize deviations between desired and observed state of the managed element and how to resolve them. This also allows detecting deviations at different granularities which can be handled by using different neural network models for generator and discriminator, coupled with the respective datasets. For instance, a deviation can be on the process level where a certain constellation of tasks indicates an undesired state. Deviations can also occur on the task level when observing undesired resource allocations or misconfigured task parameters.

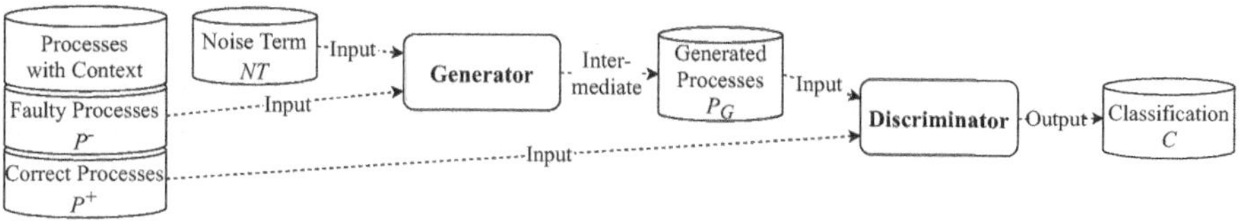

Fig. 3. Training procedure and data of the GAN architecture.

The base for learning in the ProGAN framework is given by the training data $P = P^- \cup P^+$. As illustrated in Fig. 3, the training data serves as input to the discriminator and the generator. Our approach uses traces of executed processes represented as semantic graphs (see Fig. 1) as training data where each process also contains contextual data about its environment. For instance, in smart environments this contextual data could consist of sensor data related to the process execution (e.g., states of machines, light barriers). In addition, an ontology of a smart factory (cf. [9]) could provide further data to be integrated into the learning process. In this regard, the approach benefits from the deep learning infrastructure and its ability of automatically learning patterns. This allows to include a wide range of information with a reduced expert knowledge acquisition effort, which partly limits other approaches (e.g., automated planning techniques [14] or manually modeled compensations [28]). Further, the training data is split-up into two groups of processes, i.e., processes with corresponding contextual data in which these managed elements show deviations from their desired state (P^-) and those that do not (P^+). The groups serve different purposes in the training procedure. Thereby, P^- is used as a generator input to enable the generator to learn how to resolve deviations and P^+ as a discriminator input to enable the discriminator to learn how processes without deviations are characterized.

Closely related to P is P_G. P_G has the same shape as the training data, i.e., executed processes, but has a different semantic. It stems from the generator that has the goal of reducing the deviation between desired state and actual state. Since the generator learns to generate adapted processes without deviations given data from P^-, P_G contains processes that are in line with the desired state. Besides P^-, the generator also uses additional input data NT for generating processes from P_G. NT is a noise term that is the only input of the generator according to the GAN's original definition [5] (see Sect. 2.2). We also integrate this component into our learning procedure in order to allow a certain amount of randomness for the generated data. As demonstrated in literature (e.g., [16]), this combination of contextual data – processes and the environmental state in our case – and random noise serves as a good input for suitable training generalization and enables new, unseen generated samples.

The discriminator takes P^+ and P_G as input and produces the output C. We reused the GAN's original definition [5] of C which is a binary classification of each example, indicating whether this example is from the training dataset or was created by the generator. Implicitly, this enables the discriminator to learn how processes without deviations look like, i.e., P^+, and what distinguishes them from generated processes, i.e., P_G. In a converged learning state, the discriminator should not be able to distinguish both types of processes which indicates a trained generator and a trained discriminator.

3.2 Usage of Generator and Discriminator in MAPE-K Control Loops

After being trained, the generator and discriminator can be used for different tasks regarding the MAPE-K control loop (see Sect. 2.3). The basic setup is as follows: The discriminator is used for monitoring the managed element by using sensor data as input besides the state of the process (monitor phase). Based on detected deviations, the generator is used for an in-depth analysis of the *Symptom* of the deviation and generates an adapted process to reduce the occurred deviation between the observed and the desired process state in the environment (analyze and plan phase). Afterwards, a *Workflow Management System (WfMS)* is used to execute the resulting adapted process. The joint integration of discriminator and generator in a framework to implement MAPE-K control loops distinguishes our approach from previous proposed ones where no direct integration between monitoring, analysis, and the plan phase is examined (see Sect. 2.4). The aim of this joint integration is to improve the performance of discriminator and generator with the reciprocal feedback of each other. We expect better results of this setup, compared to standalone methods for each phase that are not applied or trained in combination.

Discriminator: We apply the discriminator in the ProGAN framework for monitoring processes in their environment by using sensor data and to check whether the processes correspond to their desired state (see monitor phase). This refers to the goal of detecting deviations between the desired state and the currently observed state during process execution (similar to [22,23]). Thereby, the prediction result is a binary indication which corresponds to the *Symptom* that the current process shows a deviation (1) or not (0). Since a binary indication does not provide information on the kind of deviation observed, we only propose to use the fast inference of the discriminator in the monitor phase.

Generator: In the ProGAN framework, we propose to apply the generator in the analyze and in the plan phases. The generator is employed as a single entity to be used in both phases at once since planning an adapted process that resolves the deviation also requires the prior analysis of weaknesses. How the current deviation can be analyzed and how the desired state can be restored by an adapted process, is automatically learned during offline training of the GAN. A limitation of this single entity is that the request for change is not explicitly defined, but rather implicitly reflected in the change plan, i.e., the adapted process, at the end of the plan phase. This hinders explainability of the change plan and the performed process adaptations since no separate cause of the deviation is given. However, these problems are also faced in many other DL-based approaches, and they are already considered in research (e.g., [8]).

4 Application Scenario: Using GANs to Implement MAPE-K Control Loops

This section presents an application scenario of the ProGAN framework in the manufacturing domain. Our use case deals with structural adaptations in which tasks or whole process fragments need to be adjusted for correct execution and with substitutional adaptations in which parts of the semantic description, e.g., parameter settings of an activity, need to be modified (see [1, pp. 225–230] for more details on adaptation types). We refer to the already presented manufacturing process from Fig. 1: It is assumed that the oven that performs the `Burn` activity is not available (e.g., defective; see status "blocked" of the task node). In such a case, there may be a need for more than just a simple modification of the parameter configuration. The reason for this is that a second alternative oven is located at a different position as the planned oven. In addition to changing the parameters of which resource (in this case oven) performs the corresponding activity, other transport routes that are valid for the current situation must also be added accordingly and the semantic correctness considering the subsequent tasks must be ensured, i.e., the positions of the workpieces that are captured by the data nodes must be considered. Figure 4 depicts an excerpt of the adapted process from Fig. 1 with two inserted transportation operations before and after the burn activity to re-establish the semantic correctness (i.e., correct positioning of the workpieces) of the manufacturing process as well as the change of the resource for the `Burn` activity (both in bold font).

The training data P in this use case stems from a repository that stores previously executed processes. These processes contain contextual information on the utilization of the machines and states of the resources in the factory throughout the execution of the process. In addition, a domain ontology of the model factory [9] is provided as input data in order to make knowledge about the factory layout such as transport routes, machines, and sensor positions available. P is split-up into P^+, i.e., successfully executed processes, and P^-, i.e., not properly executed processes, according to the learning procedure in Sect. 3.1. The generator is fed with processes from P^- and the contextual information in order to generate processes, i.e., P_G, that reflect the desired state. We propose to use a definition where the generator generates an adapted process that corresponds

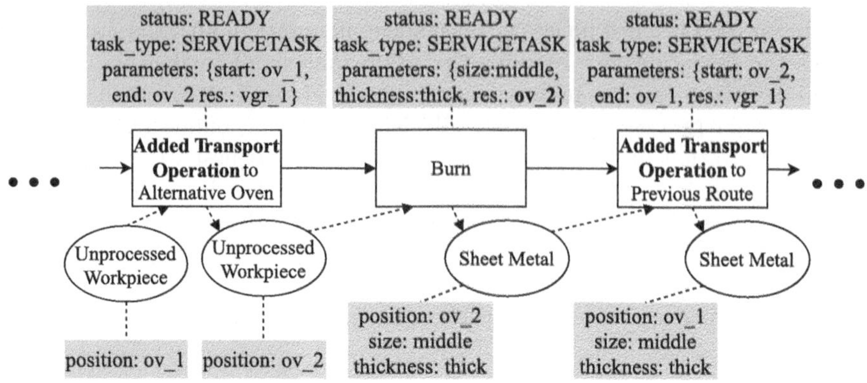

Fig. 4. Exemplary adapted process of the generator with modified task sequence.

again to the desired state by resolving the deviation. The discriminator has the goal to distinguish between processes without misconfigured tasks from P and processes with (possibly) misconfigured tasks from P_G. C is used originally as a binary indicator for each individual task.

We propose to use both components after training together in a MAPE-K control loop where the discriminator is applied in the monitor phase and the generator is applied in the analysis and plan phases. The goal of the discriminator is to monitor whether the process deviates from its desired state or not. For this purpose, the discriminator could use the provided contextual data such as the current utilization and states of the factory resources. In case a deviation during the process execution is detected, the generator analyzes the current situation and tries to solve the problem in the plan phase by generating an adapted process based on the deviated process. In this context, the generator should create an adapted process where two process fragments are inserted, i.e., the new transport routes, and the resource in the burn activity is modified, i.e., by replacing the current activity with the old parameter configuration with a new one. If the generator is not able to generate an adapted process that leads to a resolution of the deviation, a repository of process fragments resulting from inductive learning methods (see [17] for more details) can also be used by the generator for problem-solving or as additional training data.

5 Conclusion and Future Work

We presented the ProGAN framework as a first step toward an approach for using *Generative Adversarial Networks (GANs)* as an implementation of MAPE-K control loops. We apply the discriminator for business process monitoring of the managed element and its environment and the generator for enabling momentary flexibility by change in the analyze and plan phases of the MAPE-K control loop. Thereby, both components of the GAN, i.e., the generator and the discriminator, are trained together in a self-supervised way by using data in form of executed processes along with contextual data, e.g., state of machines, sensor data. We further described the application of the discriminator in the monitor phase of MAPE-K and the generator in the analyze and plan phases. Consequently, we demonstrated this application for a real-world BPM scenario from the manufacturing domain. Both the generator and the discriminator show potential of being used in similar scenarios.

In future work, we intend to further extend and implement the ProGAN framework as well as provide a technical description of certain implementation scenarios with well-defined data and functions of our GAN components. We also want to extend the discriminator to perform multi-label classifications, providing a more detailed analysis of the detected deviation which is suitable to be used in combination with the generator in the analyze and plan phases of MAPE-K. This also enables to use discriminator and generator separately after combined training. Furthermore, we also want to show the feasibility of our proposed framework in an experimental evaluation. As already shown in our demonstration, we want to focus on manufacturing processes that are used in the context of our Fischertechnik physical factory simulation model[1] at the University of Trier [9,11,13]. This allows us to evaluate against other approaches for process monitoring and our already developed methods for process adaptations [17]. Additionally, we want to compare and combine our ProGAN framework with other approaches to enable flexibility by change such as automated AI planning techniques (e.g., Use Case 2 in [11] or [14]). The combination of the predictions of the discriminator with video-based methods and complex event processing of sensor data to implement a multi-modal process monitoring tool seems to be promising [12].

References

1. Bergmann, R. (ed.): Experience Management: Foundations, Development Methodology, and Internet-Based Applications. LNCS, vol. 2432. Springer, Heidelberg (2002). https://doi.org/10.1007/3-540-45759-3
2. Bergmann, R., Gil, Y.: Similarity assessment and efficient retrieval of semantic workflows. Inf. Syst. **40**, 115–127 (2014)
3. Dumas, M., et al.: Fundamentals of Business Process Management. Springer, Heidelberg (2018). https://doi.org/10.1007/978-3-662-56509-4
4. Evermann, J., Rehse, J., Fettke, P.: Predicting process behaviour using deep learning. Decis. Support Syst. **100**, 129–140 (2017)
5. Goodfellow, I.J., et al.: Generative adversarial nets. In: Advances in Neural Information Processing Systems, vol. 27, pp. 2672–2680 (2014)
6. IBM: An architectural blueprint for autonomic computing: Autonomic Computing White Paper (2006)
7. Janiesch, C., et al.: The Internet of Things meets business process management: a manifesto. IEEE Syst. Man Cybern. Mag. **6**(4), 34–44 (2020)

[1] https://iot.uni-trier.de.

8. Keane, M.T., Kenny, E.M.: How case-based reasoning explains neural networks: a theoretical analysis of XAI using *Post-Hoc* explanation-by-example from a survey of ANN-CBR twin-systems. In: Bach, K., Marling, C. (eds.) ICCBR 2019. LNCS (LNAI), vol. 11680, pp. 155–171. Springer, Cham (2019). https://doi.org/10.1007/978-3-030-29249-2_11
9. Klein, P., Malburg, L., Bergmann, R.: FTOnto: a domain ontology for a Fischertechnik simulation production factory by reusing existing ontologies. In: Proceedings of the Conference on LWDA, vol. 2454, pp. 253–264. CEUR-WS.org (2019)
10. Kratsch, W., Manderscheid, J., Röglinger, M., Seyfried, J.: Machine learning in business process monitoring: a comparison of deep learning and classical approaches used for outcome prediction. Bus. Inf. Syst. Eng. **63**(3), 261–276 (2020). https://doi.org/10.1007/s12599-020-00645-0
11. Malburg, L., Seiger, R., Bergmann, R., Weber, B.: Using physical factory simulation models for business process management research. In: Del Río Ortega, A., Leopold, H., Santoro, F.M. (eds.) BPM 2020. LNBIP, vol. 397, pp. 95–107. Springer, Cham (2020). https://doi.org/10.1007/978-3-030-66498-5_8
12. Malburg, L., et al.: Object detection for smart factory processes by machine learning. Procedia Comput. Sci. **184**, 581–588 (2021)
13. Malburg, L., Klein, P., Bergmann, R.: Semantic web services for AI-research with physical factory simulation models in Industry 4.0. In: Proceedings of the International Conference on Innovative Intelligent Industrial Production and Logistics (IN4PL), pp. 32–43. SCITEPRESS (2020)
14. Marrella, A., Mecella, M., Sardiña, S.: Intelligent process adaptation in the SmartPM system. ACM Trans. Intell. Syst. Technol. **8**(2), 25:1–25:43 (2017)
15. Metzger, A., Kley, T., Palm, A.: Triggering proactive business process adaptations via online reinforcement learning. In: Fahland, D., Ghidini, C., Becker, J., Dumas, M. (eds.) BPM 2020. LNCS, vol. 12168, pp. 273–290. Springer, Cham (2020). https://doi.org/10.1007/978-3-030-58666-9_16
16. Mirza, M., Osindero, S.: Conditional Generative Adversarial Nets. CoRR abs/1411.1784 (2014)
17. Müller, G.: Workflow Modeling Assistance by Case-Based Reasoning. Springer, Heidelberg (2018). https://doi.org/10.1007/978-3-658-23559-8
18. Poll, R., Polyvyanyy, A., Rosemann, M., Röglinger, M., Rupprecht, L.: Process forecasting: towards proactive business process management. In: Weske, M., Montali, M., Weber, I., vom Brocke, J. (eds.) BPM 2018. LNCS, vol. 11080, pp. 496–512. Springer, Cham (2018). https://doi.org/10.1007/978-3-319-98648-7_29
19. Rama-Maneiro, E., Vidal, J.C., Lama, M.: Deep Learning for Predictive Business Process Monitoring: Review and Benchmark. CoRR abs/2009.13251 (2020)
20. Rehse, J.-R., Mehdiyev, N., Fettke, P.: Towards explainable process predictions for Industry 4.0 in the DFKI-Smart-Lego-Factory. KI - Künstliche Intelligenz **33**(2), 181–187 (2019). https://doi.org/10.1007/s13218-019-00586-1
21. Schönig, S., Ackermann, L., Jablonski, S., Ermer, A.: IoT meets BPM: a bidirectional communication architecture for IoT-aware process execution. Softw. Syst. Model. **19**(6), 1443–1459 (2020). https://doi.org/10.1007/s10270-020-00785-7
22. Seiger, R., Aßmann, U.: Consistency and synchronization for workflows in cyber-physical systems. In: Proceedings of the 10th ACM/IEEE International Conference on Cyber-Physical Systems, pp. 312–313. ACM (2019)
23. Seiger, R., Huber, S., Heisig, P., Aßmann, U.: Toward a framework for self-adaptive workflows in cyber-physical systems. Softw. Syst. Model. **18**(2), 1117–1134 (2017). https://doi.org/10.1007/s10270-017-0639-0

24. Tax, N., Verenich, I., La Rosa, M., Dumas, M.: Predictive business process monitoring with LSTM neural networks. In: Dubois, E., Pohl, K. (eds.) CAiSE 2017. LNCS, vol. 10253, pp. 477–492. Springer, Cham (2017). https://doi.org/10.1007/978-3-319-59536-8_30
25. Taymouri, F., Rosa, M.L., Erfani, S., Bozorgi, Z.D., Verenich, I.: Predictive business process monitoring via generative adversarial nets: the case of next event prediction. In: Fahland, D., Ghidini, C., Becker, J., Dumas, M. (eds.) BPM 2020. LNCS, vol. 12168, pp. 237–256. Springer, Cham (2020). https://doi.org/10.1007/978-3-030-58666-9_14
26. van der Aalst, W.M.P.: Business process management: a comprehensive survey. ISRN Softw. Eng. **2013**(1), 1–37 (2013)
27. Weinzierl, S., Dunzer, S., Zilker, S., Matzner, M.: Prescriptive business process monitoring for recommending next best actions. In: Fahland, D., Ghidini, C., Becker, J., Dumas, M. (eds.) BPM 2020. LNBIP, vol. 392, pp. 193–209. Springer, Cham (2020). https://doi.org/10.1007/978-3-030-58638-6_12
28. Wieland, M., et al.: Towards situation-aware adaptive workflows: SitOPT - a general purpose situation-aware workflow management system. In: International Conference on Pervasive Computing and Communication Workshops, pp. 32–37. IEEE (2015)

17th International Workshop on Business Process Intelligence (BPI 2021)

17th International Workshop on Business Process Intelligence (BPI 2021)

Business process intelligence (BPI) is a growing area both in industry and academia. BPI refers to the application of data- and process-mining techniques to the field of business process management. In practice, BPI is embodied in tools for managing process execution by offering several features such as analysis, prediction, monitoring, control, and optimization.

The main goal of this workshop is to promote the use and development of new techniques to support the analysis of business processes based on run-time data about the past executions of such processes. The workshop aims at discussing the current state of research and sharing practical experiences, exchanging ideas and setting up future research directions that better respond to real needs. We aim to bring together practitioners and researchers from different communities such as business process management, information systems, business administration, software engineering, artificial intelligence, process mining, and data mining who share an interest in the analysis of business processes and process-aware information systems. In a nutshell, it serves as a forum for shaping the BPI area.

The 17th edition of this workshop attracted 10 international submissions. Each paper was reviewed by at least three members of the Program Committee. From these submissions, the top five were accepted as full papers for presentation at the workshop, which was held in a hybrid format, physically in Rome but also online for remote participants. The workshop kicked-off with a keynote by Boudewijn van Dongen on the topic of "Conformance checking: A systemic view". In the talk, an interesting perspective was provided to move conformance checking from a single instance perspective to a systemic one. The five regular papers presented at the workshop and included in this volume provide a mix of novel research ideas, as described below.

Rösel, Fahrenkrog-Petersen, van der Aa, and Weidlich focus on event log anonymization. Based on learnt event embeddings, a distance measure is defined at trace level to allow semantics to be taken into account when anonymizing an event log. As such, transformations leading to swapping events that are unrelated is avoided. In an experimental analysis, the authors show that their distance-based anonymization can result in anonymized event logs that more closely match the original input compared to traditional syntactic anonymization. Next, Stein Dani, Leopold, van der Werf, Lu, Beerepoot, Koorn, and Reijers look into a completely different problem in process mining, i.e. event log extraction. Based on an in-depth literature analysis, the authors present a taxonomy of human activities that are part of event log extraction processes. By providing a deeper understanding of the human role in event log extraction, guidance is provided to future research on automating the extraction of event data. Thirdly, van der Aalst and Santos look into partial orders as a solution to address the problems of both poor timestamping quality and events subject to irrelevant orders due to timestamp proximity. Partial orders are used to exploit the uncertainty and bring in expert knowledge to (re)order events. An open-source implementation of the technique is provided in ProM. The fourth paper by Lichtenstein, Bano, and Weske introduces a novel technique for discovering case identifiers for unlabeled event logs. A particular weakness of existing techniques, namely poor performance in the case of cyclic behavior, is tackled by introducing a three-step algorithm based on event attributes. The method's effectiveness is demonstrated experimentally using two real-life event logs. Finally, Koschmider, Kaczmarek, Krause, and van Zelst review approaches to deal with outliers and noise in event logs. By first providing a classification of deviations observed in event data, and subsequently scrutinizing existing techniques and tools to verify which dimensions have already been addressed. As such, a number of gaps are identified which, together with some preliminary experimental results, provide for a nice research agenda.

As with previous editions of the workshop, we hope that the reader will find this selection of papers useful to keep track of the latest advances in the BPI area. We are looking forward to presenting more new advances in future editions of the BPI workshop.

October 2021

Andrea Burattin
Jochen De Weerdt
Marwan Hassani

Organization

Workshop Chairs

Andrea Burattin	Technical University of Denmark, Denmark
Jochen De Weerdt	KU Leuven, Belgium
Marwan Hassani	Eindhoven University of Technology, The Netherlands

Program Committee

Ahmed Awad	Cairo University, Egypt
Johannes De Smedt	KU Leuven, Belgium
Benoit Depaire	Universiteit Hasselt, Belgium
Claudio Di Ciccio	Vienna University of Economics and Business, Austria
Chiara Di Francescomarino	Fondazione Bruno Kessler, Italy
Luciano García-Bañuelos	Tecnologico de Monterrey, Mexico
Gianluigi Greco	University of Calabria, Italy
Gert Janssenswillen	Universiteit Hasselt, Belgium
Anna Kalenkova	University of Melbourne, Australia
Sander Leemans	Queensland University of Technology, Australia
Michael Leyer	University of Rostock, Germany
Fabrizio Maggi	Free University of Bozen-Bolzano, Italy
Jorge Munoz-Gama	Pontificia Universidad Católica de Chile, Chile
Marco Pegoraro	RWTH Aachen University, Germany
Pnina Soffer	University of Haifa, Israel
Boudewijn van Dongen	Eindhoven University of Technology, The Netherlands
Seppe vanden Broucke	Ghent University, Belgium
Eric Verbeek	Eindhoven University of Technology, The Netherlands
Matthias Weidlich	Humboldt-Universität zu Berlin, Germany
Hans Weigand	Tilburg University, The Netherlands
Sebastiaan Van Zelst	FIT/RWTH Aachen University, Germany
Han van der Aa	University of Mannheim, Germany
Wil van der Aalst	RWTH Aachen University, Germany

Conformance Checking: A Systemic View

Boudewijn F. van Dongen(✉)

Department of Mathematics and Computer Science, Eindhoven University of Technology, Eindhoven, The Netherlands
b.f.v.dongen@tue.nl

Abstract. Within the field of process mining, conformance checking plays an important role. While conformance checking initially mainly served as a way to explain the difference between an event log and a discovered process model, it since has evolved to a field of study in which the aim is to understand the relation between observed and modelled behaviour in the context of an abstract process.

So far, however, the focus has mainly been on the relation between an observed instance of a process and a model of that process. In other words, conformance checking considers the relation between a single observed instance of a process and a model of that process.

In this keynote, we advocate a systemic view on conformance checking. Through a series of examples, we show how a single-instance perspective would result in overlooking deviations and we provide initial ideas to resolve this and we show the theoretic framework to overcome these challenges.

1 Introduction

In the last two decades, process mining has developed as a research area focusing on the analysis of data to create insights into processes. These processes may be administrative processes such as the handling of insurance claims or production processes where parts are combined to create end products. In all processes, the day to day operations of processes differ from the designed process, i.e. the actual operational data does not conform to the designed process. That is why conformance checking has become a significant field in process mining [5]. In conformance checking, the goal is to relate reality captured in event logs with designed (or discovered) process models.

Various types of conformance checking techniques exist. Rule checking techniques [9,13], for example, are used to check if specific business rules are respected while case-replay techniques [1,3,6,12] try to identify specific deviations between modelled and observed behavior on the control flow level. More advanced techniques consider resources and data on top of the control flow [2,10]. Here, authorization (which resource can perform which activity) and data dependencies (higher-value cases take a different route) play a role. More recently, a technique was proposed to consider all perspectives at once [11].

A. Marrella and B. Weber (Eds.): BPM 2021 Workshops, LNBIP 436, pp. 61–72, 2022.
https://doi.org/10.1007/978-3-030-94343-1_5

The state-of-the-art of conformance checking uses alignments to relate the recorded execution of a process with a model of this process. An alignment (first introduced in [1], in detail explained in [5]) is a sequence of so-called moves where each move either represents a synchronous move, a log move or a model move. A synchronous move indicates that observed and modeled behaviour agree, i.e. the execution of an activity observed in the data matches the activity being performed according to the design of the process. A log move represents the fact that some observed behaviour cannot be explained by the designed process and a model move represents the fact that an observation of something that should have happened appears to be missing from the data. The latter typically indicates that the observed behaviour can be better explained if it is assumed that some data is missing.

Such alignments can be computed for individual cases in an event log using an $A^\star$ based search strategy [1,6], or by means of logic programming [4]. Furthermore, techniques exist to compute representations of classes of alignments [7] while other techniques focus on approximations of alignments [14] to increase the speed of computation or provide divide-and-conquer strategies [8]. However, all these techniques have one fundamental property in common, namely that in the end, they all reason over the so-called synchronous product (although not always instantiating this model).

The synchronous product of a process model and an observed trace is a Petri net where each of the moves explained above is represented by a transition in that Petri net. An alignment is a firing sequence of that Petri net which starts in the initial marking and reaches the final marking. By giving each type of move a weight (typically 0 for synchronous moves and 1 for non-synchronous moves) one can define an optimal alignment as a firing sequence in the synchronous product that minimizes the sum of the weights.

Common to all approaches is that they consider each case in isolation. Conformance checking for an entire event log is reduced to a case-by-case assessment of that log. In this keynote, we use a simple example to show why a systemic approach can lead to the identification of deviating behaviour that cannot be detected on a case-by-case basis.

2 Running Example

As an example, we consider a fictive production process. In this production process, a company paints red and blue products. A product comes into the paint booth and then the operator paints the product in either red or blue. After painting, another operator may check if the product was correctly painted. This check is performed at random, zero or more times. An annotated Petri net model of this very simple process is shown in Fig. 1. Please note that this is, purposely, not a typical process model in the sense that it does not have a clearly identifiable end place.

The model shows the relation between the requested color and the required path, as well as the relation between the resources used in the painting process (x) and the quality check (anybody but x).

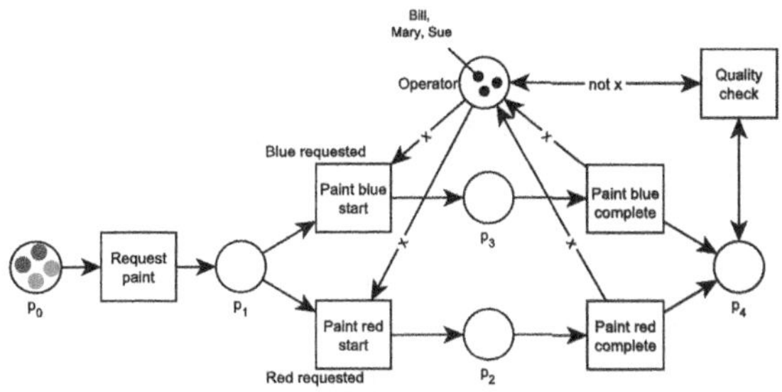

Fig. 1. Process model for our example with three resources and four cases waiting to be processed. (Color figure online)

In Table 1, we show a simple event log for this process. As common practice in event logs, they are grouped by case and contain information about the time an event was logged and the requested color which is a case attribute.

A close inspection of the event log already shows that there are some conformance issues. The first case, i.e. product $P5634$ appears to have gone through the process correctly. A request was entered to paint the product red, it was painted red by Bill and a check was performed once by Sue. In contrast, products $P5635$ and $P5636$ show deviations. In the former, the 'Request paint' activity appears to be missing and in the latter the 'Paint blue complete' activity appears to be missing. However, one could also argue that the activity 'Paint Red start' and

Table 1. Example of a log of the production process.

Product number	Event	Color	Operator	Timestamp
$P5634$	Request paint	Red		July 10, 12:31
$P5634$	Paint Red start	Red	Bill	July 10, 12:35
$P5634$	Paint Red complete	Red	Bill	July 10, 12:42
$P5634$	Quality Check	Red	Sue	July 10, 12:47
$P5635$	Paint Blue start	Blue	Sue	July 10, 12:45
$P5635$	Paint Blue complete	Blue	Sue	July 10, 12:53
$P5635$	Quality Check	Blue	Bill	July 10, 14:01
$P5636$	Request paint	Red		July 10, 12:57
$P5636$	Paint Blue start	Red	Mary	July 10, 13:15
$P5636$	Quality Check	Blue	Mary	July 10, 14:23
$P5636$	Quality Check	Blue	Sue	July 11, 8:01
$P5637$	Request paint	Blue		July 10, 12:58
$P5637$	Paint Blue start	Blue	Bill	July 10, 12:59
$P5637$	Request paint	Blue		July 10, 14:15
$P5637$	Paint Blue start	Blue	Bill	July 10, 14:15

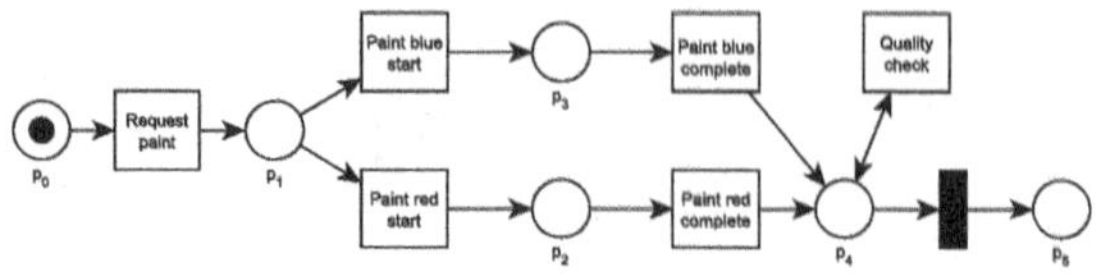

Fig. 2. A typical control-flow focused process-model version of Fig. 1

'Pains Red complete' are missing and 'Paint Blue start' was wrongly logged, since the data suggests a red product was requested, note a blue one. The last case in the log, product $P5637$ contains two request paint events and two start events for painting. The case is most likely not finished.

A traditional conformance checking technique, which focuses only on the control flow, would ignore the resource and data parts of the model and use the model shown in Fig. 2 instead. This is a classical workflow model with a clear begin and end. A so-called τ transition is included to reach the end signalling the termination of a case. This transition is not expected to be observable in an event log. Using this model results in five deviations for this log, since it produces the following alignments:

Product $P5634$:

log trace	Rp	Prs	Prc	Qc	$\gg$
model run	Rp	Prs	Prc	Qc	τ

Product $P5635$:

log trace	$\gg$	Pbs	Pbc	Qc	$\gg$
model run	Rp	Pbs	Pbc	Qc	τ

Product $P5636$:

log trace	Rp	Pbs	$\gg$	Qc	Qc	$\gg$
model run	Rp	Pbs	Pbc	Qc	Qc	τ

Product $P5637$:

log trace	Rp	Rp	Pbs	Pbs	$\gg$	$\gg$
model run	Rp	$\gg$	Pbs	$\gg$	τ	Pbc

Product $P5634$ shows a model move on the τ transition. The symbol $\gg$ is used as a 'skip' symbol indicating that at this position, something was expected but is not present. This is to be expected as this is an unobservable transition. Product $P5635$ also shows a model move on the 'Request paint' transition to start the process off. For product $P5637$, we see two log moves, i.e. the second 'Request paint' and the second 'Paint Blue start' are identified as log moves. In total, this log shows five deviations (and four model moves on τ which are not called deviations).

Rather than looking only at control flow, more advanced techniques exist for conformance checking. In particular, the data-aware techniques available today are able to detect that product $P5636$ was painted blue, while it was requested to be painted red. Such techniques explain this deviation in one of two ways depending on the parameter settings: either the product was painted in the wrong color, or the product was painted in the right color, but the log was wrong, i.e. it would produce one of the following two alignments:

Product $P5636$ painted in the wrong color:

log trace	Rp	Pbs	$\gg$	Qc	Qc	$\gg$
model run	Rp	Pbs	Pbc	Qc	Qc	τ
data view		Color $\neq$ Blue				

Product $P5636$ painted in the right color, but wrongly recorded:

log trace	Rp	Pbs	$\gg$	$\gg$	Qc	Qc	$\gg$
model run	Rp	$\gg$	Prs	Prc	Qc	Qc	τ
data view			Color = Red				

Furthermore, techniques exist to identify deviations on the resource level. Suppose, for example, that a four-eyes principle applies here, i.e. the Quality Check should not be performed by the same person who painted the product. In that case, product $P5636$ shows a deviation as Mary appears have done both the painting and the Quality Check. A typical rule-checking technique would flag this as a deviation, or, again depending on the parameter settings, would label the first quality check as a log move.

Product $P5636$ four-eyes violation:

log trace	Rp	Pbs	$\gg$	Qc	Qc	$\gg$
model run	Rp	Pbs	Pbc	Qc	Qc	τ
resource view		Mary	?	Mary shouldn't do Qc	Sue	

Product $P5636$ with log move:

log trace	Rp	Pbs	$\gg$	Qc	Qc	$\gg$
model run	Rp	Pbs	Pbc	$\gg$	Qc	τ
resource view		Mary	?	?	Sue	

With all these examples, we have shown that conformance checking techniques can be helpful in identifying deviations. However, there are some deviations in this example which cannot be caught easily with existing techniques which consider cases in isolation. Therefore, in the next section, we show that there is a more subtle problem in this example event log that calls for a systemic view on the data.

3 Alignments: A Systemic View

In the running example, the first case did not show any deviations and, when considering the control flow only, the second case shows only one deviation. However, when considering these two cases together one may see a problem. The activity 'Quality check', performed by Sue for product $P5634$ was executed at time 12:47. However, for product $P5636$, painting, by Sue, started at 12:45 and lasted until 12:53. In other words, the resource who supposedly executed the quality check for product $P5634$ was otherwise occupied at 12:47. In order to detect this, we should no longer consider the process model of Fig. 2, but we should consider the model in Fig. 1 and both cases at once.

Doing so with a focus purely on the control-flow perspective, leads to the following systemic alignments. To distinguish the case-level alignments from the systemic ones, we combine the log traces of the latter with the execution of the model in a single table (note that in all cases the trailing τ disappeared as this is not part of the model in Fig. 1).

For product $P5634$ the quality check was actually not performed, but $P5635$ was processed as logged:

log trace$P5634$	Rp	Prs	Prc	Qc				
log trace$P5635$					$\gg$	Pbs	Pbc	Qc
model run	Rp	Prs	Prc	$\gg$	Rp	Pbs	Pbc	Qc

Or, alternatively:

For product $P5634$ the quality check was performed, but painting $P5635$ was completed earlier than logged:

log trace$P5634$	Rp	Prs	Prc				Qc		
log trace$P5635$				$\gg$	Pbs	$\gg$		Pbc	Qc
model run	Rp	Prs	Prc	Rp	Pbs	Pbc	Qc	$\gg$	Qc

In both cases, one of the alignments would be considered sub-optimal from a control-flow perspective when looking at each case in isolation. However, the combination of alignments is the only way to explain the deviation seen in the event log under the assumption that Sue indeed performed all these activities. Taking the resource perspective into account as well would result in another possibility, namely:

Quality check for product $P5634$ was not done by Sue:

log trace$P5634$	Rp	Prs	Prc			Qc		
log trace$P5635$				$\gg$	Pbs		Pbc	Qc
model run	Rp	Prs	Prc	Rp	Pbs	Qc	Pbc	Qc
resource view		Bill	Bill		Sue	? but not Sue	Bill	Sue

A similar conflict exists between product $P5637$ and $P5634$. According to the event log, Bill started painting product $P5637$ blue at 12:59, but then at 14:01, Bill is seen to perform the quality check on product $P5635$ and subsequently at 14:15, product $P5637$ is requested to be painted again, a request Bill honors instantly. A logical explanation for this deviation would be that Bill realized it was break-time at 12:59 when he just started painting product $P5637$. He then decides to abort that activity and go for lunch until 14:00. In order to avoid an exceptionally long painting time to be logged, he re-requested the paint job.

While the latter would be visible from a regular alignment the former requires reasoning over two cases in relation to avoid the conclusion that Bill went for lunch for too long.

When considering all events of the log in Table 1 and the model of Fig. 1 with data and resources, the following systemic alignment can be found:

Systemic alignment for the log in Table 1 and the model in Figure 1. For readability, colors and a timeline are added.

	July 10 12:31				12:45				12:58		13:15
log trace $P5634$	Rp	Prs	Prc			Qc					
log trace $P5635$				$\gg$	Pbs		Pbc				
log trace $P5636$								Rp			Pbs
log trace $P5637$									Rp	Pbs	
model run	Rp	Prs	Prc	Rp	Pbs	Qc	Pbc	Rp	Rp	$\gg$	Pbs
resource view		Bill	Bill		Sue	not Sue	Sue			Bill	Mary
data view		color = red			color = blue						color $\neq$ blue

...

	July 10 14:01	14:15			14:23		July 11 8:00
log trace $P5634$	Qc						
log trace $P5635$				$\gg$	Qc		Qc
log trace $P5636$		Rp	Pbs			$\gg$	
log trace $P5637$							
model run	Qc	$\gg$	Pbs	Pbc	Qc	Pbc	Qc
resource view	Bill		Bill	?	Mary	?	Sue
data view							

The systemic alignment shows a total of seven deviations. Three classical model moves (in purple) and two classical log moves (in yellow). One partial synchronous move, where the event log agrees with the model on the control flow, but the indicated resource cannot be correct (in orange) and one partially synchronous move where the data perspective is wrong (in blue).

The examples above show that already for a small example, looking at the system as a whole is important and the larger the system, the more important this becomes. Examples of deviations that can only be found when looking at the system as a whole are things like:

- Privacy violations, i.e. resources who are accessing data which they are in general authorized to access, but only in the context of working on a specific case, while they are not working on that specific case.
- Time-travel violations, i.e. resources that appear to be in two places at once, or in two places after each other which would have required the resource to travel at more than it's maximum speed (think of robots with cloned ID's traveling on a grid).

- Overtaking on a conveyor, i.e. items that move over a conveyor, but suddenly depart in a different order than the order in which they entered.
- Batching violations, i.e. tasks which are executed in batches for many cases at once and which are suddenly executed only for a limited set of cases.

These, and many other types of deviations can only be identified when looking at combinations of cases and, sometimes, even combinations of process models all together in one systemic view. Unfortunately, if systems grow larger, then the computational challenges become larger too.

4 The Computational Challenge

Computing control-flow alignments for a single case is already a complex task [5]. Computing alignments for multiple cases at once is even worse in terms of complexity. While the work of [7] provides a way to compute symbolic representations of many alignments at once, these representations still focus on the individual cases, i.e. relations between cases are ignored.

An initial step in this direction is the computation of alignments for multiple, related cases by considering *joint synchronous products*. A joint synchronous product is a combination of a classical Petri net representation of the original process model and the event log as a whole. Figure 3 shows the classical Petri net representation used to get the alignments of Sect. 3. Here, two cases are represented by copies of the original model and each resource is individually represented. Note that this model becomes significantly larger if the data perspective is included or if the other two cases in the event log are included.

In a joint synchronous product the classical Petri net representation is then combined with the event log and all possible moves are encoded as transitions. Where in traditional synchronous products, the move is a pair of an event in the

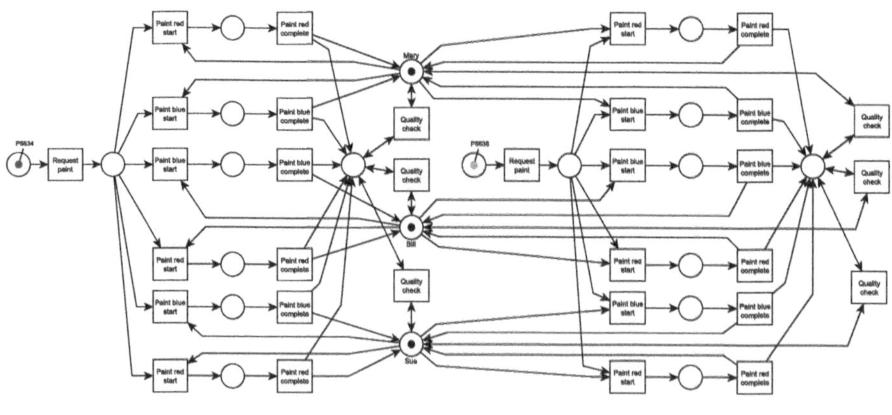

Fig. 3. A classical Petri net representation of the Fig. 1, for products $P5634$ and $P5635$, including the resource perspective, but not including the data perspective. (Color figure online)

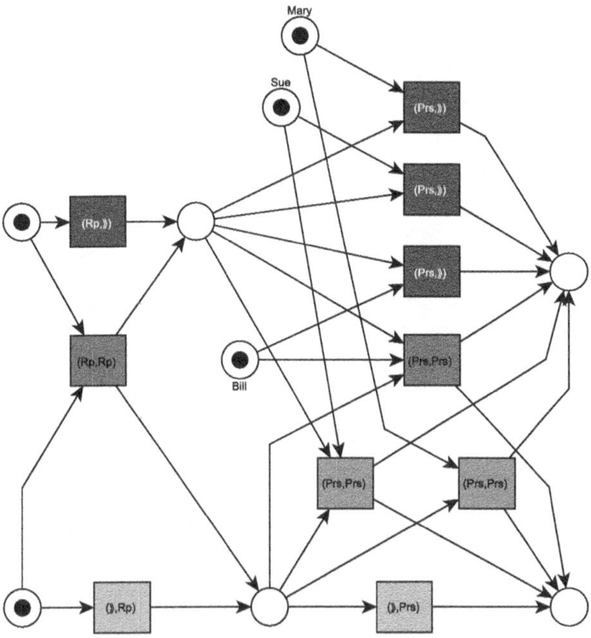

Fig. 4. Part of the joint synchronous product relating to the first two events of product $P5634$. (Color figure online)

log and a transition that fires in the model (one of which can be ≫), in a joint synchronous product, a move is a vector of all data that should be considered. In our example, this implies a vector of the product number, the event label, one of the transitions in the model, the resource involved in the transition and the requested color. However, the process of obtaining a joint synchronous product is not different from the traditional procedure for control flow [1] or the procedure for including other perspectives [11].

In Fig. 4 the start of the joint synchronous product is shown for the event log of Table 1. This snippet shows the initial 'Request paint' event from case $P5634$ and the corresponding transition with the possible synchronous move on the left. Then, on the right, for the same case, the 'Paint Red start' event is depicted, with all possible variants, i.e. on the top, three model moves for Sue, Bill and Mary. Below that the synchronous move for Bill and below that two partial synchronous moves where event and model agree, but the resource is not Bill. Finally the log move (≫, Prs) is shown on the bottom right. A similar snippet is shown for the last two events on product $P5635$ in Fig. 5. Note that the data perspective (i.e. the color) is not directly included here.

From these fragments it's easy to see that the joint synchronous product, in the general case, is huge and cannot be depicted here. Figure 6 shows the first part of the systemic alignment shown earlier as a run of the joint synchronous product where the different borders indicate the specific products of the log.

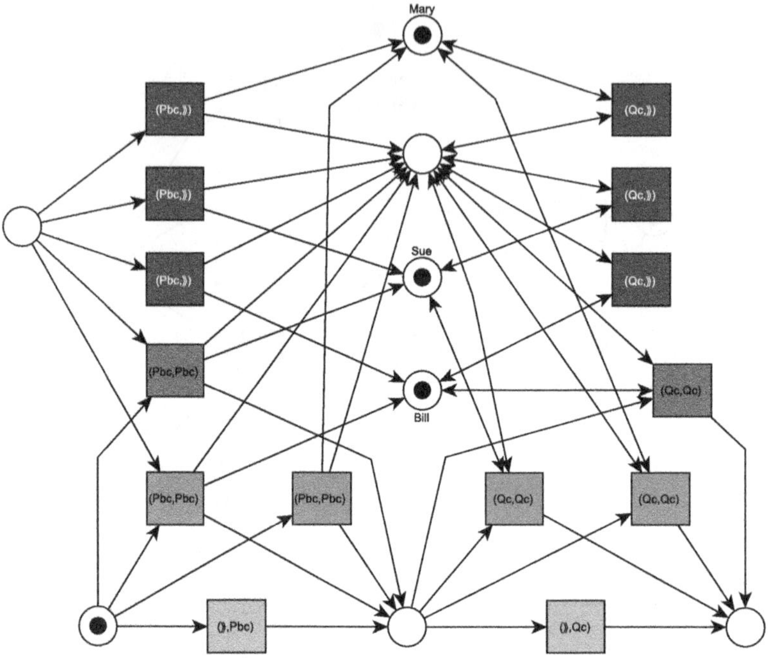

Fig. 5. Part of the joint synchronous product relating to the last two events of product *P*5635. (Color figure online)

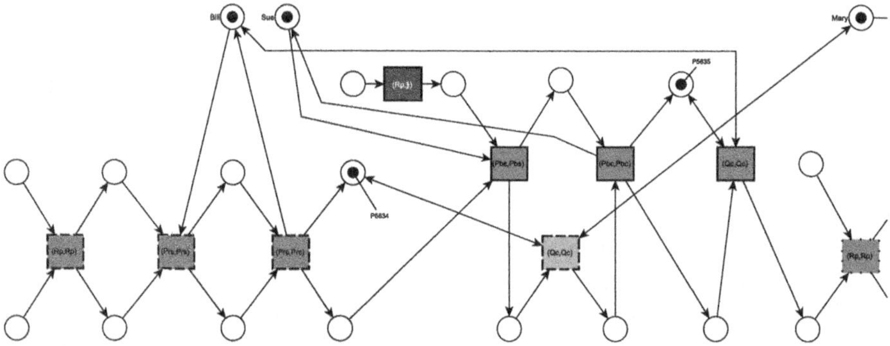

Fig. 6. First part of the optimal alignment for the event log of Table 1. This part of the joint synchronous product explains the first eight events of that log. (Color figure online)

It includes the first two products and the first event of the third product. As can be seen, there is a 'wrong resource' move for product *P*5634 which is otherwise handled fully synchronously (in orange again), while for product *P*5635 there is a model move because of the missing 'Request product'. This corresponds to the first part of the systemic alignment at the end of Sect. 3.

5 Conclusion and Future Work

In this keynote, we introduced systemic views on conformance checking. Rather than relating observed and modelled behaviour on a case-by-case basis, we showed through an example, that looking at the entire system is sometimes necessary to identify deviations. We then showed an initial approach, using a *joint synchronous product* to actually compute global alignments for an entire log including resource and data perspectives. This preliminary idea requires considerable work to operationalize as the search space for the entire system is very large. We therefore plan to look for structural reductions on this model.

References

1. van der Aalst, W.M.P., Adriansyah, A., van Dongen, B.F.: Replaying history on process models for conformance checking and performance analysis. Wiley Interdiscip. Rev. Data Min. Knowl. Discov. **2**(2), 182–192 (2012). https://doi.org/10.1002/widm.1045
2. Alizadeh, M., Lu, X., Fahland, D., Zannone, N., van der Aalst, W.M.P.: Linking data and process perspectives for conformance analysis. Comput. Secur. **73**, 172–193 (2018)
3. Berti, A., van der Aalst, W.M.P.: A novel token-based replay technique to speed up conformance checking and process enhancement. Trans. Petri Nets Other Model. Concurr. **15**, 1–26 (2021)
4. Boltenhagen, M., Chatain, T., Carmona, J.: Optimized SAT encoding of conformance checking artefacts. Computing **103**(1), 29–50 (2021)
5. Carmona, J., van Dongen, B.F., Solti, A., Weidlich, M.: Conformance Checking - Relating Processes and Models. Springer, Heidelberg (2018). https://doi.org/10.1007/978-3-319-99414-7
6. Dongen, B.F.: Efficiently computing alignments. In: Weske, M., Montali, M., Weber, I., vom Brocke, J. (eds.) BPM 2018. LNCS, vol. 11080, pp. 197–214. Springer, Cham (2018). https://doi.org/10.1007/978-3-319-98648-7_12
7. García-Bañuelos, L., van Beest, N., Dumas, M., Rosa, M.L., Mertens, W.: Complete and interpretable conformance checking of business processes. IEEE Trans. Softw. Eng. **44**(3), 262–290 (2018)
8. Lee, W.L.J., Verbeek, H.M.W., Munoz-Gama, J., van der Aalst, W.M.P., Sepúlveda, M.: Replay using recomposition: Alignment-based conformance checking in the large. In: Clarisó, R., et al. (eds.) Proceedings of the BPM Demo Track and BPM Dissertation Award co-located with 15th International Conference on Business Process Modeling (BPM 2017), Barcelona, Spain, September 13, 2017. CEUR Workshop Proceedings, vol. 1920. CEUR-WS.org (2017). http://ceur-ws.org/Vol-1920/BPM_2017_paper_157.pdf
9. Letia, I.A., Goron, A.: Model checking as support for inspecting compliance to rules in flexible processes. J. Vis. Lang. Comput. **28**, 100–121 (2015)
10. Mannhardt, F., de Leoni, M., Reijers, H.A., van der Aalst, W.M.P.: Balanced multi-perspective checking of process conformance. Computing **98**(4), 407–437 (2016). https://doi.org/10.1007/s00607-015-0441-1
11. Mozafari Mehr, A.S., de Carvalho, R.M., van Dongen, B.: Detecting privacy, data and control-flow deviations in business processes. In: Nurcan, S., Korthaus, A. (eds.) CAiSE 2021. LNBIP, vol. 424, pp. 82–91. Springer, Cham (2021). https://doi.org/10.1007/978-3-030-79108-7_10

12. Rozinat, A., van der Aalst, W.M.P.: Conformance checking of processes based on monitoring real behavior. Inf. Syst. **33**(1), 64–95 (2008)
13. Taghiabadi, E.R., Gromov, V., Fahland, D., van der Aalst, W.M.P.: Compliance checking of data-aware and resource-aware compliance requirements. In: Meersman, R., et al. (eds.) OTM 2014. LNCS, vol. 8841, pp. 237–257. Springer, Heidelberg (2014). https://doi.org/10.1007/978-3-662-45563-0_14
14. Taymouri, F., Carmona, J.: An evolutionary technique to approximate multiple optimal alignments. In: Weske, M., Montali, M., Weber, I., vom Brocke, J. (eds.) BPM 2018. LNCS, vol. 11080, pp. 215–232. Springer, Cham (2018). https://doi.org/10.1007/978-3-319-98648-7_13

A Distance Measure for Privacy-Preserving Process Mining Based on Feature Learning

Fabian Rösel[1], Stephan A. Fahrenkog-Petersen[1(✉)], Han van der Aa[2], and Matthias Weidlich[1]

[1] Humboldt-Universität zu Berlin, Berlin, Germany
{fabian.roesel,stephan.fahrenkrog-petersen, matthias.weidlich}@hu-berlin.de
[2] University of Mannheim, Mannheim, Germany
han@informatik.uni-mannheim.de

Abstract. To enable process analysis based on an event log without compromising the privacy of individuals involved in process execution, a log may be anonymized. Such anonymization strives to transform a log so that it satisfies provable privacy guarantees, while largely maintaining its utility for process analysis. Existing techniques perform anonymization using simple, syntactic measures to identify suitable transformation operations. This way, the semantics of the activities referenced by the events in a trace are neglected, potentially leading to transformations in which events of unrelated activities are merged. To avoid this and incorporate the semantics of activities during anonymization, we propose to instead incorporate a distance measure based on feature learning. Specifically, we show how embeddings of events enable the definition of a distance measure for traces to guide event log anonymization. Our experiments with real-world data indicate that anonymization using this measure, compared to a syntactic one, yields logs that are closer to the original log in various dimensions and, hence, have higher utility for process analysis.

Keywords: Privacy · Anonymization · Trace distance · Feature learning

1 Introduction

Privacy-preserving process mining [9] aims at protecting personal data, while at the same time, enabling organizations to improve their business processes. The consideration of privacy in process mining is necessary, since the event logs used as a starting point for the analysis often include information about the individual people involved in process execution. For example, for a treatment process in a hospital, an event log may include personal information about the treated patients [26]. Since the respective data was commonly not recorded for the purpose of operational improvement, process mining is often considered a secondary use of the data and, as such, strictly regulated.

A. Marrella and B. Weber (Eds.): BPM 2021 Workshops, LNBIP 436, pp. 73–85, 2022.
https://doi.org/10.1007/978-3-030-94343-1_6

In general, there are two angles to approach privacy-preserving process mining [12]: One may anonymize the event log used for process mining [13] or one may design the algorithms for process mining such that the output satisfies privacy guarantees [14].

An established strategy for anonymizing event logs is to group traces together, i.e., to merge the data of multiple executions of a process. This way, the information about a single process instance and, hence, about an individual person involved therein, is not revealed. Such an approach is realized, for instance, in PRETSA [11] and the TLKC framework [24]. In order to preserve the utility of the event log for process analysis, these anonymization algorithms incorporate a distance measure. Intuitively, by merging traces that are close to each other, the resulting log will be close to the original one. Since process mining primarily targets the analysis of the behaviour exhibited by the traces, their closeness is captured in terms of their behavioural similarity.

Yet, existing algorithms [11,24] employ simple syntactic measures, such as the Levenshtein distance. This is a limitation, since, contrary to the assumption behind the Levenshtein distance, not all events have the same closeness. Arguably, the semantics of the activities referenced in the events in terms of their context suggests that some events are closer to each other than others. Neglecting such semantics means that any event log anonymization approach may induce a higher loss in utility than what would be necessary to achieve a certain privacy guarantee.

In this paper, we therefore investigate alternative approaches to measuring the distance of traces when anonymizing event logs. We study how embeddings of events, i.e., feature vectors that provide a semantic representation of the events, may be used as a foundation for a distance measure. Specifically, we rely on the Act2Vec [6] model to learn the embeddings from the original event log, thereby capturing the context of the activities of the events. Based thereon, we design a distance measure for traces that is tailored to event log anonymization, as it indicates which traces shall be merged into which other traces when aiming at a small loss of the log's utility for process analysis. Finally, we report on evaluation experiments to shed light on the impact of the adopted distance measure on the utility of an anonymized log. For several real-world event logs, we observe that using our semantic distance measure instead of a syntactic one, yields event logs that are closer to the original log along various dimensions.

In the remainder, we first discuss essential notions and notations (Sect. 2). We then present an embedding-based trace distance measure to use in event log anonymization (Sect. 3). Finally, we report on evaluation experiments (Sect. 4), before we review related work (Sect. 5) and conclude (Sect. 6).

2 Background

Below, we provide background for our work. In Sect. 2.1, we introduce a model for event logs. Section 2.2 outlines the PRETSA algorithm, a state-of-the-art algorithm for event log anonymization. In Sect. 2.3, we review the Act2Vec model to learn a feature-based representation of the events of a trace.

Table 1. Three example traces of a request handling process.

Case	Activity	Time	Case	Activity	Time	Case	Activity	Time
1	Check Loan Req.	9:05	2	Check Loan Req.	7:37	3	Check Loan Req.	9:49
1	Negotiate rate	10:04	2	Calculate rate	7:45	3	Report fraud	10:12
1	Set up contract	10:45	2	Set up contract	8:25	3	Block account	10:16
1	Inform client	14:08	2	Mail contract	9:50	3	Inform client	11:02

2.1 Event Logs

Process Mining relies on event logs that capture the execution of business processes [28]. We consider a common model of event logs, summarized as follows.

Each step of a process is represented by an activity $a \in \mathcal{A}$, with $\mathcal{A}$ denoting the universe of all activities. Each execution of such an activity is represented as an event $e = \langle c, a, ts\rangle$, where a is the respective activity, ts is a timestamp specifying the time of activity execution, and c is a case identifier signalling the instance of the process that the activity execution was related to. Moreover, by $\mathcal{E}$, we denote the universe of all events.

All events with the same case identifier form a trace, an ordered sequence of events $\langle e_1, \ldots, e_n\rangle = t \in \mathcal{E}*$, with $t(i) \in \mathcal{E}$, $1 \leq i \leq n$ denoting the i-th event of the trace. Here, the order of events in a trace is induced by their timestamps. We write $|t|$ for the trace length, i.e., the number of its events. A finite set of traces forms an event log, denoted by $L \subseteq \mathcal{E}*$.

Three traces of an exemplary event log for a process to handle loan requests are shown in Table 1. After a request has been checked, the interest rate is negotiated or calculated automatically, a contract is set up, and the client is informed or gets the contract posted by mail. However, the check may also identify fraud, so that the account is blocked, before the client is informed about it.

2.2 Event Log Anonymization with PRETSA

Striving for privacy-preserving process mining, an event log may be anonymized to avoid disclosure of sensitive information. This approach is realized in the *pre*fix-*t*ree-based event log *sa*nitization [11], or PRETSA for short. The algorithm transforms a given log to provide privacy guarantees based on k-anonymity [27] and t-closeness [16], while aiming to maximize the utility of the anonymized log.

The general idea behind PRETSA is protect the characteristics of a trace that enable the identification of an individual person. Specifically, it considers linkage attacks that are based on activity occurrences that directly point to individual persons, or exploit sensitive attributes for which differences in the observed distributions enable conclusions on individual persons. PRETSA guards against the former attack, by adopting a notion of k-anonymity, i.e., it forms groups of traces that are undistinguishable from each other and comprise at least k members. This way, the chances of linking a trace to an individual person is limited to $1/k$. To avoid linkage attacks through sensitive attributes, PRETSA also ensures

t-closeness, i.e., it limits the difference in the distribution of a sensitive attribute of a group compared to the overall population.

To achieve the above guarantees, PRETSA transforms an event log by merging similar traces. It detects trace prefixes that violate the privacy guarantees and merges them into a similar variant. The latter is selected based on a closeness measure to limit the loss in utility caused by the transformation. However, PRETSA (and related approaches) assess the similarity of traces solely with a simple syntactic measure, i.e., the Levenshtein distance. Hence, all events of a trace are assumed to be equally close, which neglects the semantics of the respective activities, i.e., the context in which they are executed.

2.3 The Act2Vec Model

To incorporate the semantics of data elements, embeddings, i.e., feature vectors derived from a learned model, have been adopted in various domains. Most prominently, in natural language processing, an embedding can be constructed for each word of a text in order to represent its semantic meaning [19]. A model to derive the embedding of a word is learned by considering the surrounding words and, hence, captures the context in which a word typically appears.

The idea of word embeddings was recently lifted to event logs, as part of the so-called Act2Vec model [6]. This way, one may learn semantic representations of activities and, therefore, events. As in the case of word embeddings, these representations are learned from the context of an activity: If an event indicates the execution of a specific activity, the surrounding events in traces of the event log indicate which activities represent the common execution context. Consequently, closeness of activities (and hence, events) is assessed based on the closeness of their execution context, rather than other properties, such as the labels.

3 An Embedding-Based Trace Distance Measure for Event Log Anonymization

This section introduces a distance function for traces that incorporates the semantics of activities and, hence, promises to induce a lower loss of utility when used in event log anonymization. First, we discuss the intuition of the measure (Sect. 3.1), before we turn to its definition (Sect. 3.2).

3.1 Intuition

Reconsider the three traces from Table 1. When comparing the first trace (Case 1) to the others, it is clear that Case 1 is more similar to the second trace than to the third: the first two both capture the regular handling of a loan request, differing solely in the way the interest rate is determined (negotiated versus automatically calculated) and in the method used to get back to the client (informed versus per mail). The third trace (Case 3), in turn, denotes a very different scenario, in which fraud was detected, resulting in the account being blocked.

In terms of the purely syntactic Levenshtein distance, however, the second and the third trace are equally distant from the first one. Both of them include two activity executions that are in line with the first trace, whereas each of them also includes two events that are without counterpart in the first trace. Therefore, existing anonymization techniques would consider both traces to be equally suitable in a merging step, even though merging the first and the third trace could be expected to lower the utility of the resulting log more drastically than the alternative solution of merging the first and the second trace.

Instead, when the semantics of activities in terms of their execution context are incorporated, a more suitable assessment of the traces' similarity may be achieved. For instance, the "*Negotiate rate*" and "*Calculate rate*" activities can be expected to have very similar execution contexts, i.e., after the request has been checked and directly preceding the activity to set up the contract. Hence, events referring to these activities are closer than events of less related activities, such as fraud reporting or account blocking.

3.2 A Distance Measure for Traces Based on Embeddings

Feature learning based on the Act2Vec model [6] enables us to incorporate the semantics of the activities referenced in events in terms of their context of execution. The latter is induced by behavioural relations that describe common predecessors and successors of the respective activity. Since many process mining tasks are grounded in exactly these behavioural relations, preserving them as much as possible will increase a log's utility for process analysis.

Event Distance. Since embeddings encode the context of activities referenced in events, i.e., their predecessors and successors, the similarity (or the distance, respectively) of two traces can be derived from the similarity (or distance) of their individual events. Therefore, we first define a distance for events, before using it as the basis to quantify the distance of traces.

We compare two events by the Cosine similarity of their feature vectors, which is defined over a vector space V as:

$$\mathrm{cosS}: V \times V \rightarrow [-1,1], \qquad (\mathbf{v}_1, \mathbf{v}_2) \mapsto \frac{\mathbf{v}_1 \cdot \mathbf{v}_2}{||\mathbf{v}_1||_2\, ||\mathbf{v}_1||_2},$$

with $||\mathbf{v}||_2$ as the Euclidean norm of a vector $\mathbf{v}$. Note that two vectors $\mathbf{v}_1$, $\mathbf{v}_2$ are identical in direction, if $\mathrm{cosS}(\mathbf{v}_1, \mathbf{v}_2) = 1$; orthogonal, if $\mathrm{cosS}(\mathbf{v}_1, \mathbf{v}_2) = 0$, and opposing, if $\mathrm{cosS}(\mathbf{v}_1, \mathbf{v}_2) = -1$.

In the remainder, given two events $e_1, e_2 \in \mathcal{E}$, we capture their distance based on embeddings as $d_e(e_1, e_2) = 1 - \mathrm{cosS}(\mathrm{act2vec}(e_1), \mathrm{act2vec}(e_2))$.

On the Direction of a Trace Distance for Log Anonymization. To assess the distance of two traces, the pair-wise distance of the events at the respective positions is calculated and summed up. Assuming that two traces t_1 and t_2 have equal length, this idea is illustrated in Fig. 1. If two traces have a different length, the measure needs to take into account that, conceptually, events need

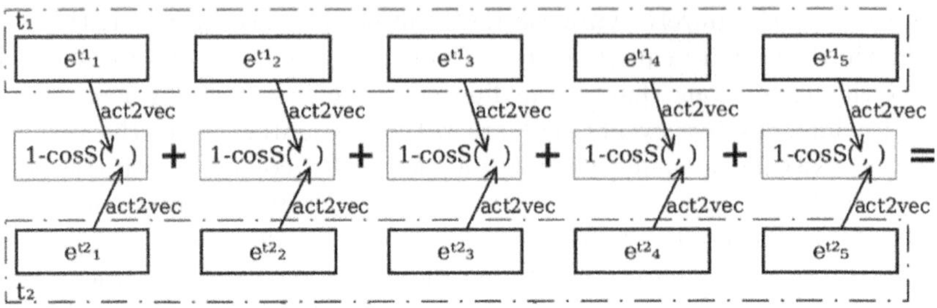

Fig. 1. Visual representation of the measure, assuming $|t_1| = |t_2|$.

to be added or removed to transform one trace into the other one. We therefore incorporate a penalty that is linear in the number of such excess events.

However, to use the distance measure to decide which traces to merge during event log anonymization, we define the penalty for excess events and, hence the trace distance measure, in an asymmetric manner. Algorithms for event log anonymization such as PRETSA merge one trace into another, which is an inherently directed operation. Here, it is desirable to avoid introducing new behaviour as part of the anonymization in order to maintain the log's utility for process analysis. Practically, PRETSA does not add new trace variants, i.e., all sequences of activity executions of traces in the anonymized log are already represented by at least one trace in the original log. As such, the algorithm guarantees that the successor relation over activities as derived from traces may only be reduced by the anonymization. The above is achieved by removing excess events of a trace when merging it into another one, which implies a potential loss of information on the context in which activities are executed. Hence, it is preferred to merge shorter traces into longer ones in order to maintain a larger part of the successor relation.

Moreover, the preservation of longer traces also helps to achieve privacy guarantees through the transformation. If traces of a certain variant, i.e., of a certain sequence of activity executions, are less than k-frequent, they have to be merged into other traces to achieve k-anonymity. Preferring the preservation of longer traces, there is a higher chance that these traces can be merged into traces that include the respective sequence of activity executions already as a prefix. In that case, no information in terms of the successor relation would be lost.

Trace Distance. Following the above arguments, we propose a directed distance measure for traces to guide event log anonymization. Given two traces t_1 and t_2, this distance incorporates that, if both traces have different lengths, merging the shorter into the longer one is preferred. The distance $d(t_1, t_2)$ is defined as:

$$d(t_1,t_2) = \begin{cases} \sum\limits_{1\leq i\leq|t_1|} d_e(t_1(i),t_2(i)) & \text{if } |t_1| = |t_2| \\ \sum\limits_{1\leq i\leq|t_2|} (d_e(t_1(i),t_2(i))) + (|t_1|-|t_2|)\cdot\rho_A & \text{if } |t_1| > |t_2| \\ \sum\limits_{1\leq i\leq|t_1|} (d_e(t_1(i),t_2(i))) + (|t_2|-|t_1|)\cdot\rho_R & \text{if } |t_1| < |t_2| \end{cases}$$

where $\rho_A, \rho_R \in \mathbb{N}$ are penalties for event addition and event removal, respectively, such that $\rho_A > 1$ and $\rho_R > \rho_A$.

As the distance measure incorporates the direction in which traces are potentially merged, the penalty ρ_R for event removal shall be larger than the one for event addition, ρ_A. Empirically, we found that a 50% increase of the penalty yielded the best results. Moreover, both penalties shall dominate the distance of any pair of individual events. Hence, we set $\rho_A = 2$ and $\rho_R = 3$ as default values.

4 Evaluation

In the following section, we present an evaluation of our approach. The aim of this evaluation is to answer the following research questions:

RQ1: Can we preserve more control-flow information by incorporating an embedding-based trace distance measure into event log anonymization?
RQ2: Do we preserve less utility in aspects unrelated to control-flow by optimizing for higher control-flow related utility?
RQ3: How do the internal properties of the distance measure impact the preserved utility?

To answer these questions, we apply our approach on the logs presented in Sect. 4.1, before Sect. 4.2 lays out the experimental setup. Section 4.3 presents and interprets the obtained results.

4.1 Dataset

We perform the evaluation on three real-world event logs. As shown in Table 2, the logs differ considerably in their domains and structuredness. The *CoSeLoG* log captures a relatively structured process, concerning the application of environmental permits, whereas the *Sepsis* log covers an unstructured process corresponding to clinical pathways from a hospital. Finally, *BPIC 2020* log captures the semi-structured reimbursement process of travel costs at a Dutch university. While all three event logs have comparable size in terms of the number of cases, and hence traces, the differences between them in terms of structuredness are clearly illustrated by the average and maximum numbers of cases per trace variant. In *CoSeLoG*, hundreds of traces may follow the same sequence of activity executions, whereas in *Sepsis*, traces often show a unique activity sequencing.

Table 2. Event log characteristics.

Event log	Cases	Variants	Avg. cases/var.	Max. cases/var.
CoSeLoG [2]	1,434	116	12.4	713
Sepsis [3]	1,050	846	1.2	35
BPIC 2020 [1]	2,099	896	2.3	206

4.2 Experimental Setup

Configuration. In our experiments, we compare the proposed method of incorporating the embedding-based distance measure against the original PRETSA using Levenshtein distance. To vary the required privacy guarantees, k-anonymity and t-closeness, we evaluate all combinations with $k = 2^i$ for $i \in [1, 8]$ and $t \in \{0.1, 0.25, 0.5, 0.75, 1.0\}$, yielding a total of 40 different settings per event log.

Evaluation Measures. To asses the utility of anonymized event logs, we use two metrics to analyse the preservation of control-flow behaviour and one metric to measure the impact on the average cycle time of each activity:

Advanced Behavioural Appropriateness. To measure the impact that anonymization has on the control-flow, we first use advanced behavioural appropriateness [25], which encodes behaviour within a log according to binary activity relations. Specifically, given a log, an activity A may {*always, sometimes, never*} indirectly {*follow, precede*} another activity B. By comparing these relations in an anonymized event log and its original counterpart, the metric yields a similarity score $\in$ [0,1], where 1 represents total equality for every relation between activity pairs in the two event logs and 0 if none of the relations are equal.

Truly Sampled Behaviour. We also use *truly sampled behaviour* [15] as a metric to measure the preservation of control-flow information. While the behavioural appropriateness considers indirectly-follows relations between activities, this measures looks at directly-follows relations and also consider their relative frequency. Specifically, it quantifies what fraction of directly-follows relations of the original log are *appropriately sampled.* It considers a directly-follows relation appropriately (or truly) sampled when, adjusted for potentially different log size, it appears approximately the same time in both logs. The *Truly Sampled Score* then is the proportion of truly sampled directly-follows relations. Therefore a Truly Sampled Score of 100% is considered optimal, while a score of 0% indicates no relation-wise similarity between the compared event logs.

Total Duration Error. Finally, we also investigate how our approach impacts the duration of an activity, which can be regarded as a sensitive attribute that shall be anonymized in accordance with a specified t-closeness guarantee. For this, we compute the total duration error. It is a simple metric based on the assumption that similar activities take a similar amount of time to execute. The error is calculated by comparing the total time it takes to execute all traces in the sanitized log to the total execution time of the original log.

Implementation. To conduct our experiments, we implemented PRETSA using the embedding-based distance measure, as well as all used evaluation measures in Python. The source code is available at GitHub[1] under MIT license.

[1] https://github.com/roeselfa/FeatureLearningBasedDistanceMetrics.

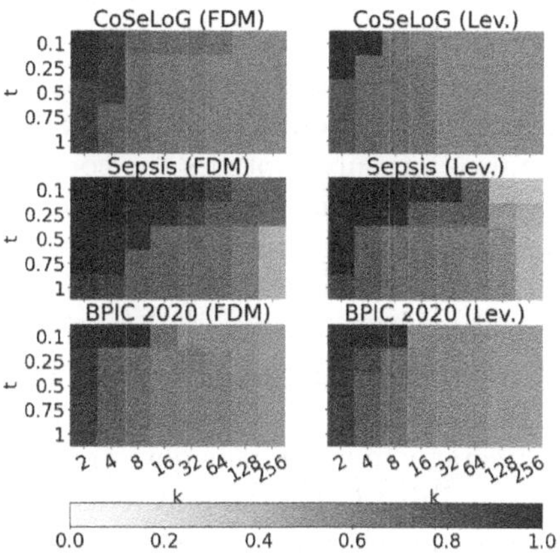

Fig. 2. Advanced behavioural appropriateness of anonymized logs.

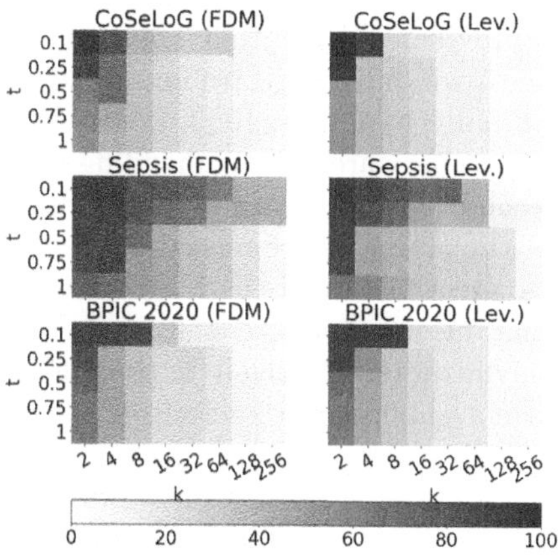

Fig. 3. Truly sampled scores of anonymized logs.

4.3 Results

To answer RQ1, we consider the evaluation metrics related to the control-flow behaviour of the log. Figure 2 illustrates that our feature-based distance measure (FDM) can improve Advanced Behavioural Appropriateness scores for certain anonymization parameters in two of the three real-life event logs. Best results were achieved using the *Sepsis* log, while *CoSeLoG* yielded partially better results. *BPIC 2020* performed approximately equal for both distance measures.

We see similar results when considering the Truly Sampled Score (see Fig. 3). Again, *Sepsis* shows the biggest improvements, for a range of k and t, whereas *CoSeLoG* improved for selected parameter combinations. No noticeable differences can be observed for *BPIC 2020*. Hence, regarding RQ1, we can say, that our embedding-based measure can preserve more control-flow utility. However, applying it gives no guarantee of achieving this aim.

Additionally, we want to investigate, if the application of the embedding-based measure harms other utility aspects of the log (RQ2). By considering the duration errors depicted in Fig. 4, we observe the following: For *Sepsis*, while results for specific scenarios vary widely, showing huge improvements in some cases and comparable declines in others, the global error neither increased, nor decreased. Rather negligible improvements were made for *CoSeLoG*, whereas both approaches are comparable for *BPIC 2020*. Concluding, in regards to RQ2, our results do

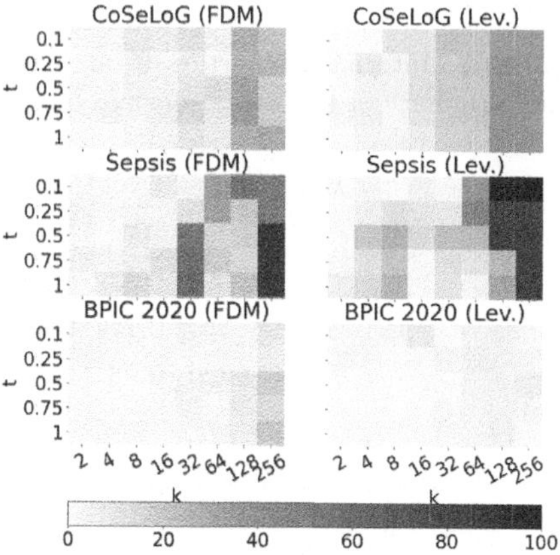

Fig. 4. Total duration error of anonymized logs.

not provide evidence that applying a distance measure focused on control-flow aspects will have negative implications for other utility aspects of the event log.

Finally, to address RQ3, we take a closer look at the internal properties of our distance measure. The embedding-based distance measure captures the context of each event in a feature vector. Similar feature vectors of two events indicate that the activities referenced in the events often appear in a similar context and thus, are semantically exchangeable to a certain extent. An event log with many similar events therefore enables the embedding-based distance measure to guide anonymization effectively in the selection of traces to merge, whereas event logs where events relate to activities of disjunct contexts, lack such possibility.

To investigate this, we first consider the average and standard deviations of the computed event distances per log, shown in Table 3. The table reveals that events in *BPIC 2020* are particularly distant from each other, with the highest average (0.93) and smallest standard deviation (0.28), whereas the *Sepsis* log has the lowest average and highest standard deviation. This reveals that for the latter, our approach using the embedding-based distance is able to have a greater impact on the anonymization procedure when compared to an approach based on the Levenshtein distance, which is indeed confirmed by the larger positive impact that our approach has for this log, when compared to the *BPIC 2020* and *CoSeLoG* logs.

Table 3 furthermore shows the median share of the x most frequent directly followers per activity (as a fraction of its total number of directly followers). This way, we measure how similar the contexts of activities are in a log. The activities of *BPIC 2020* have the most common context, backing up the high distinctiveness of them. The *Sepsis* log on the other hand, are on average noticeably closer in distance and share less neighbouring events. Allowing the measure to distinguish stronger between the closeness of activities. Consequently, we saw higher utility preservation in the previous experiments for the *Sepsis* log, compared to the Levensthein distance-based baseline. With *CoSeLoG* falling both in terms of utility preservation and internal measures between the two other logs.

Table 3. Internal properties of our measure in terms of the distance between events and the share of the top-x most frequent followers per activity (median)

Event log	Event distance		Share of top-x followers		
	Avg.	Stdev.	x = 1	x = 2	x = 3
CoSeLoG	0.83	0.29	82.8%	91.3%	99.2%
Sepsis	0.77	0.42	67.0%	82.1%	94.3%
BPIC 2020	0.93	0.28	88.8%	96.7%	99.0%

5 Related Work

Privacy-preserving process mining received much attention recently [9]. In particular, the problem of anonymizing event logs was addressed in several approaches, since logs generally show serious re-identification risks [29]. PRETSA [11] has been proposed as an algorithm to tackle this problem based on the privacy notions of k-anonymity and t-closeness. Similarly, Rafiei et al. [23,24] proposed alternative techniques to ensure a group-based privacy guarantee, inspired by k-anonymity. A similar approach to group-based privacy protection is also realized in [4]. Either way, event log anonymization relies on a distance measure to decide on which traces to merge with each other and, so far, only the Levenshtein distance as a simple syntactic measure was considered. We addressed this shortcoming with the embedding-based measure presented in this work.

Several studies also focused on differential privacy, an alternative privacy guarantee, that limits the impact one individual may have on the data. While [18] and [14] focus on specific queries performed on the event logs, the issue of publishing differential private event logs was covered by [13] and [10].

Instead of anonymizing event logs, privacy-preserving algorithms for process mining may be employed to protect the data of individual persons [12]. Here, the distributed analysis of the control-flow of event logs was explored in [7] and [17], whereas privacy-aware role mining was studied in [21].

The application of privacy-preserving process mining in the healthcare domain was studied in [20], highlighting the importance of the ability to customize the algorithms to domain-specific requirements. To foster the uptake of privacy-preserving process mining, the respective techniques have been made accessible as tools, including ELPaaS [5], Shareprom [8], and the tool described in [22].

6 Conclusion

In this paper, we presented an embedding-based trace distance measure that is tailored to event log anonymization. It leverages feature vectors learned by the Act2Vec model to assess the similarity of events and, hence, traces. Unlike syntactic distances commonly used in event log anonymization, it thereby incorporates the context in which events occur. Moreover, the measure considers differences in the trace lengths in an asymmetric manner to guide the selection of which trace to merge into which other trace as part of the anonymization. Our experimental results indicate that an embedding-based distance measure can indeed improve the results of event log anonymization, compared to the use of the Levenshtein distance. Specifically, the improvement is most pronounced for event logs that contain events of different activities that frequently appear in similar contexts.

References

1. BPI challenge 2020: Prepaid travel costs. https://data.4tu.nl/articles/dataset/BPI_Challenge_2020_Prepaid_Travel_Costs/12696722. Accessed 12 May 2020

2. Receipt phase of an environmental permit application process ('wabo'), coselog project. https://data.4tu.nl/collections/Environmental_permit_application_process_WABO_CoSeLoG_project/5065529. Accessed 11 May 2020
3. Sepsis cases - event log. https://data.4tu.nl/articles/dataset/Sepsis_Cases_-_Event_Log/12707639. Accessed 03 Apr 2020
4. Batista, E., Solanas, A.: A uniformization-based approach to preserve individuals' privacy during process mining analyses. Peer Peer Netw. Appl. **14**, 1–20 (2021). https://doi.org/10.1007/s12083-020-01059-1
5. Bauer, M., Fahrenkrog-Petersen, S.A., Koschmider, A., Mannhardt, F., van der Aa, H., Weidlich, M.: ELPaaS: event log privacy as a service. In: BPM Demos, pp. 159–163 (2019)
6. De Koninck, P., vanden Broucke, S., De Weerdt, J.: act2vec, trace2vec, log2vec, and model2vec: representation learning for business processes. In: Weske, M., Montali, M., Weber, I., vom Brocke, J. (eds.) BPM 2018. LNCS, vol. 11080, pp. 305–321. Springer, Cham (2018). https://doi.org/10.1007/978-3-319-98648-7_18
7. Elkoumy, G., Fahrenkrog-Petersen, S.A., Dumas, M., Laud, P., Pankova, A., Weidlich, M.: Secure multi-party computation for inter-organizational process mining. In: Nurcan, S., Reinhartz-Berger, I., Soffer, P., Zdravkovic, J. (eds.) BPMDS/EMMSAD -2020. LNBIP, vol. 387, pp. 166–181. Springer, Cham (2020). https://doi.org/10.1007/978-3-030-49418-6_11
8. Elkoumy, G., Fahrenkrog-Petersen, S.A., Dumas, M., Laud, P., Pankova, A., Weidlich, M.: Shareprom: a tool for privacy-preserving inter-organizational process mining. In: BPM Demos, pp. 72–76 (2020)
9. Elkoumy, G., et al.: Privacy and confidentiality in process mining-threats and research challenges. arXiv:2106.00388 (2021)
10. Elkoumy, G., Pankova, A., Dumas, M.: Mine me but don't single me out: differentially private event logs for process mining. arXiv:2103.11739 (2021)
11. Fahrenkrog-Petersen, S., van der Aa, H., Weidlich, M.: PRETSA: event log sanitization for privacy-aware process discovery. In: ICPM (2019)
12. Fahrenkrog-Petersen, S.A.: Providing privacy guarantees in process mining. In: CAiSE (Doctoral Consortium), pp. 23–30 (2019)
13. Fahrenkrog-Petersen, S.A., van der Aa, H., Weidlich, M.: PRIPEL: privacy-preserving event log publishing including contextual information. In: Fahland, D., Ghidini, C., Becker, J., Dumas, M. (eds.) BPM 2020. LNCS, vol. 12168, pp. 111–128. Springer, Cham (2020). https://doi.org/10.1007/978-3-030-58666-9_7
14. Kabierski, M., Fahrenkrog-Petersen, S.A., Weidlich, M.: Privacy-aware process performance indicators: framework and release mechanisms. In: La Rosa, M., Sadiq, S., Teniente, E. (eds.) CAiSE 2021. LNCS, vol. 12751, pp. 19–36. Springer, Cham (2021). https://doi.org/10.1007/978-3-030-79382-1_2
15. Knols, B., van der Werf, J.M.E.M.: Measuring the behavioral quality of log sampling. In: ICPM. pp. 97–104 (2019)
16. Li, N., Li, T., Venkatasubramanian, S.: t-closeness: privacy beyond k-anonymity and l-diversity. In: ICDE. IEEE (2007)
17. Liu, C., Duan, H., Zeng, Q., Zhou, M., Lu, F., Cheng, J.: Towards comprehensive support for privacy preservation cross-organization business process mining. IEEE Trans. Serv. Comput. **12**(4), 639–653 (2016)
18. Mannhardt, F., Koschmider, A., Baracaldo, N., Weidlich, M., Michael, J.: Privacy-preserving process mining. BISE **61**(5), 595–614 (2019). https://doi.org/10.1007/s12599-019-00613-3
19. Mikolov, T., Yih, W.t., Zweig, G.: Linguistic regularities in continuous space word representations. In: NAACL, pp. 746–751 (2013)

20. Pika, A., Wynn, M.T., Budiono, S., Ter Hofstede, A.H., van der Aalst, W., Reijers, H.A.: Privacy-preserving process mining in healthcare. Int. J. Environ. Res. Public Health **17**(5), 1612 (2020)
21. Rafiei, M., van der Aalst, W.M.P.: Mining roles from event logs while preserving privacy. In: Di Francescomarino, C., Dijkman, R., Zdun, U. (eds.) BPM 2019. LNBIP, vol. 362, pp. 676–689. Springer, Cham (2019). https://doi.org/10.1007/978-3-030-37453-2_54
22. Rafiei, M., van der Aalst, W.: Practical aspect of privacy-preserving data publishing in process mining. In: BPM Demos, pp. 92–96 (2020)
23. Rafiei, M., van der Aalst, W.: Group-based privacy preservation techniques for process mining. arXiv preprint arXiv:2105.11983 (2021)
24. Rafiei, M., Wagner, M., van der Aalst, W.M.P.: *TLKC*-privacy model for process mining. In: Dalpiaz, F., Zdravkovic, J., Loucopoulos, P. (eds.) RCIS 2020. LNBIP, vol. 385, pp. 398–416. Springer, Cham (2020). https://doi.org/10.1007/978-3-030-50316-1_24
25. Rozinat, A., Aalst, W.: Conformance checking of processes based on monitoring real behavior. Inf. Syst. **33**, 64–95 (2008)
26. Stefanini, A., Aloini, D., Benevento, E., Dulmin, R., Mininno, V.: Performance analysis in emergency departments: a data-driven approach. Measuring Bus. Excell. (2018)
27. Sweeney, L.: k-anonymity: a model for protecting privacy. Int. J. Uncertain. Fuzz. Knowl.-Based Syst. **10**(05), 557–570 (2002)
28. Van Der Aalst, W.: Process mining: overview and opportunities. ACM Trans. Manag. Inf. Syst. (TMIS) **3**(2), 1–17 (2012)
29. Nuñez von Voigt, S., et al.: Quantifying the re-identification risk of event logs for process mining. In: Dustdar, S., Yu, E., Salinesi, C., Rieu, D., Pant, V. (eds.) CAiSE 2020. LNCS, vol. 12127, pp. 252–267. Springer, Cham (2020). https://doi.org/10.1007/978-3-030-49435-3_16

Towards Understanding the Role of the Human in Event Log Extraction

Vinicius Stein Dani[1(✉)], Henrik Leopold[2], Jan Martijn E. M. van der Werf[1], Xixi Lu[1], Iris Beerepoot[1], Jelmer J. Koorn[1], and Hajo A. Reijers[1]

[1] Utrecht University, Princetonplein 5, 3584 CC Utrecht, The Netherlands
{v.steindani,j.m.e.m.vanderwerf,x.lu,i.m.beerepoot,j.j.koorn, h.a.reijers}@uu.nl

[2] Kühne Logistics University, Großer Grasbrook 17, 20457 Hamburg, Germany
henrik.leopold@the-klu.org

Abstract. Process mining is widely used to visualize, analyze, and improve business processes. However, often its application is hindered by the considerable preparation effort that needs to be conducted by humans. One of the key tasks required in this context is obtaining the input artifact for process mining techniques: the event log. The data that is required for building such an event log typically needs to be collected from several databases and then transformed into a suitable format. While it has become clear to both academics and practitioners that the amount of human work is substantial, there is no deep understanding of the exact activities humans need to perform. Therefore, we use this paper to develop a precise understanding of how humans are involved in event log extraction. Based on a structured literature review and qualitative data coding, we derive a taxonomy of human tasks in event log extraction. This taxonomy can serve as input for both future automation efforts, as well as for process mining methodologies.

Keywords: Process mining · Event log extraction · Human tasks

1 Introduction

Many organizations use process mining techniques to visualize, analyze, and improve their business processes. Among others, process mining has been applied in auditing [20], production planning [15], and in the healthcare domain [32]. While process mining has been shown to come with many benefits [18], its application is still often hindered by the considerable preparation effort that needs to be conducted by humans [12]. One of the key challenges in this context is to obtain the input artifact for process mining techniques, the so-called *event log*.

Depending on the specific IT landscape, the data that is required for building an event log must be collected and extracted from several databases and then transformed into an appropriate format that can be processed by available process mining tools. While a variety of automated techniques have been proposed to support organizations in obtaining event logs (e.g., [7,9,27]), human support is still required at several stages. Unfortunately, there is currently no deep understanding of the *exact activities* humans

A. Marrella and B. Weber (Eds.): BPM 2021 Workshops, LNBIP 436, pp. 86–98, 2022.
https://doi.org/10.1007/978-3-030-94343-1_7

Table 1. Excerpt from an event log of an ordering process

No	Case identifier	Event	Timestamp
1	123	Receive order	2021-05-11 11:35
2	123	Collect order items	2021-05-11 12:41
3	234	Receive order	2021-05-12 09:03
4	123	Send invoice	2021-05-12 13:49
5	234	Collect order items	2021-05-12 17:12
6	123	Ship order	2021-05-13 08:54

are performing in this context. While this may seem surprising, this can be explained by the fact that available techniques for event log extraction often focus on rather isolated technical aspects and not on the full range of tasks that are required in a practical setting.

Recognizing that the human involvement in event log extraction comes with considerable time and cost, we set out to develop a precise understanding of how humans are involved in the context of event log extraction. We believe that such an understanding is vital input for both future automation efforts as well as for process mining methodologies. Therefore, the research question of this paper is "*What are the specific manual tasks that humans perform in the context of event log extraction?*". To answer our research question, we first conduct a structured literature review targeting process mining case studies. Based on qualitative data coding, we then use the identified papers to derive a taxonomy of human tasks in event log extraction.

The remainder of the paper is organized as follows. Section 2 discusses the background of this paper. Section 3 describes our research method. Section 4 presents our taxonomy of manual tasks in even log extraction. Section 5 reflects on the implications and limitations before Sect. 6 concludes the paper.

2 Background

Process mining is a family of data analysis techniques that aims to discover, monitor, and improve organizational processes by analyzing data from so-called event logs [1]. Such event logs can be obtained from various IT systems and provide insights into how organizational processes are executed. Table 1 shows an excerpt of a simple event log of an ordering process. It shows that each entry of an event log must have at least three attributes: a case identifier, an event, and a timestamp. The event column reveals *what* happened. The timestamp column shows *when* the event occurred. The case identifier relates each event to a particular process execution, often referred to as *case*. In the example, we only observe two different cases (123 and 234).

Event logs like the one from Table 1 are a prerequisite for any available process mining technique. The process of obtaining event logs is called *event log extraction*. It is a complex and time-intensive process, which requires human involvement at several stages. The data required for constructing an event log often resides in a variety of sources. One of the main challenges is that many information systems are not process-centric and, therefore, do not record events or case identifiers explicitly. This means that we may need to identify and extract event data from a variety of different databases and transform them to a process-centric event log [12].

To provide automated support for this endeavor, many automated techniques have been developed. Recognizing the large variety of potential data sources and requirements in practice, available techniques differ considerably with respect to required inputs, their output, and also their limitations. Among others, there are extraction techniques that build on ontologies [7], redo logs [9], and database objects [27]. All these techniques require human intervention or input at some stage. Unfortunately, the exact role of the human is not always clear. Since many of these techniques address rather specific problems of event log extraction, the required human involvement can often only be understood when these techniques are applied in practice.

Given that the human involvement in event log extraction is both time and cost-intensive, we use this paper to develop a precise understanding of the respective human tasks. We argue that such an understanding is a prerequisite for further automation efforts, as well as for developing process mining methodologies. Therefore, we define our research question as: "*What are the specific manual tasks that humans perform in the context of event log extraction?*".

3 Research Method

To answer our research question, we followed a two-step approach. We first conducted a structured literature review. Based on the result, we then derived a taxonomy of human tasks via qualitative data coding. In the Sects. 3.1 and 3.2, we describe both steps in detail.

3.1 Literature Study

To identify which manual tasks have been performed in the context of event log extraction, we decided to focus on papers conducting case studies in process mining projects. In Scopus, we used the search string: ("process mining" AND "case stud*"). We included peer-reviewed papers that were published in journals or conferences between 2000 and 2020. Next, we filtered out duplicate papers, papers that were not in English, and papers that merely mentioned process mining. The search resulted in 191 papers. We identified papers that could contain tasks related to event log extraction by making a first read of the papers. This resulted in a set of 120 papers that were assessed more closely to extract the actual tasks related to manual efforts during log extraction.

As a next step in the study, we went through the 120 papers to identify human tasks related to event log extraction. To label a task as manual, we defined the following criteria:

1. It is explicitly mentioned that the task was performed manually.
2. There was no explicit counter statement that the task was performed (semi-) automatically.
3. We assume that conceptual models, such as data models and process models, were created manually unless the authors explicitly stated that they were created in an automated fashion.

With this in mind, the final set of papers from which we identified manual efforts is composed of 46 papers. On these papers we performed the coding as presented in the next section.

3.2 Coding and Taxonomy Derivation

To derive a taxonomy of manual tasks, we coded the 46 papers resulting from our literature review based on the coding of qualitative data [33]. We performed three specific steps: First, we performed hypothesis coding. This means that we devised a set of codes we expected to find in the data without actually conducting any further analysis. Second, we performed a holistic coding to identify possible categories that could emerge from and represent the data. Third, we compared both coding schemes and discussed how they relate to each other to achieve a more concise representation of the identified categories. We identified that we could directly match the holistic coding entries to our hypothesis coding, which led us to the taxonomy presented in Sect. 4.

For example, the following codes emerged from the holistic approach: i) "search into the data to select case perspective", supported by "*[...] It is therefore possible to consider the data from at least three different 'case' perspectives, i.e. an incident may be considered as a case, each patient may be considered as a case, or each response unit may be considered as a case. [...]*" [4]; and, ii) "identify, from discussion with domain expert, which case should be considered", supported by "*[...] in this case we constructed three different event logs according to the different perspectives. [...] After several discussions, it appeared that both the team flow and the document type flow represented the business process best [...]*" [10]. Both codes are related to the selection of a case, which can be summarized by our hypothesis code "Select case notion". As can be seen from these examples, we used lower-level coding in our holistic coding, and high-level coding in our hypothesis coding.

4 A Taxonomy for Manual Tasks in Event Log Extraction

In this section, we present the outcome of our research: a taxonomy for manual tasks in event log extraction. We first provide an overview of our taxonomy in Sect. 4.1. In the remainder of the section, we discuss the details of each of the five categories.

4.1 Overview

A visual representation of our taxonomy is shown in Fig. 1. It consists of five categories, represented as four squares and one rectangle. These five categories resemble the main manual tasks we identified in event log extraction. Starting from the context and scope definition, the data source need to be assessed, and attributes need to be selected. Based on this selection, an event log can be extracted from the data source, and needs to be assessed before it can be used in process mining. Note that there is no strict flow. Some tasks might be executed repeatedly or in an arbitrary order. The numbers in brackets indicate in how many papers these task categories were mentioned.

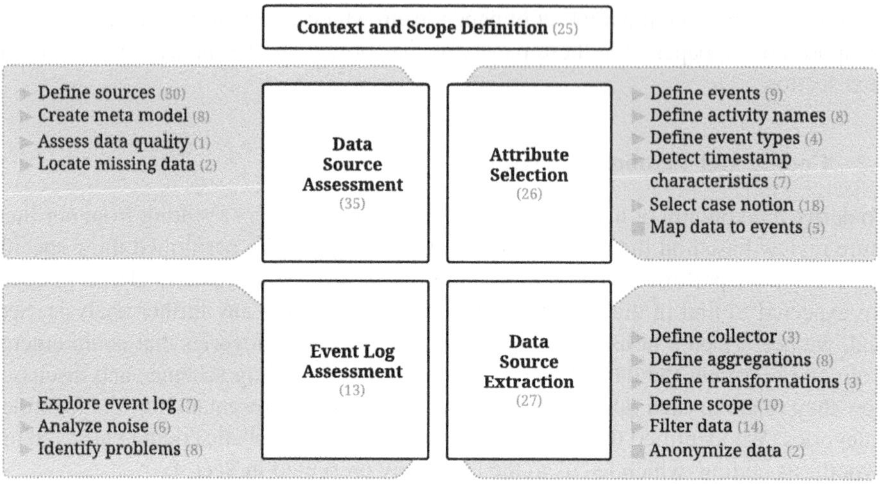

Fig. 1. Taxonomy of manual tasks in event log extraction.

4.2 Context and Scope Definition

The first step as indicated by several process mining methodologies (cf. [13]) is to define the overall context and scope (supported by 25 papers from our literature study). Naturally, this is a completely manual task that mainly consists of discussions between the process analyst and the stakeholders. The scope can vary between very exploratory to very specific. Once there is an agreement on the scope, this defines the subsequent data extraction steps. However, as pointed out in [10], the scope definition often needs to be adjusted or redefined:

> "*[...] it proved very difficult to mark out the boundaries of the process under investigation in the larger DMS [Document Management System]. Therefore, the data scope had to be fine-tuned multiple times, going through the scope adjustment loop after each inspection, until a satisfactory data set was obtained.*" [10]

This excerpt highlights that the definition of context and scope is not a one-off task, but may require repeated manual interventions.

4.3 Data Source Assessment

Event logs are extracted from data sources. As a first step, these sources need to be assessed on their usability and quality. In our literature study, we identified four task categories: reverse engineer a meta model, define sources, audit data, and locate missing data. These categories are supported by 35 papers overall.

Define Sources. Typically, there is a variety of potential data sources available. This means that a human has to understand, analyze, and decide which data sources contain the required data and, therefore, need to be considered for data extraction. That this is a complex manual task is, for example, highlighted in [36]:

> "*The selection of information on business events is an important challenge in event log extraction: while large sets of information may reduce the performance of process mining tools, too little information may affect the quality of the analysis and resulting conclusions.*" [36]

However, currently, there is no clear perspective on how this manual task can be supported or automated.

Create Meta Model. When data is extracted from multiple sources, the process analyst needs to decide on how to properly merge the data. This task can be supported by acquiring (or generating) meta models for each data source. This provides a better understanding of the different data sources [4], and allows the process analyst to already develop an understanding of how the data relates to specific process activities. In our literature study, we identified that authors leverage different meta models for this task, such as entity models [14] and hierarchy models [23].

Assess Data Quality. Prior to data extraction, there should be an assessment of the quality of the available data sources. It may be that data sources turn out to be unsuitable for process mining. An example has been reported in [22]:

> "*Clinical inspection of the data quality of the EHR [Electronic Health Record] revealed data that was considered too unreliable to use for process mining. Issues included unrecorded events and observations, recording on letters and paper records rather than the EHR, mis-diagnosis and inappropriate referrals.*" [22]

This example highlights that some manual effort is not only required for the assessment, but also for fixing the problem. In the case above, the authors decided to manually review large amounts of data, which they then manually transformed into a suitable format.

Locate Missing Data. Suppose the initially available sources do not contain all the necessary data for the process mining project. In that case, there is a need to locate the missing data in other data sources or extract it by other means. In practice, it is often necessary to obtain data managed by departments in the organization that otherwise would not be considered to take part in the process mining initiative. However, suppose the missing data is not available in *any* of the information systems of the organization. In that case, this data may be obtained by other means. In [22], for example, the authors filled the gap of missing data through interviews with domain experts. In any case, the detection and localization of missing data is so far still a manual and time-consuming effort.

4.4 Attribute Selection

Once the data sources have been assessed and selected, the next step is to select the attributes for log extraction. In this category, we identified six task categories, which are supported by 26 papers: define events, define activity names, define event types, detect timestamp characteristics, select the case notion, and map activities to events.

Define Events. One of the key tasks here is to decide which attributes will together represent the events. Attributes can be more generic towards process mining, such as resources, actors, and activities. In some case studies, highly context-specific attributes were selected. An example of such a context-specific attribute is provided in [36]:

> "*Some logistics elements (e.g., the cargo type) can be selected as the instance attributes such that the logisticians could look at the processes from a wider perspective [...] The cargo type of the instance is useful information as it has a big impact on the cargo handling processes*" [36]

The majority of the studies report on their defined events or the relevance of a proper selection of events [17,24,28] without getting much into details on the decisions related to the event definition, nor how it was performed.

Define Activity Names. Defining activity names involves different actions, such as data correlation and abstraction [12]. In [19], the authors use names already available in the data source. While in [36], to perform this task, the authors use available documents. Independently of how this task is performed, it usually still requires manual efforts and more thorough discussions within the case studies in the literature.

Define Event Types. If event types are available (e.g., start, complete, etc.), it might be needed to define which will be necessary for the analysis. In some cases [30], there are so many event types available that there is the need to trim down the data by choosing only event data from a subset of event types. Otherwise, the data would generate spaghetti-like process models.

Detect Timestamp Characteristics. Timestamps are available in data sources in many different formats, sizes, and varieties. The detection of timestamp characteristics before the data extraction can anticipate different problems, such as in [25]:

> "*[...] we do not have any information about the actual timestamps of the start and completion of the service delivered. Consequently, the ordering of events which happen on the same day do not necessarily conform with the order in which events of that day were executed.*" [25]

However, few studies reporting on the event log extraction report on the detection of timestamp characteristics in early stages [5,11]. From our findings, we identified that the studies discuss timestamp characteristics only later on in the process of event log extraction [19].

Select Case Notion. In many settings, there are several options concerning the case notion selection. A typical example could be found in the healthcare domain:

> "*It is [...] possible to consider the data from at least three different 'case' perspectives, i.e. an incident may be considered as a case, each patient may be considered as a case, or each response unit may be considered as a case.*" [5]

As also pointed out by several other authors, this decision is an important one to take and may require several analyses to understand the implications of a potential case identifier [10,17].

Map Data to Events. In this task, process analysts need to identify which tables contain data that relate to events they wish to be part of the event log. What is more, they need to define which mappings provide the intended representation of the events. Often this mapping is guided by activities from a reference model resulting from the previous task category. It can also be directly performed by the process analysts [3].

4.5 Data Source Extraction

Once the attributes have been selected, the actual event logs can be extracted. In this category, we identified six task categories: define collector, define aggregations, define transformations, define scope, filter data, and anonymize data. These categories are supported by 27 papers from our literature study altogether.

Define Collector. In some cases [5,22], the process analyst requests and receives the desired data from a data expert, who represents an involved stakeholder. In other cases, the process analyst has access to the data either via direct access to the data sources [6] or via a dump from the needed data sources [35]. The actual role of the person who collects the data is often left implicit (as e.g. in [8,26]).

Define Aggregations. When data is too fine-grained, there is often the need for the process analyst alongside the domain expert to define which will be the aggregations that need to be manually performed [16,34]. The importance of defining aggregations is reported by [31], which states that the aggregation definition leads the process analyst towards a high-level event log, which reduces the complexity of the resulting process model.

Define Transformations. Different studies report on data transformation supported by domain knowledge [21,29]. In [21], the authors report they transformed data scattered across a set of tables in a database into a unified format. In [29], the authors transformed some data elements using proxy timestamps whenever there were no previously recorded timestamps.

Transformations also refer to the decisions and manipulations performed to transform event data from non process-centric information systems into a process-centric event log. Independently of the transformations performed, it is still necessary to define and report on the data transformations to be consistent and keep documentation of what was performed over the data and why. In this way, it becomes repeatable.

Define Scope. Depending on the available data, it is necessary to define the scope further. For instance, in [17], there is too much data about a process related to many different company branches. In such a case, the authors decided to trim the data by narrowing it down to one branch to "*produce a more focused analysis*" [17].

Filter Data. After the different required definitions to perform the extraction, invariably, data filtering is performed. In many cases, the process analyst can make use of tools to support this task. However, domain knowledge and manual efforts are often driving the data filtering task [10, 17, 22]. From practice, we know that data filtering is an iterative process, which is often aligned with process discovery and conformance checking.

Anonymize Data. If the process under analysis has data linkable to a specific person, this data should be anonymized (or at least pseudonymized) to preserve privacy [2, 26, 36]. The detection of the particular attributes that may incur in links to specific persons relies on domain knowledge. Such detection is often performed manually.

4.6 Event Log Assessment

As a final category in the extraction process, the extracted event log is assessed on its quality. In this category, we identified three task categories supported by 13 papers: explore event log, analyze noise, and identify problems.

Explore Event Log. The process analyst needs to explore the extracted event log to reflect back on the previous definitions and to attempt identifying problems that might occur because of remnants of noise, incomplete or imprecise data, or even inadequate definitions. Some studies are explicit about this task, such as in [17]:

> "*The next phase of event log exploration is intended to adjust the data set according to the time and scope of the work, to identify sanitization rules, and, to identify the different analysis dimensions.*" [17]

During this task, the process analyst often leverages process discovery and conformance checking techniques to perform the exploration. However, data and scope adjustments are usually ad-hoc, and performed manually [10].

Analyze Noise. If the definitions, mapping, and filtering previously mentioned are not performed thoroughly, different types of noise in data may still be detected after data is extracted, such as incomplete cases [17], and outliers [6]. And even with thorough definitions, one can still find, for example, infrequent process variants [34]. Although there are several tools to support data filtering, domain knowledge drives the noise analysis and is usually performed by a process analyst alongside a domain expert.

Identify Problems. Once the event log is extracted, it can be assessed on potential problems. Examples of problems identified are: incomplete recordings of events [34], and incomplete cases [17]. This identification task is also driven by domain knowledge. It involves iterations between the process analyst and the domain expert, who will be redefining the scope or the events themselves, and performing data filtering.

5 Discussion

In this section, we discuss the implications and the limitations of our taxonomy for manual tasks in event log extraction.

As for the *implications* of our taxonomy, we argue that it provides an important starting point for reducing the extent of human involvement in event log extraction. We can identify two main use cases in this regard. The first use case concerns *automation*. Without exactly understanding where and how humans are involved, increasing the level of automation is hardly feasible. Our taxonomy, therefore, provides important input on the tasks that can be potentially automated. The second use case concerns *guidance*. Given the nature of some tasks (e.g. defining the scope or defining sources), it is clear that not all human tasks can be automated. However, what can still reduce the time investment and the overall effort is an increased level of guidance for manual tasks. Such guidance may include checklists or other artifacts that provide orientation and help to structure the task execution.

Naturally, our taxonomy is also subject to a number of *limitations*. The first limitation concerns the scope of our literature study. Since we exclusively targeted process mining case studies, we cannot claim to provide a complete picture. We, however, made this choice deliberately since we wanted to learn about the use of process mining in real-world environments. The second limitation concerns the use of a literature study in general. While studying the selected papers, we realized in several places that certain details are missing or not explicitly discussed. Hence, there might exist additional manual tasks that were simply not reported upon or discussed in the analyzed papers. Nonetheless, due to the extent of our literature study, we are confident that our taxonomy provides a rather comprehensive overview.

6 Conclusion

The extraction of event logs comes with substantial human effort. In this paper, we set out to develop a precise understanding of which manual tasks humans perform in the context of event log extraction. We conducted a structured literature review and applied qualitative data coding to systematically derive a taxonomy of manual tasks. Our taxonomy highlights that human work is required in various phases of the event log extraction process ranging from scope and context definition to data source assessment. As such, our taxonomy does not only provide a comprehensive overview but can also serve as input for future automation efforts and for methodological process mining support. In future research, we aim to follow up on the insights provided in this paper and develop techniques that reduce, simplify, or support human work in event log extraction.

Acknowledgments. Part of this research was funded by NWO (Netherlands Organisation for Scientific Research) project number 16672.

References

1. van der Aalst, W.M.P.: Process Mining: Data Science in Action. Springer, Heidelberg (2016). https://doi.org/10.1007/978-3-662-49851-4
2. van der Aalst, W.M.P., Guo, S., Gorissen, P.: Comparative process mining in education: an approach based on process cubes. In: Ceravolo, P., Accorsi, R., Cudre-Mauroux, P. (eds.) SIMPDA 2013. LNBIP, vol. 203, pp. 110–134. Springer, Heidelberg (2015). https://doi.org/10.1007/978-3-662-46436-6_6
3. Andrews, R., van Dun, C.G.J., Wynn, M.T., Kratsch, W., Röglinger, M.K.E., ter Hofstede, A.H.M.: Quality-informed semi-automated event log generation for process mining. Decis. Support Syst. **132**, 113265 (2020)
4. Andrews, R., et al.: Leveraging data quality to better prepare for process mining: an approach illustrated through analysing road trauma pre-hospital retrieval and transport processes in Queensland. Int. J. Environ. Res. Public Health **16**(7), 1138 (2019)
5. Andrews, R., et al.: Pre-hospital retrieval and transport of road trauma patients in Queensland. In: Daniel, F., Sheng, Q.Z., Motahari, H. (eds.) BPM 2018. LNBIP, vol. 342, pp. 199–213. Springer, Cham (2019). https://doi.org/10.1007/978-3-030-11641-5_16
6. Benevento, E., Dixit, P.M., Sani, M.F., Aloini, D., van der Aalst, W.M.P.: Evaluating the effectiveness of interactive process discovery in healthcare: a case study. In: Di Francescomarino, C., Dijkman, R., Zdun, U. (eds.) BPM 2019. LNBIP, vol. 362, pp. 508–519. Springer, Cham (2019). https://doi.org/10.1007/978-3-030-37453-2_41
7. Calvanese, D., Montali, M., Syamsiyah, A., van der Aalst, W.M.P.: Ontology-driven extraction of event logs from relational databases. In: Reichert, M., Reijers, H.A. (eds.) BPM 2015. LNBIP, vol. 256, pp. 140–153. Springer, Cham (2016). https://doi.org/10.1007/978-3-319-42887-1_12
8. Cho, M., Song, M., Yoo, S.: A systematic methodology for outpatient process analysis based on process mining. In: Ouyang, C., Jung, J.-Y. (eds.) AP-BPM 2014. LNBIP, vol. 181, pp. 31–42. Springer, Cham (2014). https://doi.org/10.1007/978-3-319-08222-6_3
9. de Murillas, E.G.L., van der Aalst, W.M.P., Reijers, H.A.: Process mining on databases: unearthing historical data from redo logs. In: Motahari-Nezhad, H.R., Recker, J., Weidlich, M. (eds.) BPM 2015. LNCS, vol. 9253, pp. 367–385. Springer, Cham (2015). https://doi.org/10.1007/978-3-319-23063-4_25
10. De Weerdt, J., Schupp, A., Vanderloock, A., Baesens, B.: Process mining for the multi-faceted analysis of business processes - a case study in a financial services organization. Comput. Ind. **64**(1), 57–67 (2013)
11. Deeva, G., De Smedt, J., De Koninck, P., De Weerdt, J.: Dropout prediction in MOOCs: a comparison between process and sequence mining. In: Teniente, E., Weidlich, M. (eds.) BPM 2017. LNBIP, vol. 308, pp. 243–255. Springer, Cham (2018). https://doi.org/10.1007/978-3-319-74030-0_18
12. Diba, K., Batoulis, K., Weidlich, M., Weske, M.: Extraction, correlation, and abstraction of event data for process mining. WIREs Data Min. Knowl. Discov. **10**(3), e1346 (2020)
13. van Eck, M.L., Lu, X., Leemans, S.J.J., van der Aalst, W.M.P.: PM^2: a process mining project methodology. In: Zdravkovic, J., Kirikova, M., Johannesson, P. (eds.) CAiSE 2015. LNCS, vol. 9097, pp. 297–313. Springer, Cham (2015). https://doi.org/10.1007/978-3-319-19069-3_19
14. Epure, E.V., Hug, C., Deneckére, R., Brinkkemper, S.: What shall I do next? In: Jarke, M., et al. (eds.) CAiSE 2014. LNCS, vol. 8484, pp. 473–487. Springer, Cham (2014). https://doi.org/10.1007/978-3-319-07881-6_32
15. Er, M., Arsad, N., Astuti, H.M., Kusumawardani, R.P., Utami, R.A.: Analysis of production planning in a global manufacturing company with process mining. J. Enterp. Inf. Manag. **31**(2), 317–337 (2018)

16. Flath, C.M., Stein, N.: Towards a data science toolbox for industrial analytics applications. Comput. Ind. **94**, 16–25 (2018)
17. dos Santos Garcia, C., Meincheim, A., Garcia Filho, F.C., Santos, E.A.P., Scalabrin, E.E.: Getting insights to improve business processes with agility: a case study using process mining. In: ICSMC, pp. 1336–1343. IEEE (2019)
18. Grisold, T., Mendling, J., Otto, M., vom Brocke, J.: Adoption, use and management of process mining in practice. Bus. Process. Manag. J. **27**(2), 369–387 (2020)
19. Gunnarsson, B.R., vanden Broucke, S.K.L.M., De Weerdt, J.: Predictive process monitoring in operational logistics: a case study in aviation. In: Di Francescomarino, C., Dijkman, R., Zdun, U. (eds.) BPM 2019. LNBIP, vol. 362, pp. 250–262. Springer, Cham (2019). https://doi.org/10.1007/978-3-030-37453-2_21
20. Jans, M., Alles, M., Vasarhelyi, M.: The case for process mining in auditing: sources of value added and areas of application. Int. J. Account. Inf. Syst. **14**(1), 1–20 (2013)
21. Jareevongpiboon, W., Janecek, P.: Ontological approach to enhance results of business process mining and analysis. Bus. Process. Manag. J. **19**(3), 459–476 (2013)
22. Johnson, O.A., Ba Dhafari, T., Kurniati, A., Fox, F., Rojas, E.: The clearpath method for care pathway process mining and simulation. In: Daniel, F., Sheng, Q.Z., Motahari, H. (eds.) BPM 2018. LNBIP, vol. 342, pp. 239–250. Springer, Cham (2019). https://doi.org/10.1007/978-3-030-11641-5_19
23. Knoll, D., Reinhart, G., Prüglmeier, M.: Enabling value stream mapping for internal logistics using multidimensional process mining. Expert Syst. Appl. **124**, 130–142 (2019)
24. Lemos, A.M., Sabino, C.C., Lima, R.M.F., Oliveira, C.A.L.: Using process mining in software development process management: a case study. In: ICSMC. IEEE (2011)
25. Mans, R.S., Schonenberg, M.H., Song, M., van der Aalst, W.M.P., Bakker, P.J.M.: Application of process mining in healthcare – a case study in a Dutch hospital. In: Fred, A., Filipe, J., Gamboa, H. (eds.) BIOSTEC 2008. CCIS, vol. 25, pp. 425–438. Springer, Heidelberg (2008). https://doi.org/10.1007/978-3-540-92219-3_32
26. Marazza, F., et al.: Comparing process models for patient populations: application in breast cancer care. In: Di Francescomarino, C., Dijkman, R., Zdun, U. (eds.) BPM 2019. LNBIP, vol. 362, pp. 496–507. Springer, Cham (2019). https://doi.org/10.1007/978-3-030-37453-2_40
27. Nooijen, E.H.J., van Dongen, B.F., Fahland, D.: Automatic discovery of data-centric and artifact-centric processes. In: La Rosa, M., Soffer, P. (eds.) BPM 2012. LNBIP, vol. 132, pp. 316–327. Springer, Heidelberg (2013). https://doi.org/10.1007/978-3-642-36285-9_36
28. Park, M., Song, M., Baek, T.H., Son, S.Y., Ha, S.J., Cho, S.W.: Workload and delay analysis in manufacturing process using process mining. In: Bae, J., Suriadi, S., Wen, L. (eds.) AP-BPM 2015. LNBIP, vol. 219, pp. 138–151. Springer, Cham (2015). https://doi.org/10.1007/978-3-319-19509-4_11
29. Partington, A., Wynn, M.T., Suriadi, S., Ouyang, C., Karnon, J.: Process mining for clinical processes: a comparative analysis of four Australian hospitals. ACM TMIS **5**(4), 1–18 (2015)
30. Potavin, J., Jongswat, N., Premchaiswadi, W.: Applying fuzzy-genetic mining in conformance and dependency relations. In: ICT & KE, pp. 228–235. IEEE (2012)
31. Rojas, E., Cifuentes, A., Burattin, A., Munoz-Gama, J., Sepúlveda, M., Capurro, D.: Performance analysis of emergency room episodes through process mining. Int. J. Environ. Res. Public **16**(7), 251–263 (2019)
32. Rojas, E., Munoz-Gama, J., Sepúlveda, M., Capurro, D.: Process mining in healthcare: a literature review. J. Biomed. Inform. **61**, 224–236 (2016)
33. Saldaña, J.: The Coding Manual for Qualitative Researchers. Sage, Thousand Oaks (2009)

34. Suriadi, S., Wynn, M.T., Ouyang, C., ter Hofstede, A.H.M., van Dijk, N.J.: Understanding process behaviours in a large insurance company in Australia: a case study. In: Salinesi, C., Norrie, M.C., Pastor, Ó. (eds.) CAiSE 2013. LNCS, vol. 7908, pp. 449–464. Springer, Heidelberg (2013). https://doi.org/10.1007/978-3-642-38709-8_29
35. Walicki, M., Ferreira, D.R.: Sequence partitioning for process mining with unlabeled event logs. Data Knowl. Eng. **70**(10), 821–841 (2011)
36. Wang, Y., Caron, F., Vanthienen, J., Huang, L., Guo, Y.: Acquiring logistics process intelligence: methodology and an application for a Chinese bulk port. Expert Syst. Appl. **41**(1), 195–209 (2014)

May I Take Your Order?

On the Interplay Between Time and Order in Process Mining

Wil M. P. van der Aalst[1,2(✉)] and Luis Santos[1]

[1] Process and Data Science (Informatik 9), RWTH Aachen University, Aachen, Germany
wvdaalst@pads.rwth-aachen.de

[2] Fraunhofer-Institut für Angewandte Informationstechnik (FIT), Sankt Augustin, Germany

Abstract. Process mining starts from event data. The ordering of events is vital for the discovery of process models. However, the timestamps of events may be unreliable or imprecise. To further complicate matters, also causally unrelated events may be ordered in time. The fact that one event is followed by another does not imply that the former causes the latter. This paper explores the relationship between time and order. Moreover, it describes an approach to preprocess event data having timestamp-related problems. This approach avoids using accidental or unreliable orders and timestamps, creates partial orders to capture uncertainty, and allows for exploiting domain knowledge to (re)order events. Optionally, the approach also generates interleavings to be able to use existing process mining techniques that cannot handle partially ordered event data. The approach has been implemented using ProM and can be applied to any event log.

Keywords: Process mining · Event data · Partial orders · Uncertainty

1 Introduction

Most process mining techniques require the events within a case to the totally ordered [2]. For example, nearly all discovery techniques convert the event log into a multiset of traces where each trace is a sequence of activities. To order the events within a case, typically timestamps are used. However, the timestamps may be unreliable or too coarse-grained. Consider, for example, a nurse taking a blood sample from a patient at 16.55 but recording this into the hospital's information system at 17.55 when her shift ends (event e_1). At 17.15, the patient's insurance company approved the operation and this was automatically recorded (event e_2). The same patient also had an X-ray in the evening, but only the date is recorded (event e_3). In this example, the real ordering of events was $\langle e_1, e_2, e_3 \rangle$, but in the event log they may appear as $\langle e_3, e_2, e_1 \rangle$. Event e_1 happened before e_2 but was recorded one hour later. Event e_3 was the last event, but because only the date was recorded, it appeared to have happened at time 00.00. Moreover, events e_1 and e_2 were fully unrelated, so why consider the temporal order? The approval was triggered by a request submitted two days before. Healthcare data are notorious for having data quality problems [10]. However, such issues can be found in any domain [3,11,15].

A. Marrella and B. Weber (Eds.): BPM 2021 Workshops, LNBIP 436, pp. 99–110, 2022.
https://doi.org/10.1007/978-3-030-94343-1_8

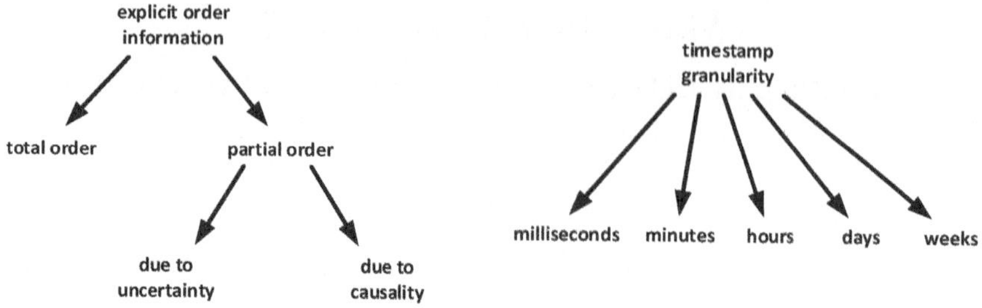

Fig. 1. We assume that events may have explicit order information (left) and have a timestamp (right). However, the ordering is partial and the timestamps can be coarse-grained.

In this paper, we assume that events are *partially ordered* and have a *timestamp* (see Fig. 1). This allows us to reason about the problems just mentioned. Given a set of events E, we assume a strict partial order $\prec_o \subseteq E \times E$. $e_1 \prec_o e_2$ means that event e_1 is before event e_2. $\pi_{time}(e_1)$ and $\pi_{time}(e_2)$ are the timestamps of both events. We assume that events are recorded at a certain granularity, e.g., milliseconds, seconds, hours, days, weeks, months, or years. Events may have more fine-grained timestamps, but we map these onto the chosen level of granularity. For example, "19-05-2021:17.15.00" and "19-05-2021:17.55" are both mapped onto "19-05-2021" when using days as a granularity.

As mentioned, next to timestamps at a selected granularity level, we also assume a partial order on events. Such a partial order can also be more coarse-grained or more fine-grained. One extreme is that the events are totally ordered, i.e., for any two events e_1 and e_2: $e_1 \prec_o e_2$ or $e_1 \succ_o e_2$. Another extreme is that that no two events are ordered ($e_1 \not\prec_o e_2$ or $e_1 \not\succ_o e_2$). The latter case (events are unordered) is similar to assuming that all events have the same coarse-grained timestamp (e.g., same year).

To better explain the problem, we consider the BPMN (Business Process Model and Notation) model shown in Fig. 2 for handling requests for compensation within an airline. Customers may request compensation for various reasons, e.g., a delayed or canceled flight. The process starts by registering the request (*reg*). After this, three checks need to be performed. The customer's ticket and history are checked for all cases, i.e., activities (ct) and (ch) need to be performed for all requests. There is a choice

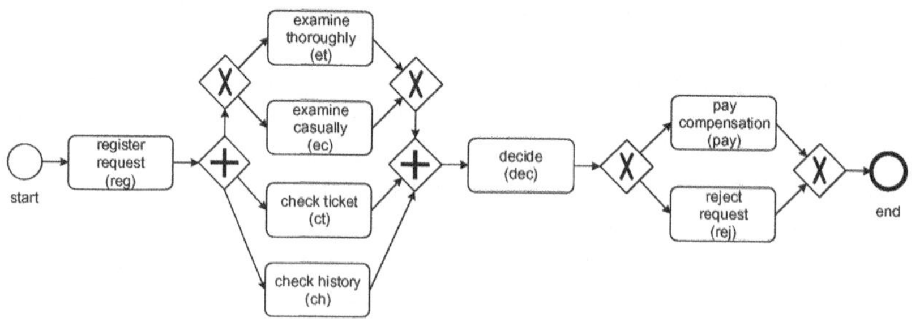

Fig. 2. A BPMN model having 4 partially-ordered runs and $2 \times 2 \times 3! = 24$ sequential runs.

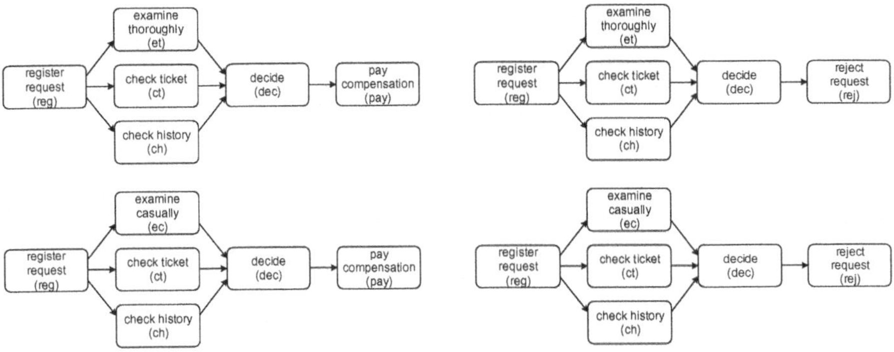

Fig. 3. The four partially-ordered runs of the BPMN model in Fig. 2.

between a thorough examination (et) and a casual examination (ec). After this, a decision is made (dec) and the request is rejected (rej) or some compensation is paid (pay). Examples of sequential runs are $\langle reg, et, ct, ch, dec, pay\rangle$, $\langle reg, ct, ec, ch, dec, rej\rangle$, and $\langle reg, ch, ct, et, dec, pay\rangle$. In total there are $2 \times 2 \times 3! = 24$ sequential runs. Note that in each run there are three concurrent activities (ct, ch, and either et or ec). These can be interleaved in $3! = 6$ ways.

There are only $2 \times 2 = 4$ partially-ordered runs. These are depicted in Fig. 3. Note that the four partially-ordered runs do not need to specify the ordering of concurrent activities. Consider the scenario with 10 concurrent activities; there is just one partially-ordered run, but there are $10! = 3628800$ sequential runs. Figure 3 helps to understand why partial orders are considered in process mining and many other analysis approaches.

Table 1. A fragment of an event log where the timestamps of events are problematic.

Event id	Properties					
	Case id	Activity	Timestamp	Resource	Cost	...
36533	9901	Register request	19-05-2021:11.02.55	Sarah	50	...
36534	9901	Check ticket	19-05-2021:13.02	John	25	...
36535	9902	Register request	19-05-2021:13.02	Sarah	50	...
36536	9902	Check history	20-05-2021:00.00.00	Pete	45	...
36537	9901	Check history	20-05-2021:00.00.00	Pete	45	...
36538	9901	Examine casually	20-05-2021:08.55.34	Mary	55	...
36539	9902	Check ticket	20-05-2021:09.11.21	John	25	...
36540	9902	Examine thoroughly	20-05-2021:10.55	Harry	55	...
36541	9901	Decide	21-05-2021	Angela	55	...
36542	9902	Decide	21-05-2021	Angela	75	...
36543	9902	Reject request	22-05-2021:14.12.45	Sarah	20	...
36544	9901	Pay compensation	22-05-2021:16.52.37	Sarah	150	...
...	...	...	...	...	...	...

Table 1 shows a fragment of an event log corresponding to the BPMN model in Fig. 2. Process discovery techniques aim to learn a process model based on such data. If we assume the events to be sorted based on the identifier in the first column, then case 9901 corresponds to sequential run $\langle reg, ct, ch, ec, dec, pay \rangle$ and case 9902 corresponds to sequential run $\langle reg, ch, ct, et, dec, rej \rangle$. However, a closer inspection of the timestamp column suggests that there several problems. Some events have a precision in seconds, others in minutes, or even days. There are also timestamps of the form "20-05-2021:00.00.00" which suggests that times are sometimes rounded to days. Moreover, we may know that some timestamps show the time of recording and not the actual event. At the same time, we may know that the registration activity (reg) always happens before the check activities.

When timestamps are unreliable or imprecise, like in Table 1, we cannot use them as-is. One approach is to make the timestamps more coarse-grained (e.g., just consider the day). This automatically leads to partially ordered traces. Moreover, there may be explicit information that reveals explicit causal relations. For example, when the concurrent activities do not share any information. We may know that ct, ch, ec, and et use only data collected in ref, but that dec uses the outcomes of the three checks. Such causal dependencies can be derived based on data-flow analysis or explicit domain knowledge, e.g., a payment is always preceded by a decision. As Fig. 1 shows, partial orders can be used to express either *uncertainty* or *explicit causality* (i.e., partial orders have a dual interpretation).

Next to discussing the relationship between time and order, we present a concrete preprocessing approach implemented in ProM (contained in the *PartialOrderVisualizer* package that can be downloaded from `promtools.org`). The approach uses a *time aggregator* and *tiebreaker* to create a *partially-ordered* event log. Moreover, it is possible to create a k*-sequentialization* of the partially-ordered event log to be able to apply conventional approaches.

The remainder of this paper is organized as follows. Section 2 presents related work and Sect. 3 provides a theoretical foundation to reason about the relationship between time and order. Section 4 presents our preprocessing approach, followed by implementation details and an example (Sect. 5). Section 6 concludes the paper.

2 Related Work

For an overview of process mining techniques, we refer to [2]. See [5] for conformance checking and [14] for large-scale applications in organizations such as BMW, Uber, Siemens, EDP, ABB, Bosch, and Telekom.

Recently, many papers on data quality in process mining were published [3, 11, 15]. Earlier [2, 10] already provided a framework for data quality issues and guidelines for logging. Timestamp-related data quality problems are seen as one of the main roadblocks in process mining.

Explicit uncertainty is considered in the work of Pegoraro et al. [12, 13]. Event logs are annotated with explicit uncertainty and this information is used when discovering process models or checking conformance. For example, timestamps of events have upper and lower bounds and conformance checking yields optimistic and pessimistic bounds for the actual fitness.

Partial orders are a well-studied topic in modeling checking and concurrency theory. Partially-ordered causal runs are part of the standard true-concurrency semantics of Petri net [7]. See [4] for an example of a synthesis technique using partial orders as input. There are a few process mining techniques that start from partial orders, e.g., the conformance checking technique in [9] and the process discovery technique in [8].

In [1] techniques for partial order resolution are presented. These aim to convert a strict weak ordering into a probability distribution over all corresponding total orders. In [6] also the "same-timestamp problem" is addressed, again aiming at creating total orders. It is impossible to list all partial-order-based approaches here. Moreover, the goal of this paper is not to present new conformance checking or process discovery techniques. Instead, we provide a framework to reason about the relation between order and time, and the corresponding challenges.

In this paper, we focus on the preprocessing of event data while using standard process mining techniques. The main contribution is a discussion on the interplay between time and ordering and a concrete preprocessing tool implemented in ProM. Obviously, our framework can be combined with existing partial-order-based techniques such as [1,4,6,8,9].

3 On the Interplay Between Time and Order in Process Mining

In this section, we define event logs that may have both explicit ordering information and timestamps (possibly rounded to hours, days, or weeks). We relate such event logs to the simplified event logs typically used as input for process discovery.

3.1 Event Logs with Time and Order

We first define universes for events, attribute names, values, activities, timestamps, and attribute name-value mappings. Attribute name-value mappings will be used to assign at least a case, activity, and timestamp to each event.

Definition 1 (Universes). *$\mathcal{E}$ is the universe of event identifiers. $\mathcal{N}$ is the universe of attribute names with $\{case, act, time\} \subseteq \mathcal{N}$, $\mathcal{V}$ is the universe of attribute values, $\mathcal{C} \subseteq \mathcal{V}$ is the universe of case identifiers, $\mathcal{A} \subseteq \mathcal{V}$ is the universe of activity names, $\mathcal{T} \subseteq \mathcal{V}$ is the universe of totally-ordered timestamps, and $\mathcal{M} \subseteq \mathcal{N} \not\rightarrow \mathcal{V}$ is the universe of attribute name-value mappings such that for any $m \in \mathcal{M}$: $\{case, act, time\} \subseteq dom(m)$, $m(case) \in \mathcal{C}$, $m(act) \in \mathcal{A}$, and $m(time) \in \mathcal{T}$. For any $n \in \mathcal{N}$ we write $m(n) = \perp$ if $n \notin dom(m)$.*

The properties of an event are described by an attribute name-value mapping that provides at least a case identifier, activity name, and timestamp. Moreover, events may have an explicit order next to timestamp information.

Definition 2 (Event Log). *An event log* $L = (E, \pi, \prec_o)$ *consists of a set of events* $E \subseteq \mathcal{E}$, *a mapping* $\pi \in E \to \mathcal{M}$,[1] *and* $\prec_o \subseteq E \times E$ *such that* $(E, \prec_o)$ *is a strict partial order (i.e., irreflexive, transitive, and asymmetric).*[2]

Table 1 shows a fragment of a larger event log. Consider the first event in the table: $e = 36533$, $\pi_{case}(e) = 9901$, $\pi_{act}(e) = \textit{register request}$, $\pi_{time}(e) =$ 19-05-2021:11.02.55, $\pi_{resource}(e) = \textit{Sarah}$, and $\pi_{cost}(e) = 50$. Table 1 does not define an explicit order. Possible interpretations are that $\prec_o = \emptyset$ (no order) or a total order based on the order in the table, i.e., $e_1 \prec_o e_2$ if the row corresponding to e_1 appears before the row corresponding to e_2. However, $\prec_o$ may also be based on domain knowledge or data-flow analysis (events can only use a data value produced by an earlier event).

Definition 3 (Notations). *Let* $L = (E, \pi, \prec_o)$ *be an event log.*

- $A(L) = \{\pi_{act}(e) \mid e \in E\}$ *are the activities in* L, $C(L) = \{\pi_{case}(e) \mid e \in E\}$ *are the cases in* L, *and* $E\restriction_c = \{e \in E \mid \pi_{case}(e) = c\}$ *are the events of case* $c \in C(L)$.
- $\prec_t = \{(e_1, e_2) \in E \times E \mid \pi_{time}(e_1) < \pi_{time}(e_2)\}$ *is the strict partial order based on the timestamps,*
- $\prec_{ot} = \prec_o \cup \prec_t$ *is the union of the strict partial orders* $\prec_o$ *and* $\prec_t$.
- *If two events* $e_1, e_2 \in E$ *are unordered with respect to* $\prec_o$ *(i.e.,* $e_1 \not\prec_o e_2$ *and* $e_1 \not\succ_o e_2$*), we write* $e_1 \sim_o e_2$. *Similarly,* $e_1 \sim_t e_2 \Leftrightarrow e_1 \not\prec_t e_2 \wedge e_1 \not\succ_t e_2$, *and* $e_1 \sim_{ot} e_2 \Leftrightarrow e_1 \not\prec_{ot} e_2 \wedge e_1 \not\succ_{ot} e_2$.

It is easy to verify that also $(E, \prec_t)$ is a strict partial order (i.e., irreflexive, transitive, and asymmetric). $\sim_o$, $\sim_t$, and $\sim_{ot}$ are reflexive and symmetric by construction. Note that $e_1 \sim_t e_2$ if an only if $\pi_{time}(e_1) = \pi_{time}(e_2)$.

3.2 Consistency

The relation $\prec_{ot}$, which combines $\prec_o$ and $\prec_t$, does not need to be a strict partial order. For example, e_1 happens before e_2 according to $\prec_o$, but e_2 happens before e_1 according to $\prec_t$. Because both ordering relations disagree, $\prec_{ot}$ is not asymmetric. Therefore, we introduce the notion of *consistency*.

Definition 4 (Consistent). *An event log* $L = (E, \pi, \prec_o)$ *is consistent if for any* $e_1, e_2 \in E$: $e_1 \prec_o e_2$ *implies* $\pi_{time}(e_1) \leq \pi_{time}(e_2)$.

This can also be formulated as follows (using transposition): $e_1 \not\prec_o e_2$ or $e_1 \not\succ_t e_2$, for any $e_1, e_2 \in E$. Hence, it is impossible that $e_1 \prec_o e_2$ and $e_1 \succ_t e_2$ hold at the same time. Since both orderings are not conflicting and $\prec_t$ is also a strict weak ordering, the combination yields a strict partial order.

[1] We use the shorthand $\pi_n(e) = \pi(e)(n)$. Note that $\pi_{case}(e)$, $\pi_{act}(e)$, and $\pi_{time}(e)$ denote the case, activity, and timestamp of an event $e \in E$.

[2] For any $e, e_1, e_2, e_3 \in E$: $e \not\prec_o e$ (irreflexivity), if $e_1 \prec_o e_2$ and $e_2 \prec_o e_3$, then $e_1 \prec_o e_3$ (transitivity), and if $e_1 \prec_o e_2$, then $e_2 \not\prec_o e_1$ (asymmetry).

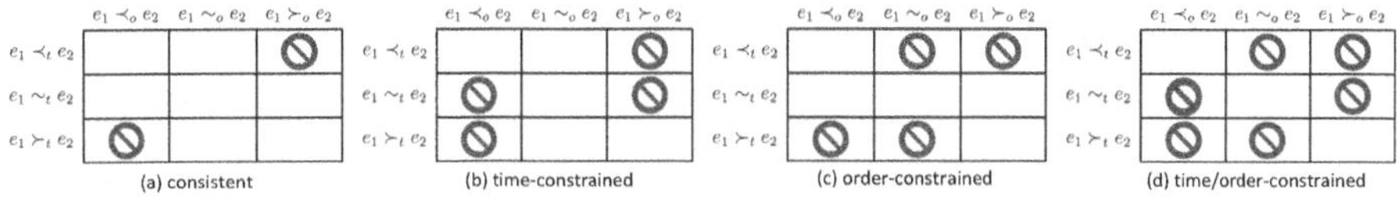

Fig. 4. Possible combinations of order and time relations between two events e_1 and e_2 assuming that the event log is (a) consistent, (b) time-constrained ($\prec_o \subseteq \prec_t$), (c) order-constrained ($\prec_t \subseteq \prec_o$), and (d) time-constrained and order-constrained ($\prec_t = \prec_o$).

Proposition 1 (Consistency Implies Strict Partial Ordering). *Let $L = (E, \pi, \prec_o)$ be an event log. $(E, \prec_t)$ is a strict weak ordering (i.e., a strict partial order with negative transitivity*[3]*), and $(E, \prec_{ot})$ is a strict partial order if L is consistent.*

Proof. $(E, \prec_t)$ is irreflexive, transitive, and asymmetric by construction. Remains to show that negative transitivity holds. Assume that $e_1 \not\prec_t e_2$ and $e_2 \not\prec_t e_3$, i.e., $\pi_{time}(e_1) \geq \pi_{time}(e_2)$ and $\pi_{time}(e_2) \geq \pi_{time}(e_3)$. Hence, $\pi_{time}(e_1) \geq \pi_{time}(e_3)$, i.e., $e_1 \not\prec_t e_3$. Therefore, $(E, \prec_t)$ is a strict weak ordering.

Next, assume that L is consistent. We show that $(E, \prec_{ot})$ is a strict partial order, i.e., for any $e, e_1, e_2, e_3 \in E$: $e \not\prec_{ot} e$ (irreflexivity), if $e_1 \prec_{ot} e_2$ and $e_2 \prec_{ot} e_3$, then $e_1 \prec_{ot} e_3$ (transitivity), and if $e_1 \prec_{ot} e_2$, then $e_2 \not\prec_{ot} e_1$ (asymmetry). Because $\prec_{ot} = \prec_o \cup \prec_t$, irreflexivity follows from $e \not\prec_o e$ and $e \not\prec_t e$. Asymmetry follows directly from consistency: It is impossible that both $e_1 \prec_o e_2$ and $e_1 \succ_t e_2$ hold, so no cycles are introduced. Transitivity relies on the fact that that negative transitivity holds for $\prec_t$. One can use case distinction using the following four cases. (1) $e_1 \prec_t e_2 \ \wedge \ e_2 \prec_t e_3 \Rightarrow e_1 \prec_t e_3 \Rightarrow e_1 \prec_{ot} e_3$. (2) $e_1 \prec_o e_2 \ \wedge \ e_2 \prec_o e_3 \Rightarrow e_1 \prec_o e_3 \Rightarrow e_1 \prec_{ot} e_3$. (3) Assume $e_1 \prec_t e_2 \wedge \ e_2 \prec_o e_3 \wedge \ e_2 \not\prec_t e_3$. Using consistency, we know that $e_2 \not\succ_t e_3$, hence $e_2 \sim_t e_3$. Since $e_1 \prec_t e_2$ and $e_2 \sim_t e_3$, also $e_1 \prec_t e_3$. (If $e_1 \not\prec_t e_3$, then negative transitivity implies $e_1 \not\prec_t e_3 \wedge e_3 \not\prec_t e_2 \Rightarrow e_1 \not\prec_t e_2$ leading to a contradiction.) Since $e_1 \prec_t e_3$, also $e_1 \prec_{ot} e_3$. (4) Assume $e_1 \prec_o e_2 \ \wedge \ e_2 \prec_t e_3 \ \wedge \ e_1 \not\prec_t e_2$. Consistency implies $e_1 \not\succ_t e_2$, hence $e_1 \sim_t e_2$. Since $e_1 \sim_t e_2$ and $e_2 \prec_t e_3$, also $e_1 \prec_t e_3$ and $e_1 \prec_{ot} e_3$. Hence, in all four cases transitivity holds, thus completing the proof.

Both $\prec_o$ and $\prec_t$ order events. L is called time-constrained if $\prec_t$ is at least as strict as $\prec_o$, i.e., $e_1 \prec_o e_2$ implies $e_1 \prec_t e_2$. L is order-constrained if $e_1 \prec_t e_2$ implies $e_1 \prec_o e_2$. Figure 4 illustrates these notions.

3.3 Simplified Event Logs

The basic process discovery techniques assume linear traces and only consider the activity names. Therefore, we connect the more involved event log notion $L = (E, \pi, \prec_o)$ (Definition 2) to simplified event logs and standard discovery techniques.

Definition 5 (Simplified Event Log, Process Model, and Discovery Technique). *A trace $\sigma = \langle a_1, a_2, \ldots, a_n \rangle \in \mathcal{A}^*$ is a sequence of activities. $S \in \mathcal{B}(\mathcal{A}^*)$ is a simplified*

[3] Recall that negative transitivity means that if $e_1 \not\prec_t e_2$ and $e_2 \not\prec_t e_3$, then $e_1 \not\prec_t e_3$. In a strict weak ordering, incomparability is transitive, i.e., $e_1 \sim_t e_2 \ \wedge \ e_2 \sim_t e_3 \Rightarrow e_1 \sim_t e_3$.

event log, i.e., a multiset of traces. A process model $M \subseteq \mathcal{A}^*$ is a set of traces. A discovery function $disc \in \mathcal{B}(\mathcal{A}^*) \rightarrow \mathcal{P}(\mathcal{A}^*)$ maps an event log onto a process model.

We abstract from the process model notations (e.g., BPMN or Petri nets) and focus on the modeled behavior. This allows us to define a model as a set of possible traces $M \subseteq \mathcal{A}^*$. $M = disc(S)$ is the process model discovered from simplified event log S. A simplified event log is a multiset of traces, e.g., $S = [\langle reg, ct, ch, ec, dec, pay\rangle^3, \langle reg, ch, ct, et, dec, rej\rangle^2]$ contains five traces.

Definition 6 (Sequential Runs). *Let $L = (E, \pi, \prec_o)$ be a consistent event log. For any case $c \in C(L)$, $\sigma = \langle a_1, a_2, \ldots, a_n\rangle \in \mathcal{A}^*$ is a sequential run of c if there is a bijection $f \in \{1, 2, \ldots n\} \rightarrow E\lceil_c$ such that $a_i = \pi_{act}(f(i))$ for any $1 \leq i \leq n$ and $e_i \not\succ_{ot} e_j$ for any $1 \leq i < j \leq n$. $seqr_L(c) \subseteq \mathcal{A}^*$ are all sequential runs of case c.*

$\sigma \in seqr_L(c)$ is a trace where each activity refers to an event of case c in such a way that there is a one-to-one correspondence between the elements of σ and $E\lceil_c$, and the order does not contradict the combined ordering $\prec_{ot}$. Given a partial order, there may be many linearizations, i.e., total orders that are compatible. In a k-sequentialization of L, we pick k linearizations for each case.

Definition 7 (k-Sequentialization of an Event Log). *Let $L = (E, \pi, \prec_o)$ be a consistent event log. $S = [\sigma_1, \sigma_2, \ldots \sigma_n] \in \mathcal{B}(\mathcal{A}^*)$ is a k-sequentialization of L if (1) there is function $f \in \{1, 2, \ldots n\} \rightarrow L(C)$ such that $\sigma_i \in seqr_L(f(i))$ for any $1 \leq i \leq n$, and (2) $|\{i \in \{1, 2, \ldots, n\} \mid f(i) = c\}| = k$ for any $c \in C(L)$. $seql_k(L) \subseteq \mathcal{B}(\mathcal{A}^*)$ are all possible k-sequentializations of L.*

Definition 7 shows how event log L can be converted into a simplified event log $S \in seql_k(L)$. Each case in L corresponds to k linearizations in S. We leave it open how the linearizations are selected. This can be probabilistic or deterministic. (In our implementation, all linearizations are sampled from $seqr_L(c)$ using equal probabilities).

4 What if Timestamps Are Imprecise?

As described in the introduction, timestamps may be imprecise or partially incorrect. Therefore, we provide transformations of the event log, making time more coarse granular, e.g., all events on the same day have the same timestamp.

Definition 8 (Time Aggregator). *$ta \in \mathcal{T} \rightarrow \mathcal{T}$ is a time aggregator if for any $t_1, t_2 \in \mathcal{T}$ such that $t_1 < t_2$: $ta(t_1) \leq ta(t_2)$.*

For example, *ta*(19-05-2021:13.02) = *ta*(19-05-2021:17.55) = 19-05-2021:00.00 if a time granularity of days is used, i.e., all timestamps on the same day are mapped onto the same value. By making time more coarse-grained, more events become unordered. These may still be ordered by $\prec_o$, e.g., based on data-flow analysis. Next to $\prec_o$, we may use domain knowledge in the form of a so-called *tiebreaker* to optionally order events having identical coarse-grained timestamps.

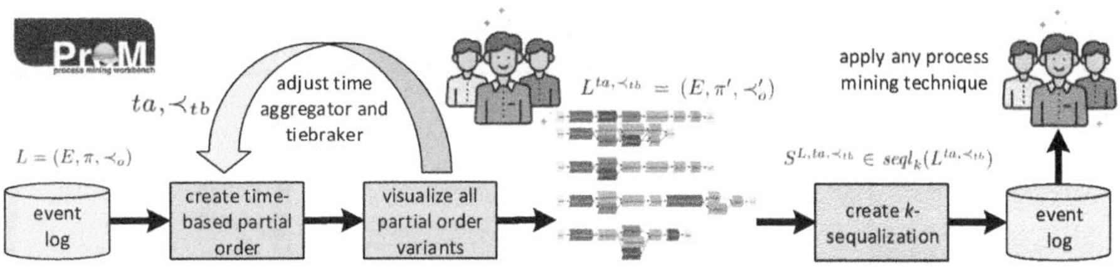

Fig. 5. Overview of the functionality of the *PartialOrderVisualizer* package. The user can change the time granularity and modify the tiebreaker using domain knowledge. Cases with the same partial order are grouped into partial-order variants. These can be sorted and inspected. At any point in time, it is possible to create a regular event log (using k-sequentialization).

Definition 9 (Tiebreaker). *A tiebreaker* $\prec_{tb} \subseteq \mathcal{A} \times \mathcal{A}$ *is a strict partial order used to order activities having the same aggregated timestamp.*

A tiebreaker adds causal dependencies between events that have the same coarse-granular timestamps and belong to the same case. Using time aggregator ta and tiebreaker $\prec_{tb}$, we can create a new event log $L^{ta,\prec_{tb}}$.

Definition 10 (Preprocessing). *Let* $L = (E, \pi, \prec_o)$ *be a consistent event log,* $ta \in \mathcal{T} \rightarrow \mathcal{T}$ *a time aggregator, and* $\prec_{tb} \subseteq \mathcal{A} \times \mathcal{A}$ *a tiebreaker.* $L^{ta,\prec_{tb}} = (E, \pi', \prec'_o)$ *is the event log after applying the time aggregator ta and tiebreaker* $\prec_{tb}$ *such that* $\pi'_n(e) = \pi_n(e)$ *for any* $e \in E$ *and* $n \in \mathcal{N} \setminus \{time\}$, $\pi'_{time}(e) = ta(\pi_{time}(e))$ *for any* $e \in E$, *and* $\prec'_o = \prec_o \cup \{(e_1, e_2) \in E \times E \mid \pi_{case}(e_1) = \pi_{case}(e_2) \wedge \pi'_{time}(e_1) = \pi'_{time}(e_2) \wedge \pi_{act}(e_1) \prec_{tb} \pi_{act}(e_2)\}$.

As long as the tiebreaker $\prec_{tb}$ does not contradict $\prec_o$, $\prec'_o$ is a partial order and $L^{ta,\prec_{tb}} = (E, \pi', \prec'_o)$ is a consistent event log.

Hence, we can compute a k-sequentialization of $L^{ta,\prec_{tb}}$ and produce a simplified event log $S^{L,ta,\prec_{tb}} \in seql_k(L^{ta,\prec_{tb}})$. As mentioned before, we leave it open how sequential runs are selected. Using the simplified preprocessed event log, we can apply any process discovery technique to obtain process model $M^{L,ta,\prec_{tb}} = disc(S^{L,ta,\prec_{tb}})$.

5 Implementation

The new ProM package *PartialOrderVisualizer* implements the approach described in Sect. 4 and can be downloaded as part of ProM's nightly builds (http://www.promtools.org/doku.php?id=nightly). It has been implemented as a so-called "visualizer" and can be selected by choosing *Explore Partial Orders (Variants)* in the pull-down menu. The visualizer can be applied to any event log.

Figure 5 shows the main components of the *PartialOrderVisualizer*. Based on time aggregator ta and tiebreaker $\prec_{tb}$ the partial orders are computed and visualized. Initially, ta is set to hours and $\prec_{tb} = \emptyset$. The user can experiment with the time granularity and add ordering constraints. One can inspect partial order variants and details of the corresponding cases. Moreover, the user can create a k-sequentialization of $L^{ta,\prec_{tb}}$. The resulting event log can be analyzed using classical process mining techniques.

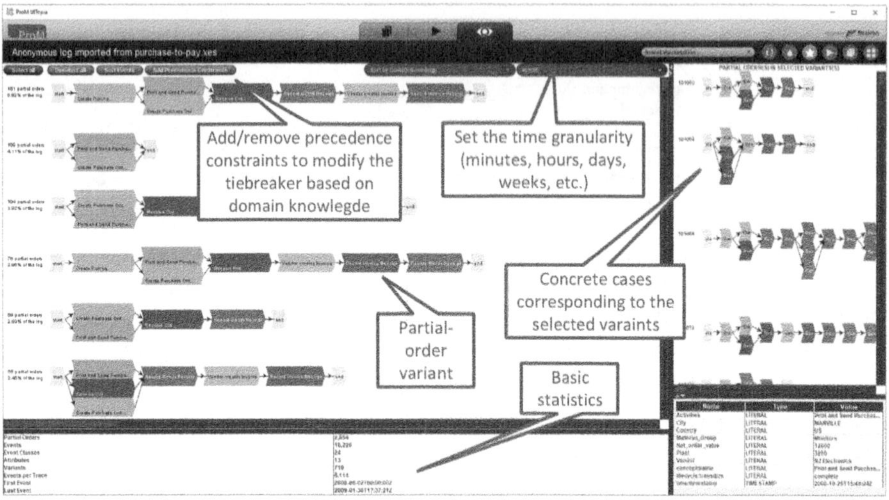

Fig. 6. Data from a Purchase-to-Pay (P2P) process visualized using *PartialOrderVisualizer*. What can be seen is that activities *Create Purchase Order Item* and *Print and Send Purchase Order* often happen in the same hour.

Figure 6 shows the *PartialOrderVisualizer* for the event data of a Purchase-to-Pay (P2P) process with 2,654 cases and 16,226 events. There are 685 trace variants in the original event log. The view shown in Fig. 6 uses a time granularity of one hour leading to a similar number of partial-order variants (i.e., 710). However, one can clearly see that some activities often happen within the same hour. It is possible to add ordering information to make these sequential (if desired). Note that seeing which activities happen in the same period is valuable and provides new insights.

Figure 7 shows the *PartialOrderVisualizer* for the same event data, but now using a time granularity of a week. One can see that more events become unordered, because they happen in the same week. Using this view, we generated a new event log replicating each partially-ordered case 10 times. As expected, this new event log has 26,540 cases and 162,260 events. Interestingly, there are now 8,864 trace variants.

The *PartialOrderVisualizer* has been applied to a range of event logs, e.g., we have used it to analyze the treatment of Covid-19 patients at Uniklinik RWTH Aachen. In this Covid-19 dataset, only the dates are reliable. Naïvely using the ordering in the event log or the recorded timestamps leads to incorrect conclusions.

For the Covid-19 dataset it takes just a few seconds. For a larger data sets like the well-known road fines event log[4], which has over 560.000 events and 150.000 cases, it takes around 10 s (using for example the day, hour, minute, and second abstractions).

What is interesting is that in many applications the number of partially-ordered variants temporarily goes up when coarsening the time granularity. However, by definition,

[4] Road Traffic Fine Management Process, 4TU.ResearchData, https://doi.org/10.4121/uuid:270fd440-1057-4fb9-89a9-b699b47990f5.

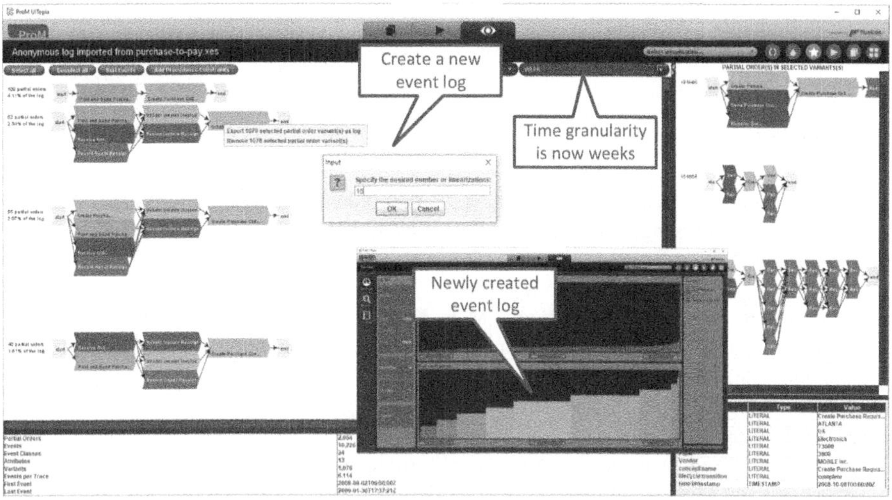

Fig. 7. Another view on the same P2P dataset now using a time granularity of a week. A new event log was created using a k-sequentialization (with $k = 10$).

the number of partially-ordered variants is the smallest when all events are mapped onto the same time period.

6 Conclusion

The contribution of this paper is twofold. On the one hand, we discussed the possible interplay between time and order using an event log definition where events have timestamps and may be ordered by some partial order. Since (rounded) timestamps provide a strict weak ordering, the combination is a partial order (provided that the event log is consistent). On the order hand, we described a new way to preprocess event logs using a time aggregator ta and a tiebreaker $\prec_{tb}$. The time aggregator makes the timestamps more coarse-grained to avoid accidental event ordering due to imprecise or partially incorrect timestamps. The tiebreaker can be used to order events that have the same coarse-grained timestamp (e.g., date). The preprocessed event log can be created and explored using the new *PartialOrderVisualizer* implemented in ProM. Viewing the event log at different time scales provides novel insights and helps to counteract data quality problems. Moreover, the *PartialOrderVisualizer* can also generate a k-sequentialization of the partially-ordered event log. This allows for the application of regular process mining techniques (e.g., discovery and conformance checking).

Although some process mining techniques have been developed for partially-ordered event logs, we feel that more research is needed. Partial orders may be the result of uncertainty or explicit causal information. These need to be treated differently. We also plan to integrate frequent item-set mining into our approach. The fact that certain combinations of activities often happen in the same period can be used to create event logs where high-level events refer to sets of low-level events.

Acknowledgments. We thank the Alexander von Humboldt (AvH) Stiftung and the NHR Center for Computational Engineering Sciences (NHR4CES) for supporting our research.

References

1. van der Aa, H., Leopold, H., Weidlich, M.: Partial order resolution of event logs for process conformance checking. Decis. Support Syst. **136**, 113347 (2020)
2. van der Aalst, W.M.P.: Process Mining: Data Science in Action. Springer, Heidelberg (2016). https://doi.org/10.1007/978-3-662-49851-4
3. Andrews, R., van Dun, C.G.J., Wynn, M.T., Kratsch, W., Röglinger, M.K.E., ter Hofstede, A.H.M.: Quality-informed semi-automated event log generation for process mining. Decis. Support Syst. **132**, 113265 (2020)
4. Bergenthum, R., Desel, J., Mauser, S., Lorenz, R.: Synthesis of petri nets from term based representations of infinite partial languages. Fund. Inform. **95**(1), 187–217 (2009)
5. Carmona, J., van Dongen, B., Solti, A., Weidlich, M.: Conformance Checking: Relating Processes and Models. Springer, Cham (2018). https://doi.org/10.1007/978-3-319-99414-7
6. Conforti, R., La Rosa, M., ter Hofstede, A.H.M., Augusto, A.: Automatic repair of same-timestamp errors in business process event logs. In: Fahland, D., Ghidini, C., Becker, J., Dumas, M. (eds.) BPM 2020. LNCS, vol. 12168, pp. 327–345. Springer, Cham (2020). https://doi.org/10.1007/978-3-030-58666-9_19
7. Desel, J.: Validation of process models by construction of process nets. In: van der Aalst, W., Desel, J., Oberweis, A. (eds.) Business Process Management. LNCS, vol. 1806, pp. 110–128. Springer, Heidelberg (2000). https://doi.org/10.1007/3-540-45594-9_8
8. van Dongen, B.F., Desel, J., van der Aalst, W.M.P.: Aggregating causal runs into workflow nets. In: Jensen, K., van der Aalst, W.M., Ajmone Marsan, M., Franceschinis, G., Kleijn, J., Kristensen, L.M. (eds.) Transactions on Petri Nets and Other Models of Concurrency VI. LNCS, vol. 7400, pp. 334–363. Springer, Heidelberg (2012). https://doi.org/10.1007/978-3-642-35179-2_14
9. Lu, X., Fahland, D., van der Aalst, W.M.P.: Conformance checking based on partially ordered event data. In: Fournier, F., Mendling, J. (eds.) BPM 2014. LNBIP, vol. 202, pp. 75–88. Springer, Cham (2015). https://doi.org/10.1007/978-3-319-15895-2_7
10. Mans, R., van der Aalst, W.M.P., Vanwersch, R.: Process Mining in Healthcare: Evaluating and Exploiting Operational Healthcare Processes. Springer Briefs in Business Process Management, Springer, Cham (2015). https://doi.org/10.1007/978-3-319-16071-9
11. Martin, N.: Data quality in process mining. In: Fernandez-Llatas, C. (ed.) Interactive Process Mining in Healthcare. Health Informatics, pp. 53–79. Springer, Cham (2021). https://doi.org/10.1007/978-3-030-53993-1_5
12. Pegoraro, M., van der Aalst, W.M.P.: Mining uncertain event data in process mining. In: Carmona, J., Jans, M., La Rosa, M. (eds.) International Conference on Process Mining (ICPM 2019), Aachen, Germany, pp. 89–96. IEEE Computer Society (2019)
13. Pegoraro, M., Uysal, M.S., van der Aalst, W.M.P.: Efficient construction of behavior graphs for uncertain event data. In: Abramowicz, W., Klein, G. (eds.) BIS 2020. LNBIP, vol. 389, pp. 76–88. Springer, Cham (2020). https://doi.org/10.1007/978-3-030-53337-3_6
14. Reinkemeyer, L.: Process Mining in Action: Principles, Use Cases and Outlook. Springer, Cham (2020). https://doi.org/10.1007/978-3-030-40172-6
15. Suriadi, S., Andrews, R., ter Hofstede, A.H.M., Wynn, M.T.: Event log imperfection patterns for process mining: towards a systematic approach to cleaning event logs. Inf. Syst. **64**, 132–150 (2017)

Attribute-Driven Case Notion Discovery for Unlabeled Event Logs

Tom Lichtenstein(✉), Dorina Bano, and Mathias Weske

Hasso Plattner Institute, University of Potsdam, Potsdam, Germany
Tom.Lichtenstein@student.hpi.de, {Dorina.Bano,Mathias.Weske}@hpi.de

Abstract. Event logs can be analyzed using various process mining techniques (e.g., process discovery) to obtain valuable information about the actual behavior of business process executions. Typically, these techniques rely on the presence of a case identifier linking events to process instances. However, if the process involves information systems that do not record events in a process-oriented manner, a clear case identifier may be missing, resulting in an unlabeled event log. While some approaches already address the challenge of inferring case identifiers for unlabeled event logs, most of them provide limited support for cyclic behavior without additional inputs. This paper proposes a three-step approach to correlate events with case identifiers for unlabeled event logs originating from processes with cyclic behavior. While evaluating the accuracy of our approach with two real-world event logs (MIMIC-IV and Road Traffic Fine Management), we show that our approach, compared to the existing ones, detects cyclic behavior and correlates events closer to the original process instances without additional inputs.

Keywords: Process mining · Unlabeled event log · Case notion discovery

1 Introduction

Event logs extracted from historical data of business process executions contain valuable information about the actual process behavior. Process mining techniques allow us to retrieve the information from execution logs. For example, process models can be discovered, process conformance can be checked, and process improvement opportunities can be identified [3]. To allow the process mining experts to apply different process mining techniques on the event data contained in the event log, each event must refer to a particular *case* (i.e., process instance) in which it occurred. The correlation between an event and its process instance is usually represented by a common attribute, referred to as a *case identifier* (i.e., hereinafter referred to as *case ID*). The assignment of case IDs to events – the *case notion* – determines one perspective from which an event log can be analyzed [15].

Process-aware information systems already manage their data in a process-oriented way which simplifies the extraction of events in the required format [2].

A. Marrella and B. Weber (Eds.): BPM 2021 Workshops, LNBIP 436, pp. 111–122, 2022.
https://doi.org/10.1007/978-3-030-94343-1_9

However, if event data was recorded by process-unaware information systems or by humans, some events may not be associated with a case ID. This leads to so-called *unlabeled event logs* that lack a clear case notion [5]. Some approaches already address the challenge of correlating events with process instances by focusing on different aspects to determine a case [5,7,8,11]. Nonetheless, to adequately handle unlabeled event logs originating from processes with cyclic behavior (i.e., containing loops), additional process information (e.g., process models) is often required.

Since this information is not always available, in this paper, we propose a three-step approach that addresses the challenge of correlating events with process instances using the unlabeled event log as the single source of information. We argue that it is possible to derive the case IDs and the cyclic behavior (i.e., if it exists) from an unlabeled event log. To achieve this, we are considering attributes contained in event logs.

The remainder of this paper is organized as follows: Sect. 2 explains the underlying concepts and a running example, Sect. 3 briefly reviews research in this area, and Sect. 4 presents a case notion detection approach which is then evaluated in Sect. 5. Finally, the findings are summarized in Sect. 6.

2 Preliminaries and Running Example

This section presents the basic concepts of an event log as well as a running example that is used to explain the approach proposed in Sect. 4.

Event Log. An event log contains events extracted from various information systems involved in a business process. Usually, event logs are stored according to the *eXtensible Event Stream* (XES) standard [1]. Each contained event is required to have attributes describing (i) the *activity* performed, (ii) the *timestamp* of its occurrence and (iii) the *case* in which it occurred. Additional information related to the events can be stored in optional attributes (e.g., resource, department) [5]. An individual case is represented by a *trace* which consists of a sequence of events having the same case ID [15]. Traces with the same sequence of activities in their events are referred to as a *variant*. However, if an event log contains events that are not explicitly associated with case IDs, they are categorized as *unlabeled* event logs.

Running Example. Figure 1 shows an event log example extracted from the Medical Information Mart for Intensive Care (MIMIC)-IV database [12], which describes the treatment process of 100 patients with acute kidney failure in a hospital. The process starts with each patient being admitted normally or to the intensive care unit (ICU). In the first case, various medications are administered or prescribed. For the cases with ICU admission, several ICU-related procedures are performed. Afterward, each patient is discharged. As a patient may be admitted multiple times, each process instance contains complex cyclic behavior. However, the extracted event log is unlabeled, as it does not have an explicit case identifier assigned to each event. This event log serves as an example to describe the approach presented in this paper.

Activity	Time	subject_id	hadm_id	location	drug	val	...
Added Patient	13:45	14824					
Admitted Patient	13:57	14824	84273	EMERG.			
Administered Drug	14:13				Bisacodyl		
Admitted Patient	15:04	22153	23962	REHAB			
Prescribed Drug	15:42				Bisacodyl	10	

Fig. 1. Excerpt of an unlabeled event log representing an Acute Kidney Failure Treatment process extracted from the MIMIC-IV Medical Database [12]

3 Related Work

Inferring case IDs for unlabeled event logs has been a topic in process mining for several years. Some existing techniques addressing the problem of correlating events with case IDs are presented below.

Bayomie et al. [7] discuss how to derive case IDs of an unlabeled event log or data stream in real-time using a decision tree learning approach. In addition to the unlabeled event log, the approach requires a process model and heuristics data describing the behavior of the process. For each event, a set of case IDs is derived using the decision tree combined with the additional process information. Thus, this approach generates several case notions simultaneously and returns a set of labeled event logs. The approach is extended in [5] to support event logs with cyclic behavior. Nevertheless, it still requires a process model and heuristics data as input. In [6] the authors propose another approach called *EC-SA*, which is based on probabilistic optimization. The approach requires as input the unlabeled event log and the process model. Differently from their solution, we consider the unlabeled event log as a single source of information to derive the case IDs and discover the cyclic process models.

Another approach, proposed by Burattin et al. [8], assumes that the case IDs of unlabeled event logs can be derived from a set of optional attributes. In their approach, sets of process instance-defining attributes are specified using heuristics, such as considering attributes with specific data types. The sets are chained together based on common attribute values, and the resulting chains allow the correlation of events with case IDs. While this approach does not depend on a reference model, it relies on the value similarity of the corresponding attributes within a process instance, which may not apply to instances with cyclic behavior.

In [11] Ferreira et al. propose an iterative expectation-maximization procedure based on a first-order Markov chain. Their approach estimates a probabilistic model from a control flow perspective by considering only the sequence of activities. Using a greedy algorithm, they further attempt to find a local maximum in a likelihood function regarding the probabilistic model to determine the case notion. However, unlike our work, the approach cannot handle loops appropriately as every detected trace can only contain each activity once.

4 Attribute-Driven Case Notion Discovery

Our goal is to discover a case notion for an unlabeled event log by correlating its events with process instances. Since business processes often involve various entities (e.g., actors, resources, or locations) [16], and widespread information systems (e.g., relational databases) tend to follow an entity-oriented design [9], we assume that unlabeled event logs containing optional attributes allow the identification of underlying entities involved in the process executions. By identifying the references between the entities and attributes, the event log can be divided into event log chunks, each associated with a class of entities. Based on the event log chunks, the approach aims to identify sub-process instances for the underlying entities, which are then merged into process instances according to the relationships between the entities, resulting in a case notion. Since a process instance can contain events of multiple entities of the same entity class, loops can be uncovered, while cyclic behavior within a sub-process instance can be detected based on attribute value similarity.

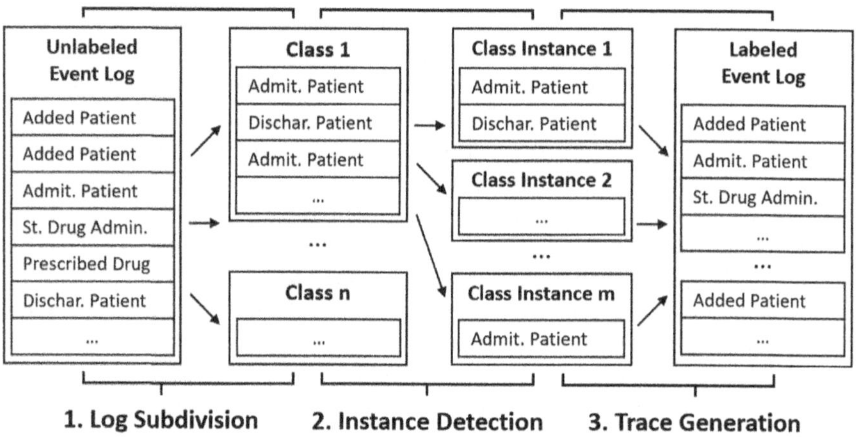

Fig. 2. Conceptual depiction of the steps involved in the attribute-driven case notion discovery approach including the artifacts generated after each step

To realize this, we are following a three-step approach (cf. Fig. 2): *Event Log Subdivision*, *Entity Class Instance Estimation* and *Trace Generation*. The first step takes an unlabeled event log as input and discovers a data model, which determines the division of the unlabeled event log into multiple event log chunks (i.e., each data model class determines an event log chunk). The second step estimates the sub-process instances based on attribute value similarity and the transition probability between activities for each chunk. The final step generates the event log traces. A detailed explanation of each step is provided in the following.

4.1 Event Log Subdivision

The event log subdivision step takes as input an unlabeled event log and divides it into smaller chunks by considering the attributes associated with the events. To achieve this, first, we are discovering the data model from the unlabeled event log, and then the events are assigned to the data model classes.

Data Model Detection. To discover a data model from an unlabeled event log, we follow the approach proposed by Bano et al. [4]. The first step of this approach is to derive an *activity-attribute relationship diagram* (i.e., A2A diagram), representing the number of access relations between all activities and optional attributes in an event log. The second step takes the A2A diagram as input and groups the attributes into data model classes according to a predefined set of rules. A class could cover the attributes for either all patients, admissions, or medications in the running example.

Event Assignment. After grouping the attributes into the data model classes, the next step is to assign the events (i.e., taken from the unlabeled event log) to these classes. Therefore, for each event, it is checked which attributes of the discovered data model classes it accesses (i.e., the event writes an attribute value). An event is assigned to a data model class when it accesses at least one of the data model class attributes. It should be noted that the same event can also be assigned to multiple data model classes if it simultaneously accesses attributes of different classes.

4.2 Class Instance Detection

The second step of our approach aims to arrange the events associated with the data model classes into instances of the respective class, referred to as *class instances* (CI). Based on the running example (explained in Sect. 2), a class instance (i.e., entity) can be a patient, admission, or medication. Thus, the events assigned to a class instance represent a sub-process instance. Below, the assignment procedure is described for a single data model class, as it is performed for each class individually.

For the class instance detection step, we consider the iterative expectation-maximization approach presented in [11] as a basis. The approach assigns each event to the most appropriate source sequence (i.e., class instance) using a transition matrix, capturing the transition probabilities between activities. As a starting point, the first event is assigned to an initial class instance. Afterward, each event is assigned to the class instance with the highest transition probability (i.e., considering the activity of the current event and the activity of the last assigned event of the class instance), or a new instance is created for the event. After each iteration, the transition matrix is updated according to the order of activities within the class instances. The class instance detection step is repeated until the assignment of events to instances is equal to a previous iteration.

The unlabeled event log also provides additional process-related information in the form of optional attributes. Therefore, we extend the class instance selection by comparing attribute values between an event and the class instance. The instance assignment procedure first considers an assignment corresponding to similar attribute values before an assignment based on transition probabilities occurs. The extension is explained in more detail in the following paragraphs.

Similarity-Based Instance Assignment. For the similarity-based instance assignment, the class instances are extended with attribute sets, including the corresponding data model class attributes. The attribute values of a class instance are updated every time an event is assigned to it according to the values of the event's attributes with the same names.

For each event to be assigned, a *similarity score* is calculated for all available class instances, indicating the degree of affiliation between an event and a class instance. More specifically, the score evaluates the matching attributes between the event and the instance, which both have the same name and value. However, each attribute within a class might have a different significance for determining affiliation. For example, an identical Boolean or timestamp value is not as significant as an identical patient identifier or drug name when determining a suitable instance for an event. The importance of each attribute is also taken into account in [8], where attributes are excluded based on assumptions made by domain experts about their data type. In contrast, our approach evaluates the significance of attributes without requiring additional domain knowledge, as this may not be available in process discovery scenarios. To achieve this, we are introducing the *similarity weight* concept, which weights each attribute based on the number of times an event accesses it and its values. Let A be the set of optional attributes included in the unlabeled event log. The similarity weight $w_s(a)$ for an attribute $a \in A$ is composed out of two metrics $comp(a)$ (completeness) and $part(a)$ (partitioning) and defined as follows:

$$w_s(a) = comp(a) \cdot part(a) = \frac{|E_a|}{|E|} \cdot e^{-\alpha \cdot (\frac{|\hat{V_a}|}{|V_a|} - 0.5)^2} \tag{1}$$

Let E be the set of events assigned to the data model class having $a \in A$ as an attribute, the first metric $comp(a)$ results from the number of events $|E_a|$ that access a relative to the total number of events of the data model class $|E|$. Also, let V_a be a multiset of all values assigned to a in E_a and let $\hat{V_a}$ be the set of unique values in V_a. The second metric $part(a)$ determines the number of unique values $|\hat{V_a}|$ concerning the total number of values $|V_a|$, representing the partitioning of the events in a chunk that could be achieved with the attribute. Since we expect attributes with a low or high count of unique values to be less significant for indicating affiliation (as it might be the case for Boolean or timestamp values), we chose the function shown in Eq. 1 to produce a partition score $\in (0, 1]$ as a generic interpretation for the ratio. The score decreases as the ratio gets further away from the mean. Here, $\alpha \in \mathbb{R}_0^+$ determines the gradient of the tangent line at the inflection point. Figure 3 visualizes the parameters and concepts used for the metrics to determine the similarity weights.

Since the degree of affiliation between an event and a class instance depends on several attributes and the similarity weight is determined for each attribute independently, all similarity weights of a data model class are considered

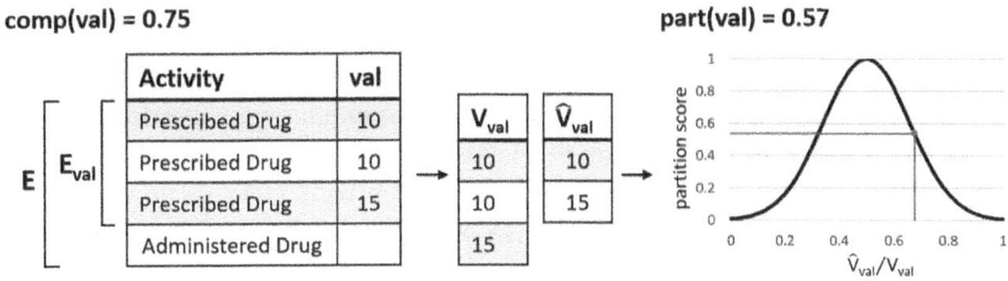

Fig. 3. Considered parameters for the completeness and partitioning metrics to determine the similarity weights using the val attribute as an example ($\alpha = 20$)

relative to each other for the similarity score as illustrated in Fig. 4. Therefore, the similarity score results from the sum of the relative similarity weights of the matching attributes between an event and a class instance. Figure 4 illustrates the determination of similarity scores for an event and two exemplary class instances. Although each class instance has two attribute values matching the event, the first class instance receives a higher similarity value because the similarity weights rank its matching attributes as more relevant. The event is then assigned to the class instance with the highest similarity score. To avoid events being assigned to class instances with low similarity, the score must exceed a threshold $t \in [0, 1]$ which is set in advance by the domain user. If there is no suitable class instance, then the assignment is carried out according to the expectation-maximization approach presented in [11].

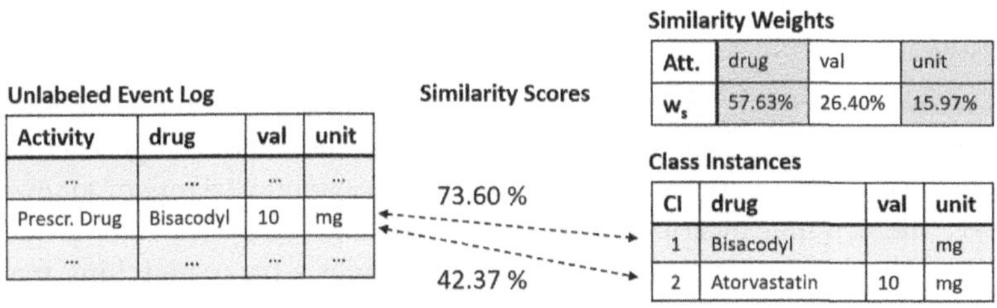

Fig. 4. Similarity score determination for an event and two class instances having different attribute values assigned considering the similarity weights of the attributes

4.3 Trace Generation

The last step of the approach composes the class instances into event log traces. To achieve this, the domain user must select a *root class* by having a specific business goal in mind. The root class must have a class instance related to the first event of the unlabeled event log. Next, the class instances are assigned to traces using an approach inspired by the concept proposed by Ferreira et al. [11].

Starting with an empty set of traces, the first event of the unlabeled event log creates the initial trace. If an event is added to a trace, the trace tracks all class instances related to that event. Each subsequent event is added to a trace that already tracks a class instance related to the event. If this is not possible and the event is related to a class instance of the root class, a new trace is created. Otherwise, the event assignment to a trace is estimated using a transition matrix that captures the transition probabilities between activities with their associated data model classes. The trace generation step is repeated until the assignments to traces correspond to a previous iteration. Furthermore, the transition matrix is updated after each iteration according to the events contained in the traces. Finally, the resulting labeled event log is serialized according to the XES standard [1].

5 Evaluation

To evaluate our approach, we use two real-live event logs. The first log is extracted from the MIMIC-IV database [12] and represents the acute kidney failure treatment introduced in Sect. 2. The second event log, called Road Traffic Fine Management [13] contains information regarding the road traffic fines processed by an Italian municipality. Both event logs were transformed to unlabeled event logs by removing their case ID attribute. In this section, we first introduce a metric that will be used for evaluation. Afterward, the two event logs are described in more detail, followed by a comparison of the estimation accuracy between our approach and the Multiple Instance Model (MIM) approach[1] presented in [11]. An implementation of our approach is made available on GitHub[2].

Evaluation Metric. To determine the accuracy of our approach, we are adapting the G-score metric introduced in [11] which calculates the geometric mean between the detected variants and the initially intended variants of an event log. However, the original metric only matches variants with the same sequence of activities in both event logs. In contrast, we consider a fuzzy matching technique for comparing the estimated variants with the original ones while allowing for partial mismatches in the activity sequences due to the complexity of the traces. For example, since one treatment may involve hundreds of events, an estimated trace that differs by a single event from the original trace would be considered a mismatch by the original G-score. Different from that, our metric still assesses the similarity between the traces of the variants. To achieve this, we are using the normalized Levenshtein distance metric (d_{NGLD}) proposed by Yujian et al. [14]. This metric evaluates the distance between two sequences returning a normalized result $\in [0, 1]$. Therefore, the similarity results from complementing the metric to 1.

[1] https://github.com/diogoff/unlabelled-event-logs.
[2] https://github.com/tom-lichtenstein/attribute-driven-case-notion-discovery.

The consideration of matching variants with similar but not identical activity sequences implies the need to extend the original G-score formula. The extended metric is referred to as G_{sim}-score in the following, and its result can be interpreted as the percentage of similarity between two event logs. To define the G_{sim}-score for two labeled event logs L and L' with different case notions, let $\mathcal{T}$ be the set of all traces contained in L and Υ be the set of all distinct variants of $\mathcal{T}$. Likewise, let $\mathcal{T}'$ be the set of traces in L' and Υ' be the set of variants in $\mathcal{T}'$. For each variant $\upsilon \in \Upsilon$, a set S_υ is created that contains all variants $\upsilon' \in \Upsilon'$, that have the highest similarity with υ according to $1 - d_{NGLD}(\upsilon, \upsilon')$ with regard to the activity sequences. If in addition, $p(\upsilon)$ returns the number of traces $\mathcal{T}_\upsilon \subseteq \mathcal{T}$ matching υ with respect to the total number of traces $|\mathcal{T}|$, and in the same way $q(\upsilon')$ returns the ratio of traces $\mathcal{T}'_\upsilon \subseteq \mathcal{T}'$ matching υ' relative to $|\mathcal{T}'|$, the G_{sim}-score can be defined as follows:

$$G_{sim}(L \parallel L') \triangleq \sum_{\upsilon \in \Upsilon} \sqrt{p(\upsilon) \cdot \sum_{\upsilon' \in S_\upsilon} (q(\upsilon') \cdot (1 - d_{NGLD}(\upsilon, \upsilon')))} \tag{2}$$

Estimation Accuracy Analysis on Real-Life Event Logs. To evaluate the approach with the acute kidney failure treatment process, we consider multiple unlabeled event logs, each containing events extracted from a different set of MIMIC-IV database tables. The unlabeled event logs are associated with a *complexity level* (CL) (i.e., illustrated in Fig. 5) which corresponds to the number of database tables involved in the extraction. On each CL, the event log contains the events from all tables of the previous levels plus an additional table. The data model depicted in Fig. 5 references the additional table for each CL. All event logs concern the treatment process of 100 patients, while each patient's treatment is initially intended as a process instance. In terms of cyclic behavior, the process instances contain simple loops from CL 2 and nested loops from CL 4. In addition, we are investigating how the foreign key attribute values coming from the database can affect the assignments. Therefore, we consider two types of event logs: one containing attribute values derived from foreign keys (Type A); and another without attribute values derived from foreign keys (Type B).

A comparison between our approach (for Type A and B event logs) and the MIM approach considering the G_{sim}-score is illustrated in Fig. 6. The results reveal that overall our approach is more accurate than the MIM approach. When considering unlabeled event logs containing attributes derived from foreign keys (Type A), the approach achieves results above 80% for all tested scenarios despite nested loops. In contrast, the MIM approach already falls below this mark at CL 2. Figure 7 shows the basic behavior detected in the event logs labeled by both approaches using a process discovery algorithm from the process mining workbench ProM [10]. It can be seen that the labeled event log generated by our approach contains loops in contrast to the labeled event log generated by the MIM approach, which does not contain loops. However, looking at the results of our approach applied to the unlabeled event logs without attributes derived from foreign keys (Type B), the accuracy drops significantly at CL 3. In this

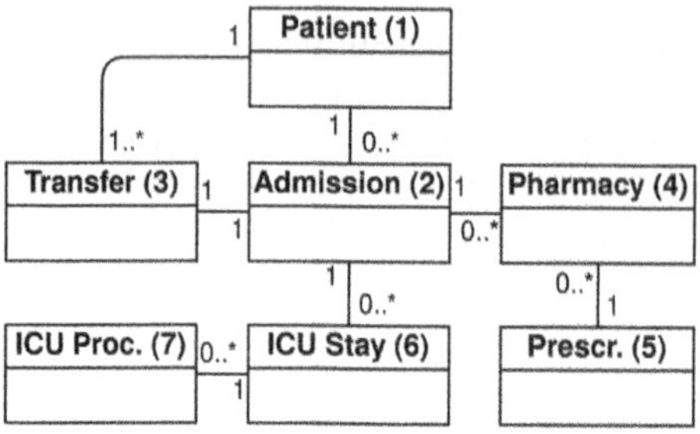

CL	Events	Activities
1	100	1
2	586	3
3	2281	5
4	4633	7
5	7104	9
6	7222	11
7	53124	13

Fig. 5. Excerpt of the MIMIC-IV data model including the tables considered for each complexity level represented as classes and the resulting properties for the event logs

scenario, the approach relies more on probability than on similarity, while the probability-based assignment is adversely affected by loops.

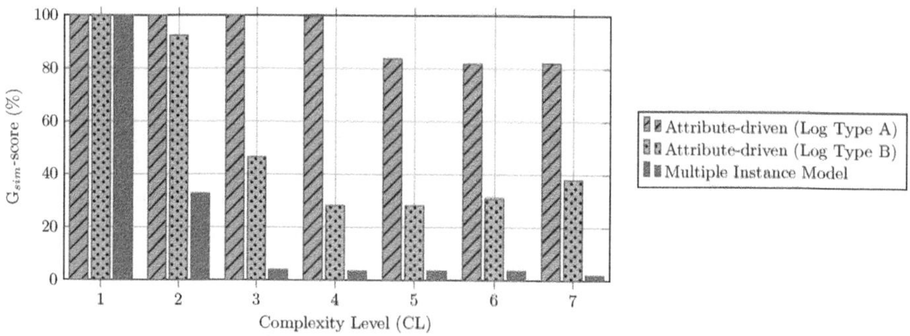

Fig. 6. Accuracy comparison of the case notion estimations of the attribute-based approach (similarity-based assignment threshold set to 70%) and the MIM approach. Considered are the G_{sim}-scores resulting from their application to multiple unlabeled event logs at different levels of complexity extracted from an Acute Kidney Failure Treatment process. Higher values show better accuracy.

The Road Traffic Fine Management event log is the second event log considered in this evaluation. It contains 21,954 events with 8 different activities and 5,421 traces having up to 13 optional attributes. Since the event log does not have to be extracted, we consider only the given structure without the case identifier attribute. In this case, our approach achieves a G_{sim}-score of 65.30%, whereas the MIM approach achieves 22.29%, indicating a nearly threefold accuracy.

Discussion. According to the G_{sim}-score results for the two examined real-life event logs, our approach achieves similar or higher accuracy estimating the initially intended case notion compared to the MIM approach. However, unlike the MIM approach, our approach relies on optional attributes for each event. Moreover, the fewer connections exist between the attributes in the unlabeled

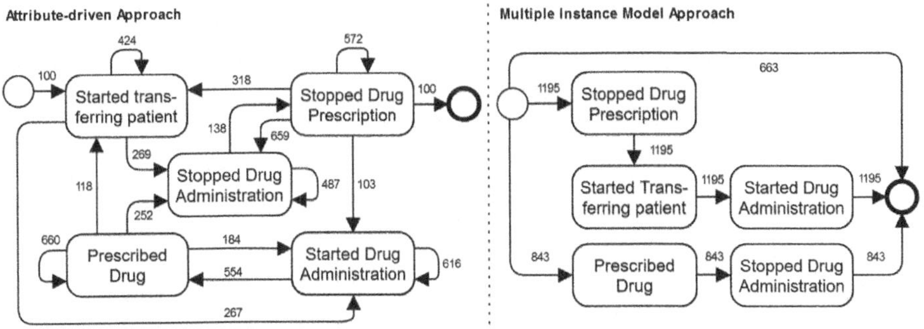

Fig. 7. Process models generated for the resulting labeled event logs (Type A; CL 6) of the attribute-driven and the MIM approach using the Directly Follows Graph from ProM (Activities: 0.4; Paths: 0.35)

event log, the more the approach must rely on probability-based assignments, which cannot capture loops. Nevertheless, for unlabeled event logs containing loops extracted with entity-related information, the approach allows an adequate estimation of a case notion without requiring further domain-related context information.

6 Conclusion

This paper proposes a novel approach addressing the challenge of correlating events with case IDs for unlabeled event logs. In addition, we show that it is possible to discover a process model with cyclic behavior by just considering an unlabeled event log as a single source of information. The approach splits the unlabeled event log into event log chunks using a data model derived from the log to detect sub-process instances. The detection considers the attribute similarity and activity transition probabilities. Eventually, related sub-process instances are merged into traces, generating a case notion for the unlabeled event log. The correlation accuracy of the presented approach is compared with an existing approach using two real-life event logs. The results show that our approach is more accurate while being able to capture cyclic behavior. As future work, we aim to extend the similarity weight calculation by considering additional properties of the attribute values. Furthermore, the conditions for applying the approach could be relaxed by supporting events without optional attributes.

References

1. IEEE standard for extensible event stream (XES) for achieving interoperability in event logs and event streams. IEEE Std 1849–2016, pp. 1–50 (2016). https://doi.org/10.1109/IEEESTD.2016.7740858

2. van der Aalst, W.M.P.: Process-aware information systems: lessons to be learned from process mining. Trans. Petri Nets Model. Concurr. **2**, 1–26 (2009). https://doi.org/10.1007/978-3-642-00899-3_1
3. van der Aalst, W.M.P.: Data science in action. In: Process Mining, pp. 3–23. Springer, Heidelberg (2016). https://doi.org/10.1007/978-3-662-49851-4_1
4. Bano, D., Weske, M.: Discovering data models from event logs. In: Dobbie, G., Frank, U., Kappel, G., Liddle, S.W., Mayr, H.C. (eds.) ER 2020. LNCS, vol. 12400, pp. 62–76. Springer, Cham (2020). https://doi.org/10.1007/978-3-030-62522-1_5
5. Bayomie, D., Awad, A., Ezat, E.: Correlating unlabeled events from cyclic business processes execution. In: Nurcan, S., Soffer, P., Bajec, M., Eder, J. (eds.) CAiSE 2016. LNCS, vol. 9694, pp. 274–289. Springer, Cham (2016). https://doi.org/10.1007/978-3-319-39696-5_17
6. Bayomie, D., Di Ciccio, C., La Rosa, M., Mendling, J.: A probabilistic approach to event-case correlation for process mining. In: Laender, A.H.F., Pernici, B., Lim, E.-P., de Oliveira, J.P.M. (eds.) ER 2019. LNCS, vol. 11788, pp. 136–152. Springer, Cham (2019). https://doi.org/10.1007/978-3-030-33223-5_12
7. Bayomie, D., Helal, I.M.A., Awad, A., Ezat, E., ElBastawissi, A.: Deducing case ids for unlabeled event logs. In: Reichert, M., Reijers, H.A. (eds.) BPM 2015. LNBIP, vol. 256, pp. 242–254. Springer, Cham (2016). https://doi.org/10.1007/978-3-319-42887-1_20
8. Burattin, A., Vigo, R.: A framework for semi-automated process instance discovery from decorative attributes. In: 2011 IEEE Symposium on Computational Intelligence and Data Mining (CIDM), pp. 176–183. IEEE (2011)
9. Chen, P.P.S.: The entity-relationship model-toward a unified view of data. ACM Trans. Database Syst. (TODS) **1**(1), 9–36 (1976)
10. van Dongen, B.F., de Medeiros, A.K.A., Verbeek, H.M.W., Weijters, A.J.M.M., van der Aalst, W.M.P.: The ProM framework: a new era in process mining tool support. In: Ciardo, G., Darondeau, P. (eds.) ICATPN 2005. LNCS, vol. 3536, pp. 444–454. Springer, Heidelberg (2005). https://doi.org/10.1007/11494744_25
11. Ferreira, D.R., Gillblad, D.: Discovering process models from unlabelled event logs. In: Dayal, U., Eder, J., Koehler, J., Reijers, H.A. (eds.) BPM 2009. LNCS, vol. 5701, pp. 143–158. Springer, Heidelberg (2009). https://doi.org/10.1007/978-3-642-03848-8_11
12. Johnson, A., Bulgarelli, L., Pollard, T., Horng, S., Celi, L.A., Mark, R.: MIMIC-IV (2020). https://doi.org/10.13026/A3WN-HQ05, https://physionet.org/content/mimiciv/0.4/
13. de Leoni, M.M., Mannhardt, F.: Road Traffic Fine Management Process (2015). 2, https://doi.org/10.4121/uuid:270fd440-1057-4fb9-89a9-b699b47990f5, https://data.4tu.nl/articles/dataset/Road_Traffic_Fine_Management_Process/12683249
14. Li, Y., Liu, B.: A normalized levenshtein distance metric. IEEE Trans. Pattern Anal. Mach. Intell. **29**(6), 1091–1095 (2007). https://doi.org/10.1109/TPAMI.2007.1078
15. de Murillas, E.G.L., Reijers, H.A., van der Aalst, W.M.P.: Case notion discovery and recommendation: automated event log building on databases. Knowl. Inf. Syst. **62**(7), 2539–2575 (2019). https://doi.org/10.1007/s10115-019-01430-6
16. Weske, M.: Business Process Management - Concepts, Languages, Architectures, 3rd edn. Springer, Heidelberg (2019)

Demystifying Noise and Outliers in Event Logs: Review and Future Directions

Agnes Koschmider[1(✉)], Kay Kaczmarek[1], Mathias Krause[1], and Sebastiaan J. van Zelst[2,3]

[1] Group Process Analytics, Kiel University, Kiel, Germany
ak@informatik.uni-kiel.de,{stu96465,stu113088}@mail.uni-kiel.de
[2] Fraunhofer Institute for Applied Information Technology, Fraunhofer Gesellschaft, Sankt Augustin, Germany
sebastiaan.van.zelst@fit.fraunhofer.de
[3] Chair of Process and Data Science, RWTH Aachen University, Aachen, Germany

Abstract. Various process mining techniques exist, e.g., techniques that automatically discover a descriptive model of the execution of a process, based on event data. Whereas the premise of process mining is clear, i.e., as witnessed by the tremendous growth of the field, data quality issues often hamper the direct applicability of process mining techniques. Several authors have studied data quality issues in process mining, yet, these works primarily propose data pre-processing techniques. An overarching study of the nature of data quality issues, the types of available techniques, and the general possibilities of (semi)-automated outlier/noise detection methods is missing. Therefore, in this paper, we propose a first attempt to structure and study the field of outlier/noise detection in process mining and understand to what degree knowledge on noise and outliers from other domains could advance the process mining field. We do so by answering three central research questions, covering various aspects related to (semi)-automated outlier/noise detection.

1 Introduction

Process mining [1] techniques are able to, largely automatically, derive valuable insights in the execution of a process, based on recorded *event data*. For example, techniques exist that automatically discover a process model based on event data, i.e., referred to as *automated process discovery algorithms* [2]. The premise of process mining is promising, yet, several practical issues, primarily related to *outlier and noise detection*, hamper the direct application of process mining techniques on the recorded event data. As such, various techniques to pre-process and filter event data have been proposed in process mining, e.g., [3–5]. However, a clear overview and categorization of these techniques, as well as the concepts of *noise* and *outliers* have not been proposed.

In statistics, outliers are defined as "*high measurements where the value is some standard deviation above the average*" [6]. In data engineering, outliers, commonly referred to as "*anomalies*", refer to "*something that is out of range*".

A. Marrella and B. Weber (Eds.): BPM 2021 Workshops, LNBIP 436, pp. 123–135, 2022.
https://doi.org/10.1007/978-3-030-94343-1_10

This can, on the one hand, point to insignificant data or, on the other hand, to interesting and useful information about the underlying system [7]. Hence, distinguishing the essence of outliers in terms of undesired or unwanted behavior versus surprisingly correct and informative data is of particular interest for the quality of process discovery. However, studies on the interrelation between outliers and noise have received little interest. We consider outliers in event logs as *divergent data* that is out of range of behavior that we expect, while noise is "semantic", i.e., it refers to, e.g., erroneous data recordings. Both outlier and noise detection are essential for process discovery since they might negatively influence the usefulness of the discovered process model [8]. For example, incorrect logging of event positions causes problems for process discovery algorithms to discover the correct control flow (i.e., relationships between events may be inferred that do not exist in reality). Consequently, a clear understanding of noise and outliers bridges the gap between detecting errors in event data recordings versus unexpected or even unwanted, yet, correclty logged behavior.

While several approaches exist to handle outliers in event logs and to filter noise within the process discovery algorithms [4,9,10], we aim to understand to which degree the potential of outlier and noise detection techniques from other fields have been exploited for process mining. To provide an answer, we formulate the following three research questions: **RQ1**: What types of outlier and noise exist for event logs? **RQ2**: How are outliers and noise detected in event logs? **RQ3**: What are potential research directions for future research in outlier and noise detection techniques in process mining? Equipped with such knowledge, we aim to define an outlier and noise detection model that quantifies outliers and noise in event logs.

Against this background, this paper is structured as follows. Section 2 classifies outlier, noise, and "normal" behavior in an event log aiming to answer **RQ 1**. Relying on this classification, a literature search has been conducted with the purpose of addressing **RQ 2**. The results are summarized in Sect. 3. To further understand the capacities of noise and outlier techniques for event logs, corresponding tools were analyzed. Based on the literature search results and the tool analysis, potential research directions are formulated and summarized in Sect. 5 referring to **RQ 3**. The paper ends with a conclusion.

2 Noise vs. Outliers

Techniques for outlier detection have been suggested to a large extent for data analysis in general [7], whereas a clear conceptualization of outlier and noise detection for event logs is lacking. Instead, noise and outliers are often used synonymously. Figure 1 illustrates how outliers and noise in an event log can be considered and differentiated from normal behavior. An outlier in event data, is a behavior that is out of range of the behavior that we expect. It is something that differs considerably from all or most other behavior. According to Fig. 1 these are points laying outside the thin, dotted rectangle. Noise is any undesirable or unwanted value of a case or event attribute. Behavior without any outlier or

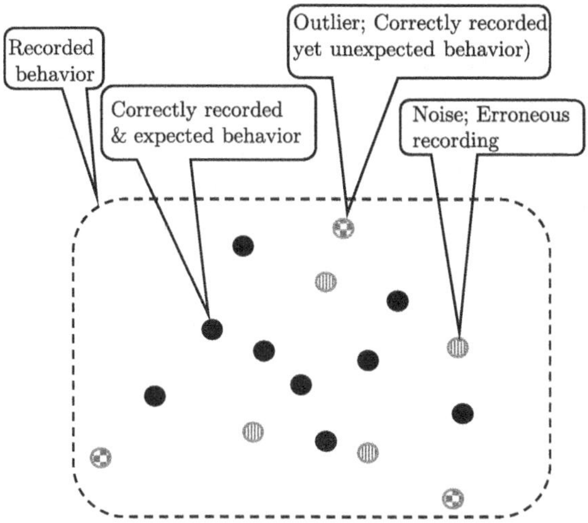

Fig. 1. Difference between outliers, noise and normal behavior in an event log.

noise is considered "normative behavior" and is represented as a black dot in Fig. 1.

To understand outliers in event logs, we use an existing classification from the literature [11] and map it for our purpose on outliers in event logs:

- *Point as outlier*: this is the simplest type of ouliers and plenty of literature exist how to detect them, for an overview we refer to [12]. In the context of an event log *point as outlier* would be a trace with activities significantly deviating from the rest. E.g., in a log $L = \{\langle A, B, C, E, F\rangle^{50}, \langle A, B, C, D, E, F\rangle^{100}, \langle A, A, A, B, C, E, F\rangle^{2}\}$ the third trace is the only trace with three As and points to an outlier. Note that although this trace is classified having an outlier, this outlier might point to interesting observations.
- *Context as outlier*: this category of outliers, also referred to as conditional anomalies, requires a notion of context. A contextual outlier is given if an observation is uncommon in a certain context but not unexpected in another context. A contextual outlier is a trace that significantly depends on a context dimension. To structure context dimensions for our purpose, we apply context dimensions as they have being defined for conceptual models [13], which are: personal & social, task, environmental and spatial-temporal. For instance, the process for oncology in the daytime consists of 14 activities while the process for intensive care in the night has 8 activities with two loops. In this example time is a contextual factor (it refers to the spatial-temporal dimension) and thus traces with 8 activities are annotated with spatial-temporal attribute.
- *Subsequence as outlier*: in the context of an event log a subsequence outlier means that a subset of the trace deviates significantly from the rest, even if the individual activities in the subset may not be outliers. For example in a log $L = \{\langle A, B, C, E, F, G\rangle^{30}, \langle A, B, C, D, E, F, G\rangle^{25}, \langle A, B, C\rangle^{2}\}$ the last

trace is an outlier since there are only a few traces of it although the behavior A, B, C is a regular behavior.

Common techniques to detect outliers are density-based or distance-based clustering [14]. These algorithms build a model (e.g. a statistical, probabilistic model) describing the normal behavior, and consider as outliers all data points which deviate from this model [14]. In the context of event logs also sequence-based anomaly detection, contiguous subsequence-based anomaly detection and pattern frequency-based anomaly detection are common techniques, since they consider particular aspects of event logs for outlier detection [15].

To understand and classify noise, we browsed the literature on data quality for databases and machine learning [16–19] and event log imperfection patterns [20]. Accordingly, we adopt the classification of attribute and class noise and map it for our purpose:

- *Attribute noise*: Arises when an imprecision, incompletion or an error is introduced to one or more attributes. Attribute noise can be totally unpredictable i.e., random, or simply a low variation with respect to the correct value [16]. In the literature the following types of attribute noise are distinguished [16]: (1) erroneous attribute values, (2) missing or don't know attribute values and (3) incomplete attributes or don't care values [16]. An event log might contain erroneous attribute values due to a logging error that recorded identical timestamp for different events (see Table 1). Missing attribute values might arise due to e.g., faults in sensor devices and they are shown by a missing entry in an event log. Incomplete attributes might occur due to irregularities in sampling like that the data may not be available or it was not considered important.
- *Class noise*: this category refers to the semantics of attributes (i.e., mislabeled or contradictory activity labels). The event log imperfection patterns [21] describe examples for class noise like the patterns Synonymous Labels and Homonymous Label.

Figure 2 shows in which steps attribute and class noise can be found in the process from raw (sensor) event data to the discovered process model and also where outliers affect this process. This figure might lay the foundation to conceptualize a holistic detection technique for outliers and noise. Although attribute and class noise can already be found in the raw (sensor) event data, the classification if this noise is completely random or not is determined at event log level. Class noise is also detected when abstracting or aggregating events to activities since the semantics of attributes is disambiguated at this stage. Outliers are detected either at the aggregation level or during process discovery.

To understand how outliers and noise have been addressed for process mining, a literature search was conduced. The results are summarizes in the next section.

Table 1. An example event log, adapted from [22], with attribute noise (highlighted in red) due to identical timestamps and missing values that were not recorded for resources.

Case id	*Timestamp*	*Activity*	*Resource*	*Transactional*	*Cost*	...
12373	30-7-2019 11.12	register request	Bas	complete	50	...
12374	30-7-2019 11.32	register request	–	start	50	...
12374	30-7-2019 11.44	register request	Agnes	complete	50	...
12373	30-7-2019 11.44	check ticket	–	start	100	...
⋮	⋮	⋮	⋮	⋮	⋮	...

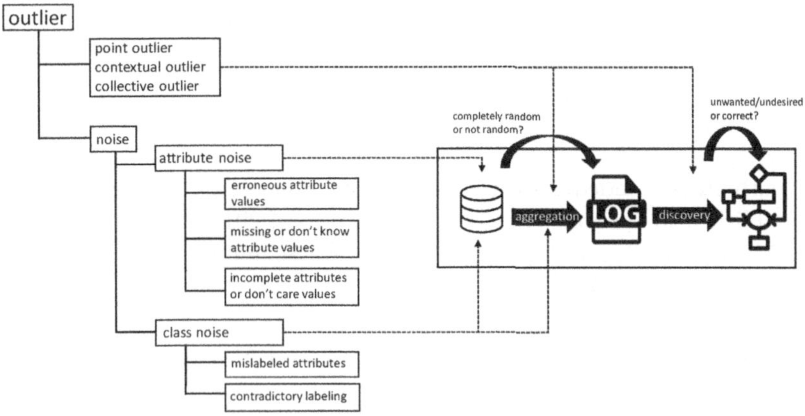

Fig. 2. Classification of outliers and noise and its relevance for the steps from raw data to the discovered process model.

3 Literature Search

This section describes how the literature search was conducted. We searched the ACM, IEEE and Springer research databases. We also used Google Scholar to find appropriate literature by browsing the citations of related publications already found in the scientific databases. Second, we conducted a backward search to find more appropriate publications cited in papers of the first search round. To find synonyms for the term *outlier* we checked all related papers found by Springer having this term in the title. The result list was cross-checked with related papers on event log quality [20,23] and one highly cited paper on outlier detection for process discovery [3]. We used the following query to identify appropriate publications:

```
{(''process mining'' OR ''event log'')  AND
(''outlier'' OR ''anomaly'' OR ''infrequent behavior'' OR ''noise''
OR ''abnormal behavior'') AND ''process mining''}
```

Table 2. Type of Outlier and Noise techniques suggested for event logs.

Outlier	
Point	chaotic activity [26] spurious event [4,10,27]
Contextual	personal & social (e.g., resource) [28] task (e.g., control-flow): [5,9,25,29–33] environmental (e.g., training data): [34–36] spatial-temporal (e.g., time): [31]
Noise	
Attribute Noise	erroneous attribute values: [24,25,37,38] missing attributes values [39] incomplete attribute values [3,25,40]
Class Noise	synonymous label [24,38,41,42] homonymous label [24]

Eventually, we ended up with 24 relevant publications. The 24 publications were published in the following years: 2008 (4.5%), 2015 (9.1%), 2017 (13.6%), 2018 (31.8%), 2019 (31.8%), 2020 (9.1%). These numbers indicate a recent increasing interest in the topic.

3.1 Literature Result: Type of Outliers and Noise

Table 2 shows the results of the literature search. Referring to **RQ1** (What types of outlier and noise exist for event logs?) the following observations can be given. The main focus has been put on control-flow aspects like probability and frequency of traces. Some works exist on point as outlier detection. Only one paper has been found that detects attribute and class noise, see [24] and one paper [25] addressing attribute noise and context as outlier. No paper was found for subsequence as outlier.

3.2 Literature Result: Outlier Detection Techniques

To answer **RQ2** (How are outliers and noise detected in event logs?) we analyzed the related literature with respect to the detection techniques used to identify noise and outliers. Table 3 summarizes the results. It shows that unsupervised learning is the most common technique. According to our analysis only one approach (see reference [25]) discusses the characteristics of outliers (i.e., is this undesired, unwanted behavior vs. surprisingly correct and informative data).

4 Analysis of Tools for Noise and Outlier Filtering for Event Logs

The intention of this analysis is to identify the efficiency of available tools and to understand if noise and outliers are kept properly apart. We first

Table 3. Outlier detection techniques used for process mining

Article	Type	Year	Type of Supervision			Technique	
			Superv.	Semi-s.	Unsuperv.	Method	Explain?
Ghionna et al. [34]	-O-	2008			x	Distance	
Wang et al. [41]	-N-	2015			x	Branch and Bound	
Cheng et al. [36]	-O-	2015	x			Rule based	
Nolle et al. [35]	-O-	2016			x	Deep Learning	
Conforti et al. [3]	-N-	2017			x	Log Automaton	
Chapela-Campa et al. [32]	-O-	2017			x	Pattern	
Mannhardt et al. [33]	-O-	2017			x	Density	
Fani Sani et al. [5]	-O-	2018			x	Rule based	
Fani Sani et al. (b) [29]	-O-	2018			x	Covering Probability	
Fani Sani et al. (c) [9]	-O-	2018			x	Density	
van Zelst et al. [10]	-O-	2018			x	Probabilistic Automata	
Conforti et al. [37]	-N-	2018			x	Automata	
Nolle et al. [31]	-O-	2018		x		Deep Learning	
Tax et al. [26]	-O-	2019			x	Entropy,Statistical	
Böhmer et al. [28]	-O-	2019			x	Rule based	
Sun et al. [27]	-O-	2019			x	Density	
Sadeghianasl et al. [38]	-N-	2019		x		Density	
Chapela-Campa et al. [30]	-O-	2019			x	Pattern, Abstraction	
Nguyen et al. [39]	-N-	2019			x	Machine Learning	
Niels et al. [25]	-O-N-	2019			x	Heuristics	x
Sadeghianasl et al. [42]	-N-	2019			x	MST clustering	
Sarno et al. [40]	-N-	2020			x	Rule based	
van Zelst et al. [4]	-O-	2020			x	Probabilistic Automata	

narrowed our analysis on noise detection in event logs. For this purpose, we generated an event log with 1000 traces based on the Petri net shown in Fig. 3[1]. To add noise to the event log, we use the ProM[2] plugin "Add Noise to log filter" with a noise threshold of 20%. A manual inspection, however, shows that the changed event log has only 135 out of 1000 noise-free traces (=13,5%). Due to this inappropriate result, we decided to manually add (attribute) noise[3]. Based on the Petri net shown in Fig. 3 we generated a log with $L = \{\langle A,D,H,K\rangle^{68}, \langle A,C,N,H,K\rangle^{22}, \langle A,C,G,L,M,J\rangle^{12}, \langle A,C,G,M,L,J\rangle^{3}, \langle A,B,F,I\rangle^{25}, \langle A,B,F,E,F,I\rangle^{4}, \langle A,B,F,E,F,E,F,I\rangle^{1}\}$. To understand whether available tools also detect outliers and not only noise, we inserted varying trace frequency and control-flow errors into the event log. Additionally, we assign resources to activities, which is relevant when analyzing noise like missing attribute values or incomplete assignment of resources. For this purpose, we assign ResGr1 to activity A, ResGr2 to activities B,E,F,I, ResGr3 to activities C, G, L, M, J and ResGr4 to activities D, N, H, K. Normal behavior are all

[1] To generate the event log we used the Petri Net-based Event Log Generator http://processmining.be/loggenerator/.

[2] https://www.promtools.org/doku.php.

[3] Available tools do not resolve synonyms nor homonyms. Therefore we restricted our analysis only to attribute noise.

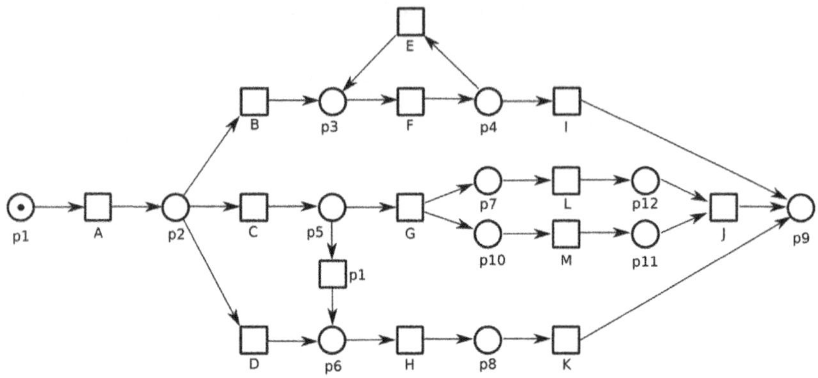

Fig. 3. Petri net used to generate a synthetic event log.

traces that can be replayed on the Petri net of Fig. 3. Besides, we define two control-flow rules: 1) activity E may only be performed a maximum of three times, 2) the order in which activities M and L are performed is not important. However, activity G must precede both activities and J must succeed. Abnormal behavior is any behavior that deviates from our defined behavior according to our Petri net. For this event log we manually inspected these four tools:

- The process mining tool Disco[4]: here filtering of specific attributes is not possible nor filtering by frequency within a trace or the frequency of subsequently followed activities.
- Filter events based on attribute values (See footnote 2): this tool only allows to filter events from traces, but not the incorrectly assigned resources.
- Filter Event Log (See footnote 2): although this tool has plenty of noise filtering techniques, some errors were found in this tool. For instance, filtering the order in which activities have to be executed does not work appropriately.
- Filter/Edit Attributes of an Event Log (See footnote 2): the filtering technique is limited to renaming of attributes rather than noise or outlier detection.

Summarizing, our analysis shows that available noise filtering tools do not appropriately filter noise. Although they aim to identify outliers (i.e., point as ouliers) there is ample potential to improve the filtering techniques e.g., in terms of erroneous order of activities. We are convinced that our classification of outliers and noise could be seen an appropriate foundation for revision. In future we plan to compare available tools for outlier detection with emphasis on control-flow in order to give more insights into the quality of the detection techniques.

5 Future Research Directions

Based on our different analysis, we see the following future research directions responding to **RQ3**:

[4] https://fluxicon.com/disco/.

Correlation Between Noise and Outliers on Process Quality: No paper exists analyzing the influence of noise and outliers on process discovery (i.e., Is it more beneficial, first to find noise and then outliers or vice versa? How does different attributes impact the process quality? or How does attribute and class noise detection impact each other?) In terms of data quality and classification accuracy for machine learning, plenty of experiments were conducted to analyze the mutual impact of attribute and class noise [16,18,19]. Related literature shows that attribute noise depends on the correlation between attribute and class noise. The higher the correlation, the more negative impact the attribute noise may bring [16]. Noisy attributes that have a weak correlation with class label could be eliminated, while noisy attributes with a high correlation with the class label may be subject to data cleansing. Our analysis shows that such analysis have not been conducted for event logs. To provide a solution, attributes could be ranked according to their sensitivity from most to least noisy [18] and then to apply polishing techniques instead of noise filtering for highly noise sensitive attributes [19].

Explainability: Our analysis shows that only little attention has been given to the explainability of outlier and noise detection. As shown in Table 3 existing approaches do not discuss the characteristics of outliers like interesting vs. unwanted behavior, nor if the attribute noise is random. However, this classification is essential for appropriate filtering. A thorough analysis of explainability, also calls for experimental settings on event log quality of the complete data pipeline. This calls for a holistic approach evaluating the impact of noise and outlier and its mutual impact on process discovery.

6 Related Work

A partial view on outlier detection for process mining has been addressed by [43]. This paper, however, mainly focuses on outlier detection in business process runtime behavior and sets a different focus on this topic. Complementary to our paper are approaches on data quality issues in event logs [20,23]. The data quality frameworks defined in [20,23] classify quality issues, however, they do not provide guidance as how to discover noise and outliers for process mining, nor do they provide a mechanism to discuss to which extent an event log is affected by noise and outliers (i.e., how is the quality of an event log affected by attribute noise?). Quality issues in an event log can be identified by event log imperfection patterns http://www.workflowpatterns.com/patterns/logimperfection/, which are mainly related to noise and disregard outliers.

Some process discovery algorithms like the Inductive Miner has embedded filtering mechanisms to deal with some types of outliers. According to [44] infrequent behavior is considered as paths that are taken infrequently, or traces that only differ by occurrence of infrequent activities. The directly-follows graph (DFG) is filtered until it only contains most frequent edges or the mainstream behavior (often called as happy path). When applying this approach, however, it has been identified that the DFG may be misleading [45]. To overcome limitations of the directly-follows graph in terms of appropriately filtering infrequent behavior, the eventually-follows graph (EFG) has been suggested [44].

To the best of our knowledge this is the first literature review on outlier and noise detection for process mining.

7 Conclusion and Future Directions

It is generally acknowledged that outlier detection is a challenging task. Anyway, handling outliers is an essential task in order to identify desired vs. correct deviations. Within the field of process mining, data quality, i.e., the presence of outliers and/or noise, is one of the most urgent problems hampering the direct application of process mining techniques on event data. Whereas a significant amount of research has been conducted toward (algorithmic) techniques for outlier/noise identification and removal, an overall viewpoint on the matter is lacking. In this paper, on the basis of existing literature on general data quality issues, we provide a structured qualification of the notions of outliers and noise in event data. Furthermore, we provide an overview and classification of the different existing techniques developed for the purpose of outlier/noise detection and removal for process discovery. Future research avenues in this field are (1) discussions about the characteristics of outliers (i.e., undesired, unwanted behavior vs. surprisingly correct and informative data), (2) holistic approach to analyze attribute and class noise, while class noise has received little attention yet, (3) sophisticated detection techniques for event logs (e.g., filtering by specific attributes, their frequency within a trace or the frequency of subsequently followed activities).

References

1. van der Aalst, W.M.P.: Process Mining - Data Science in Action, Second edn. Springer, Heidelberg (2016)
2. Augusto, A., et al.: Automated discovery of process models from event logs: review and benchmark. IEEE Trans. Knowl. Data Eng. **31**(4), 686–705 (2019)
3. Conforti, R., Rosa, M.L., ter Hofstede, A.H.M.: Filtering out infrequent behavior from business process event logs. IEEE TKDE **29**(2), 300–314 (2017)
4. van Zelst, S.J., Sani, M.F., Ostovar, A., Conforti, R., Rosa, M.L.: Detection and removal of infrequent behavior from event streams of business processes. Inf. Syst. **90**, 101451 (2020)
5. Fani Sani, M., van Zelst, S.J., van der Aalst, W.M.P.: Applying sequence mining for outlier detection in process mining. In: Panetto, H., Debruyne, C., Proper, H.A., Ardagna, C.A., Roman, D., Meersman, R. (eds.) OTM 2018. LNCS, vol. 11230, pp. 98–116. Springer, Cham (2018). https://doi.org/10.1007/978-3-030-02671-4_6
6. Freedman, D.: Statistical Models: Theory and Practice. Cambridge University Press, Cambridge (2005)
7. Ord, K.: Outliers in statistical data, 3rd edition, (john wiley & sons, chichester). Int. J. Forecast. **12**(1), 175–176 (1996)

8. Ghionna, L., Greco, G., Guzzo, A., Pontieri, L.: Outlier detection techniques for process mining applications. In: An, A., Matwin, S., Raś, Z.W., Ślęzak, D. (eds.) ISMIS 2008. LNCS (LNAI), vol. 4994, pp. 150–159. Springer, Heidelberg (2008). https://doi.org/10.1007/978-3-540-68123-6_17
9. Sani, M.F., van Zelst, S.J., van der Aalst, W.M.P.: Improving process discovery results by filtering outliers using conditional behavioural probabilities. In: Teniente, E., Weidlich, M. (eds.) BPM 2017. LNBIP, vol. 308, pp. 216–229. Springer, Cham (2018). https://doi.org/10.1007/978-3-319-74030-0_16
10. van Zelst, S.J., Sani, M.F., Ostovar, A., Conforti, R., Rosa, M.L.: Filtering spurious events from event streams of business processes. In: CAiSE 2018, Proceedings, pp. 35–52 (2018)
11. Chandola, V., Banerjee, A., Kumar, V.: Anomaly detection: a survey. ACM Comput. Surv. **41**(3) (2009)
12. Gupta, M., Gao, J., Aggarwal, C.C., Han, J.: Outlier detection for temporal data: a survey. IEEE TKDE **26**(9), 2250–2267 (2014)
13. Koschmider, A., Mannhardt, F., Heuser, T.: On the contextualization of event-activity mappings. In: BPM 2018 International Workshops, pp. 445–457 (2018)
14. Aggarwal, C.C.: Outlier Analysis. 2nd edn. Springer, Heidelberg (2016)
15. Chandola, V., Banerjee, A., Kumar, V.: Anomaly detection for discrete sequences: a survey. IEEE Trans. Knowl. Data Eng. **24**(5), 823–839 (2012)
16. Zhu, X., Wu, X.: Class noise vs. attribute noise: a quantitative study of their impacts. Artif. Intell. Rev. **22**(3), 177–210 (2004)
17. SáEz, J.A., Galar, M., Luengo, J., Herrera, F.: Tackling the problem of classification with noisy data using multiple classifier systems: analysis of the performance and robustness. Inf. Sci. **247**, 1–20 (2013)
18. Khoshgoftaar, T.M., Van Hulse, J.: Empirical case studies in attribute noise detection. IEEE Trans. Syst. Man Cybern. **39**(4), 379–388 (2009)
19. Gupta, S., Gupta, A.: Dealing with noise problem in machine learning data-sets: a systematic review. Procedia Comput. Sci. **161**, 466–474 (2019). The Fifth Information Systems International Conference
20. Dixit, P.M., et al.: Detection and interactive repair of event ordering imperfection in process logs. In: Krogstie, J., Reijers, H.A. (eds.) CAiSE 2018. LNCS, vol. 10816, pp. 274–290. Springer, Cham (2018). https://doi.org/10.1007/978-3-319-91563-0_17
21. Andrews, R., Suriadi, S., Ouyang, C., Poppe, E.: Towards event log querying for data quality. In: Panetto, H., Debruyne, C., Proper, H.A., Ardagna, C.A., Roman, D., Meersman, R. (eds.) OTM 2018. LNCS, vol. 11229, pp. 116–134. Springer, Cham (2018). https://doi.org/10.1007/978-3-030-02610-3_7
22. van Zelst, S.J., Mannhardt, F., de Leoni, M., Koschmider, A.: Event abstraction in process mining - literature review and taxonomy. Granul. Comput. (2020)
23. Bose, R.P.J.C., Mans, R.S., van der Aalst, W.M.P.: Wanna improve process mining results? In: 2013 IEEE Symposium on Computational Intelligence and Data Mining (CIDM), pp. 127–134 (2013)
24. Ziolkowski, T., Brandt, L., Koschmider, A.: Elogqp: an event log quality pointer. In: ZEUS 2021. Volume 2839 of CEUR Workshop Proceedings, pp. 42–45. CEUR-WS.org (2021)
25. Martin, N., Martinez-Millana, A., Valdivieso, B., Fernandez-Llatas, C.: Interactive data cleaning for process mining: a case study of an outpatient clinic's appointment system, pp. 532–544, September 2019

26. Tax, N., Sidorova, N., van der Aalst, W.M.P.: Discovering more precise process models from event logs by filtering out chaotic activities. J. Intell. Inf. Syst. **52**(1), 107–139 (2018). https://doi.org/10.1007/s10844-018-0507-6
27. Sun, X., Hou, W., Yu, D., Wang, J., Pan, J.: Filtering out noise logs for process modelling based on event dependency. In: ICWS 2019, pp. 388–392. IEEE (2019)
28. Böhmer, K., Rinderle-Ma, S.: Mining association rules for anomaly detection in dynamic process runtime behavior and explaining the root cause to users. Inf. Syst. (2019)
29. Fani Sani, M., van Zelst, S.J., van der Aalst, W.M.P.: Repairing outlier behaviour in event logs. In: Abramowicz, W., Paschke, A. (eds.) BIS 2018. LNBIP, vol. 320, pp. 115–131. Springer, Cham (2018). https://doi.org/10.1007/978-3-319-93931-5_9
30. Chapela-Campa, D., Mucientes, M., Lama, M.: Simplification of complex process models by abstracting infrequent behaviour, pp. 415–430, October 2019
31. Nolle, T., Seeliger, A., Mühlhäuser, M.: Binet: multivariate business process anomaly detection using deep learning. In: BPM 2018, Proceedings, pp. 271–287 (2018)
32. Chapela-Campa, D., Mucientes, M., Lama, M.: Discovering infrequent behavioral patterns in process models. In: Carmona, J., Engels, G., Kumar, A. (eds.) BPM 2017. LNCS, vol. 10445, pp. 324–340. Springer, Cham (2017). https://doi.org/10.1007/978-3-319-65000-5_19
33. Mannhardt, F., De Leoni, M., Reijers, H.A., van der Aalst, W.M.P.: Data-driven process discovery - revealing conditional infrequent behavior from event logs. In: CAiSE 2017, Proceedings, pp. 545–560 (2017)
34. Ghionna, L., Greco, G., Guzzo, A., Pontieri, L.: Outlier detection techniques for process mining applications. In: An, A., Matwin, S., Raś, Z.W., Ślęzak, D. (eds.) ISMIS 2008. LNCS (LNAI), vol. 4994, pp. 150–159. Springer, Heidelberg (2008). https://doi.org/10.1007/978-3-540-68123-6_17
35. Nolle, T., Seeliger, A., Mühlhäuser, M.: Unsupervised anomaly detection in noisy business process event logs using denoising autoencoders. In: Calders, T., Ceci, M., Malerba, D. (eds.) DS 2016. LNCS (LNAI), vol. 9956, pp. 442–456. Springer, Cham (2016). https://doi.org/10.1007/978-3-319-46307-0_28
36. Cheng, H.J., Kumar, A.: Process mining on noisy logs - can log sanitization help to improve performance? Decis. Support Syst. **79**, 138–149 (2015)
37. Conforti, R., La Rosa, M., ter Hofstede, A.: Timestamp repair for business process event logs. Technical report (2018)
38. Sadeghianasl, S., ter Hofstede, A.H.M., Wynn, M.T., Suriadi, S.: A contextual approach to detecting synonymous and polluted activity labels in process event logs. In: Panetto, H., Debruyne, C., Hepp, M., Lewis, D., Ardagna, C.A., Meersman, R. (eds.) OTM 2019. LNCS, vol. 11877, pp. 76–94. Springer, Cham (2019). https://doi.org/10.1007/978-3-030-33246-4_5
39. Nguyen, H.T.C., Comuzzi, M.: Event log reconstruction using autoencoders. In: Liu, X., et al. (eds.) ICSOC 2018. LNCS, vol. 11434, pp. 335–350. Springer, Cham (2019). https://doi.org/10.1007/978-3-030-17642-6_28
40. Sarno, R., Sinaga, F., Sungkono, K.: Anomaly detection in business processes using process mining and fuzzy association rule learning. J. Big Data **7** (2020)
41. Wang, J., Song, S., Lin, X., Zhu, X., Pei, J.: Cleaning structured event logs: a graph repair approach. In: Proceedings - International Conference on Data Engineering 2015, pp. 30–41, May 2015
42. Sadeghianasl, S., ter Hofstede, A.H.M., Wynn, M.T., Suriadi, S.: A contextual approach to detecting synonymous and polluted activity labels in process event logs.

In: Panetto, H., Debruyne, C., Hepp, M., Lewis, D., Ardagna, C.A., Meersman, R. (eds.) OTM 2019. LNCS, vol. 11877, pp. 76–94. Springer, Cham (2019). https://doi.org/10.1007/978-3-030-33246-4_5
43. Böhmer, K., Rinderle-Ma, S.: Anomaly detection in business process runtime behavior - challenges and limitations. CoRR abs/1705.06659 (2017)
44. Leemans, S.J.J., Fahland, D., van der Aalst, W.M.P.: Discovering block-structured process models from incomplete event logs. In: Ciardo, G., Kindler, E. (eds.) PETRI NETS 2014. LNCS, vol. 8489, pp. 91–110. Springer, Cham (2014). https://doi.org/10.1007/978-3-319-07734-5_6
45. van der Aalst, W.: A practitioner's guide to process mining: limitations of the directly-follows graph. Procedia Comput. Sci. **164**, 321–328 (2019). CENTERIS 2019

1st International Workshop on BPM Governance for and Beyond Digital Transformation (BPMGOV 2021)

1st International Workshop on BPM Governance for and Beyond Digital Transformation (BPMGOV 2021)

With this workshop we would like to contribute to the strengthening of the managerial-oriented part of the BPM conference. The main goal is to provide a discussion platform concerning BPM governance open questions that are valid both in the context of digital transformation and also beyond.

BPM governance denotes the establishment of appropriate and transparent accountability in terms of roles and responsibilities for different levels of BPM and has been recognized as one of the core capabilities and critical success factors for BPM. Moreover, BPM governance coexists in organizations and should be integrated with corporate governance and IT governance.

To deal with the challenges of digital transformation, organizations face clearly defining the role of Chief Digital Officer and its connections to other roles, among which are also process owners, BPM offices, IT departments, CIOs, and others who aim to improve organizational performance. In the era of digital business models, the relations within a business ecosystem, i.e., business partners, customers, and government institutions should be also taken into account when talking about BPM governance. When it comes to digital transformation, customers' journeys and their behavior and habits play a crucial role in organizational processes' governance and redesign, thus shifting the focus of BPM governance from mainly internally oriented to externally oriented. BPM governance issues are equally important for private and public sector organizations. Despite its importance, BPM governance is still under-researched and not deployed enough in practice.

We are convinced that the first edition of our workshop contributed to the partial filling of this gap and the development of a reliable body of evidence in this field. We received eight high-quality submissions, the top four of which were selected for presentation and publication.

Tomasz Helbin and Amy Van Looy (Ghent University, Belgium) took up the topic of BPM ambidexterity in the public sector based on a case study of EU institutions. They explore the critical enablers for the success of BPM initiatives in this sector.

Piotr Sliż, Jędrzej Siciński, Paweł Antonowicz, and Robert Bęben (all authors from the University of Gdansk, Poland) contributed to deepening the understanding of BPM governance in the public sector. In their research factors supporting and hindering implementation of BPM governance at a public university are addressed.

An ambidextrous BPM environment is also acknowledged by Tahir Ahmad and Amy Van Looy (Ghent University, Belgium) in their research designed to translate the original task-technology fit theory into a process-technology fit theory.

Ivona Jovanović, Dragana Stojanović, Dragoslav Slović, Ivan Tomašević, Barbara Simeunović, and Milica Maričić form a team of researchers from the University of Belgrade (Serbia), who explored Business Process Improvement factors in gaining a competitive advantage in Serbian firms.

We would like to thank the authors and Program Committee members for their contribution to the 1st edition of the International Workshop on BPM Governance For and Beyond Digital Transformation. We are firmly convinced that the BPMGOV workshop will help strengthen the existing body of BPM governance knowledge.

We recommend reading these selected papers to the wider community of BPM researchers, practitioners, and enthusiasts.

September 2021

Vesna Bosilj Vukšić
Renata Gabryelczyk
Mojca Indihar Štemberger
Andrea Kő

Organization

Program Chairs

Vesna Bosilj Vukšić	University of Zagreb, Croatia
Renata Gabryelczyk	University of Warsaw, Poland
Mojca Indihar Stemberger	University of Ljubljana, Slovenia
Andrea Kő	Corvinus University of Budapest, Hungary

Program Committee

Agnieszka Bitkowska	Warsaw University of Technology, Poland
Edyta Brzychczy	AGH University of Science and Technology, Poland
Silvia Inês Dallavalle de Pádua	University of Sao Paulo, Brazil
Peter Fettke	German Research Center for Artificial Intelligence, Germany
Daniela Garbin Praničević	University of Split, Croatia
Tomislav Hernaus	University of Zagreb, Croatia
Krzysztof Kluza	AGH University of Science and Technology, Poland
Milanović-Glavan Ljubica	University of Zagreb, Croatia
Ivan Matić	University of Split, Croatia
Andrzej Niesler	Wroclaw University of Economics and Business, Poland
Gregor Polančič	University of Maribor, Slovenia
Natalia Potoczek	Polish Academy of Sciences, Poland
Dragana Stojanović	University of Belgrade, Serbia
Dalia Suša Vugec	University of Zagreb, Croatia
Martina Tomicic Furjan	University of Zagreb, Croatia

Process Innovation in the EU Public Sector Through the Lens of BPM Ambidexterity

Tomasz Helbin(✉) and Amy Van Looy

Faculty of Economics and Business Administration, Department of Business Informatics and Operations Management, Ghent University, Tweekerkenstraat 2, 9000 Ghent, Belgium
Tomasz.Helbin@Ugent.be, Amy.VanLooy@UGent.be

Abstract. Public sector organizations need to innovate their processes incrementally and radically to face political challenges. BPM ambidexterity, which is defined as the simultaneous pursuit of incremental and radical process innovation, provides a relevant lens to analyze this. Since there are only a few studies on ambidexterity in the public sector, we have conducted an embedded case study at an EU institution. We confirm that process characteristics, and the resulting culture and structure are critical enablers for the success of BPM initiatives. Departments with more mature practices for process optimization also turn out to be more open to radical process change. Moreover, temporal ambidexterity is the most common strategy to resolve the tension between incremental and radical innovation, while departments with a higher process orientation also show preference for contextual ambidexterity. We end with public sector guidelines on how to develop organizational capabilities in incremental and radical process innovation.

Keywords: Business process management · Organizational ambidexterity · Public sector · Innovation

1 Introduction

Public sector organizations have to respond to multiple challenges, such as the climate crisis and the need for green transition, the digital disruption of multiple industries, geopolitical instabilities, as well as the need to deliver savings and efficiencies in administration. Business Process Management (BPM) plays a critical role in improving organizations' efficiency and competitiveness, through the optimization of internal processes, overcoming of organizational silos and reduction of waste [1]. However, the discipline has focused to a limited extent on disruptive process improvements [2] and, as argued by some researchers, may not be fit for the challenges of today [3].

As indicated by [4, 5], there are very limited guidelines for practitioners on how to simultaneously manage incremental and disruptive process improvements, and resolve the resulting tensions, which is defined as BPM ambidexterity [6]. Moreover, the contextual factors impacting BPM ambidexterity are not fully researched, in particular the broader Business Process Orientation [7], which covers the organizational structure and

A. Marrella and B. Weber (Eds.): BPM 2021 Workshops, LNBIP 436, pp. 141–152, 2022.
https://doi.org/10.1007/978-3-030-94343-1_11

culture enabling BPM. Hence, our objective is to understand how EU public organizations improve their business processes [8]. The research questions are based on prior work [4], aiming to explore various contextual dimensions of BPM ambidexterity:

RQ1. What are the triggers and obstacles for incremental and radical process improvement in the EU public sector?
RQ2. What practices exist for business process ambidexterity (i.e., exploitation and exploration) in the EU public sector?
RQ3. What are the contextual factors influencing different levels of process improvement activities in the EU public sector?
RQ4. How can tensions be resolved between exploration and exploitation in EU public sector?

We conducted a single but embedded in-depth case study at an EU institution. The main implication for theory is deepening our understanding of BPM ambidexterity in the public sector, the contextual factors underpinning its growth, as well as the preferred strategies to resolve ambidextrous tension. We also provide guidelines for public sector practitioners on how to mature BPM ambidexterity.

We proceed as follows: Sect. 2 provides the theoretical background for the study. Next, Sect. 3 describes the methodology, followed by Sect. 4 covering the results of the case study. Section 5 provides a discussion, to conclude in Sect. 6.

2 Literature

2.1 Business Process Management Ambidexterity

BPM can be defined as *"the art and science of overseeing how work is performed in an organization to ensure consistent outcomes and to take advantage of improvement opportunities"* [6 p. 1]. Historically, the discipline was focused on control and rigor, delivering process efficiency, optimization and operational excellence, but not necessarily process exploration and radical innovation [9]. As argued by [10], digital innovation can today obliterate entire business processes, putting into question the BPM's premise of optimizing them. In this context, organizations need to maintain the organizational capabilities to both incrementally and radically improve business processes – which is referred to as BPM ambidexterity [4]. As widely described in academic literature of organizational ambidexterity [11], organizations are advised to devise strategies to resolve the inherent tension between exploration and exploitation through various strategies, which can be grouped into a few categories: 1) structural (creating 'start-up' entities focusing on exploration, structurally separate from the rest of the organization), 2) temporal (strategically focusing the whole business unit on either exploration or exploitation in different timeframes), and 3) contextual (enabling a unit to pursue simultaneously exploration and exploitation) [12]. More recently, there has been evidence that incremental and radical process improvement may be paradoxically reinforcing each other [12, 13]. BPM ambidexterity can be then visualized in a quadrant of focus for exploitative and explorative BPM (Fig. 1).

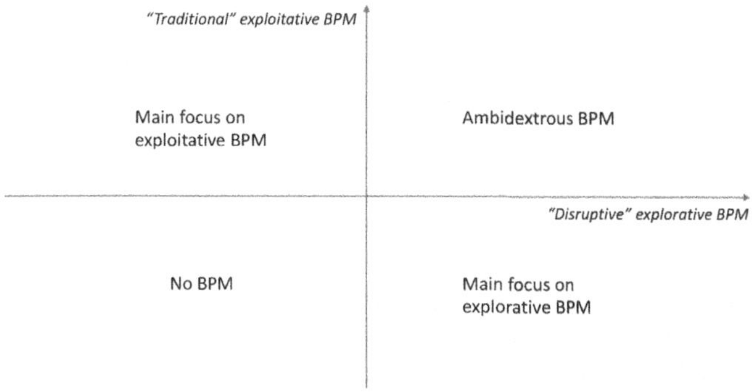

Fig. 1. The quadrant of explorative and exploitative BPM (source: own elaboration).

The successful delivery of BPM initiatives is dependent on several contextual factors related to the characteristics of the organization, its environment, and the process itself [15]. A holistic capability model has been delivered by [7], covering both the process lifecycle capabilities (i.e., modelling, deploying, optimizing, managing) and the broader organizational capabilities enabling BPM, which are defined as Business Process Orientation (BPO) (i.e., including structure and culture). Most recently, [16] have established that a culture supportive of BPM mediates the impact of BPM methods on process performance. Nevertheless, there is still a gap regarding whether a culture supportive of process optimization also supports radical process change [4]. Similarly, there are few guidelines on how to align organizational structures to deliver BPM ambidexterity [5].

2.2 Process Innovation in the Public Sector

While BPM is applied in many organizations of different types, BPM studies mainly focus on large and private organizations [17]. Nonetheless, public sector organizations also profit from improving their processes and from digital process innovation, to better serve their customers (as tax payers). As argued by [15], context awareness is one of the ten principles of good BPM, and [18] provide evidence that the maturity of processes is generally lower in the public sector, which does not face the competitive pressure of the private sector. The vision of process innovation in the public sector has two historical antecedents dating to the 1990s: (1) 'New Public Management' (NPM) [19] and (2) Business Process Reengineering (BPR) [20]. NPM was founded on the themes of disaggregation, competition and incentivization, extending private sector management practices to the public sector, thus creating market-based relationships between departments, and leading to significant contractual outsourcing of services. Previously, BPR already demanded a radical redesign of processes in the early 1990s, with the view of significantly increasing the process efficiency [21]. From the early 2000s onwards, BPM continued to be present in the public sector as a strategic enabler of eGovernment initiatives [22], and, more recently, in digital transformation and digital strategy initiatives [23]. As argued by [24], there are fundamental differences between BPM in the public and private sector, such as differences in the value creation concept (i.e., which includes less profitability concerns in the public sector) and the organizational culture (i.e., which is typically less oriented towards innovation, change and more risk averse). The present paper intends to add an ambidexterity view to this perspective to better understand BPM contingencies in the public sector.

3 Methodology

We conducted a single, but embedded in-depth case study [23, 24] at an EU institution, covering six departments, eights interviews and nine respondents. The units of analysis were departments within this organization [26]. The design was explorative in nature [27], with the objective to better understand the contextual factors influencing the exploitative and exploratory activities for BPM ambidexterity. This case study design allowed us to control certain contextual variables (in particular the environmental and organizational dimensions such as market competitiveness, uncertainty and industry), while at the same time enabling pattern matching between the embedded units of the analysis (i.e., the departments) to identify differences in ambidexterity approaches.

3.1 Case Description

The organization was selected as the first one from several planned for further research, as one of the authors had in-depth access to data and the organization showcased an interesting journey in the adoption of BPM in the public sector. Historically, a major corporate initiative to transform the organization to a process-oriented one took place in early 2010s, driven by the need to deliver tax-saving efficiencies and inter-departmental synergies. This initiative led to a successful optimization and standardization of some core processes (e.g., grant management, procurement), however BPM was deprioritized and discontinued in other departments. Moreover, the organization decided to let BPM re-appear in 2021 as an enabler of the organization's digital strategy, leading to the setup of a Business Process Automation center of excellence. This new endeavor can serve as a reference to assess its coming efforts, for which lessons from the past become crucial as well. Looking at the organization from a contextual point of view [15], we can define the four BPM-related contextual dimensions as visualized in Table 1, explaining which are stable for the entire organization and which are variable across the departments.

3.2 Data Collection

The data collection was guided by data triangulation [25], including interviews, corporate documents (i.e., related to BPM implementations and in-house training materials), wiki reviews, and participation in specific events (e.g., meetings of the Business Process Automation center of excellence and BPM in-house training events). The respondents were EU employees, namely members of the BPM competence centers, process owners and staff of IT departments working on BPM, covering various business domains (Table 2). The departments were selected following the first interview in the center of excellence, to cover higher and lower process orientation (i.e., covering most mature departments in terms of BPM, as well departments with lower levels of process maturity but with significant IT investments).

Table 1. Contextual dimensions [15] of the studied organization.

Goal dimension	Both exploration and exploitation are in scope of this study
Process dimension	There is a variability between the departments with regard to the value contribution (i.e., core and supportive), repetitiveness, knowledge intensity, creativity, and variability
Organization dimension	The organization is large-sized (i.e., 32,000 employees) with high organizational resources. There is variability in the process-oriented culture between the departments
Environment dimension	The level of competitiveness and uncertainty is relatively low (i.e., compared to the private sector)

The interview questionnaire is available online: https://www.dropbox.com/s/fihaocorkg2ns2m/Case%20study%20questions.docx?dl=0. The interviews were carried out from January 2020 until April 2021. The average interview time was 50 min. In total, we collected 127 pages of transcripts, which were analyzed via text analysis using codes or nodes [28]. As the case study was explorative, open coding was carried out by one researcher and validated by another [29]. The first set of questions focused on the traditional exploitative changes, whereas the second set focused on the disruptive, explorative ones, providing the interviewees with clear contextual examples of various dimensions of incremental and radical process change (e.g., by inquiring about whether and how the interviewees dealt with process improvements versus complete process change) [7].

Table 2. Departments interviewed

Departments	Respondent	Role
A	A1	PIT
A	A2	PIT
B	B1	PIT
C	C1	PE
D	D1	PE
D	D2	PE
E	E1	PE
E	E2	PO
F	F1	PE

PIT – Process Improvement Team; PE – Process Executer (IT); PO – Process Owner

3.3 Validity and Reliability

To mitigate risks to validity, we used multiple sources of evidence by means of embedded units (construct validity). We also used multiple measurement frameworks (e.g., the measurement instrument with capabilities for managing business processes by [7], the maturity model for business process orientation by [30], and the model for radical process innovation by [13]) to run the interviews (internal validity). Finally, we followed a case study protocol to build a case study database with a chain of evidence, including all event logs, interview recordings, transcripts and saved documents and communications (reliability) [31].

4 Results

4.1 Triggers and Obstacles for Process Improvement (RQ1)

When asked explicitly, most interviewees initially responded that the public sector is not the place where extensive innovation takes place, which was also mentioned in previous literature on BPM in the public sector [24]. However, after a deeper inspection, it became evident that both incremental and radical innovation had been realized across all studied departments, often in unexpected forms (e.g., by imitating innovation activities from the private sector, or finding creative ways to deal with bureaucracy).

Table 3. Triggers and obstacles for BPM ambidexterity

	Incremental process improvements	Disruptive process improvements
Triggers	• Broad range of examples driven by the business needs and IT opportunities (A, B, C, D, E, F) • Accepted as needed, to keep process automation aligned with business needs (A, B, C, D, E, F) • Bottom-up driven by user needs (A, C, D, E, F)	• Political mandate and legal basis needed for disruptive process change (C, D, E, F) • Top-down vision of management (B, C, D, E) • Opportunities of AI for fraud detection (D, E) • Cross-unit/cross-department coordination needed (A, B, D, E) • Environmental factors (e.g. the COVID pandemic) (E)
Obstacles	• Some stakeholders benefit from status quo (change management) (A, B) • Culture based on legal compliance (B, F)	• Legal basis as the main constraint (C, D, E, F) • Organizational norms and rituals (A, B) • Internal users expect continuity (A, B, E, F)

Department codes are in parentheses.

In general, incremental process improvements were mostly realized as a bottom-up activity, which was often seen as a necessity to keep the processes relevant (B1), and to respond to users' needs. On the other hand, radical process changes were clearly a top-down activity, driven by changes to the legal basis of the departments' activities or by top managers' vision. Disruptive process improvements were perceived as most successful when delivered holistically to the end-to-end process in a cross-departmental way, instead of being done "locally" by a single team. The obstacles to incremental process improvement were often cultural, since some stakeholders were seen to benefit from the status quo. Alternatively, the main obstacle to disruptive innovation appeared to be the constraints of the legal basis, organizational norms and rituals which were adopted over the years, as well as the fact that several end-users prefer continuity of service. The observed triggers and obstacles are visualized in Table 3.

4.2 Ambidexterity Practices Linked to Maturity (RQ2)

Interestingly, the practices for business process exploitation and exploration differed between the departments with higher and lower maturity levels of process orientation [30]. For the purpose of this study, we used the conceptual model for process orientation [30] together with the initial in-depth input from the corporate BPM center of excellence to establish the historical context and to group departments into more and less mature ones. For process exploration, the main success factor for driving process orientation was establishing governance and leadership for BPM on the business side. In the most successful department (E), the process team was reporting directly to the Deputy Director General. This was clearly linked and reinforced by a departmental culture supportive of performance efficiency. On the other hand, the less mature departments had BPM activities primarily driven by IT or isolated process teams, and their culture was usually oriented towards legal compliance, leading to BPM projects stalling beyond the process modelling phase.

When it comes to process exploration and disruptive change, we observed the prerequisite of a clearly scoped and predefined "innovation windows" (D1) where, within the mandate of the department, an opportunity can be created to radically improve business processes and deliver on the political objective of the department. For instance, several departments used AI techniques for fraud detection in various big data sets. The different ambidexterity practices are visualized in Table 4.

4.3 Contextual Factors (RQ3)

Regarding the contextual factors influencing incremental and radical process improvements, we identified two interrelated factors: (1) the core business of the department and (2) the supporting organizational culture [15]. Processes which were: less variable, dealing with financial aspects and thus often audited, and externally focused seemed paired with a culture of efficiency (e.g., in the departments dealing with fund management). This contrasted with departments dealing with processes that showed more variability, were relatively more knowledge intensive or not dealing specifically with financial aspects (e.g., foreign aid, legislation, HR), which were linked to a culture of legal compliance. In such departments, radical digital innovation was typically seen as desirable, however not strategically enabling the goals of the departments. As explained in Sect. 4.2, a culture of efficiency turned out to be a strong trigger of process orientation.

Table 4. Ambidexterity practices for more and less mature departments

	More mature departments	Less mature departments
Exploitation	• Process management driven by the business, and enshrined in the legal basis ("business process driven by design in the regulation itself") (E) • IT systems built on top of business processes (C, D, E) • Culture supportive of performance efficiency (fund management) (C, D, E) • Explicit BPM lifecycle: optimizing processes before automating (D, E) • Organizational structures aligned with end-to-end processes (D, E)	• Process management driven by IT, or isolated teams (A, B, F) • IT systems built on legal basis, business intelligence hardcoded in the IT systems (A, B) • Culture focused on legal compliance (B, F) • Process modelling phase doesn't go beyond process optimization (A, B, F)
Exploration	• Innovation window: use of AI for fraud detection, pattern recognition (D, E) • Radical innovation is embedded in the process teams (contextual ambidexterity) (C, D, E)	• Separate radical innovation units created (structural ambidexterity) (B)

Moreover, departments with a strong process orientation embedded in the organizational culture and structures, were able to deliver radical change much faster, leveraging the underlying processes and structure (e.g. by radically shortening processes), and therefore were more willing to engage in radical change. Finally, some external environmental contextual factors linked to political priorities were the most important driver of radical process improvement (e.g., priorities linked to COVID).

4.4 Solutions to Ambidexterity Tensions (RQ4)

As explained in the Sect. 2, organizations can employ structural or contextual strategies to resolve the tension between exploration and exploitation. In the studied organization, the more process-oriented departments preferred contextual strategies for BPM ambidexterity. This was explained by the need to motivate staff with varied and exciting tasks (D1), and to avoid siloed initiatives with limited impact (E1). We observed an overall greater acceptance of radical change in process-oriented departments. On the other hand, the less process-oriented departments rather preferred to separate structurally radical innovation activities. At the same time, we argue that for the whole of public sector, disruptive process change is most often enabled by external, political or legal changes, which happen rarely, and are separated by periods of stability, and incremental process improvement – which can be defined as temporal ambidexterity.

5 Discussion

This in-depth case study can be analyzed through multiple lenses. First, looking from the Business Process Ambidexterity point of view [4], we can perform a plotting of the studied departments on the dimensions of incremental and radical process improvements (see Fig. 1 – Sect. 2.1). Considering the contextual factors of the public sector, we argue that not all quadrants are achievable or desirable for a public organization, as compared to the private sector:

(1) The ‘no BPM’ quadrant is in the long-term unsustainable, as confirmed by our respondents. In practice, it may lead to a disruptive, radical process change when the processes become too misaligned with the business needs.
(2) The solely ‘main focus on explorative BPM’ state is not desirable for any private or public organization (with the exception of start-ups), as the operational backbone is a prerequisite for successful scaling-up of radical innovations [9].
(3) The ‘main focus on exploitative BPM’ quadrant may be the target state for some departments, since the public sector is less driven by competitive forces. The political drive to deliver efficient, digitalized and user-centric services may indicate that this is not a sustainable state for most departments and units.
(4) ‘Ambidextrous BPM’ is the target state for several departments. We argue that temporal ambidexterity is the most common form in the public sector, as most radical process change is driven by external factors such as legislative changes which happen rarely (e.g., every 7 years for financial legislation). On the other hand, this quadrant remains the preferred state for private organizations, where market forces and increasing digital disruption typically require a combination of incremental and radical process change [32].

The plotting of departments is visualized in Fig. 2, where the values of individual dimensions and the positioning of departments correspond to the concepts of {low, high}. Our studied organization has no department with established processes for explorative or disruptive BPM yet. However, as shown in Sect. 4.2, two departments already have a culture and structure that is more supportive of the traditional, exploitative BPM. This, in turn, provides the ground work for an easier implementation of radical process change, driven by external contextual factors. As argued by [14], process optimization may enable radical process innovation through the growth of the organization’s resources, knowledge capacity and managerial vision, which has been confirmed by a respondent in the department with higher maturity for process optimization (E2).

If we look at the case organization’s journey, the current political priorities (e.g., Green Deal, Europe fit for the Digital Age) will necessitate radical changes in the working methods, so as to deliver greening and digitalization of the institution. Centers of excellence have been set up for innovation, and business process automation with a greater focus on both these capabilities is expected in the coming years.

Based on our interviews, we have derived the following recommendations for public organizations willing to grow in terms of business process orientation (arrow A):

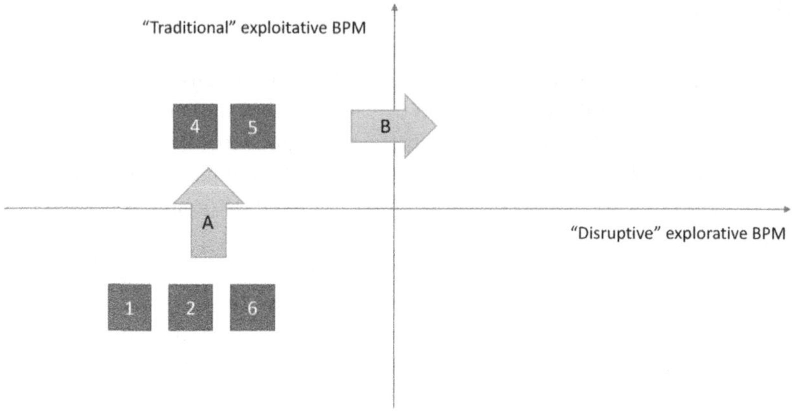

Fig. 2. Mapping of departments on an axis of incremental and radical process improvements

(1) Ensure that the BPM and Digital Transformation activities are driven by the business (not IT), with especially top management support and an appropriate organizational structure;
(2) Ensure early value creation by focusing on a single critical core process with a view of automating it, rather than investing in multiple parallel initiatives. This can be supported by Lean and agile practices in delivering digital solutions based on processes;
(3) Ensure that IT systems are built on business processes by establishing the process owner role in IT project delivery;
(4) Promote a culture of transparency, openness and efficiency.

Radical and disruptive changes in the EU public sector are driven by environmental (political) forces [33]. Practitioners willing to grow organizational capabilities in that domain should consider the following (arrow B):

(1) Growing the incremental process orientation of the organization will help deliver radical changes faster, and with greater organizational acceptance;
(2) Within the mandate and the legal basis of the organization, establish opportunities where radical innovation may deliver significant benefits to the organization's political objectives (e.g., fraud detection using machine learning, automation of processes to support GDPR compliance)

Finally, as argued recently by several researchers [33], contextual factors play a decisive role in the success and effectiveness of BPM initiatives. This was reiterated in our study, where greater business process orientation was linked with cultures and structures supportive of BPM. Based on this case study, we confirm that a culture supportive to BPM also creates a greater acceptance and readiness for radical change [16], which goes in line with the arguments of [14] that managerial vision and resulting culture focusing on optimization, may drive openness towards exploration. Interestingly, we have not observed an 'exploitation bias' [34], which would lead exploitation-oriented organizations to be less disruptive.

6 Conclusion

In this in-depth case study, we have contributed to the body of knowledge by providing a unique view on BPM ambidexterity in the public sector, with particular focus on the contextual factors impacting incremental and radical process innovation. We theorize that a structure and culture supportive of incremental innovation may enable radical innovation. The main implication is a set of guidelines for practitioners, linking capabilities in incremental process innovation with radical process innovation. The main limitation is that the case study only covers a single public-sector organization, with a limited set of departments. We are therefore planning to expand the study to other EU public institutions to validate and broaden the understanding of those contextual factors (including business-unit characteristics), and to provide guidelines to public organizations to further develop their capabilities in process innovation.

References

1. Dijkman, R., Lammers, S.V., de Jong, A.: Properties that influence business process management maturity and its effect on organizational performance. Inf. Syst. Front. **18**(4), 717–734 (2016). https://doi.org/10.1007/s10796-015-9554-5
2. Kohlborn, T., Mueller, O., Poeppelbuss, J., Roeglinger, M.: Interview with michael rosemann on ambidextrous business process management. Bus. Process Manag. J. **20**(4), 634–638 (2014)
3. Beverungen, D., et al.: Seven paradoxes of business process management in a hyper-connected world. Bus. Inf. Syst. Eng. **63**(2), 145–156 (2021)
4. Helbin, T., Van Looy, A.: Is business process management (BPM) ready for ambidexterity - conceptualization, implementation guidelines and research agenda. Sustainability **13**(4), 1–25 (2021)
5. Grisold, T., Vom Brocke, J., Groß, S., Mendling, J., Röglinger, M., Stelzl, K.: Digital innovation and business process management: opportunities and challenges as perceived by practitioners. Commun. Assoc. Inf. Syst. (2021)
6. Benner, M.J., Tushman, M.L.: Exploration, exploitation and process management: the productivity dilemma revisited. Acad. Manag. J. **28**(2), 238–256 (2003)
7. Van Looy, A., De Backer, M., Poels, G.: A conceptual framework and classification of capability areas for business process maturity. Enterp. Inf. Syst. **8**(2), 188–224 (2014)
8. Dumas, M., La Rosa, M., Mendling, J., Reijers, H.A.: Fundamentals of Business Process Management. Springer, Heidelberg (2013). https://doi.org/10.1007/978-3-642-33143-5
9. Stelzl, K.: Organizational ambidexterity – exploring and exploiting the role of business process and project portfolio management (2020)
10. Mendling, J., Pentland, B., Recker, J.: Building a complementary agenda for business process management and digital innovation. Eur. J. Inf. Syst. **29**(3), 208–219 (2020)
11. Lavie, D., Stettner, U., Tushman, M.L.: Exploration and exploitation within and across organizations. Acad. Manag. Ann. **4**(1), 1941–6067 (2010)
12. Raisch, S., Birkinshaw, J.: Organizational ambidexterity: antecedents, outcomes, and moderators. J. Manage. **34**(3), 375–409 (2008)

13. Ng, S.C.H., Rungtusanatham, J.M., Zhao, X., Lee, T.S.: Examining process management via the lens of exploitation and exploration: reconceptualization and scale development. Int. J. Prod. Econ. **163**, 1–15 (2015)
14. Berente, N., Lee, J.: How process improvement efforts can drive organisational innovativeness. Technol. Anal. Strateg. Manag. **26**(4), 417–433 (2014)
15. vom Brocke, J., Zelt, S., Schmiedel, T.: On the role of context in business process management. Int. J. Inf. Manage. **36**(3), 486–495 (2016)
16. Schmiedel, T., Recker, J., Vom Brocke, J.: The relation between BPM culture, BPM methods, and process performance: evidence from quantitative field studies. Inf. Manag. **57**, 103175 (2020)
17. Van Looy, A., Van den Bergh, J.: The effect of organization size and sector on adopting business process management. Bus. Inf. Syst. Eng. **60**(6), 479–491 (2017). https://doi.org/10.1007/s12599-017-0491-3
18. Van Der Aalst, W.M.P., La Rosa, M., Santoro, F.M.: Business process management: don't forget to improve the process! Bus. Inf. Syst. Eng. **58**(1), 1–6 (2016)
19. Hood, C.: A Public Management for all Seasons? Public Adm. **69**, 3–19 (1991)
20. Hammer, M.: Reengineering work: don't automate, obliterate. Harv. Bus. Rev. **68**(4), 104–112 (1990)
21. Halachmi, A., Bovaird, T.: Process reengineering in the public sector: learning some private sector lessons. Technovation **17**(5), 227–235 (1997)
22. Stemberger, M.I., Jaklic, J.: Towards E-government by business process change—a methodology for public sector. Int. J. Inf. Manage. **27**(4), 221–232 (2007)
23. Weerakkody, V., Janssen, M., Dwivedi, Y.K.: Transformational change and business process reengineering (BPR): lessons from the British and Dutch public sector. Gov. Inf. Q. **28**(3), 320–328 (2011)
24. Bandara, W., Gable, G.G., Rosemann, M.: Factors and measures of business process modelling: model building through a multiple case study. Eur. J. Inf. Syst. **14**(4), 347–360 (2005)
25. Recker, J.: Scientific Reseearch in Information Systems (2005)
26. Yin, R.: Case Study Research and Applications: Design and Methods (2017)
27. Benbasat, I., Goldstein, D.K., Mead, M.: The case research strategy in studies of information systems. MIS Q. **11**(3), 369–386 (1987)
28. Darke, P., Shanks, G., Broadbent, M.: Successfully completing case study research: combining rigour, relevance and pragmatism. Inf. Syst. J. **8**(4), 273–289 (1998)
29. Bhattacherjee, A.: Introduction to Research, Social Science Research: Principles, Methods, and Practices (2012)
30. McCormack, K.P.: Business process orientation: do you have it? Qual. Prog. **34**(1), 51–58 (2001)
31. Noble, H., Smith, J.: Issues of validity and reliability in qualitative research. Evid. Based. Nurs. **18**(2), 34–35 (2015)
32. Benner, M.J., Tushman, M.L.: Reflections on the 2013 decade award - 'exploitation, exploration, and process management: the productivity dilemma revisited' ten years later. Acad. Manag. Rev. **40**(4), 497–514 (2015)
33. Manning, N., Lau, E.: Public management reforms across OECD countries. In: Public Management and Governance, pp. 65–80 (2015)
34. Moreno Luzon, M.D., Gil Marques, M., Arteaga, F.: Driving organisational ambidexterity through process management. The key role of cultural change. Total Qual. Manag. Bus. Excell. **25**(9–10), 1026–1038 (2014)

The BPM Governance Supporting Factors and Implementation Barriers – The Experience of a Public University

Piotr Sliż[(✉)], Jędrzej Siciński, Paweł Antonowicz, and Robert Bęben

Faculty of Management, University of Gdansk, Gdańsk, Poland
{piotr.sliz,jedrzej.sicinski,pawel.antonowicz,
robert.beben}@ug.edu.pl

Abstract. The positive impact associated with the implementation of process solutions in private-sector organizations has been signaled in the literature on the subject. From the cognitive perspective, assessment of the vulnerability of BPM and BPM Governance implementation in public sector organizations, with particular emphasis on public universities, is of significance. The research gap, meaning the small number of publications presenting implementation of BPM Governance elements in these organizations, needs to be underlined here. The article's originality lies in the focus on describing the empirical experience associated with strategy reconfiguration and resulting from the implementation of characteristic process organization solutions at a higher education institution. The main aim of this paper is to present the factors supporting and hindering implementation of BPM Governance at a public university. As a result of the research carried out using the methods of systematic literature review and participant observation, a catalog of the factors supporting and rigidifying the implementation of BPM Governance elements was developed. The article additionally describes the Authors' experience in identifying the university stakeholders, the processes architecture, and the formalization of selected processes using authorial IT tools.

Keywords: BPM governance · BPM initiatives · Public university · Public organization

1 Introduction

Public universities and their activity constitute an important element of a society's civilizational development. Accordingly, a university's involvement in the development of world science plays an important role from the perspective of the university stakeholders, as it can allow them to discount the benefits resulting from maintaining such development. In the context of the challenges faced by public universities in Poland, it is crucial to answer the question of whether the current organizational system enables achievement of such a goal. One of the systemic solutions increasing an organization's flexibility in responding to exogenous and endogenous factors entails implementation of Business Process Management (BPM) and the resulting focus of the organization's

A. Marrella and B. Weber (Eds.): BPM 2021 Workshops, LNBIP 436, pp. 153–165, 2022.
https://doi.org/10.1007/978-3-030-94343-1_12

activities on processes by identifying them as the key objects of its structure. This results from the cross-functional and the inter-business nature of BPM [1, 2]. In the aftermath of the dynamic changes taking place in the university environment, such structural features as centralization, specialization and formalization have expanded in a relatively short time, which has rigidified processes by transforming them into procedures. Since the changes were adaptive in character, and the functioning system, including the structural solutions, were modified in an evolutionary and fragmentary manner (in various areas), the efficacy of the organizations' operating systems decreased. The review of BPM and BPM Governance literature, carried out for the purpose of this article, allowed identification of the research gaps (RG). RG1: Small quantity of publications analyzing the implementation of BPM and BPM Governance in public universities. RG2: Scarcity of well-established in the scientific literature theory, illustrating concepts, as well as design and reconfiguration methods for university structural solutions aimed at flexibility (processization). RG3: Scarcity of publications presenting implementation of the integrated information and communication systems supporting the strategic and operational dimensions of university management. RG4: Lack of a set of good practices regarding BPM implementation at public universities. The research problem has been formulated as four research questions (RQ). RQ1: How can BPM and BPM Governance be integrated in a public university at the stage of its strategy and BMP-implementation-aimed activity formulation? RQ2: Which of the BPM Governance elements identified can be implemented at a public university? RQ3: What are the barriers and limitations to BPMG implementation at a public university? RQ4: What steps should be taken, considering the reality of the organizational system functioning at universities, in order to discount the processization outcome using BPM Governance? The article mainly aims at presenting the factors supporting and hindering implementation of BPM Governance at a public university. Secondarily, partial objectives, intersecting at three planes: the epistemological (EO), the methodological (MO), and the utilitarian (UO), were outlined as a result of the research problem posed and the main objective formulated: EO1: Identification of BPM Governance elements and assessment of their possible implementation at a public higher education institution. EO2: Outlining the framework of common-ground BSC, BPM and BPM Governance implementation at a public university. MO: Presentation of BPM and BPM Governance implementation at a public university. UO1: Presentation of a catalog of recommendations for BPM implementation at a public university. In order to achieve the objectives formulated, the methods of systematic subject literature review and participant observation were used. UO2: Development, based on the set of the supporting and hindering factors presented, of utilitarian recommendations regarding reconfiguration of a public university's organizational structure.

2 BMP and BPM Governance in the Public Sector - Theoretical Background

The years of neglect, including the belittling of the role implementation of new management concepts in public organizations plays, have led to an accumulation of a full spectrum of disturbances within the sphere of governance, causing emergence of an aura of 'silent' acceptance for the defective dimension of the organizational structure along with the amassing of a wide spectrum of pathologies, the catalog of which begins, inter alia, with bureaucracy and suppressed, centralized decision-making [3]. The managerial deficiencies that had accumulated over the years in public universities were, in many cases, relatively effectively eliminated by the dominant competitive position of this type of organizations. Private sector organizations, unlike the public sector, focus their activity on obtaining and maintaining competitive advantage in terms of profit maximization [4]. As long as private educational entities did not pose a serious threat to the public entities in Poland, the motivation for pensive and systemic changes in the art of university management was rather modest [5]. The polarity reversal occurred as a result of the dynamic increase in the importance of private universities as well as due to the unfavorable demographic trends and the COVID-19 pandemic, which have virtualized science globally, causing traditional education to be put on hold in at least 120 countries [6]. The aggregated impact of this chain of events has accelerated what might have seemed inevitable, namely the increasingly pronounced tendency to revise the ineffective management concepts applied at public universities. The very idea of process management (BPM) is a concept offering particular streamlining potential. This approach, in recent years, has been gaining increasing approval within the broadly understood public sector, as a potential way to boost the agility and the ability to compete on the progressively decentralized markets [7]. The relatively low popularization of BPM at public entities (particularly at universities) results from numerous doubts and from the generally unjustified fear of potential implementation difficulties. The research diagnosing the BPM implementation potential, carried out in Brazilian state entities, proved that the factor hindering implementation entails, inter alia, the lower-level (linear) personnel's poor awareness, the general unpreparedness, and the lack of willingness to understand the benefits arising from reconfiguring the existing status-quo into a process-oriented dimension [8].

When assessing the BPM implementation potential in public organizations, attention should be given to the aspect public and private entities have in common, i.e., the six core elements of BPM, which affect successful implementation of the process approach [9]. Reflecting on the conclusions formulated by M. Rosemann and J. vom Brocke, the Authors' observations follow. The six factors that are fundamental to the BPM concept constitute the common denominator for both public and private entities. Due to this underlying similarity, public organizations should not be stigmatized as those unable to fully implement BPM. This observation should induce revision of the opinion treating public entities, virtually in a top-down manner, as impermeable and 'difficult' objects of potential processization. This mental simplification is, according to the Authors, a view that impoverishes diffusion of scientific and economic practice. This contributes to the slowing down in public entity (here: university) improvement, and thus to this sector's low propensity for implementation of the process approach, which in turn results in a decrease in the quality of the services provided.

The growing interest in this method of management is currently noticeable among the representatives of both the academic community and the business practice [7, 10]. Despite the raising interest in the BPM method, signs of problems associated with the incomprehension of this organizational management method can be noticed [2, 11]. Such a state of the matter results in unsuccessful attempts to implement BPM, which may result from the lack of understanding of the BPM framework assumptions, as well as from the orientation of management activities on the ad hoc and atomized initiatives that are aimed at improvement of individual processes or their stages, without viewing the problem through the prism of the entire organization [12, 13].

As emphasized in the works [1, 2, 7, 14–18], BPM Governance is identified as a critical factor determining the success of the BPM initiatives carried out in an organization. A.F.L. Santana et al., based on a systematic literature review of the works [14, 16, 18–24], propounded a catalog of BPM elements, which includes: business process standards, business process roles and responsibility, business process objectives, tasks, assessment methods, control methods, governance structures, as well as architecture and infrastructure [7].

Summing up, the fact that the research questions formulated are clearly related to the needs reported in the contemporary literature on the subject constitutes the strong point of the study. These needs are related, among others, to the issue of BPM implementation outside the business sector, since its application potential is noticeable, inter alia, also in in the field of education [25].

3 Materials and Methods

3.1 Research Object

The object of the empirical research is one of the largest public universities in Poland. Its organizational structure includes 11 faculties, representing 21 scientific disciplines that are subject to periodic parametric evaluation for scientific quality (excellence) assessment purposes. The university's structure is also expanded by satellite units, including its own library, a publishing house, and special purpose vehicles established to commercialize the research workers' knowledge. The university employs nearly 3 000 persons, 60% of which are research and teaching personnel. The university's strategic goals have been defined within four perspectives addressing its mission via assurance and development of: (1) the highest level of university education; (2) the highest quality of scientific research and innovation; (3) openness, social responsibility and the university's commitment to, inter alia, cooperation with the socio-economic environment; and (4) the highest level of university managerial efficiency. The university's management activity defined necessitated, in particular, implementation of a management methodology that would grant highest priority to process management.

The study entailed a non-random sampling technique, incorporating deliberate choice, used for selecting the study participants. 230 university employees took part in the procedure. The selection criterion primarily concerned the competences and the knowledge associated with the activities carried out as part of the processes identified.

3.2 Characteristics of BPM and BPM Governance Implementation at a Higher Education Institution

The article's empirical layer presents the stages of process solution implementation, using the example of a multi-faculty university in Poland, as well as outlines the directions of further activities aimed at increasing the level of process maturity and BPM

Governance implementation. The propounded and described BPM and BPMG implementation methodology propounded and described has been formulated based on the current state of the work on modifying the university's strategy and implementing the BPM solutions and the BPMG elements identified as the factors supporting its processization. Appropriately, the project team responsible for the BPM implementation formulated 8 transition stages constituting the authorial BPM implementation methodology propounded:

1. Formulation of a new strategy for the application of a system of measures and the execution of BSC implementation,
2. Identification of the university process's stakeholders,
3. Verification, in cooperation with the process implementers, of the university process architecture identified and the grouping of the processes into X areas,
4. Identification of the degree of digitization, the improvement potential, and the complexity (number of the functional areas involved in the process implementation) of the university's processes. Identification of processes requiring reconfiguration based on the evaluation criteria adopted,
5. Recreation of the current organizational structure and roles, as well as assessment of the capacity for reconfiguration of the structural solution towards processes,
6. Analysis of the identified processes' trajectory through the functional areas (the university's organizational units),
7. External and internal identification of the suppliers and the clients (recipients) in the processes identified,
8. Formalization of the process group selected.

The above description of the activities derives from the consultations and the workshops carried out by the Authors at the university under the supervision of the Rectoral team. It constitutes a proposal indicating the activities aimed at gradual (evolutionary) processization of a public organization. The sequence of actions adopted in such fashion allows, inter alia, to mitigate the barriers associated with the excess of the administrative structures and the human factor's resistance, as well as to gradually change the organizational culture. It, of course, entails a multi-year process. This process has been - owing to the co-development, in cooperation with the representatives of all organizational units, of new process solutions - strengthened through the use of the participatory model. The study carried out was of action research character (it combined research with broad reflection on the practice) [26].

4 Results and Discussion

4.1 Characteristics of BPM, BOMG and BSC Implementation Stages

Formulation a New a New Strategy for the Organization's Processization and the Implementation of a Measure System

When formulating a new strategy for a public higher education institution, one should take the barriers resulting from the current system of functioning (including the process architecture) into account, including the degree of centralization, formalization, and organizational-structure specialization. This means that reconfiguration of the strategic solutions calls for trainings and employee activation in the terms of their comprehension of the changes being introduced. Conversely, within the dimension of the functioning system, the main element of which is the process, it is important to consider the role of the stakeholders who, through the structure of their expectations, should have real impact on the shape of the process effect, and indirectly on the course thereof. Considering the public space of universities, this mainly applies to the clients, in internal terms. In order to perform such an assessment, it is, at first, necessary to: diagnose the current state of organization (e.g., through the so-called mapping and prioritization of the processes being implemented); identify and assess the capacity for the use of the good practices developed in this area; take the goals and the expectations (internal and external) that have been set for the organization into account.

Identification of the University's Process Stakeholders/Stakeholder Engagement
In order to effectively use the knowledge accumulated in the organization, with the aim to increase the level of the process recipients' satisfaction with the way their needs are met, a diversified group of stakeholders was invited to participate in the diagnostic and design work. As not to omit any of the important processes being implemented at the organization, a list of processes was developed in cooperation with intentionally selected (indicated by the Rectoral team) representatives of various organizational units, who had the knowledge of and experience with the activities carried out at the university. The analysis of the processes being carried out was preceded by a thorough breakdown of the process stakeholders, so as to enable both the process recipients and the implementers, as well as any other persons who have impact on the course of a given process, to be invited to participate in the assessment. Due to the specificity of higher education institutions' activity, identification of the processes implemented by the unit was carried out with regard to the following eleven areas: (1) scientific corpus management; (2) teaching body management; (3) education of the future; (4) finance, administration and budgeting; (5) infrastructure management; (6) human resources; (7) client – the students; (8) client – the society; (9) client – the economy; (10) promotion; (11) internationalization.

Process Identification and Prioritization at the University Under Examination
The process identification stage was divided into three substages. In the first stage, the project team attempted to identify the main and the auxiliary processes implemented at the University of Gdańsk. An attempt was then made to assess the degree of interaction between the processes within the architecture identified. For this purpose, a matrix was created, in which the impact of the process A outputs on the process B inputs was assessed. In this substage, a 5-point measurement scale was used (0 - a given process has no impact on another; 1 – a given process has little impact on another; 2 – a given process has average impact on another; 3 – a given process has above-average impact on another; 4 – a given process has significant impact on another; 5 – a given process has great impact on another). Subsequently, based on the data obtained, an attempt was made to identify those process pairs, which are characterized by the highest dependency. This procedure allowed for initial reconstruction of the process architecture and enabled discussion about the blending of processes or the division thereof into smaller ones. By analyzing

the anonymized expert assessments, which determine the value of individual process pairs, the processes that may be of the greatest significance for the implementation of other processes were identified. Simultaneously, a multi-criteria method of process evaluation was developed as part of the diagnostics, incorporating the following criteria: (1) the number of employees involved in the implementation of a given process, the process significance, measured by the number of the persons it directly concerns; (2) the degree of process formalization; (3) the process digitization potential, with particular emphasis on electronic document circulation; (4) the level of the domain solution's independence from the system solutions; (5) the potential to improve the processes, the paste of change implementation (achievement of an acceptable level); (6) the number of the organizational units involved in the process implementation.

Process Formalization

The stage associated with the formalization of the processes identified was preceded by identification of the group of the processes characterized by the highest potential for digitization and implementation of electronic document circulation, and by the highest complexity, defined by the number of the functional areas implementing a process. Consequently, two processes were qualified for further work: the handling and settlement of the purchases made under and outside the provisions of the Public Procurement Law (P1), as well as the University-wide planning and allocation of the teaching loads (P2). It should be noted here that during the work, attempts were made to reconfigure the processes into an end-to-end category. Both the functional area representatives who declared implementation of these processes as well as the process recipients (external and internal clients) indicated were invited to participate in this stage. First, the real process proceedings were recreated using a BOC Adonis tool. Following the study results, a set of supporting factors and those hindering the implementation of selected BPM Governance elements at the university under examination is presented in Table 1.

As indicated in Table 1, implementation of BPM Governance solutions in a public university is primarily limited by the lack of prior activities that would be aimed at implementing BPM. On the structural solution level, despite the high activation of the university employees and the recognition of the necessity to implement changes, as confirmed by the study, that would enable a flexible response to the external and internal factors, the highly hierarchical functional organizational structure constitutes a limiting factor. The lack of previous activities aimed at BPM implementation did not result in the focus on the effects of the processes and their compliance, in terms of external and internal aspects, with the client's needs. What is more, the faculties' high autonomy causes BPM implementation at a multi-faculty university to be problematic, due to the lack of full strategic managerial support for BMP. Faculties maximize their own goals at the operational and tactical levels, rather than the strategic level of the entire university. The activities associated with the verification of the research-generated effects (scientific) and the didactic processes were usually carried out as a result of the inspections conducted by the decision-making bodies (e.g., the ministry). This has made the necessity to implement these processes in accordance with predefined standards quite obvious, which in turn resulted in an increase in the level of process standardization and in occasional measurement thereof (measurements performed for the needs of the

Table 1. Factors supporting and hindering BPM Governance implementation at a public university

Elements of BPMG	Supporting factors	Hindering factors
Business process standards	• University authorities' openness to the transfer of the good BPM and BPMG practices derived from the private sector • Comprehension of the process standardization level's gradeability that is dependent on the prerogatives of the implementers	• Lack of BPM training • Incorrect understanding of the concept of a process and its identification with such notions as a procedure or a function
Business process roles and responsibility	• Activation of employees in the context of the process improvements proposed • Comprehension of the shortcomings resulting from the high hierarchization of the university's organizational structure	• High level of specialization • The phenomenon of area sub-optimization in relation to the organization • Failure to identify the customer-supplier relationship in internal approach
Business process objectives	• The implementation of process solutions has been included in the university's new strategy	• The strategic goals did not concern the processes implemented previously • Lack of a measurement system for university processes
Assessment methods	• Application of the knowledge accumulated in the organization enables elaboration of the indicators reflecting the essence of the changes taking place in the organization	• Lack of measurement of the added value generated in the processes
Control methods	• The search for solutions aimed at achievement of the requirements designed by decision-making bodies (the Ministry, etc.)	• Lack of identified management processes • Lack of a self-diagnosis mechanism

(continued)

Table 1. *(continued)*

Elements of BPMG	Supporting factors	Hindering factors
Governance structures	• Support for interdisciplinary forms of work • Employees' openness for sharing the knowledge regarding the activities implemented	• The main object of organizational structure building is departments, rather than teams • Lack of a Process Management Office unit and organizational process roles (e.g. process owner) • High structure hierarchy
Architecture and infrastructure	• Implementation of an ERP tool • Recognition of the need for digitization • Preparedness for adaptation of the IT department (the role of programmers) to the current university processization activities	• Diffused decision-making centers use domain-specific IT tools that are not complementary to other systems

Source: own elaboration.

controlling institutions). One important factor supporting the BPM Governance implementation entailed the understanding of the need for digitization and an ERP class tool implementation on the part of the university authorities and employees. It should be emphasized here that, within the IT dimension, activity should be undertaken to replace the scattered IT tool and develop a university management IT system. Figure 1 shows the actual cognizance of the processes at the university under examination.

Fragmentary perception of the processes is noticeable on the part of the implementers. It should be underlined here that the functional organizational structure constitutes an important hindering factor in the BPM and BPM Governance implementation. Figure 2 shows the work objective, aimed at changing the scope of the implementers' process identification, taking a holistic outlook on the entire process into account, with focus on identifying the suppliers and recipients, in external and internal terms, including the understanding of the significance of the process outputs generated, which in turn power other processes in the organization.

The plan for further actions assumes conduction of trainings, modification towards reduction of the functional areas implementing processes, increased prerogative of the process implementers, and digitization of the activities characterized by a high level of repeatability and standardization (e.g., document approval in the process of public procurement, using electronic document circulation).

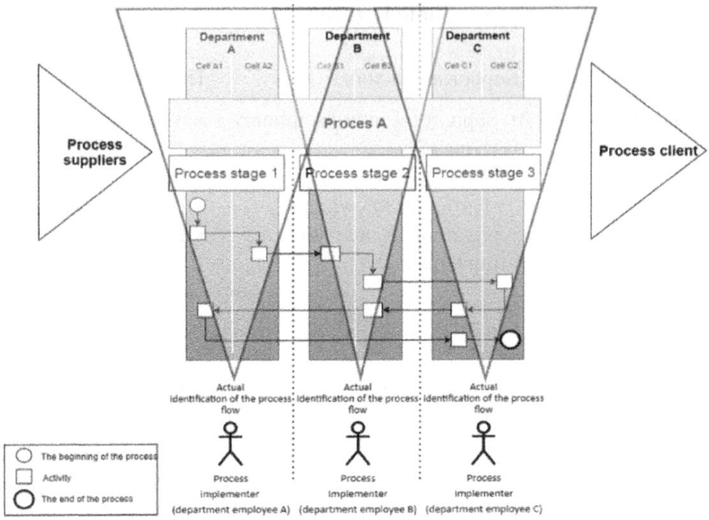

Fig. 1. Diagram of the deficiencies in the current management formula against the implementation of process organization elements

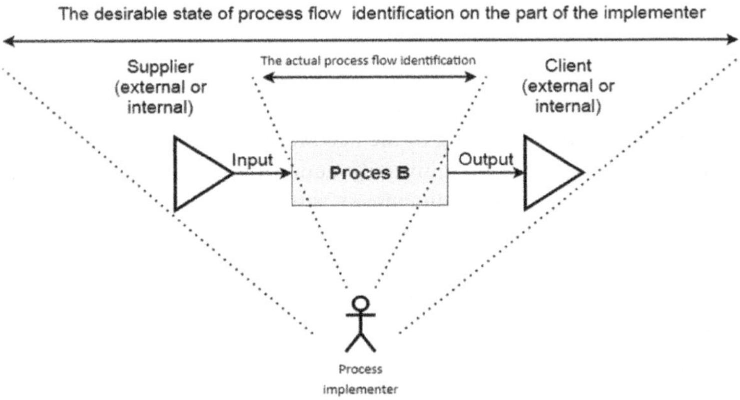

Fig. 2. Identification of the actual and the desired cognizance of a process and its suppliers and recipients

5 Conclusion

The study's partial aim was to prove that these limitations are often apparent, because the similarities existing between private and public entities (in terms of the BPM approach application) are not characterized by such a high dispersion. According to the Authors, the element that binds the public and the private spheres, in terms of process approach applicability, entails the Six Core Elements of the BPM model. The Authors' experience with BPM management and BPM Governance implementation, described in this article, has been outlined within the scope of five generalizing conclusions: First, identification and reduction of the proceedings in the processes (elimination of double-executed activities and elimination of excessive decision gates) allows for exclusion of many intermediate decision-making links, which reduces the number of the decisions made in the organization, which, in turn, positively affects the process runtime. Secondly, the phenomenon of responsibility dispersion (and avoidance) can be observed in public entities. In order to minimize the responsibility for potential errors, processes are often migrated between the cells, which repeatedly approve individual stages of implementation - this causes the blurring of the responsibility for errors, in the event of problems. The last conclusion: recognition of the need to parallelly implement IT and communication techniques as well as ERP-class tools, with the purpose of data collection and exploration, including electronic document circulation. This means that achievement of a state in which conscious implementation of an ERP tool follows from the actions aimed at increasing the level of the university's process maturity. The above-presented results of the empirical procedure reflecting the activities that have been carried out so far with the aim of implementing BPM and BPM Governance at universities exhibit, just as other studies, certain limitations. The lack of a project activity stage that would be aimed at identification of the degree of a higher education institution's exposure to BPM and BPM Governance implementation has been classified as such a limitation. The limitation identified constituted an incentive for designation, within the area of the issues discussed, of a new research direction associated with the design of a descriptive model of maturity, adapted to the specificity of the organization analyzed and enabling objectified assessment (based on the symptoms of processability) of the current degree of maturity. Despite the fact that this stage was not taken into account prior to the process identification and formalization in the organization under analysis, in the Authors' opinion, the supporting and hindering factors identified in this article allow for better formulation of such a model. What is more, the Author's previous experience with BPM implementation at a public university has delineated another research direction associated with implementation of BPM ambidexterity assumptions, including consideration of processes in operational and exploratory distribution. Within the context of social implication, the fact that the solution proposed pertains to a wide range of beneficiaries constitutes an undoubted advantage. The changes developed through dialogue lead to an optimization of the domain solutions, which in turn results in the reduction of costs or an increased work efficiency and quality. The general society, which in many countries (including Poland) finances the higher education sector, thus benefits from the increased efficiency and quality of university operation. The internal stakeholders, such as the employees, who improve the conditions of their work by co-developing new solutions and ameliorate the processes in which they participate, must be taken into consideration as well. As such, a culture of cooperation develops, which can also bear fruit in other areas. Other beneficiaries here are the stakeholders, i.e., the students, business environment institutions, entrepreneurs whose perspective (as the recipients of the processes reconfigured) constitutes the key element of the changes designed.

References

1. Rosemann, M., vom Brocke, J.: The six core elements of business process management. In: vom Brocke, J., Rosemann, M. (eds.) Handbook on Business Process Management 1. IHIS, pp. 107–122. Springer, Heidelberg (2010). https://doi.org/10.1007/978-3-642-00416-2_5
2. Markus, M.L., Jacobson, D.D.: Business process governance. In: vom Brocke, J., Rosemann, M. (eds.) Handbook on Business Process Management 2, 1st edn., pp. 201–222. Springer, Heidelberg (2010). https://doi.org/10.1007/978-3-642-01982-1_10
3. Martin, B.R.: What's happening to our universities? Prometheus **34**(1), 7–24 (2016)
4. Colley, C.L., Doyle, J.L., Logan, G.W., Stettinius, W.: What is Corporate Governance? McGraw-Hill Companies Inc., USA (2005)
5. Woźniak, P.: Podejście procesowe w systemie zarządzania uczelnią. Konferencja Innowacje w Zarządzaniu i Inżynierii Produkcji, Polskie Towarzystwo Zarządzania produkcją (2014). http://ptzp.org.pl/files/konferencje/kzz/artyk_pdf_2014/T2/t2_481.pdf. Accessed 26 Mar 2021
6. Shahzad, A., Hassan, R., Aremu, A.Y., Hussain, A., Lodhi, R.N.: Effects of COVID-19 in E-learning on higher education institution students: the group comparison between male and female. Qual. Quant. 1–22 (2020)
7. Santana, A.F.L., Alves, C., Santos, H.R.M., Felix, A.L.C.: BPM governance: an exploratory study in public organizations. In: Halpin, T., et al. (eds.) BPMDS/EMMSAD-2011. LNBIP, vol. 81, pp. 46–60. Springer, Heidelberg (2011). https://doi.org/10.1007/978-3-642-21759-3_4
8. Valenca, G., Alves, C.F., Santana, A.F.L., de Oliveira, J.A.P., Santos, H.R.M.: Understanding the adoption of BPM governance in brazilian public sector. In: ECIS 2013 Completed Research, p. 56 (2013)
9. Rosemann, M., vom Brocke, J.: The six core elements of business process management. In: vom Brocke, J., Rosemann, M. (eds.) Handbook on Business Process Management 1. IHIS, 2nd edn., pp. 105–122. Springer, Heidelberg (2015). https://doi.org/10.1007/978-3-642-45100-3_5
10. Syed, R., Bandara, W., French, E., Stewart, G.: Getting it right! Critical success factors of BPM in the public sector: a systematic literature review. Australas. J. Inf. Syst. **22**, 1–39 (2018)
11. Niehaves, B., Plattfaut, R., Budde, M., Becker, J.: Business process governance: theorizing and empirical application. In: Proceedings of the Seventeenth Americas Conference on Information Systems (2011)
12. Trkman, P.: The critical success factors of business process management. Int. J. Inf. Manag. **30**(2), 125–134 (2010)
13. de Bruin, T., Rosemann, M.: Using the Delphi technique to identify BPM capability areas. In: ACIS 2007 Proceedings, p. 42 (2007)
14. Korhonen, J.: On the lookout for organizational effectiveness – requisite control structure in BPM governance. In: 1st International Workshop on BPM Governance – WoGo (2007)
15. Bandara, W., Indulska, M., Sadiq, S., Chong, S., Rosemann, M., Green, P.: Major issues in business process management: an expert perspective. Department technical report. School of Information Technology and Electrical Engineering, The University of Queensland (2007)
16. Jeston, J., Nelis, J.: Business Process Management – Practical Guidelines to Successful Implementations, 2nd edn. Elsevier/Butterworth-Heinemann (2008)
17. Ravesteyn, P., Batenburg, R.: Surveying the critical success factors of BPM-systems implementation. Bus. Process. Manag. J. **16**(03), 492–507 (2010)
18. Rosemann, M., de Bruin, T.: Towards a business process management maturity model. In: Proceedings of the 13th European Conference on Information Systems, ECIS (2005)

19. Harmon, P.: BPM governance. BPTrends **3**(3) (2005). http://www.bptrends.com/publicationfiles/bptemailadvisor020805.pdf. Accessed 26 Mar 2021
20. Harmon, P: Best practices in the governance of business process management. BPTrends 1–23. http://www.bptrends.com/publicationfiles/05-25-05BPT1HrTalkforIQPCBPMConf-Harmon.pdf. Accessed 26 Mar 2021
21. Kirchmer, M.: Business process governance for MPE. In: Kirchmer, M. (ed.) High Performance Through Process Excellence From Strategy to Operations, pp. 69–84. Springer, Heidelberg (2009). https://doi.org/10.1007/978-3-540-77825-7_5
22. Spanyi, A.: Business process management governance. In: vom Brocke, J., Rosemann, M. (eds.) International Handbook on Business Process Management 2 - Strategic Alignment, Governance, People and Culture, pp. 223–238. Springer, Heidelberg (2010). https://doi.org/10.1007/978-3-642-01982-1_11
23. Richardson, C.: Process governance best practices: building a BPM center of excellence. BPTrends (2006)
24. Paim, R., Nunes, V., Pinho, B., Santoro, F., Cappelli, C., Baião, F.A.: Structuring a process management center of excellence. In: Proceedings of the 2009 ACM Symposium on Applied Computing - SAC 2009, p. 281. ACM Press, New York (2009)
25. Tuček, D., Basl, P.J.: Using BPM principles to increase the efficiency of processes in higher education in the CR, p. 47 (2011)
26. Somek, B.: Action Research: A Methodology for Change and Development, p. 6. McGraw-Hill Education, Open University Press, England (2006)

About a Process-Technology Fit for Process Improvements in an Ambidextrous Environment

Tahir Ahmad(✉) and Amy Van Looy

Faculty of Economics and Business Administration, Department of Business Informatics and Operations Management, Ghent University, Tweekerkenstraat 2, 9000 Ghent, Belgium
{Tahir.Ahmad,Amy.VanLooy}@UGent.be

Abstract. Since business process management (BPM) can become more disruptive due to new technologies, both explorative and exploitative BPM approaches are being combined to improve business processes. This combination of approaches refers to the notion of ambidexterity. The question, however, remains which emerging technology is appropriate for which type of process improvement. Therefore, this research is designed to translate the original task-technology fit (TTF) theory into a process-technology fit (PTF) theory by also acknowledging an ambidextrous BPM environment. We have derived an initial theoretical model from existing theories and frameworks. Afterwards, we have conducted an expert panel with a content analysis of 19 interviews. Our findings demonstrate ambidextrous BPM practices, and uncover what it means when processes fit or do not fit certain technologies. Novelty resides in translating TTF into PTF and BPM ambidexterity, and so contributing to BPM's theoretical underpinnings. In addition, we provide more insight into related practical issues.

Keywords: Process-technology fit · Process improvement · Process innovation · Digital innovation · Ambidexterity

1 Introduction

Since emerging technologies (e.g., artificial intelligence or AI, blockchain, Internet of Things or IoT) have the potential to influence the nature of business processes, an organization's information technology (IT) adoption becomes a critical factor for business process management (BPM) success [1]. This IT adoption can be within a business process (i.e., for specific process tasks) or in the entire organization (i.e., a process portfolio) with disruptive business models. An IT adoption, however, heavily relies on the technological characteristics or features related to particular technologies. For instance, specific technologies can have a different degree of technology disruptiveness. Nonetheless, since IT can bring innovation into the BPM discipline (i.e., leading to radical and/or continuous process improvements), new technologies enable an organization to rethink its legacy to new ways of performing business processes [2].

Previously, the task-technology fit (TTF) theory has been used to understand a fit between IT and process tasks, but not for aligning emerging technologies with BPM.

A. Marrella and B. Weber (Eds.): BPM 2021 Workshops, LNBIP 436, pp. 166–178, 2022.
https://doi.org/10.1007/978-3-030-94343-1_13

More specifically, BPM is getting increasingly digitalized and is adopting more agile methods, but BPM practitioners are not necessarily accepting such new technologies [3]. Because these technologies open new gateways for BPM practitioners to improve their business process outputs, there is a dire need of timely selecting the best fitting technology to optimize a process improvement's efficiency. Secondly, an alternative perspective that profits from finding a fit between IT and BPM is called ambidexterity [4]. From an ambidexterity point of view, BPM requires both exploitative and explorative BPM approaches for finding a balance between incremental process changes and more drastic process innovations [5]. Nonetheless, finding the appropriate technological benefits in an ambidextrous BPM environment requires further investigation as well. To address both gaps, we intend to answer the following research questions:

RQ1: Which factors explain a process-technology fit?
RQ2: How can BPM ambidexterity facilitate a process-technology fit?

Our purpose is to translate the task-technology fit theory to the BPM discipline in two ways, namely (1) by converting the task characteristics into process characteristics, and (2) by respectively positioning incremental and drastic process improvements as exploitative and explorative efforts in the ambidexterity theory. Therefore, we use a qualitative research approach by conducting expert panel interviews (i.e., experts with experience of process innovation projects). The findings are based on a content and conceptual analysis to develop the related nodes and themes. This work has scientific relevance by aligning the constructs with the dominant frameworks and theories in order to strengthen BPM's foundation. Furthermore, BPM practitioners can profit from our findings and the process improvement expertise of our expert panel.

2 Research Background and Theoretical Framework

The notion of BPM has been derived from the business process re-engineering and workflow management waves in the 1990s, as well as the earlier quality management tradition with total quality management (TQM). The contemporary BPM literature offers several maturity models, success factors and principles of BPM [1]. In addition, BPM typically follows a life cycle approach with dedicated methods and techniques, which starts from process identification and process discovery, followed by process analysis, process design, process implementation, and process monitoring and controlling [6]. Although BPM's main focus has been on process modelling, mapping and ontologies, it is now being used across diverse sectors such as healthcare and human resources [7], as well as across diverse trends such as green BPM [8] and agile BPM.

The notion of digital innovation is based on creativity and disruptive business models, and refers to the innovation of products, services, business processes and business model by making use of IT's full potential. Product and service innovations help organizations to sustain in competitive business environments. This brings us to the concept of business process innovation, which combines the concepts of BPM and digital innovation. It refers to new (e.g., smart) business processes that are particularly designed and executed to bring change in the organizational way of working [9]. Since emerging technologies can be used across the different BPM life cycle phases, new business models are being developed such as Uber and Airbnb [10]. These are only a few examples of how digital process innovation can change an organization's way of working and its strategic business ideas. Thus, business process innovation is generally seen as a promising factor for stimulating sustainable growth in the market by aligning emerging technologies with business activities on the operational level.

Subsequently, we explain the theoretical lenses underlying our research model.

2.1 Task-Technology Fit Theory

TTF claims that business performance will be enhanced if IT capabilities match with specific tasks, or vice versa, if task properties match with the characteristics of the supporting IT [11]. Since each business process comprises many tasks and requires an alignment with supporting technologies, technology adoption (i.e., by organizations) and acceptance (i.e., by employees) are key factors in achieving a task-technology fit [12]. Task non-routineness, interdependence and receptiveness are examples of task characteristics. On the other hand, technology characteristics are dependent on the information systems or software used [11]. As such, we derived the factors of process characteristics, technology characteristics and fit to be used in our study. BPM capabilities also mediate positively between ambidexterity and process performance [13].

2.2 Ambidexterity Theory

Many of today's organizations are shifting from older technologies and traditional process-related methods and techniques to novel (e.g., more disruptive or agile) alternatives. Since rapid changes are difficult to achieve, the optimal option is to combine the strategies of exploitation and exploration. This idea is documented in the ambidexterity theory, which also defines incompatibilities between explorative and exploitative tasks [14]. Explorative BPM helps realize new business process models based on disruptive technologies (e.g., bitcoins), whereas exploitative BPM relies on continuity in existing paths (e.g., traditional BPM methods and techniques for Six Sigma or Lean). When an organization balances both strategies, it is said to be an ambidextrous organization [2]. We derived ambidexterity as an additional factor in our study, next to the process and technology characteristics. The readiness of business processes for accepting ambidexterity is vital [15], as ambidexterity enhances overall performance [16]. [17] suggested to prioritize strategies and to find a balance between explorative and exploitative BPM.

2.3 Business Process Management (BPM) Context Factors

Since the application and scope of BPM can vary across sectors and departments, [18] have described four dimensions of circumstantial or contextual factors related to BPM, namely related to: (1) goals, (2) processes, (3) environments and (4) organizations. First, BPM success strongly depends upon the tasks to be performed and the organizational goals, which can aim towards innovation and/or incremental improvements. Secondly, process characteristics include repetitiveness, knowledge-intensity, creativeness, value contribution, interdependence and variability. Environmental factors can also affect a process-technology fit by means of competitors and stakeholders. Since the adoption of new technologies can help gain a competitive advantage, it is important to scan the business environment and develop a market portfolio. Hence, risk and uncertainty are influential factors within an organization's business environment to achieve a successful BPM implementation [19]. Finally, BPM success is based on organizational factors, such as size and resources.

2.4 Other Theories Related to 'Fit'

To gain more insight into the 'fit' factor, we have looked at technology acceptance theories. First, [20] presented the initial technology acceptance model (TAM), which has been used abundantly in information systems (IS) research. Especially the perceived ease of use (P.E.O.U.) and perceived usefulness (P.U.) are important behavioral factors from a user perspective, such as BPM practitioners, to achieve a process-technology fit. Alternatively, IT characteristics such as complexity, compatibility and relative advantage are defined in technology adoption theories [21], while technological readiness can predict whether an organization's technical capacity is ready to adopt a new technology [21]. Furthermore, we have relied on the ten principles to realize BPM success [1]. While context awareness has been discussed in Sect. 2.3, especially the principles of continuity and technology appropriateness are additional indicators to find the best fitting technology for a specific business process because they literally refer to different strategies for incremental and radical process improvements by IT.

2.5 Performance Metrics (or Fit Outcomes)

Although process performance indicators depend upon the business process type (e.g., task intensity) and organizational context (e.g., size and structure), it is generally accepted that process performance can be measured on four criteria based on cost, time, flexibility and quality [22]. Similarly, we use a generally accepted framework to measure organizational performance, namely the balanced scorecard (BSC) [23].

2.6 Initial Research Model

Based on this background, we have derived an initial research model (Fig. 1).

While process and technology characteristics are highly important to find a process-technology fit, Fig. 1 adds ambidextrous BPM to refine this interplay. Environmental and organizational factors have moderation effects (Sect. 2.3). Finally, PTF affects process performance and organizational performance, in line with BSC.

3 Methodology

We conducted an expert panel to explore the impact of factors (RQ1) related to our initial research model (Fig. 1) and the added value of BPM ambidexterity in PTF (RQ2).

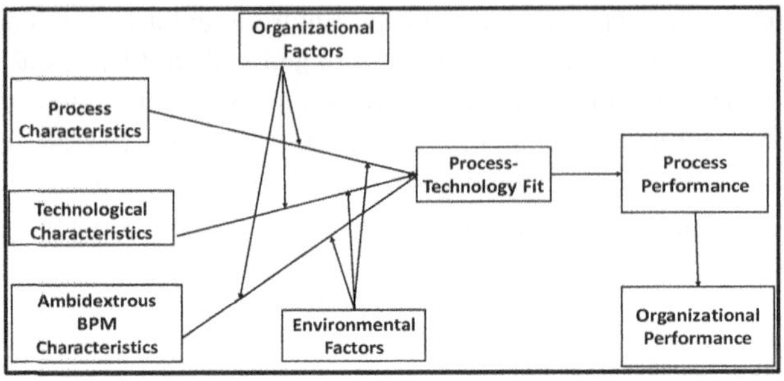

Fig. 1. Our initial research model for a Process-Technology Fit (PTF).

3.1 Selection of Experts

We organized 21 interviews with international BPM consultants, of which two incomplete interviews were dropped. The final sample thus contained 19 interviews. We did not target a specific industry or location, because our focus was on the themes rather than the industry. In a purely purposive sampling, we selected only BPM consultants because they deal with multiple clients and they are typically more aware of the emerging technologies to be used in a BPM environment. Our expert profiles included business process analysts, business process specialists, business process managers and business process re-engineering managers. More information about the expert profiles is available at: https://1drv.ms/x/s!Ahs8XtPFzYl1j1l9jttmp5AoDWJL?e=v9GpRV/.

The BPM consultants were found through LinkedIn profiles and then contacted for a skype meeting. For reasons of generalizability, our experts belonged to both developed countries (i.e., in Western Europe, Eastern Europe, United States) and developing countries (i.e., in Asia). We assigned an anonymous ID per respondent (Rxx).

3.2 Operationalization of Variables

Based on Fig. 1, we developed a questionnaire based on the eight main constructs and in accordance to the prior theories and frameworks of Sect. 2. The eight constructs were further operationalized across 40 sub-dimensions observed in the literature. The questionnaire contained two sections, namely starting with five demographic questions, followed by 45 questions related to the sub-dimensions. The questions were developed on the basis of describing the conceptual meaning, and finding the occurrence of the constructs. The questionnaire is available at: https://1drv.ms/w/s!Ahs8XtPFzYl1j1bFwhZn35uToeix?e=cSUDph/.

3.3 Data Analysis

All interviews were recorded and fully transcribed [24]. In total, 344 nodes were developed against the sub-dimensions (i.e., first round as a content analysis), which we subsequently linked to themes (i.e., second round as a conceptual analysis). In the first round, the 344 nodes were coded manually by looking at each response separately. In the second round, all nodes were converted into higher-level themes using a conceptual analysis, during which presence and frequency were registered in summary tables.

4 Results

4.1 Themes for Process and Technology Characteristics

Table 1 shows the sub-dimensions of the process and technology characteristics.

Table 1. Process characteristics and technological characteristics (expert count).

Repetitive tasks(14) Non-repetitive tasks(5)	Knowledge-intensive (15) Not knowledge-intensive (4)
Creative (17) Non-creative (2)	Value contribution in core process (1), support process (5), and both (2)
Variability (19)	Interdependency (19)
Complexity exist (14) Complexity does not exist (5)	Relative advantage (17) No relative advantage (2)
Security/privacy/trust issues (08) No issues (11)	Disruptive (14) Non-disruptive (5)

First, the six process-related characteristics were: repetitiveness, knowledge intensity, creativeness, value contribution, variability and interdependence. The experts highlighted few instances when a process was *non-repetitive* in nature, and those instances were related to continuous process improvement. Next, the experts discussed various aspects of *knowledge intensity,* such as the creation of logs and high-level process maps for their workflows. Knowledge-intensive BPM was required from a sustainability and testimonial market perspective. Further on, *creativity* was seen as vital for a business process' success. Creative ideas seemed highly useful for re-design and automation (R9, R7, R17). Ideally, such creative ideas should be at the start of a process innovation project, which can be obtained through brainstorming sessions and consulting handbooks or scholarly work. The question about *value contribution* was related to aligning core and support processes with the business strategy, for which the experts also differentiated between human resource management processes and other management processes as they create a link between core and support processes (R4). *Task variety* seemed to exist in every business process, albeit mainly dependent on user requirements and specifications. For instance, information-based tasks typically contain more variations (R5, R9, R18, R13). For process *interdependency*, it was mentioned that front-end processes are often linked to back-end processes, and that process tasks depend on higher-level, end-to-end business processes. For instance, process architectures show more interdependency (R8), whereas quality control processes were mentioned as more independent processes (R1).Secondly, the four sub-dimensions regarding the technology-related factors were: complexity, relative advantage, security/privacy/trust issues and technology disruptiveness. *Complexity* of technology was rather positioned as how users apply a technology in diverse ways (R18). For instance, user acceptance was esteemed meaningful, and a change in technology can be a reason of increased perceived complexity in IT systems (R12, R13, R16, R17, R2). Alternatively, R13 argued that technology can become more complex in the later innovation stages when constraints become more visible. It was also observed that business process modelling engines can already be experienced as complex for some BPM users. Next, service level agreements and legal agreements were mentioned for ensuring *security and privacy in technologies.* The experts repeatedly ensured that they do not worry too much about the security or privacy issues of new technologies, given their technical background. Next, when a single technology is not providing the expected advantage, a combination of technologies can be used to achieve a *relative advantage* (R16). AI and machine learning seemed to create significant differences for predictions and analytics. Finally, we received mix comments for *technology disruptiveness.* Disruptiveness especially occurred with AI, machine learning, and blockchains (R1, R12, R18).

4.2 Themes for Ambidextrous BPM

Table 2 continues with the eight sub-dimensions of ambidextrous BPM: innovation type, trigger, solution, techniques, focus, risk management, structure and leadership.

Table 2. Ambidextrous BPM characteristics and sub-dimensions (expert count).

Incremental Innovation (8), radical (2), or both (9)	Problem-driven (4), opportunity-based (4), or both (11)
Existing legacy (7), clean slate (3), or both(9)	Best Practices (3), openness (4), or both (12)
Efficiency(8), flexibility(4), or both(9)	Uncertainty (1), certainty (8), or both (8)
Formal structure (9), adaptive (7), or both (3)	Bureaucratic (2), visionary entrepreneurship (15), or both (2)

The experts agreed that emerging technologies enable both *innovation types*, i.e., incremental and radical process improvements, and depend upon software engineering methodologies like agile or waterfall (R4, R5). Regarding the *innovation triggers*, problem solving helped in retaining the customers whereas new opportunities helped to grab customers (R10, R12, R13). The innovation triggers seemed to depend upon the client type and the nature of a problem or a particular business segment. For instance, an organization with ample financial resources has more capabilities to grab new opportunities. When thinking about *innovation solutions*, the experts inclined more towards ambidexterity again. Innovation solutions also seemed to depend upon the existing organizational and IT structures, as well as the division and the strategy of the organization. For instance, niche market organizations have the choice of being a startup with the help of a compatible environment (R7, R12, R15). Regarding *innovation techniques*, the experts recommended to use an agile methodology to experience more openness. Nonetheless, innovation techniques depend upon the customer requirements and budget. For instance, creative ideas can demand more investments but ensure a higher return on investment (R7, R12). Next, *risk management* during process innovation projects were considered critical. For instance, even for relatively lower risks, some issues can grow in the later project stages. Also the project's integration phase typically contains more risks (R7, R8, R18). Regarding the *innovation focus*, half of the experts emphasized both efficiency and flexibility, while four experts did not agree with flexibility (i.e., in the sense that any adaptation can be done in the process). Several experts argued that the focus should rather be on efficiency, policies, rules, and procedures (R1, R3, R4, R7, R8, R9, R10, R12, R13, R14). E.g., after-sales services were observed as difficult to remain flexible. Next, the *innovation structure* or layout of ongoing process innovation projects received mixed comments, with only three experts referring to the hybrid philosophy. Instead, nine experts mentioned that people are mostly working in formal structures. Agile methods were again mentioned to trigger more adaptive project structures (R3, R5, R6, R13, R16). Overall, most comments related to an innovation structure with typical management levels such as top, middle and operational levels. Finally, *leadership* styles were rather seen as a personality trait. When leaders are highly involved in low-level activities, they also do regular follow ups and better motivate their staff (R2, R4, R13, R18).

4.3 Themes for Process-Technology Fit

The seven sub-dimensions for a process-technology fit are shown in Table 3.

Table 3. Process-technology fit and sub-dimensions (expert count).

Compatible (17), incompatible (2)	With continuity (16), not willing to continue (3)
Top management support (15), without (4)	Achieving goals (17), or not (2)
Technology appropriateness (19)	

Most experts claimed that business processes are increasingly aligned with IT, while older software versions are not necessarily *compatible* with new IT systems. Nonetheless, only when compatibility is experienced between the process and a technology, organizations will *continue* with the digital process innovation. Innovative technologies were also considered as useful for continuous improvement and for showing the advantages to top managers in order to convince them to invest more. In contrast, new technologies also come with new costs, expenses, training requirements, and knowledge (R8, R9, R10). Regarding *technology appropriateness,* BPM practitioners need to decide whether a specific technology is appropriate for a certain process before starting the implementation project phase (R7, R12). Technologies like IoT and AI were frequently mentioned for their potential to be integrated with core processes. Technology appropriateness is difficult to discover at the start of a project, since benefits rather become visible later on. Alternatively, conflicting thoughts emerged about *top management support.* Some experts felt that a direct manager is more supportive than a top manager, and other experts argued that top managers are mostly concerned about performance outputs and do not necessarily care about what is going on at the operational level (R2, R4, R7, R8, R16). Nonetheless, top management support was mainly seen in the form of financial approvals or a positive attitude towards innovation. Next, organizational *goals* are usually monitored through key performance indicators, which also need to be achieved by process innovation projects. Especially the financial and quality-oriented goals were mentioned. While the experts did not doubt about the *usefulness (P.U.)* factor, more comments were received on the *P.E.O.U.* factor. In standardized software tools, such as SAP solutions or salesforce, the interfaces are not easily customized. Many experts emphasized that user-friendly interfaces matter a lot, and that software developers can make them more user friendly by changing colors, button styles and fields (R1, R5, R18). Most experts were convinced that BPM users perceive a technology as easy once they get user-friendly interfaces.

4.4 Themes for Mediating and Performance Variables

The reference count for other sub-dimensions are illustrated in Table 4.

Table 4. Mediators and performance outcomes (expert count).

With high market competition (15), or without (4)	Uncertainty in sector (19)
Size matters (9), or size does not really matter (9)	Culture matters (18), or culture does not really matter (1)
Sufficient resources (13), insufficient resources (6)	Technologically ready (17), or not ready (2)
Process cost savings (14), or not(5)	Time savings (17), or not (2)
Flexibility gains (19)	Quality gains (19)
Financial gains (16), non-financial gains (3)	Learning and growth (17), or not much (2)
Customer satisfaction (19)	

All experts agreed that the environmental factors were highly influential to find a process-technology fit, and they especially agreed upon sector and uncertainty. For instance, the oil and gas sector has other core processes than the software sector. Market uncertainty also impacts, for instance, on changes in political and economic conditions (R1, R3). Some experts illustrated this impact for non-profit organizations which do not face a highly competitive market, as compared to manufacturing. Also the organizational culture was esteemed highly influential for the fit because certain actions and decisions can be derived from it. A digital culture is more supportive and adaptive in terms of technology adoption (R3, R14). While mixed opinions were observed for organization size, most respondents agreed that many organizations have sufficient resources to start a digital process innovation project. Finally, process and organizational performance seemed to be highly affected when finding a process-technology fit.

5 Discussion

This study has described the factors (i.e., process-related, technological, ambidextrous and fit) involved in a Process-Technology Fit (Sects. 4.1, 4.2 and 4.3), for addressing

RQ1. Similarly, the mediating factors have been highlighted in Sect. 4.4. We have replied to RQ2 by determining ambidextrous factors supporting PTF in Sect. 4.2.

The need to find a technology appropriate for a specific task has been discussed since the 1990s [11]. However, in today's digital economy with high-speed internet, business processes are reshaping in every iteration of the BPM's life cycle. The expert opinions in our study have addressed all components of Fig. 1, during which a deeper understanding of 'Fit' and 'No Fit' situations are acquired. This understanding is visualized in Fig. 2 by portraying two balances (i.e., between process and technology, and between exploitation and exploration). Figure 2 illustrates that 'Fit' situations will directly lead to increased process performance, while 'No Fit' situations will lead to equal or (more frequently) decreased process performance.

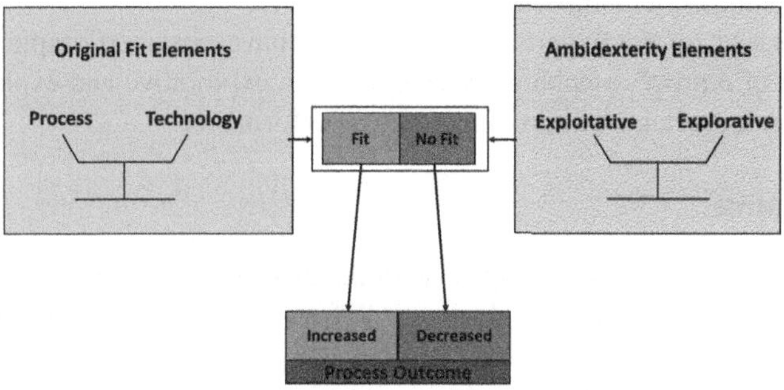

Fig. 2. Explaining the notions of "Fit" and "No Fit" for Process-Technology.

We have particularly observed many instances in which the experts stated that a process-technology fit results in increased process performance. For instance, R7 highlighted the importance of a 'Fit' situation as follows: *"When IT fits our portfolio, we just stop with adding a new technology. We all should align to the policy"*. Furthermore, R2 described a 'Fit' advice: "*Provide to clients the solutions of IT and AI for reasons of analytics and statistics, so that they can decide in the future to improve with these analytics*", indicating that AI is suitable for doing analytics. Another example is given by R8: "*Our technology is suited for some of our tasks, which are not operational and not related to manufacturing. It is suited for internal communication and doing these kind of work*". Consequently, the notion of a process-technology fit is situational, and can create a big difference among organizations and processes.

Next, evidence has been provided regarding the ambidexterity balance added to the PTF theory, namely on the right-hand side of Fig. 2. With the rapid boom of emerging technologies and business process advancements, many experts have emphasized a hybrid approach that combines exploitative elements with explorative elements. For instance, R5 explained that: "*when we are using BPM, innovation can be incremental and radical. It depends on your design and the selected methodology for using agile or waterfall.*" Another example was given by R10: "*Problem solving helps in retaining the customer and new opportunities are needed to grab the customer. So both are in a continuous process.*", indicating that both improvement triggers are important. Moreover, R15 continued with: "*We prefer to think out-of-the-box, but within the brackets of the project activities because all objectives and activities of the project are predefined. So, it is a combination of both."*

Subsequently, we discuss some additional lessons learned from our expert panel, in addition to Fig. 1 and Fig. 2. First, many experts have emphasized that the process' nature and user requirements highly impact on the process design and modeling phase. Secondly, resources (such as financial, technological or specialized people resources) have not necessarily been considered as major issues anymore in medium and large-sized organizations, whereas especially a lack of creativity among business process handlers (e.g., process developers, process designers, and process managers) appears to more seriously limit BPM exploration. Thirdly, we have uncovered that some sub-dimensions as mentioned in prior studies show overlaps in our initial PTF theory. E.g., although the literature can differentiate between process variability and repetitiveness, our panel suggests to combine them into one sub-dimension.

In sum, based on the experts' comments, we have observed ample evidence for the relevance of a process-technology fit in both an explorative and exploitative BPM approach, which is vital for improving process performance.

6 Conclusion

Since the BPM dynamics are changing with the growth of new technologies, we have started to translate the task-technology fit (TTF) theory into a process-technology fit (PTF) theory based on an expert panel of BPM consultants.

In RQ1, we have observed which process characteristics and technological characteristics are most relevant for finding a fit. By aligning the theoretical assumptions with the experts' practical experience, we have gained insight into the notion of a 'fit' between business processes and IT, and we specifically differentiated the 'fit' and 'no fit' elements for process improvements. For instance, we have found that agile methods were often recommended by the experts (e.g., agile software engineering for business intelligence projects). In terms of new technologies, it appeared that especially artificial intelligence has been perceived as highly favorable in BPM environments nowadays. Nonetheless, the experts have confirmed that practitioners are still urgently looking for better graphical user interfaces.

In RQ2, we have positioned the process-technology fit idea under the umbrella of ambidexterity by also observing the seemingly conflicting forces of ambidextrous BPM. The expert panel has confirmed that an ambidextrous BPM approach is highly supportive. It appeared that an explorative BPM approach (i.e., including openness and flexibility) has not yet been fully adopted but that most organizations rather focus on using new technologies for exploitative process improvements with explorative process improvements being seen as future ambitions.

A major advantage of our expert panel is that BPM consultants have relevant expertise across multiple process innovation projects, which has allowed us to better understand the research model. Nonetheless, we acknowledge the fact that the themes have been developed on the basis of expert opinions instead of being tested in a generalizable way. Therefore, in follow-up research, we will first look at multiple case studies to compare particular process innovation projects more deeply, followed by a survey with BPM users to contribute to our theory building and testing efforts.

References

1. vom Brocke, J., Schmiedel, T., Recker, J., Trkman, P., Mertens, W., Viaene, S.: Ten principles of good business process management. BPM J. **20**, 530–548 (2014)
2. Rosemann, M.: Proposals for future BPM research directions. In: Ouyang, C., Jung, J.-Y. (eds.) AP-BPM 2014. LNBIP, vol. 181, pp. 1–15. Springer, Cham (2014). https://doi.org/10.1007/978-3-319-08222-6_1
3. Badakhshan, P., Conboy, K., Grisold, T., vom Brocke, J.: Agile business process management: a systematic literature review and an integrated framework (2019)
4. Heckmann, C.S., Maedche, A.: IT ambidexterity for business processes: the importance of balance. Bus. Process. Manag. J. **24**, 862–881 (2018)
5. Binci, D., Belisari, S., Appolloni, A.: BPM and change management: an ambidextrous perspective. Bus. Process. Manag. J. **26**, 1–23 (2019)
6. Marrella, A.: Automated planning for business process management. J. Data Semant. **8**(2), 79–98 (2018). https://doi.org/10.1007/s13740-018-0096-0
7. Van Looy, A., Shafagatova, A.: Business process performance measurement: a structured literature review of indicators, measures and metrics (2016)
8. Couckuyt, D., Van Looy, A.: Green BPM as a business-oriented discipline: a systematic mapping study and research agenda. Sustainability **11**, 4200 (2019)
9. Schniederjans, D.G.: Business process innovation on quality and supply chains. Bus. Process. Manag. J. **24**, 635–651 (2018)
10. Anwar, S.T.: Growing global in the sharing economy: lessons from Uber and Airbnb. Glob. Bus. Organ. Excell. **37**, 59–68 (2018)
11. Goodhue, D.L., Thompson, R.L.: Task-technology fit and individual performance. MIS Q. **19**, 213 (1995)
12. Fichman, R.G.: Information Technology Diffusion: A Review of Empirical Research (1992)
13. Ferraris, A., Monge, F., Mueller, J.: Ambidextrous IT capabilities and business process performance: an empirical analysis. BPM J. **24**, 1091–1109 (2018)
14. March, J.G.: Exploration and exploitation in organizational learning. Organ. Sci. **2**, 71–87 (1991)
15. Helbin, T., Van Looy, A.: Is business process management (BPM) ready for ambidexterity? Conceptualization, implementation guidelines and research agenda. Sustainability (Switzerland) **13**, 1–25 (2021)
16. Del Giudice, M., Soto-Acosta, P., Carayannis, E., Scuotto, V.: Emerging perspectives on business process management (BPM): IT-based processes and ambidextrous organizations, theory and practice. BPM J. **24**, 1070–1076 (2018)
17. Afflerbach, P., Frank, L.: Customer experience versus process efficiency: towards an analytical framework about ambidextrous BPM. In: ICIS 2016, vol. 4801 (2016)
18. Brocke, J.V., Zelt, S., Schmiedel, T.: On the role of context in business process management. Int. J. Inf. Manag. **36**, 486–495 (2015)
19. Zelt, S., Recker, J., Schmiedel, T., vom Brocke, J.: A theory of contingent business process management. Bus. Process. Manag. J. **25**, 1291–1316 (2018)
20. Davis, F.D., Bagozzi, R.P., Warshaw, P.R.: User acceptance of computer technology: a comparison of two theoretical models. Manag. Sci. **35**, 982–1003 (1989)
21. Low, C., Chen, Y., Wu, M.: Understanding the determinants of cloud computing adoption. Ind. Manag. Data Syst. **111**, 1006–1023 (2011)
22. Dumas, M., La Rosa, M., Mendling, J., Reijers, H.A.: Introduction to business process management. In: Dumas, M., La Rosa, M., Mendling, J., Reijers, H.A. (eds.) Fundamentals of Business Process Management, pp. 1–31. Springer, Heidelberg (2013). https://doi.org/10.1007/978-3-642-33143-5_1

23. Kaplan, R.S., Norton, D.P.: The balanced scorecard: measures that drive performance. Harvard Bus. Rev. **70**(1), 71–79 (1992)
24. Recker, J.: Scientific Research in Information Systems. Springer, Heidelberg (2013). https://doi.org/10.1007/978-3-642-30048-6

Exploring the Key Factors of Process Improvement that Drive Competitive Advantage: A Case of Serbia

Ivona Jovanović(✉), Dragana Stojanović, Dragoslav Slović, Ivan Tomašević, Barbara Simeunović, and Milica Maričić

Faculty of Organizational Sciences, University of Belgrade, Belgrade, Serbia
{ivona.jovanovic,dragana.stojanovic,dragoslav.slovic, ivan.tomasevic,barbara.simeunovic,milica.maricic}@fon.bg.ac.rs

Abstract. The aim of this study is to define the set of BPI factors that contribute the most to the success of gaining competitive advantage in the company. This paper aims to investigate the linkages between BPI in context of BPM and competitive advantage, and whether BPI practices ensure that companies can accomplish desired results. Structural equation modeling (SEM) was conducted to identify key factors that contribute to the success of BPI, which in turn drives competitive advantage on the market. An online survey was conducted and data has been collected from the Serbian companies from different sectors. The results highlight the impact BPI has on competitive advantage in companies and shows which factors significantly contribute to success of BPI in context of leadership and employees. To achieve desired results, the focus has to be on strong support of the leadership and good communication between employees. The study highlights significance of setting short- and long-term objectives, so that employees can understand the need for BPI projects, as well as their position in them. This study can help practitioners to be more successful in achieving BPI goals, to understand the nature of process improvement and to transform their process knowledge into competitive advantage.

Keywords: BPI · Competitive advantage · Success · SEM analysis

1 Introduction

With high level of competitiveness in the global market, in order to gain a competitive advantage, it becomes increasingly important for companies to improve their process performance. The companies take process improvement initiatives in order to eliminate the deficiencies identified in their performance and to maintain competitive advantage [23]. Although the implementation of process improvement in practice is growing over the years for enhancement of company competitiveness, studies shows that about 80% of process improvement initiatives were unsuccessful [3, 5, 30]. The main reason for failure is the complexity of implementation, because it requires resources, time and effort [9], as companies are impatient to see results and achieve success in a short

A. Marrella and B. Weber (Eds.): BPM 2021 Workshops, LNBIP 436, pp. 179–191, 2022.
https://doi.org/10.1007/978-3-030-94343-1_14

time [28, 34]. Since the competitive advantage is one of the main sources for achieving continuous improvement of business processes [1], it is important for companies to focus on the different aspects of competition strategies and consider what can lead to success for process improvement. Company should enhance both their people and processes to balance with the changing business environment and to maintain their competitive advantage [19, 22, 24]. If the company does not improve performance with competition strategies, their competitors will [4].

Although there are researches on how process improvement effects on performance of the company, there has been a lack of practical studies about its success in companies in Serbia. Previous research [28] shows a lack of strategic focus of management for process improvements and their implementation on the leadership level, while process improvement initiatives are focused mostly on operational performance. Process improvement must be adopted through alignment with organizational structure and people empowerment [23]. It is important to establish the balance within employees and higher management in process improvement initiatives in the company. If the employees do not have basic skills and knowledge to support process improvement initiatives, or the management does not participate in improvements, the effects may be unsatisfactory [4].

Therefore, this research investigate which process improvement initiatives can be a sustainable practice which help companies in Serbia to achieve competitive advantage and desired results. This is achieved by the assessment of the relationship between constructs among which are business process improvement (BPI), competitive advantage and goals of BPI, through the application structured equation modeling (SEM) analysis. According to Ketchen et al. [18], the researchers mostly lean on analysis techniques such as regression analysis and ANOVA, so that there is a limited extent on analysis techniques such as SEM.

The paper is organized as follows: next section is based on theoretical background about BPI as a part of BPM initiatives, and third section describes the research methodology that was used for the purpose of this paper. Results are given in fourth section, and discussion and conclusion in the fifth section.

2 Theoretical Background

Business Process Improvement (BPI) is "a systematic approach which can help a company to optimize its elemental processes to achieve more efficient results" [26]. BPI can be accomplished by changing the elements of a business process so that it increases the fulfillment of organizational goals and enhances the business process performance of the company [13]. BPI as a part of business process management (BPM) represents an important part of managing the effectiveness and efficiency of organization processes [7, 32] and creating a sustainable competitive advantage in companies and organizations [8, 23]. Also, it includes Lean/Six Sigma as one of the main approaches for implementing process approach, which embraces an organizational transformation throughout the company [14]. Different empirical studies confirmed the positive correlation between process improvement and business success [2, 22, 23], and have shown that companies can achieve competitive advantage through BPI [29]. Process-oriented companies are better in achieving desired results [31], and in achieving a competitive advantage compared to others [34]. The main challenge remains how to implement process principles of process approach onto an organization's operations [31].

As employees become more specialized and as leadership cannot have knowledge about various disciplines, a need for making process improvement projects across the company is rising, so it increases the need for employees involvement in process improvement projects [26]. Therefore, the selection of important process improvement initiatives cannot depend only on the status and roles of individuals or groups of individuals in the company. According to Abderrazak & Kacem [2], the success of the company varies depending on the commitment of individuals to the company, management and employees of the company, so process improvement still is an important issue for top management [13]. The company must ensure that its leadership and employees have appropriate knowledge for the process approach, so they can use it for improvement. For process improvement is necessary to establish communications within employees, because it enables the company to work together as one unit in gaining their efficiency [23].

On the other side, Kumar et al. [17] highlight the significance of process improvement as a critical element for driving customer satisfaction, but states that majority of companies are focusing only upon quality improvement issues. Also, customer satisfaction can be identified as one of the basic drivers for process improvement initiatives in companies in Serbia [27]. Based on similar researches in this field, this study identifies key factors for the success of process improvement, from the leadership and employees aspects, and align these factors with companies' outcomes. Leading practices in companies in Serbia, this paper identifies the competencies which company should have and which steps should take in order to strengthen competitive advantage efficiency.

3 Research Methodology

The purpose of this research is to demonstrate the influence of BPM through BPI on competitive advantage in companies in Serbia. Besides that, it was interesting to see which key factors of BPI contribute the most to fulfillment of BPI goals and competitive advantage. We expect that the fulfillment of the goals have a positive association with the competitive advantage.

For the purpose of the study, three research questions were defined (Table 1):

Table 1. Research questions

Q1	Which leadership and employees factors in BPI context contribute the most to competitive advantage of companies in Serbia?
Q2	Does BPI have a positive impact on fulfillment of BPI goals in companies in Serbia?
Q3	Does BPI have a positive impact on success of BPM program in companies in Serbia?

In order to answer on research questions and collect data for this model, survey research was conducted. Questionnaire was a part of a wider research about BPM and

Lean Transformation practice. Questions about BPM practice were divided into four categories: (1) general questions about company and respondents (7 questions), (2) questions about process maturity, Process Performance Index (12 questions), and (3) BPM and BPI practice (59 questions) and, (4) questions about goals fulfillment and competitive advantage (22 questions). For these categories, several constructs were defined, each consisting of several questions. The survey questions included single or multiple choices, but several questions had an open form to better perceive the view of the respondent. The Convenience Sampling, as a nonprobability sampling method, was applied in this research. Participation was voluntary and anonymous and respondents were informed that the data would be used for research purposes only. The proposed model is presented in the Fig. 1.

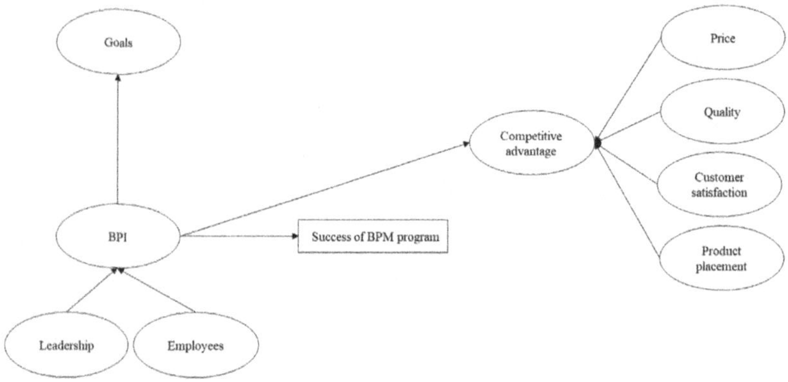

Fig. 1. Proposed model

BPI. The BPI was measured through two dimensions: leadership and employees. The leadership and employees dimensions are given in the Table 2. It was requested from participants to express the extent of their agreement with the offered statements about process improvement. This part of the model observes the role of BPI from the leadership and employees perspectives, in order to indicate the set of factors whose mutual influence can lead to increase of competitive advantage of the company.

The Likert scale was used for all variables, on a five-point rating scale varying from 1 (strongly disagree) to 5 (strongly agree). The following statements are adapted from Wolf & Harmon [33], in accordance with similar research in the field [16, 25].

Competitive Advantage. The respondents were asked to indicate the extent to which they agree with different statements about competitive advantage of their company. It was measured through four dimensions: price, quality, customer satisfaction and product placement (Table 3). These measurable quality characteristics of product are important for setting the process improvement strategy. According to Kumar et al. [17] and Stojanović et al. [27] the customer satisfaction represents an important indicator of process improvement success, so as the product quality [20]. Trihatmoko et al. [30] highlight

Table 2. The BPI statements (Source: Adapted from Wolf & Harmon [33])

BPI	Items no	Statements
Leadership	B1	Senior leadership provides visible support (budget, actions, etc.) for process improvement
	B2	Senior leaders are decisive with respect to organizational goals, priorities and strategies concerning the process improvement
	B3	We have the ability to identify roadblocks to process improvement
	B4	We have the ability to overcome roadblocks to process improvement
	B5	We set short- and long-term objectives for process improvement
	B6	The company has a process improvement plan that is put in writing
	B7	Process improvement plans tend to become fully deployed
Employees	B8	Everyone is aware of the process improvement mission and vision
	B9	We have successfully obtained support for BPI from our employees
	B10	Employees understand the need for process improvement
	B11	Employees understand how they fit into the process impromevent
	B12	The roles of all employees in the process improvement have been communicated
	B13	The outcomes and benefits of process improvement are not well communicated
	B14	Our annual objectives reflect the company's long-term focus on process improvement

product placement and innovation of the products as an important phenomenon in business market competition, while the company needs to consider product price for gaining a competitive advantage [10].

BPI Goals Fulfillment. The respondents evaluated the extent up to which their company fulfilled different BPI objectives. The questions are adapted from Marshall [21] and include following goals: eliminating waste, reducing cost, improving organizational capabilities, improving competitive position of the organization, and improving financial and operational performance. In order to achieve its goals, for organizations it is necessary to identify process improvement activities and to understand the nature of a business.

Table 3. The competitive advantage statements (Source: Adapted from Marshall [21])

Price	We offer competitive prices
	We are able to offer prices as low or lower than our competitors
Quality	We are able to compete based on quality
	We offer products that are very durable to our customers
	We provide reliable products in quality
Customer satisfaction	We offer products that are highly reliable
	We deliver customer order on time (Just-in-time)
	We alter our product offerings to meet client needs
Product placement	We respond well to customer demand for "new" features
	We are first in the market in introducing new products
	We have time-to-market lower than industry average

Success of BPM Program. As BPI represents an important part BPM initiatives, it was interesting to measure how the mutual influence of leadership and employees factors increases the effectiveness of the BPM program, which should be part of strategic targets of the company. In this part of the research, the respondents were asked to indicate their opinion about how much the business process improvement program is successful in their company.

In order to answer the research questions, in this paper was used structural equation modeling (SEM) analysis. The SEM analysis is a multivariate statistical analysis which is based on the principles of factor analysis and multiple regression, which allow the construction of latent variables and exploration of their mutual relations. Also, latent variables of higher order - hierarchical latent variables can be defined [6]. The results of the SEM analysis should indicate how well the collected data fits the proposed model. In order to test the validity of the proposed model, we used the Chi-square statistics, Comparative fit index (CFI), Tucker-Lewis Index (TLI) and Root mean square error of approximation (RMSEA). The survey results were analyzed using SPSS 25 and AMOS 22.

SEM analysis was previously investigated in the field of business improvement factors for process success. For example, Ghlichlee and Bayat [12] employed it in analysing employees' engagement and customer-oriented behaviors as one of the main factors affecting business performance, while Li et al. [19] used it to identify the effects of BPI and human resource enhancement on operational performance in manufacturing organizations.

4 Results

4.1 Sample Caracteristic

In order to confirm the influence of BPI initiatives on competitive advantage, an online survey was conducted. A questionnaire and letter invitation to participate were sent to

700 potential participants from Serbian and foreign companies. In total, 82 responses have been received, representing 11.71% response rate. According to Wong et al. [35], the response rate from studies in this field vary from 6.7% to 20%, so the response rate in this research can be confirmed as adequate.

Among 82 respondents, 6.1% stated that their position is general manager, 7.3% work as executive officers, 36.6% work as business department managers, business analyst makes 3.7% of respondents, consultant 15.8%, researcher 2.4%, process specialist/process analyst 24.4%, while 3.7% of respondents stated that the position they work on was not offered by the survey. The 62% respondents stated that they work in large companies (62%) with 250 employees or more. When it comes to business field that the company is operating in, the 42.7% respondents work in manufacturing companies, 42.7% in service oriented companies, while 14.6% are working in the company in both industries. About one third of respondents stated that they were involved in business process improvements for more than two years (32.9%), and 14.6% of respondents were engaged five years or more on process improvements. When it comes to the current interest in process improvements, most respondents stated that BPI is as a part of everyday business (67.1%) and 20.7% respondents stated that the interest for implementation is growing. The respondents indicated that the average length of BPM projects in their company were between 6 and 12 months (30.5%) and about one quarter of respondents stated that average length of projects was between 3 and 6 months (28%).

4.2 Data Analysis and Results

To evaluate the search model with SEM analysis, the initial step was to inspect the reliability of the proposed scales and to calculate the internal consistency. The metrics used in this research is the Cronbach's alpha [11]. The obtained Cronbach's alpha and the number of items per scale are presented in Table 4. The internal consistency on the first-level of constructs ranges from 0.743 (Product placement) to 0.913 (Leadership). On the second-level constructs, the internal consistency is 0.882 (Competitive advantage) and 0.947 (BPI). The calculated metrics are above the 0.7 threshold, which indicates that the proposed constructs measure a particular aspect. Therefore, the conclusion is that the data is suitable for SEM analysis.

The initial model had relatively low fit of the data to the model: Chi-square = 958.011, df = 429, CFI = 0.750, TLI = 0.729, RMSEA = 0.123. This indicated that there was place to re-specify the model. There was no need to alter the structure of latent variables, as all measured variables were statistically significant. Also, all the hypothesized relations within the latent variables/constructs were statistically significant (|C.R. – Critical ratio| > 1.96). Therefore, to alter our model, modification indices were used.

The final model had better fit to the data: Chi-square = 728.864, df = 417, CFI = 0.852, TLI = 0.835, RMSEA = 0.096. Taking into account the model complexity and the sample size, although the values of the chosen model fit are close to the thresholds (CFI > 0.9, TLI > 0.8, RMSEA < 0.1 – as suggested by Hsu [15]) we can conclude that the model is satisfactory, so we continue with the interpretation of the results. The assessment of the outer model is given in Table 5.

Table 4. Cronbach's Alfa per first-level and second-level Construct and the Number of Items per Scale

	Competitive advantage				Goals	BPI	
	Price	Quality	Customer satisfaction	Product placement	/	Leadership	Employees
Items	2	3	2	3	6	7	7
(first-level construct)	0.766	0.910	0.879	0.743	0.873	0.913	0.904
(second-level construct)	0.882				/	0.947	

Table 5. Assessment of the outer model: construct, predictors, obtained standardised and unstandardized coefficients, and Critical Ratio (C.R.)

Construct	Predictors	UnStd Coeff	Std Coeff	C.R.
BPI	Leadership	1.000	1.000	/
	Employees	1.415	0.913	5.202**
CA	Price	1.000	1.000	/
	Quality	1.767	1.000	4.165**
	Customer satisfaction	1.519	1.000	3.915**
	Product placement	1.417	0.784	3.754**

Note: **$p < 0.01$, C.R. – critical ratio

The proposed model is a hierarchical latent variable model [6], because the construct BPI and competitive advantage are second-order latent variables. To additionally explore the created model, each second-order latent variable was inspected. The obtained results led us to raise the following two observations:

1. The results indicate that first-order constructs "Leadership" and "Employees" have a statistically significant direct effect on BPI, which indicates that if the predictors leadership or employees increase, it will increase the value of the construct BPI.
2. The results indicate that all predictors have a statistically positive impact on competitive advantage, meaning that if the company focus on any of the observed aspects, it will increase competitive advantage. The results show that Quality has the highest obtained unstandardized coefficient 1.767, meaning if the company products/services quality increases, the competitive advantage in company increases. Other predictors, such as customer satisfaction, also have a positive impact on competitive advantage. If the company focus on customer satisfaction, they will have higher success on achieving competitive advantage, which confirms research conducted by Kumar et al. [17].

The assessment of the inner model indicates that all coefficients are statistically significant and positive (Table 6).

Table 6. Assessment of the inner model: construct, predictors, obtained standardised and unstandardized coefficients, Critical Ratio (C.R.), and the R square of the construct

Construct	Predictors	UnStd Coeff	Std Coeff	C.R.	R^2
Goals	BPI	0.417	0.618	3.231	0.382
CA	BPI	0.358	0.470	2.932	0.220
Success of BPM program	BPI	0.900	0.665	4.682	0.442

Note: **$p < 0.01$, C.R. – critical ratio

Taking a closer look at the model of BPI influence, it can be concluded that BPI has a statistically significant positive impact on fulfillment of BPI goals. The value of the unstandardized regression coefficient is 0.417 in absolute value indicating a moderate effect, so if the value of the BPI increases, the fulfillment of the goals will increase. The R square of this construct is 0.382, meaning that 38.2% of variability of Goals is explained by BPI. Having in mind that there is only one explanatory variable in the model, we can conclude that the model is of good quality.

According to the the value of unstandardized regression coefficient, BPI has a statistically significant positive impact on competitive advantage, which means if the BPI increase, the competitive advantage will increase. The R square of this construct is 0.220, meaning that 22.0% of variability of competitive advantage is explained by BPI.

The results show that the regression coefficient between BPI and success of BPM program is 0.900, which indicates a statistically positive impact of BPI on the success of BPM program. The R square of this construct is the highest, 0.442, which means that 44.2% of variability of BPM program success is explained by BPI.

To exert more reliability and practical implication to this study, further research was conducted. Since the BPI proved to have a direct impact on competitive advantage and fulfillment of BPI goals, we strived to explore the structure of the first-order latent variables Leadership and Employees which are used to quantify BPI in order to answer the research question "Which leadership and employees factors in BPI context contribute the most to competitive advantage of companies in Serbia?" The results are given in the following Table 7.

According to the results, all presented factors are statistically significant for BPI initiatives. The highest value of the regression coefficients for leadership have items B1, B5 and B6, meaning if practitioners want to increase the effectiveness of their process improvement, the main focus should be on:

- Leadership which provides visible support (budget, actions, etc.) for process improvement;
- Setting short- and long-term objectives for process improvement;
- Process improvement plan that is put in writing.

Table 7. Assessment of the first-level constructs: obtained standardized, unstandardized coefficients, and Critical Ratio (C.R.)

First-level construct	Items no	UnStd Coeff	Std Coeff	C.R.
Leadership	B1	1.423	0.791	5.186**
	B2	1.289	0.766	5.104**
	B3	1.354	0.819	5.295**
	B4	1.375	0.799	5.230**
	B5	1.621	0.860	5.431**
	B6	1.531	0.766	5.142**
	B7	1.000	0.791	/
Employees	B8	1.000	0.840	/
	B9	0.887	0.778	8.449**
	B10	0.956	0.858	9.895**
	B11	0.941	0.885	10.434**
	B12	1.029	0.836	9.461**
	B13	0.374	0.308	2.777**
	B14	0.900	0.833	9.406**

Note: **$p < 0.01$, C.R. – critical ratio

In order to increase the process improvement initiatives in context of employees, the results indicate that the focus of the company should be on the following factors:

- The roles of all employees in the process improvement need to be communicated;
- Employees must understand the need for process improvement;
- Employees need to understand how they fit into the process improvement.

The factor that has the lowest regression coefficients for BPI is that the outcomes and benefits of process improvement are not well communicated, and because of the formulation of the question, it can also be an important factor which highlights good communication among employees and leadership. Increased focus on these factors should bring greater success rate of BPI projects. Based on the results of study, it can be concluded that it is very important for everyone, from management to every employee, to participate in the process improvement.

5 Discussion and Conclusion

This research observes the role of BPI within a perspective of business process management, indicating the set of factors whose common influence can lead to increase of competitive advantage of the company. As the competition is growing in the market, the company needs to focus on process improvement strategies.

The structural equation modeling (SEM) analysis confirmed that the BPI is positively related to competitive advantage and success of BPI goals in the company. Also, it recognised that the BPI can contribute to the success of the basic elements of competitive advantage, through price, quality, customer satisfaction and product placement. In order to determine how BPI can help companies to achieve competitive advantage, and to additionally explore the research questions, in-depth analysis of the first-order latent constructs was conducted. Based on the obtained results, it can be concluded that it is very important that the process improvement in the company has strong support of the leadership. The majority of respondents assign department managers to be responsible for process improvements and their difficulties and obstacles [28]. Based on this assumption, the results show that it is important to increase the involvement of leadership in BPI initiatives and that not all responsibilities should be on department managers. A lack of support from the leadership can lead to unsatisfactory outcomes of BPI initiatives and failure. Also, the lack of knowledge of process improvement could be the reason for failure, but with formal training of employees in process redesign and process analysis, companies could experience success in BPI which leads to better results. The most respondents stated that they were involved in business process improvements for more than two years and that process improvement is a part of everyday business in the company. This indicates process awareness for companies in Serbia and shows that companies in Serbia are actively engaged to process improvement.

According to the results, one of the key factors for BPI is the setting of short- and long-term objectives, but sometimes in the company the goals can be imposed without a final explanation. This is directly related to the statement that employees must understand the need, outcomes and benefits of process improvement in their company. When the goals of improvement are clearly defined and when it is properly communicated to the employees, it will contribute to the growth of BPI projects success. The companies that undertook more serious process improvement not only on a leadership level, but also in accordance with employees, had more success in BPM projects.

The findings in this paper have practical implications of BPI practice through represented recommendations for BPI practitioners for the companies that operate in Eastern Europe. The contribution is that the key factors for the success of BPI are recognized, so the company does not have to identify them, but only work on their improvement. That will allow companies to better understand the nature of process improvement and how they can overcome problems with management and employees.

The limitation in this research is the response rate and that the data were collected from companies in just one country. The research is also important for companies outside Serbia, so one of the future directions could be the inclusion of other countries in order to generalize the conclusions more easily. The managerial implications of improving processes in this study are that managers should emphasize different things according to their roles and be more dedicated to BPI projects.

Future research should include a larger-scale study in Serbia, which will consider additional variables, such as the influence of the international environment on the development of BPI initiatives. Also, expended research should include clustering of companies and testing of connections per cluster. It would give verifiability and universality to the model and would have a potentially better comparability with previous studies.

Using a different formulation of BPM process improvement criterias (cost, time, quality, flexibility), the research should get a more complete and targeted coverage.

References

1. Abbas, S., Hosein, D.: Application of analytic network process in selection of six sigma projects. Int. J. Ind. Prod. Res. **20**(4), 157–164 (2010)
2. Abderrazak, B., Kacem, T.: Investigating the impact of the process approach on the performance of the company - empirical study of the Moroccan context. Int. J. Adv. Res. (IJAR) **8**(12), 321–337 (2020)
3. Abdolvand, N., Albadvi, A., Ferdowsi, Z.: Assessing readiness for business process reengineering. Bus. Process. Manag. J. **14**(4), 497–511 (2008)
4. Andersen, B.: Business Process Improvement Toolbox. Quality Press (2007)
5. Bai, C., Sarkis, J.: A grey-based DEMATEL model for evaluating business process management critical success factors. Int. J. Prod. Econ. **146**(1), 281–292 (2014)
6. Becker, J.-M., Klein, K., Wetzels, M.: Hierarchical latent variable models in PLS-SEM: guidelines for using reflective-formative type models. Long Range Plan. **45**(5–6), 359–394 (2012)
7. Beimborn, D., Joachim, N.: The joint impact of service-oriented architectures and business process management on business process quality: an empirical evaluation and comparison. Inf. Syst. E-Bus. Manag. **9**(3), 333–362 (2011)
8. Broadbent, M., Weill, P., St. Clair, D.: The implications of information technology infrastructure for business process redesign. MIS Q. **23**(2), 159–182 (1999)
9. Buh, B., Kovačič, A., Indihar Štemberger, M.: Critical success factors for different stages of business process management adoption–a case study. Econ. Res. - Ekonomska Istraživanja **28**(1), 243–258 (2015)
10. Chen, C.-H., Lu, C.-L.: Optimum profit model based on order quantity, product price, and process quality level. Expert Syst. Appl. **38**(6), 7886–7893 (2011)
11. Cronbach, L.J.: Coefficient Alpha and the internal structure of tests. Psychometrika **16**(3), 297–334 (1951)
12. Ghlichlee, B., Bayat, F.: Frontline employees' engagement and business performance: the mediating role of customer-oriented behaviors. Manag. Res. Rev. **44**(2), 290–317 (2020)
13. Griesberger, P., Leist, S., Zellner, G.: Analysis of techniques for business process improvement. In: Proceedings of 19th European Conference on Information Systems (ECIS) (2011). https://aisel.aisnet.org/ecis2011/20/. Accessed 5 May 2021
14. Harmon, P.: The scope and evolution of business process management. Handb. Bus. Process Manag. **1**, 37–81 (2010)
15. Hsu, L.: Work motivation, job burnout, and employment aspiration in hospitality and tourism students – an exploration using the self-determination theory. J. Hosp. Leis. Sport Tour. Educ. **13**, 180–189 (2013)
16. Jae, Y.M., Byeong, W.K., Young, H.S., Choon, S.L., Hyungjoon, M.: An integrated method for business process improvement. Int. J. Innov. Comput. Inf. Control **8**(7(B)), 5237–5250 (2012)
17. Kumar, V., Smart, P.A., Maddern, H., Maull, R.S.: Alternative perspectives on service quality and customer satisfaction: the role of BPM. Int. J. Serv. Ind. Manag. **19**(2), 176–187 (2008)
18. Ketchen, D.J., Boyd, B.K., Bergh, D.D.: Research methodology in strategic management: past accomplishments and future challenges. Organ. Res. Methods **11**(4), 643–658 (2008)
19. Li, Y., Wang, Q., Wang, Z., Chen, L.: Improving business processes or human resources? The performance implications and contingencies. Ind. Manag. Data Syst. **121**(7), 1577–1598 (2021)

20. Maddern, H., Maull, R., Smart, P.A.: Customer satisfaction and service quality in UK financial services. Int. J. Oper. Prod. Manag. **27**(9), 999–1019 (2007)
21. Marshall, D.A.: Lean transformation: overcoming the challenges, managing performance, and sustaining success. Theses and Dissertations–Marketing and Supply Chain (2014)
22. McCormack, K., et al.: A global investigation of key turning points in business process maturity. Bus. Process. Manag. J. **15**(5), 792–815 (2009)
23. Nadarajah, D., Kadir, A.S.L.S.: A review of the importance of business process management in achieving sustainable competitive advantage. TQM J. **26**(5), 522–531 (2014)
24. Niehaves, B., Poeppelbuss, J., Plattfaut, R., Becker, J.: BPM capability development – a matter of contingencies. Bus. Process. Manag. J. **20**(1), 90–106 (2014)
25. Ravensteyn, P., Batenburg, R.: Surveying the critical success factors of BPMsystems implementation. Bus. Process. Manag. J. **16**(3), 492–507 (2010)
26. von Rosing, M., Scheer, A.-W., von Sheer, H.: The Complete Business Process Handbook: Body of Knowledge from Process Modeling to BPM, vol. 1 (2015)
27. Stojanović, D., Jovanović, I., Slović, D., Tomašević, I., Simeunović, B.: BPM adoption in serbian companies. In: Di Ciccio, C., et al. (eds.) BPM 2019. LNBIP, vol. 361, pp. 324–327. Springer, Cham (2019). https://doi.org/10.1007/978-3-030-30429-4_23
28. Stojanović, D., Simeunović, B., Tomašević, I., Jovanović, I.: The influence of business process prioritization on success of BPM adoption. Smart Gov. Reg. Cities 235–258 (2019)
29. Stojanović, D., Tomašević, I., Slović, D., Gošnik, D., Suklan, J., Kavčič, K.: BPM in transition economies: joint empirical experience of Slovenia and Serbia. Econ. Res. - Ekonomska Istraživanja **30**(1), 1237–1256 (2017)
30. Trihatmoko, R.A., Mulyani, R., Lukviarman, N.: Product placement strategy in the business market competition: studies of fast moving consumer goods. Bus. Manag. Horizon **6**(1), 150–161 (2018)
31. Trkman, P.: The critical success factors of business process management. Int. J. Inf. Manag. **30**(2), 125–134 (2010)
32. van der Aalst, W.M.P., ter Hofstede, A.H.M., Weske, M.: Business process management: a survey. In: van der Aalst, W.M.P., Weske, M. (eds.) BPM 2003. LNCS, vol. 2678, pp. 1–12. Springer, Heidelberg (2003). https://doi.org/10.1007/3-540-44895-0_1
33. Wolf, C., Harmon, P.: The State of business process management (2012). http://www.bptrends.com/bpt/wp-content/surveys/2012-_BPT%20SURVEY-3-12-12-CW-PH.pdf. Accessed 3 May 2021
34. Wong, W.P.: Business-process management: a proposed framework for future research. Total Qual. Manag. Bus. Excell. **24**(5–6), 719–732 (2013)
35. Wong, W., Ahmad, N., Nasurdin, A., Mohamad, M.: The impact on external environmental on business process management and organizational performance. Serv. Bus. **8**(4), 559–586 (2014)

1st International Workshop on Business Process Management and Routine Dynamics (BPM&RD 2021)

1st International Workshop on Business Process Management and Routine Dynamics (BPM&RD 2021)

Business process management (BPM) and routine dynamics (RD) are both interested in how organizations carry out work [1]. Research has portrayed BPM and RD as "two islands of process research" that could profit tremendously from one another [2, 3]. To date, however, conversations in both fields largely take place independently of one another. To facilitate exchange among researchers from both fields, we organized and held the 1st International Workshop on Business Process Management and Routine Dynamics in conjunction with the 19th International Conference on Business Process Management in Rome, Italy. The workshop attracted a total of six submissions. Each submission was reviewed by at least two members of the Program Committee. We accepted four papers for presentation at the workshop and invited one of the submissions as a panel keynote.

The workshop was organized in two sessions. The first session was comprised of the paper presentations. The paper by Haase et al. [4] discussed how modeling notations can be expanded to model creative processes. Mahringer [5] showed how heat mapping can be used to detect interesting insights into process behavior. The paper by Kim et al. [6] used digital trace data to predict process structure after a disruption. Finally, the paper by Plattfaut and Borghoff [7] reported on an ongoing action research study on process innovation.

The second session of the workshop was dedicated to a panel on "Digital Twins of Organizations". Based on their paper, Markus Becker and Brian Pentland [8] posed the question "Digital Twin of an Organization: Are you serious?". Afterward, Barbara Weber approached and reflected on the topic from a technical perspective. As a final discussant, Kalle Lyytinen presented his ideas, taking a rather philosophical stance. In what followed, we had a question and answer session before moving into a more open discussion among the workshop participants.

We wish to express our gratitude to all those who made this first iteration of the workshop a success. In particular, we thank the Program Committee for providing high-quality reviews and the workshop chairs for their support regarding organizational matters. Lastly, we want to thank the authors and all workshop participants who contributed to making the workshop a lively discussion.

For future iterations of the workshop, we encourage submissions that capitalize on digital trace data to examine dynamics of organizational processes [3, 9], techniques to make such dynamics visible, and other kinds of thought-provoking papers that help strengthen the connection between BPM and RD.

November 2021

Bastian Wurm
Thomas Grisold
Waldemar Kremser
Jan Mendling

References

1. Mendling, J., Berente, N., Seidel, S., Grisold, T.: Pluralism and pragmatism in the information systems field: the case of research on business processes and organizational routines. Data Base Adv. Inf. Syst. Forthcom. (2021)
2. Pentland, B.T., Vaast, E., Wolf, R.: Theorizing process dynamics with directed graphs: a diachronic analysis of digital trace data. MIS Q. **45**, 967–984 (2020)
3. Wurm, B., Grisold, T., Mendling, J., vom Brocke, J.: Business process management and routine dynamics. In: Cambridge Handbook of Routine Dynamics. Cambridge University Press (2021)
4. Haase, J., Thim, C., Bender, B.: Expanding modeling notations: requirements for creative process modeling. In: Marrella, A., Weber, B. (eds.) BPM 2021 Workshop Proceedings. LNBIP. Springer, Berlin (2021)
5. Mahringer, C.A.: Analyzing digital trace data to promote discovery – the case of heatmapping. In: Marrella, A., Weber, B. (eds.) BPM 2021 Workshop Proceedings. LNBIP. Springer, Berlin (2021)
6. Kim, I., Frank, K.A., Wolf, J.R., Pentland, B.T.: Predicting process structure after a disruption: an example from the clinical documentation process. In: Marrella, A., Weber, B. (eds.) BPM 2021 Workshop Proceedings. LNBIP. Springer, Berlin (2021)
7. Plattfaut, R., Borghoff, V.: Capabilities for digital process innovation: results of an ongoing action research study. In: Marrella, A., Weber, B. (eds.) BPM 2021 Workshop Proceedings. LNBIP. Springer, Berlin (2021)
8. Becker, M.C., Pentland, B.T.: Digital twin of an organization: are you serious ? In: Marrella, A., Weber, B. (eds.) BPM 2021 Workshop Proceedings. LNBIP. Springer, Berlin (2021)
9. Grisold, T., Wurm, B., Mendling, J., vom Brocke, J.: Using process mining to support theorizing about change in organizations. In: 53rd Hawaiian International Conference on System Sciences (HICSS 2020) (2020)

Organization

Workshop Chairs

Bastian Wurm	Vienna University of Economics and Business, Austria
Thomas Grisold	University of Liechtenstein, Liechtenstein
Waldemar Kremser	Radboud University, The Netherlands
Jan Mendling	Humboldt-Universität zu Berlin, Germany

Program Committee

Wil van der Aalst	RHTW Aachen University, Germany
Adriano Augusto	University of Melbourne, Australia
Markus Becker	University of Southern Denmark, Denmark
Iris Beerepoot	Utrecht University, The Netherlands
Nicholas Berente	University of Notre Dame, USA
Maria Booth	Stockholm School of Economics, Sweden
Thorvald Hærem	Norwegian Business School, Norway
Kalle Lyytinen	Case Western Reserve University, USA
Christian Mahringer	University of Stuttgart, Germany
Martin Matzner	University of Erlangen-Nürnberg, Germany
Brian Pentland	Michigan State University, USA
Jan Recker	University of Hamburg, Germany
Christoph Rosenkranz	University of Cologne, Germany
Stefan Seidel	University of Liechtenstein, Liechtenstein
Inge van de Weerd	Utrecht University, The Netherlands
Verena Wolf	University of Paderborn, Germany
Youngjin Yoo	Case Western Reserve University, USA

Expanding Modeling Notations: Requirements for Creative Process Modeling

Jennifer Haase[1,2(✉)], Christof Thim[1], and Benedict Bender[1,2]

[1] Chair of Business Informatics, University of Potsdam, Potsdam, Germany
jennifer.haase@wi.uni-potsdam.de

[2] Weizenbaum-Institute for a Digitized Society, Berlin, Germany

Abstract. Creativity is a common aspect of business processes and thus needs a proper representation through process modeling notations. However, creative processes constitute highly flexible process elements, as new and unforeseeable outcome is developed. This presents a challenge for modeling languages. Current methods representing creative-intensive work are rather less able to capture creative specifics which are relevant to successfully run and manage these processes. We outline the concept of creative-intensive processes and present an example from a game design process in order to derive critical process aspects relevant for its modeling. Six aspects are detected, with first and foremost: process flexibility, as well as temporal uncertainty, experience, types of creative problems, phases of the creative process and individual criteria. By first analyzing what aspects of creative work modeling notations already cover, we further discuss which modeling extensions need to be developed to better represent creativity within business processes. We argue that a proper representation of creative work would not just improve the management of those processes, but can further enable process actors to more efficiently run these creative processes and adjust them to better fit to the creative needs.

Keywords: Modeling · Requirements · Pockets of creativity · Creative process

1 Introduction

Creativity is a natural aspect of knowledge work. In many knowledge intensive jobs, daily tasks include dealing with open and complex problems and demand the creation of prior unknown ideas, solutions and products [1]. Recent work which applied a routine perspective at creative performance shows a high degree of creativity within every day routine tasks [2, 3]. Thus, noteworthy levels of creativity can be achieved through repeated patterns of action – routines [4]. Such repeating process patterns can be made visible and manageable through modeling. However, creative processes – a set of linked tasked to develop something new and useful [5] – pose a certain challenge to common modeling techniques, as creativity requires unforeseeable process elements to be achieved.

Several methods do exist to capture creative tasks using standardized modeling techniques [6, 7]. Process segments containing some form of creativity can be indicated by an

A. Marrella and B. Weber (Eds.): BPM 2021 Workshops, LNBIP 436, pp. 197–208, 2022.
https://doi.org/10.1007/978-3-030-94343-1_15

extra symbol for creative work [6], potential process variation is signalled by thickened lines [8] or creative tasks are modelled as an unstructured case [9]. For these modeling approaches, we argue that they capture the creative process flow only approximately. The aspects relevant for creative work, like individual expertise, process phases and types of creative work, are yet unable to become visible with the modeling approaches existing so far. In this paper, we aim at a closer look at the requirements for creativity to be modeled. What must a modelling technique consider to represent the important aspects of creative processes? By dissecting the characteristics of creative work, we derive the components relevant for a detailed modeling approach. We further present the already available possibilities to model a detailed creative process based on existing modeling methods, as well as necessary modeling features which need to be developed in future work.

2 Understanding Creative Work

In order to model creative work, the specifics of what makes a process creative needs to be understood. In general, processes can be differentiated in their knowledge-intensity needed to perform the process [10]. When business processes are knowledge-intensive, like in product-development or strategy planning, the sequence of such processes is harder to predict, many parallel workflows are possible, and often documentation about the exact process flow is scarce. Thus, creative processes could be included to knowledge-intensive processes (KiP, [11]), and to differentiate any KiP from a creative one, an extension was made to specify creative-intensive processes (CiP). “A CiP is either a single creative activity that cannot be further broken down or a business process that contains at least one creative activity" [12, p. 3]. Even when creativity is understood as the overall result of a creative project [13], the actual tasks performed can be broken down into flexible, in which ideas are generated or worked upon, as well as more straight forward processes, which follow a predefined work-flow. Thus, even a core creative task like brainstorming is accompanied by more straight forward aspects, like writing down the ideas. This is captured under the concept of *Pockets of Creativity* (PoC, [14]).

2.1 Example of a Level Design from a Game Design Process

To illustrate a CiP, one approach for a level design from an overall game design process is graphically presented. Game design as a creative process was already analyzed from a routine perspective [15, 16], in which the dynamic and complex features of CiPs were also found. CiPs are characterized mainly by flexibility, as well as temporal and conditional uncertainty, experience, type of creative problem at hand, process phases and specific creator attributes. Based on an interview with an experienced gamer designer, the steps from a level design process were recorded and discussed (see Fig. 1). The process of level design includes the development of a games' environment, with the local features like landscapes, paths for the characters to use, obstacles and tokens [17]. Within the interview, a detailed level design process was identified, based on the latest process he went through for a third person action game. Three main steps were performed, which were processed one after the other. For each step, a set of tasks was identified, partly performed repeatedly, with some repetitions in altered form compared to the first process run. Especially the first task, the development of a paper plan, was characterized by many sub-processes, which were performed once in a logical order (solid arrows), but were performed in some different order in the following iterations (dashed arrows). Overall, within this stage, the highest level of flexibility could be seen compared to the overall process, as the subprocess steps are highly iterative, characterized by the game designer's experience from previous projects and in which he could play with new ideas. When the paper plan was set, the ideas for the level design were rebuilt in a second process phase using building blocks. First, a color and shape code was set for the 3D elements to model all elements of the game (like landscapes, characters, tokens), which were positioned according to the paper plan. Then, the positioning was checked, improved, also based on play-imitations with the character figures. The adapted design was tracked using photographs and short videos of the scenery. By iteratively checking all perspectives and possible movements, the designs functionality was improved. The goal of this phase is twofold: to check whether the prior paper plan works out in "reality" and to improve the setting and incorporate possibly new and interesting features. These steps of this review process were first carried out alone, then again with colleagues. The suggestions for improvement from those colleagues were collected with post-it's on the 3D elements, and later incorporated. As a result of the second step, the finished design was captured with pictures and videos. Next, the design was transferred to the virtual world, by using a game engine to start programming the actual game. The designer used virtual 3D objects as placeholders and by developing the virtual level, created a list of affordances for the programmers in parallel. At the end, the level design is finished when the programmers are fully instructed on what to implement.

2.2 Characteristics of Creative-Intensive Processes

Drawing upon the explanations from the level design process, combined with literature about the creative process' specifications, we aim to discuss main characteristics of the creative process deeming important for our modeling plan.

Flexibility. As a creative process leads to something new and useful, the exact path how to reach such a new idea or product is unclear in advance and different for process repetitions. This can only be achieved when the process includes some form of flexibility for the PoCs to be realized. Flexibility from a process perspective is closely related to the concept of agility. For the sake of clarification, agile processes describe a form of process management, which is characterized by short iterations between execution and evaluation, high level of collaboration and thus, high levels of work process flexibility [18]. In context of creative work, agile methods can be used and are often applied to structure creative projects, especially for software development [19], however, an agile process per se does not necessarily have to be a creative one. Focusing on the process at task execution level, an agile approach relies on flexibility, which is why we focus solely on the concept of flexibility.

The paper plan development from our case interview represents a PoC. Here, flexibility can be found within the iterations of tasks, and the time and perseverance spend

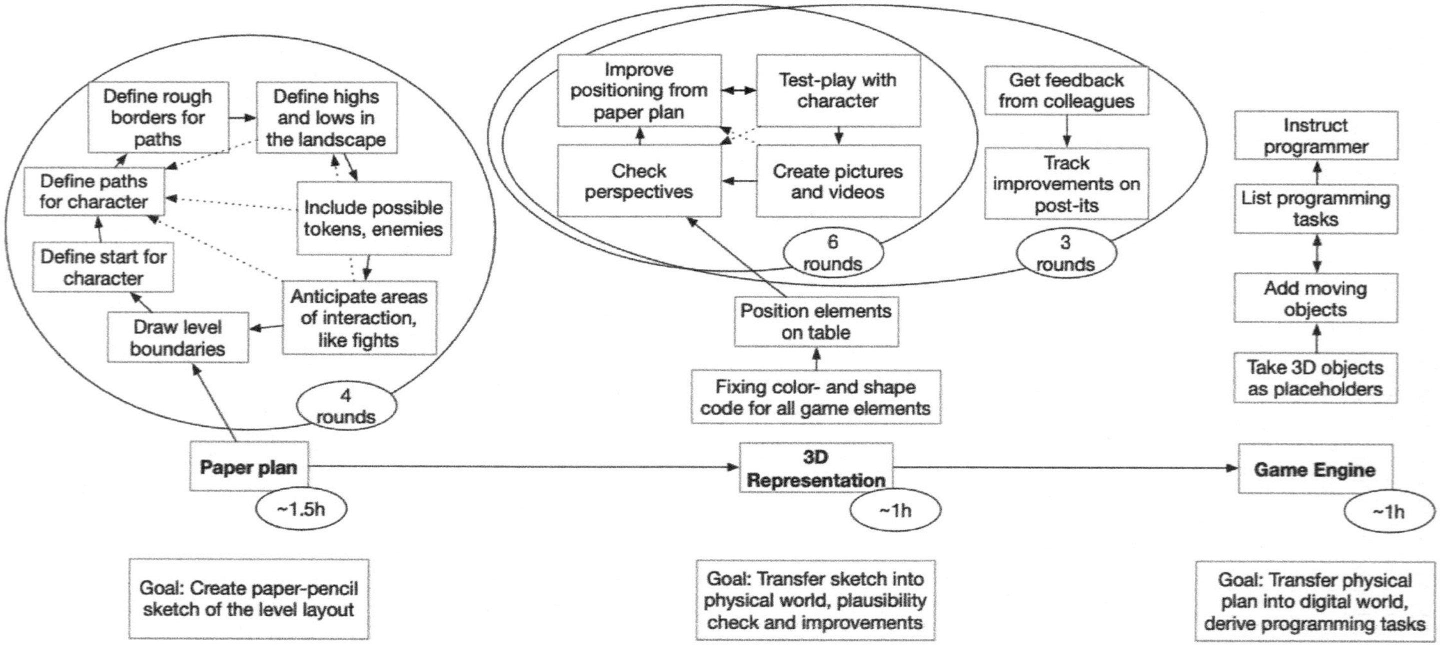

Fig. 1. Graphical representation of a level design process for a third person action game

by the game designer. In line with that, prevailing models describing organizational creativity emphasize the importance of flexibility at the structural level: individual freedom in terms of autonomy and decision making, as well as cooperation between teams and organizational units [20]. As the precise creative process unfolds through the performance, the actors need the freedom to choose when to perform which subprocesses, in what order, when to integrate third persons and other resources. Although some processes within the overall game design process are often repeatedly performed, those adaptations can lead to highly dynamic process repetition, with much different numbers of actions and thus differing degrees of complexity [20].

Temporal and Conditional Uncertainty. The aspect of flexibility leads to a challenge in planning the time needed for a process step. As subprocesses and activities can be repeated and iterated several times, the time needed for the overall process to be performed can differ hugely for process repetitions. In the level design case, the time taken on an instance level could be clearly reported by the game designer. However, based on past experience, he also reported that the same process sequence can take quite different amounts of time, depending on the task complexity or subjective evaluation of the achieved results. As mentioned for flexibility, the actor performing the process needs relative freedom concerning the time taken for a process step. Granting this is an elementary structural requirement for creative processes [21].

Experience. Another aspect we discussed in the interview is the knowledge and experience that is needed within the process. Following psychological approaches explaining creative performance, mainly three aspects are relevant to achieve creativity: domain knowledge, creative skills and task-motivation [20]. The basic idea is, that in order to create something of meaning, and to be able to solve a creative task within a domain, the person needs minimal knowledge about that domain [22]. However, for some tasks, a specific expert knowledge is crucial (like in the design-process, the software knowledge of the game engine to model the virtual scenery), whereas for other tasks, much less specific domain knowledge is necessary (like subjective evaluations concerning the functionality of the level design). For some modeling cases, it can be useful to indicate what kind and level of expertise is needed to successfully perform a process.

Type of Problem. Following the logic of different expertise needed within the creative process, creative tasks differ according to the type of problem at hand. For open problem task, ideation processes are required to find many and wide associations, often gathered in some form of brainstorming sessions [23]. In comparison, closed problem spaces require convergent thinking, as a form of complex problem solving and rely much higher on explicit domain knowledge [24]. Thus, differentiating those types of creative activities can improve the understanding of expertise needed for a successful process performance. Software supporting creative group work also differentiates between ideation and convergent thinking, to better support the different accompanied process needs (e.g., CreativeFlow, [25]).

Process Phases. Building on the differentiation between creative problems, creative work can usually be divided into distinguishable phases that can be found across domains [26]. Starting with a creative problem, early process steps are characterized by ideation processes, whereas later steps rely more on convergent thinking and elaboration of the idea [23]. In our case, especially the first step, the creation of the paper plan, was characterize by a very high level of ideation, as there are less many boundaries which narrow the problem space. Whereas later phases, like the development of the game-engine, the creative act relies mostly in translating an idea into concrete programming tasks. This is in so far relevant, as convergent thinking tasks are closer to KiPs than to CiPs and require thus less flexible management.

Creator Attributes. Drawing again on the psychological understanding for successful creative performance, two of the three main aspects – creative skills and task-motivation [20] – are yet to be addressed. Overall, creative process performance is highly influenced by the individuals' ideation skills [23], self-evaluation in terms of their own creative skills [27], and thus, their motivation to engage in a creative task [20]. In comparison to KiPs, for which a certain type and level of knowledge is needed, the individual preconditions necessary to perform a creative process successfully are located less objectively within the person, but depend strongly on the interaction of the person with the situation at hand [28].

3 Modelling Creative Work

Modelling business processes is a way to explicate enacted organizational routines. Here, we aim to explain why such an approach would be highly beneficial for the specific content of creative work. The core idea that visualized and recorded processes increase the understanding of processes' specifics, interdependencies, and possible improvements also hold up for creative work. Modeling constitutes means for the process' performance, management and information system support. The latter is eased by the formal description which can be interpreted by information systems [29]. For the management, process modelling generates an instrument to define expectations towards process and task performance. From this perspective, diverse measures can follow a thorough process understanding: resource allocation (like budget, equipment, time, employees), efforts enhancing creative output (through access to knowledge, communication practices, training of employees and teams), and risk management (through the implementation of review cycles with competent employees) [30]. The management of creative work requires especially the management of knowledge and individual creative skills. A thorough process understanding can improve the actor's performance by raising the understanding of when in the process (what type of) creativity will be best needed. We know from the conceptual proximity to KiP, as well as from psychological research on individual creative work: it requires specialist knowledge, creative competencies and motivation to act [20]. Following on research about "reflective talk" of organizational routines, the explication of shared mental models can improve the work flow and supports process adaptations [31]. Thus, modelling creative working routines can be one way to enable the employees involved to share their mental process models to better align their routine understanding with each other.

3.1 Modeling Languages Capturing Creative Work

Several ways already exist to model creative processes, since creativity is a typical component of knowledge work. However, classical modeling techniques capture standardized work patterns, in which creativity and flexibility are less focused upon. As the prior analysis has shown, the relevant aspects of creative work are rather more difficult to visualize. Equally to modelling approaches for KiPs, CiPs are well defined on a generic process level, but become complex and unstructured at finer granularity. As such, flexibility is the key aspect required to model creative processes. Thus, existing modeling approaches were searched and analyzed in how they reflect creativity, as well as flexibility aspects of processes. The approaches which we found differ in their aim to either capture directly PoCs, process variability and flexibility.

Modeling Pockets of Creativity. A limited amount of research has tried to map creativity directly. For example, the Architecture of Integrated Information Systems (ARIS) was extended to capture PoCs. A specific PoC-Symbol was introduced to represent a creative subtask within a workflow. This PoC can then be further described and specified on an associated PoC-sheet [6].

Modeling Process Variability. Variability of processes refers to the presentation of several alternative processes for the same subject matter [32]. This is technically possible with all notation languages, as the same process could be repeatedly modeled to show the nuanced differences between the processes for each variant, or for typical differences between variants. A thorough analysis and comparison of different modeling techniques capturing methods for process variability was performed by La Rosa and colleagues [9].

Modeling Process Flexibility. A set of methods have been developed to model some degree of flexibility into the process. Different forms of flexibility within processes are distinguished, with under-specification (accounts for hard to predict in advance processes, when actual execution of a process becomes clear through the performance) being most relevant for creative processes. Under-specification can be visualized by placeholders. In context of creative tasks, such placeholders could be used for PoCs, in which the detailed performance becomes apparent only through the performance. YAWL [7] is the only modeling language able to visualize flexibility by under-specification [33]. YAWL builds on petri-nets and includes placeholders for tasks, which's specifics are unknown before the performance [7].

Another quite different approach focuses on *flexibility in built-time* by modeling the intentions of a task as well as strategies to reach those, but not the actual task performance [34]. Here, processes are represented in modules (MAPs), from which each contains a source intention, a target intension and a strategy. Thus, for each MAP an entry state is defined, as well as a final state, plus a strategy to reach it. This method allows for a high level of process flexibility as well as a more detailed representation within those MAPs, compared to the YAWL approach. In a similar line, the Business Process Modeling Notation got extended to account for such processes which are rather unpredictable and evolve through enactment (case management model and notation, CMMN, [35]). CMMN was primarily invented to represent case management which require human intervention. Here, within the overall process flow it is known when such a case will arise, as well as what kind of tasks will be required. Although recommended for the modeling of creative tasks [30], CMMN is commonly seen as a solution for specific cases within an overall standardized process, as well as for exception handling [36], thus rather less prone for processes with high levels of flexibility and in-situ adaptations.

3.2 Modeling Extensions to Capture Creative Work

Based on the derived features of the creative process, we discuss the current modeling methods in their abilities to capture creativity, and highlight the aspects that need further development of modeling tools. For flexibility, as the most important feature to model creative work, several methods exist to represent e.g., under-specification and task-iterations. The variability approach could be applied in the creative work context especially for a higher-level process analysis like stage-gate-models, as they are often used in game design [20]. However, for a more detailed analysis of repeated creative performance, the variability approach does not fit well. A creative process can hardly be predefined as it establishes through the enactment. For a more nuanced approach, process aspects of flexibility *within* each process need to modeled. As such, the representations of cases (like CMMN), placeholders (YAWL) and PoCs (ARIS adaptation) gives an approximation to the visualization of creative tasks. However, as these keep the creative tasks within black boxes, those approaches do not allow for the representation of creativity-related particularities.

Modeling PoCs as an entity representing creative tasks might be sufficient for most modeling purposes. However, in some cases a more comprehensive understanding of the process will be necessary and useful for the management and improvement of creative output. Adequate modeling methods are required to illustrate the important aspects of creative work. We see four strategies by which modeling tools could be extended to better represent creative work (in detail, see Tab. 1). First, creative work is better represented on a sub-task level, in which the elements of a PoC are modeled. This could be done with existing methods, however even when creative *intensive* processes are modeled, this is rarely done (compare [6]). Second, creative work differs on what kind of task – open vs. closed – they address. Differentiating the kind of creative task could help to better determine what kind of expertise is needed, as well as the time probably needed for a sub-task. Third, complex creative tasks are performed in phases.

Indicating the creative process phases could help to better manage associated challenges and risks (compare [38]). Fourth, relevant individual (expertise, creative skills, motivation) and situational aspects (like external support, resources, time allocation) have a huge influence on the outcome of the creative performance. Especially expertise in terms of skills and knowledge can be highly relevant for the performance of a creative task by individuals and teams. Including these aspects in a model can significantly increase the visibility and possible improvement of these aspects.

Table 1. Overview of creative process aspects with current modeling methods and the derived modeling extensions.

Aspect	Current modeling approaches	Proposed modeling extensions
Flexibility	• Modelling parallel processes (variability) • Signalling for iterative process-part and/or tasks • Signalling underspecified tasks	• Flexibility of creative work is already well represented on a higher process level
Temporal uncertainty	• Indication whether a process comes with time-pressure (e.g., ARIS, [6])	• Differentiation between types of creative problems could indicate where time is better to predict
Experience	• Required knowledge/expertise can be assigned to tasks in form of "requirements" (e.g. Knowledge Modelling Description Language, KMDL [37])	• Differentiation between types of creative problems could indicate where what level of (non)expertise is needed to ideate and/or solve complex problems
Types of creative problems	• PoC vs. "normal" standardized process	• Differentiation between types of PoC: addressing open vs. closed problems
Phases of creative process	• Indirectly, as tasks of an overall creative process can be modelled	• Differentiation between phases of CiP: mainly ideation vs. elaboration [34]
Individual criteria	• Individual-related competences can be assigned to tasks in form of "requirements" (e.g., KMDL, [37])	• Indicating external factors that affect the specific creative process

4 Conclusion

An effective management of creative business processes is essential, as creative outcomes contribute significantly to organizational development and product innovations. Process modeling is an effective tool to visualize and manage creative business processes, which includes also the adequate representation of creative work. So far, some modeling methods annotate creative aspects within business processes, but keep the actual creative tasks in black-boxes. Often enough, modeling creativity on this level suffices for the modeling reason. If, however, influencing factors are to be modeled in addition, to better assign and plan creative process tasks [39], a more detailed process representation is necessary. This taps into a main dilemma of modeling creative work: modeling approaches aim to precisely describe processes, which in the case of creative work cannot be predicted *precisely* in advance. A middle ground needs to be found between the simplified "black-boxing" approach of creative tasks done so far and its actual performative complexity. A more detailed representation of creative work can be beneficial for the operative management of processes, especially in terms of resource allocation (time, budget, staff) and risk management. Process transparency can be enabling for creative work, in the sense that actors can be better trained in their skills and knowledge when it is known what precisely the process (step) needs. Also, organizational studies could show that an explicit examination of the creative process at work and its characteristics and challenges helps actors to improve their performances, as well as strengthens their understanding for future creative work [40]. Further, this effects can also be found for teams, as the explication of how the creative work should be best performed in a team can align the shared mental models of the team members, with positive impact on their overall creative output [21]. Overall, modeling important peculiarities of creative work can benefit their management and performance, plausibly resulting in higher quality – meaning more creative – work.

The presented analysis of current modeling methods is bound to the most prominent ones discussed in the literature. Also, our approach to analyze creative work using a case from video game design, combined with general statements on the part of the literature, limits the generalizability of our results. We aimed to discuss creative work from a general point of view, which comes with the downside of possibly missing important aspects from various domains [22]. However, our analysis can be used as a starting point to extend explicit modeling methods for the representation of creative work processes.

Acknowledgements. This work was supported by the German Federal Ministry of Education and Research (BMBF), grant number 16DII127 (Weizenbaum-Institute). We like to thank Friedrich Schadow for his in-depth input about level design within digital game design. We further like to thank Stefan Seidel and Thomas Grisold on their conceptual input during the data acquisition.

References

1. Eppler, M.J., Seifried, P., Röpnack, A.: Improving knowledge intensive processes through an enterprise knowledge medium. In: Meckel, M., Schmid, B.F. (Eds.) Kommunikationsmanagement im Wandel: Beiträge aus 10 Jahren, pp. 371–389. Gabler, Wiesbaden (2008)
2. Sonenshein, S.: Routines and creativity: from dualism to duality. Organ. Sci. **27**(3), 739–758 (2016). https://doi.org/10.1287/orsc.2016.1044
3. Deken, F., Carlile, P.R., Berends, H., Lauche, K.: Generating novelty through interdependent routines: a process model of routine Work. Organ. Sci. **27**(3), 659–677 (2016)
4. Pentland, B.T., Hærem, T.: Organizational routines as patterns of action: Implications for org Behavior. Ann. Rev. Organ. Psychol. Organ. Behav. **2**(1), 465–487 (2015)
5. Runco, M.A., Jaeger, G.J.: The standard definition of creativity. Creat. Res. J. **24**(1), 92–96 (2012). https://doi.org/10.1080/10400419.2012.650092
6. Karow, M., Reul, D.: Modellierung kreativitätsintensiver Prozesse – Methodische Unterstützung der Analyse und Dokumentation. In: Becker, J., Schwaderlapp, W., Seidel, S. (eds.) Management kreativitätsintensiver Prozesse: Theorien, Methoden, Software und deren Anwendung in der Fernsehindustrie, pp. 39–69. Springer, Heidelberg (2009). https://doi.org/10.1007/978-3-642-21678-7_3

7. van der Aalst, W.M.P., ter Hofstede, A.H.M.: YAWL: yet another workflow language. Inf. Syst. **30**(4), 245–275 (2005). https://doi.org/10.1016/j.is.2004.02.002
8. Gottschalk, F., Van Der Aalst, W.M., Jansen-Vullers, M.H., La Rosa, M.: Configurable workflow models. Int. J. Coop. Inf. Syst. **17**(2), 177–221 (2008)
9. La Rosa, M., Van der Aalst, W.M., Dumas, M., Milani, F.P.: Business process variability modeling: a survey. ACM Comput. Surv. CSUR **50**(1), 1–45 (2017)
10. Schmelzer, H.J., Sesselmann, W.: Geschäftsprozessmanagement in der Praxis: Kunden zufrieden stellen - Produktivität steigern - Wert erhöhen. Carl Hanser Verlag (2020)
11. Aureli, S., Giampaoli, D., Ciambotti, M., Bontis, N.: Key factors that improve knowledge-intensive business processes which lead to competitive advantage. Bus. Process Manag. J. **25**(1), 126–143 (2019). https://doi.org/10.1108/BPMJ-06-2017-0168
12. Seidel, S.: A business process management perspective on creativity: creativity intensive processes. In: AMCIS 2007 Proceedings, pp. 417–423 (2007)
13. Obstfeld, D.: Creative projects: a less routine approach toward getting new things done. Organ. Sci. **23**(6), 1571–1592 (2012). https://doi.org/10.1287/orsc.1110.0706
14. Seidel, S., Muller-Wienbergen, F., Rosemann, M.: Pockets of creativity in business processes. Commun. Assoc. Inf. Syst. **27**(1), 415–436 (2010)
15. Goh, K.T., Pentland, B.T.: From actions to paths to patterning: toward a dynamic theory of patterning in routines. Acad. Manage. J. **62**(6), 1901–1929 (2019)
16. Cohendet, P.S., Simon, L.O.: Always playable: recombining routines for creative efficiency at ubisoft montreal's video game studio. Organ. Sci. **27**(3), 614–632 (2016)
17. Schell, J.: The Art of Game Design: A Book of Lenses. CRC Press, Boco Raton (2008)
18. Bider, I., Jalali, A.: Agile business process development: why, how and when—applying Nonaka's theory of knowledge transformation to business process development. IseB **14**(4), 693–731 (2014). https://doi.org/10.1007/s10257-014-0256-1
19. Meyer, B.: Agile!: The Good, the Hype and the Ugly. Springer, Heidelberg (2014). https://doi.org/10.1007/978-3-319-05155-0
20. Amabile, T.M.: Componential theory of creativity. Harv. Bus. Sch. **12**(96), 1–10 (2012)
21. Martins, E.C., Terblanche, F.: Building organisational culture that stimulates creativity and innovation. Eur. J. Innov. Manag. **6**(1), 64–74 (2003). https://doi.org/10.1108/14601060310456337
22. Baer, J.: The importance of domain-specific expertise in creativity. Roeper Rev. **37**(3), 165–178 (2015). https://doi.org/10.1080/02783193.2015.1047480
23. Runco, M.: Div thinking, creativity and ideation. Camb. Handb. Creat. 413–446 (2010)
24. Cropley, A.: In praise of convergent thinking. Creat. Res. J. **18**(3), 391–404 (2006)
25. Becker, J., Heide, T., Breuker, D., Voigt, M.: Evaluating groupware for creative group processes: the case study of CreativeFlow. In: AMCIS Proceedings, Detroit, pp. 1–8 (2011)
26. Botella, M., Lubart, T.: From dynamic processes to a dynamic creative process. In: Beghetto, R.A., Corazza, G.E. (eds.) Dynamic Perspectives on Creativity. CTAE, vol. 4, pp. 261–278. Springer, Cham (2019). https://doi.org/10.1007/978-3-319-99163-4_15
27. Haase, J., Hoff, E.V., Hanel, P.H.P., Innes-Ker, Å.: A meta-analysis of the relation between creative self-efficacy and different creativity measurements. Creat. Res. J. **30**(1), 1–16 (2018). https://doi.org/10.1080/10400419.2018.1411436
28. Zhang, Z., Bartol, K.M.: Linking empowering leadership and employee creativity: the influence of psychological empowerment, intrinsic motivation, and creative process engagement. Acad. Manag. J. **53**(1), 107–128 (2010). https://doi.org/10.5465/amj.2010.48037118
29. Recker, J., Mendling, J.: The state of the art of business process management research as published in the BPM conference. Bus. Inf. Syst. Eng. **58**(1), 55–72 (2015). https://doi.org/10.1007/s12599-015-0411-3

30. Cham, R.K.: Creativity management–the new challenge for business process management. In: Khullar, L., Kavishwar, S., Bhutda, P. (eds.) Enterprise Value Creation Through Creative Management Practices, Nagpur, pp. 69–79 (2015)
31. Dittrich, K., Guérard, S., Seidl, D.: Talking about routines: the role of reflective talk in routine change. Organ. Sci. **27**(3), 678–697 (2016)
32. Valença, G., Alves, C., Alves, V., Niu, N.: A systematic mapping study on business process variability. Int. J. Comput. Sci. Inf. Technol. **5**(1), 1–21 (2013)
33. Schonenberg, H., Mans, R., Russell, N., Mulyar, N., Aalst, W.: Process flexibility: a survey of contemporary approaches. In: Dietz, J.L.G., Albani, A., Barjis, J. (eds.) CIAO!/EOMAS -2008. LNBIP, vol. 10, pp. 16–30. Springer, Heidelberg (2008). https://doi.org/10.1007/978-3-540-68644-6_2
34. Sadler-Smith, E.: Wallas' four-stage model of the creative process: more than meets the eye? Creat. Res. J. **27**(4), 342–352 (2015). https://doi.org/10.1080/10400419.2015.1087277
35. Freund, J., Rücker, B.: Praxishandbuch BPMN: mit Einführung in CMMN und DMN. Carl Hanser Verlag GmbH Co KG (2016)
36. Marrella. A., Lespérance, Y.: Towards a goal-oriented framework for the automatic synthesis of underspecified activities in dynamic processes. In: IEEE 6th International Conference on Service-Oriented Computing and Applications, pp. 361–365 (2013)
37. Gronau, N., Müller, C., Korf, R.: KMDL-capturing, analysing and improving knowledge-intensive business processes. J. Univers. Comput. Sci. **11**(4), 452–472 (2005)
38. Vargas-Hernández, J.G.: Modeling risk and innovation management. J. Compet. Stud. **19**(3/4), 45–57 (2011)
39. Waples, E.P., Friedrich, T.L.: Managing creative performance: important strategies for leaders of creative efforts. Adv. Dev. Hum. Resour. **13**(3), 366–385 (2011)
40. Byrge, C., Hansen, S.: Course in New Thinking in higher education: Enhancing creativity through the means of training, theory and workshop. Probl. Educ. 21st Century, **51**(51), 18–32 (2013)

Analyzing Digital Trace Data to Promote Discovery – The Case of Heatmapping

Christian A. Mahringer[1,2](✉)

[1] University of Stuttgart, Stuttgart, Germany
christian.mahringer@bwi.uni-stuttgart.de
[2] Heidelberg Academy of Sciences and Humanities, Heidelberg, Germany

Abstract. Business Process Management and Routine Dynamics are two streams of research that both explore process. To this end, Business Process Management has developed a rich array of methods that can be used to analyze digital trace data. Routine Dynamics has put less emphasis on the analysis of digital trace data, but it has advanced a methodological approach that promotes discovery, i.e., the process that actors perform and experience as they develop novel insights. This paper argues that the analysis of digital trace data can promote the process of discovery. It uses heatmapping as a specific example to show how analyzing digital trace data can promote discovery. The paper thus emphasizes a specific way how Business Process Management and Routine Dynamics can fertilize each other.

Keywords: Business process management · Digital trace data · Discovery · Heatmaps · Methodology · Organizational routines · Process analysis · Routine dynamics · Visualization

1 Introduction

Organizational processes have gained vast attention by scholars in different scientific fields. Two streams of research in which process has played a key role are Routine Dynamics and Business Process Management. Routine Dynamics (RD) is a stream of research that takes a processual perspective on organizational routines [1, 2], and focuses on the role of repetitive action patterns in organizations. These scholars seek to understand how routines matter for central organizational phenomena, such as change [3, 4] and innovation [5]. Similarly, Business Process Management (BPM) is concerned with the identification, analysis, redesign, implementation, and monitoring of business processes [6]. In this vein, these two research streams have been described as "two islands of process research" [7, 8].

However, from a methodological point of view, we can observe some differences between these islands. On one hand, BPM has developed sophisticated methods to analyze digital trace data. Trace data consist of a series of discrete events that are ordered sequentially [9]. Due to the increasing use of software tools in organizations, digital trace data are more and more available, providing many opportunities to study organizational processes. While the analysis of digital trace data also plays a role in RD research [10], most scholars in this domain have conducted ethnographic studies. These ethnographic studies typically follow a process of discovery [11–13]. Rather than focusing on

A. Marrella and B. Weber (Eds.): BPM 2021 Workshops, LNBIP 436, pp. 209–220, 2022.
https://doi.org/10.1007/978-3-030-94343-1_16

the application of a rigorous scientific method aiming at the generation of universally "valid" insights, scholars strive to follow surprises and hunches and, thus, deviate from pre-established conceptions.

In this paper, I argue that joining the forces of BPM and RD can lead to methodological advances in analyzing digital trace data. Business Process Management provides a rich array of methods that can be used to analyze digital trace data, whereas Routine Dynamics can contribute a methodological orientation towards discovery. This paper suggests that it is valuable to bring these aspects together, i.e., promoting discovery through the analysis of digital trace data. To illustrate this argument, the paper draws on "heatmapping" as a specific method of analyzing digital trace data. A heatmap is a graphical visualization of a matrix that assigns different colors to areas according to a certain logic. For instance, higher values can take darker colors while lower values take lighter colors. In this vein, heatmaps can visualize patterns in the data and facilitate the discovery process of the researcher. In this paper, heatmapping serves as a "boundary object" to show how both streams of research can join their forces.

The data set used for illustration is derived from a larger study of software development teams in a medium sized high-tech manufacturing company, which develop software tools to control complex cabin-like machines [14, 15]. The software development teams enact routines based on the Scrum framework, which splits software development into distinct phases of two weeks (i.e., sprints), and defines events, rules, and roles that support the sprints [16]. The study combines ethnographic field work (i.e., observation of the teams in conjunction with interviews and documents) with digital trace data extracted from a software tool that the teams use to organize their work. This paper uses a sub-set of the digital trace data that covers a period of two years. The data set consists of approximately 41.000 actions. The aim of this paper is not to produce novel insights into organizational processes, but to illustrate how the analysis of digital trace can promote discovery.

2 Discovery in Routine Dynamics and Business Process Management Research

From a methodological point of view, detailed, ethnographic studies play a key role in RD research. Scholars have performed such ethnographic studies in a variety of different contexts such as student housing organizations [17], start-ups [18] or artificial intelligence labs [19]. In this vein, Dittrich [11] identifies 45 different ethnographic papers in the RD field. The analysis of ethnographic data does not only entail the application of suitable methods to pre-specified questions, but it involves a process of discovery [13]. By discovery I refer to the process that actors perform and experience as they develop novel insights. Dittrich [11, p. 3] characterizes the process of discovery, stating that "[e]thnographers are open to surprises in the field and they work with these surprising observations to develop new ideas, concepts and/ or explanations." As scholars engage with a phenomenon, they see things that they did not expect, and they are subsequently drawn towards these surprising or puzzling things.

While most Routine Dynamics scholars have performed ethnographic studies, some scholars have also analyzed (digital) trace data. Pentland and Feldman [20] introduce the notion of narrative networks. In a narrative network, the nodes represent actions, and the edges represent sequential relations between those actions. Narrative networks can be used to compare organizational processes across time and space. While narrative networks can be used to analyze ethnographic data [18, 21], they are also useful for analyzing (digital) trace data [22–24]. Based on the notion of narrative networks, Goh and Pentland [25] analyze the paths of how routines are and could be performed. Identifying such paths enables unpacking the dynamics of a process by tracing how paths change over time [26].

In Business Process Management the analysis of digital trace data has been a more central focus. As Wurm, Grisold, Mendling and vom Brocke [7] note, this stream of research is closely connected to the emergence and development of software tools that accumulate digital trace data. As a consequence, a variety of methods, such as algorithms [27] or process modelling [28], has been developed that can be used to analyze digital trace data. BPM research, however, has put less emphasis on how using such methods can drive the process of discovery as defined above. In BPM, the term "discovery" seems to be used in a different sense. The business process management lifecycle [6], for instance, includes "process discovery" as a key phase, but here discovery means the identification of "descriptive information about how the process is performed" [29, p. 5494], and not the process of following hunches and surprises. In BPM, discovery refers to the technical methods that can inductively be used to identify a "unique underlying process model" [30, p. 56] in a complex data set, while in RD discovery refers to the process that actors perform and experience as they develop novel insights, which is shaped by hunches and surprises. This paper follows the latter definition of the term. Notably, both aspects are interrelated, as BPM methods can be used to promote the discovery process as conceived in RD.

Visualizations play a key role in analyzing digital trace data. Even though puzzles can become apparent from observation per se (e.g., observing a situation that does not fit the theory during fieldwork) or coding [12], visualizations are useful because they can bring a specific point of view to the foreground and reduce complexity. While visualizations can also be constructed from ethnographic data, they are a key method of identifying patterns in processes represented by digital trace data. Goh and Pentland [25], for instance, visualize how paths change over the sprints of an agile project. This enabled them to identify different "phases of patterning" in the project. Similarly, BPM scholars have developed various visualizations such as different kinds of process diagrams [31], petri nets [32] or heatmaps [33–36].

3 Using Heatmapping to Promote Discovery

3.1 Heatmapping

A heatmap is a graphical visualization of a matrix that assigns different colors to areas according to a certain logic. For instance, higher values attain darker colors while lower values attain lighter colors. Figure 1 shows such a heatmap. In this example, the matrix represents transition probabilities between actions that make up the Scrum routines of a software development team. The library (i.e., the set of actions considered) consists of 51 distinct actions. Each cell represents the probability that a specific action X follows another action Y. As shown in the legend on the right-hand side, transition probabilities of 10% and below are colored light blue, probabilities of 30% and above are colored dark blue, and probabilities between 10 and 30% are distributed on the color spectrum ranging from light to dark blue.

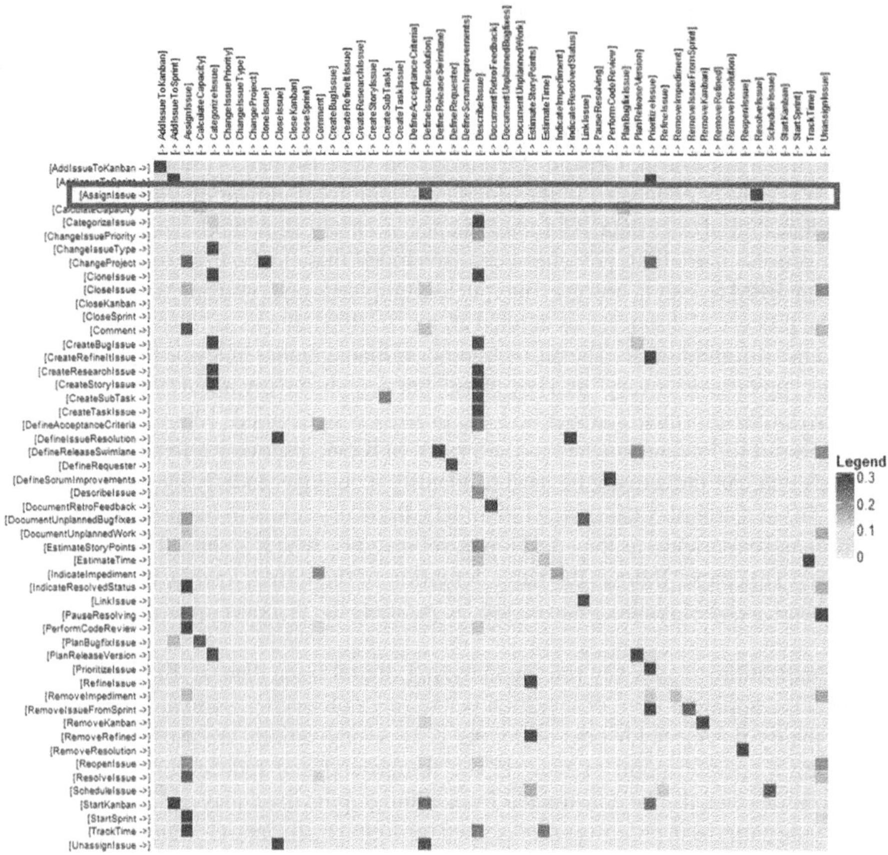

Fig. 1. Heatmap of the transition matrix of software development routines

For instance, the red-framed row indicates actions that follow the assignment of an issue (e.g., a software bug or a new feature requirement) to a specific software developer in the software tool. The heatmap shows that there are two actions that typically follow assigning, i.e., developing a resolution for the issue and indicating that the issue has been resolved. This insight can facilitate discovery, as we strive to understand these patterns. In this case, the reason why assigning an issue was followed by resolving it was that the developers assigned their own account to an issue before they started to resolve it, so that other developers knew that the issue was "reserved". This prevented that multiple developers simultaneously worked on the same issue. When a developer was done resolving an issue, he assigned a tester that set the resolution of the issue to "fixed" if it satisfied his requirements. This explains why the assignment of an actor (in this case a tester) was oftentimes followed by the indication of the resolution.

3.2 How to Use Heatmapping to Promote Discovery?

As the prior example reveals, heatmapping (as any other type of visualization) requires a profound sensitivity towards the data. In the following, I describe how to use heatmapping in studying organizational processes. Complementary instructions for the analysis of digital trace data have been provided elsewhere [6, 10, 37].

Step 1: Defining the Lexicon. The first step in the analysis is the definition of the set of actions to be considered (i.e., lexicon). The 51 distinct actions shown above are a choice of the researcher. I selected a lexicon that closely resembles how the actors in the field see the Scrum routines. This was possible due to the deep contextualization of the digital trace data through fieldwork and interviews. A larger lexicon has more potential to facilitate the discovery process, because it provides more opportunities to ask questions about the relationships between actions. However, when we compare multiple heatmaps across time and space this process can easily be overwhelming. A smaller heatmap reduces complexity and helps to mitigate this challenge. Hence, there is a need to balance richness and complexity in a way that it promotes discovery.

Step 2: Calculating the Transition Matrix. Heatmapping requires a matrix that can be transformed into a heatmap visualization. I used a transition matrix that shows how actions follow each other. To compute the matrix I used the "TraMineR" package in R [38], and to create the heatmaps I used the "ComplexHeatmap" package [39]. However, we could also apply heatmapping to other applications, such as a matrix that shows how different actors or roles interact with each other. Yeshchenko et al. [33], for example, use heatmaps to detect process drift. Burattin, Kaiser, Neurauter and Weber [35] use heatmaps to visualize eye tracking results in process modelling.

Another important decision is whether to use transition probabilities or absolute transitions. Absolute transitions sum up how often an action X follows an action Y, while transition probabilities represent the probability that an action X follows an action Y. Transition probabilities can be biased through rare values. For example, when an action only occurs one times in the data set, there is only one other action that follows, leading to a transition probability of 100%. On the other side, absolute transitions make it difficult to compare processes across time and space when the compared matrices differ substantially in the total number of actions.

Figure 2 shows two heatmaps for the same data set. The heatmap on the left-hand side is based on transition probabilities and the heatmap on the right-hand side is based on absolute transitions. As shown by the red frame on the left-hand side, rare values—in this case changing the project assigned to an issue, which only occurs 4 times in the data set—signify high transitions probabilities. On the right-hand side, the absolute count of describing an issue (n = 2928) leads to highlights that do not occur according to probabilities. However, we can also use these sensitivities to ask new questions about the process, for instance, whether transition probabilities of low or high frequency events are significant for process dynamics.

Step 3: Defining the Color Scheme. The definition of the color scheme is another crucial step in the process, because the heatmap visualization is influenced by the color thresholds. Figure 3 illustrates three different thresholds for the same process. In the heatmap on the left-hand side, colors become darker starting from transition probabilities of 1%, in the middle 10%, and on the left-hand side 20%. A low threshold is

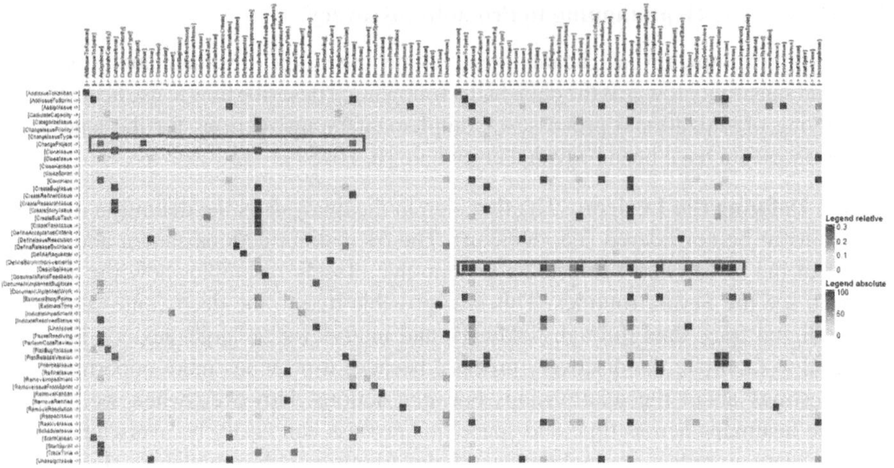

Fig. 2. Comparison of heatmaps based on transition probabilities (left) and absolute transitions (right)

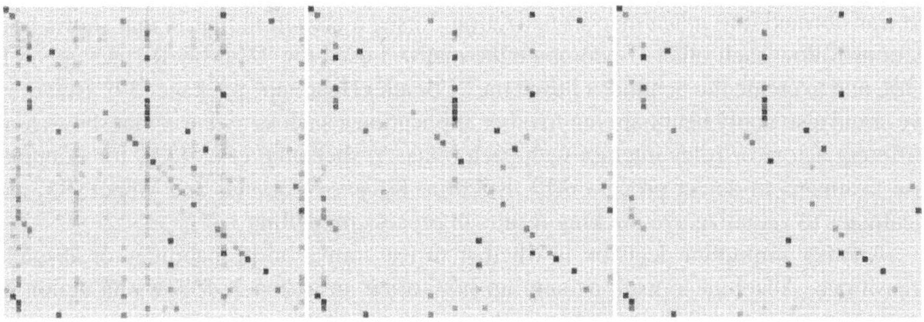

Fig. 3. Heatmaps with different color thresholds (from left to right: 1%, 10%, 20%)

sensitive towards small transition probabilities but can also be overwhelmingly complex. A higher threshold supports the identification of more typical transitions in the action pattern. Here, heatmapping plays out one of its core strengths because it is possible to adjust the sensitivity of the heatmap in a way that accounts for the question at hand. I selected only one color with different shades, which is helpful to highlight frequent transitions, but it is also possible to use a color gradient across multiple colors, which may lead to different insights.

Step 4: Comparing Across Time and Space. Finally, we can compare heatmaps across time and space to generate new insights into process dynamics. For example, we could split the data set into different time frames and create heatmaps for each time frame. Subsequently, it is possible to compare how the heatmaps (and thus the process) change over time. In this sense, heatmapping can be used to better understand how processes change over time—a key question in routine dynamics research [40, 41].

Alternatively, we can compare how a process is enacted in different places. This can be relevant for different questions such as the role of context for routine enactment [42] or routine replication [43, 44]. BPM scholars have also emphasized the need of considering

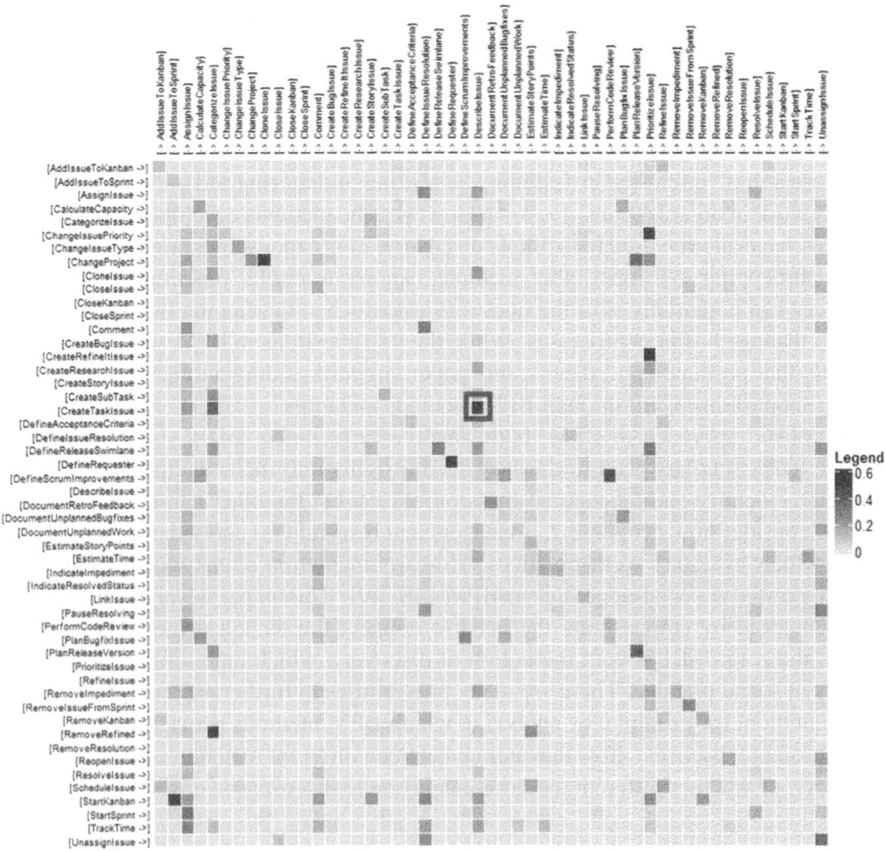

Fig. 4. Differences in transition probabilities between two teams

context [7, 45]. Figure 4 shows the differences between transition probabilities of an old and a new software development team. The matrix reveals that the routines in both teams were not similar, even though the novel team replicated the routines of the old team. We can use the heatmap to examine how the routines in both teams differed. For instance, the red highlight in the figure signifies a difference in how creating a task issue (a specific kind of issue as opposed to bugs and stories) was followed by describing the issue. Following this surprising puzzle revealed that only the new team used this specific type of issue. This leads to further questions, such as why this was the case. In this vein, the heatmap drives discovery.

4 Conclusion

This paper suggests that there is ample potential for research on Business Process Management and Routine Dynamics to learn from each other. It has focused on a specific aspect, that is, the combination of methods for analyzing digital trace data with a methodological orientation toward discovery. Put differently, this paper suggests the fusion of BPM methods and RD methodology. In this vein, it addresses scholars that have emphasized the synergetic potential of both streams of research [7, 29, 46, 47].

4.1 Promoting Discovery in Business Process Management

BPM research is "interested in designing, analyzing, monitoring and improving business process work in organizations" [7, p. 4]. However, the methods developed in this stream of research can also be used to facilitate discovery. This paper shows how this is possible with heatmapping, but most other methods can serve similar functions. Considering discovery can potentially shift the questions asked when developing new methods and tools to manage business processes. While the key questions in BPM seem to focus on methods that produce valid results, this paper suggests adding another question: how can we support discovery through BPM methods and tools?

A holistic approach to BPM, both, involves a technical and a social dimension. On one hand, scholars need to develop deep insights into methods and tools that can be used to identify and manage business processes. On the other hand, BPM scholars need to consider how people (e.g., managers and researchers) apply these methods and tools in their everyday work practices. This argumentation aligns with scholars who have suggested a social approach to BPM [48, 49]. This paper advances this line of argumentation, by emphasizing the need to connect the technical methods that have been at the core of BPM with the notion of discovery, which represents a process that unfolds in social practice. There are many possibilities to further expand on this line of argumentation, for example, by analyzing which BPM methods managers and scholars use to promote discovery, how they iterate between different methods as they engage in discovery, or whether different methods can promote discovery in different ways.

4.2 Enriching the Repertoire of Methods in Routine Dynamics

Discovery is a key process in RD research, but it has mostly been emphasized in the context of ethnographic studies. This paper suggest that the analysis of digital trace data can also be a driver of discovery; like coding field notes and interviews it is possible to apply methods for the analysis of digital trace data. The resulting metrics and visualizations can then lead to new questions that lead to the application of different methods, propelling the discovery process forward. The analysis of digital trace data can most likely lead to different questions than the analysis of fieldnotes and interview data because digital trace data oftentimes extend across longer periods of time and across multiple locations [10]. When we apply multiple methods (e.g., analysis of digital trace data, field notes, and interview data) in conjunction, it is possible to further reinforce the momentum of discovery.

RD research has developed some methods that can help analyze (digital) trace data in a way that promotes discovery, such as narrative networks [20, 22], path-based analysis [25], or analysis based on clock- and event-time [50]. BPM research also provides many tools that could be used to analyze digital trace data in a way that promotes discovery [6, 31, 33]. This paper complements this set of methods by adding heatmapping to the repertoire of possible methods.

A common way of visualizing the results are network graphs, that display the nodes and edges of a narrative network. However, there are notable differences between network graphs and heatmaps. Network graphs play out their strengths when we strive to understand the overall structure of a routine pattern and how this structure changes over time, or when we want to get a visual impression of the complexity of a routine. By contrast, heatmaps (how they have been applied in this paper) foreground differences in transitions between actions. This can be useful for Routine Dynamics research, because (a) heatmaps help to see more likely transitions (i.e., which action does typically come next?), (b) the nuances of color differences help generate a more fine-grained picture of the routine pattern, and (c) heatmaps also visualize transitions that have not happened, i.e., the "in-between" or "white" space [23]. Moreover, heatmaps could be used beyond visualizing transitions between actions. For example, we could visualize transitions between routines or degrees of variation of routines in a routine ecology. Depending on how heatmaps are applied, they may guide the discovery process in different directions.

4.3 Driving Discovery Through Visualization

This paper has elaborated heatmapping as a specific way of analyzing and visualizing organizational processes to promote the process of discovery. The specific examples from the data reveal that discovery depends on how visualizations are crafted. Hence, visualizations play an important role for the discovery process, that goes beyond structuring and representation of results. This argument echoes Feldman [51], who argues that making process visible requires different approaches to visualization. She also explains that different visualizations offer different points of view on a phenomenon and may influence theorizing in different ways. This paper adds to this line of argumentation by emphasizing that visualizations can play a fundamental role in driving discovery.

Acknowledgements. This research has been funded by the Heidelberg Academy of Sciences and Humanities. I am grateful for the support of the editors of this issue, for the feedback of Thomas Grisold on prior versions of this paper, the helpful comments of three anonymous reviewers, and the access and support provided by the host company in which the data has been collected. I also thank Brian Pentland, who has inspired me to approach the analysis of digital trace data.

References

1. Feldman, M.S.: Routines as process. In: Howard-Grenville, J., Rerup, C., Langley, A., Tsoukas, H. (eds.) Organizational Routines. How They are Created, Maintained, and Changed, pp. 23–46. Oxford University Press, Oxford (2016)
2. Howard-Grenville, J., Rerup, C.: A process perspective on organizational routines. In: Langley, A., Tsoukas, H. (eds.) The Sage Handbook of Organization Process Studies, pp. 323–337. Sage, Thousand Oaks (2017). https://dx.doi.org/10.4135/9781473957954.n20
3. Feldman, M.S., Pentland, B.T., D'Adderio, L., Dittrich, K., Rerup, C., Seidl, D.: What is routine dynamics? In: Feldman, M.S., Pentland, B.T., D'Adderio, L., Dittrich, K., Rerup, C., Seidl, D. (eds.) Cambridge Handbook of Routine Dynamics, pp. 1–18. Cambridge University Press, Cambridge (2021). https://doi.org/10.1017/9781108993340.002
4. Feldman, M.S., D'Adderio, L., Dittrich, K., Jarzabkowski, P.: Routine dynamics in action: replication and transformation. Res. Sociol. Organ. **61**, 1–10 (2019)

5. Deken, F., Sele, K.: Innovation work and routine dynamics. In: Feldman, M.S., Pentland, B.T., D'Adderio, L., Dittrich, K., Rerup, C., Seidl, D. (eds.) Cambridge Handbook of Routine Dynamics, pp. 288–300. Cambridge University Press, Cambridge (2021). https://doi.org/10.1017/9781108993340.025
6. Dumas, M., La Rosa, M., Mendling, J., Reijers, H.A.: Fundamentals of Business Process Management. Springer, Heidelberg (2018). https://doi.org/10.1007/978-3-662-56509-4
7. Wurm, B., Grisold, T., Mendling, J., vom Brocke, J.: Business process management and routine dynamics. In: Feldman, M.S., Pentland, B.T., D'Adderio, L., Dittrich, K., Rerup, C., Seidl, D. (eds.) Cambridge Handbook of Routine Dynamics, pp. 513–524. Cambridge University Press, Cambridge (2021). https://doi.org/10.1017/9781108993340.042
8. Pentland, B.T., Vaast, E., Ryan Wolf, J.: Theorizing process dynamics with directed graphs: a diachronic analysis of digital trace data. MIS Q. **45**, 967–984 (2021). https://doi.org/10.25300/MISQ/2021/15360
9. Cornwell, B.: Social Sequence Analysis: Methods and Applications. Cambridge University Press, New York (2015). https://doi.org/10.1017/CBO9781316212530
10. Mahringer, C.A., Pentland, B.T.: Sequence analysis in routine dynamics. In: Feldman, M.S., Pentland, B.T., D'Adderio, L., Dittrich, K., Rerup, C., Seidl, D. (eds.) Cambridge Handbook of Routine Dynamics, pp. 172–183. Cambridge University Press, Cambridge (2021). https://doi.org/10.1017/9781108993340.015
11. Dittrich, K.: Ethnography and routine dynamics. In: Feldman, M.S., Pentland, B.T., D'Adderio, L., Dittrich, K., Rerup, C., Seidl, D. (eds.) Cambridge Handbook of Routine Dynamics, pp. 103–129. Cambridge University Press, Cambridge (2021). https://doi.org/10.1017/9781108993340.011
12. Locke, K., Feldman, M.S., Golden-Biddle, K.: Discovery, validation, and live coding. In: Elsbach, K.D., Kramer, R.M. (eds.) Handbook of Qualitative Organizational Research, pp. 371–379. Routledge, New York (2015)
13. Locke, K.: Field research practice in management and organization studies: reclaiming its tradition of discovery. Acad. Manag. Ann. **5**, 613–652 (2011). https://doi.org/10.5465/19416520.2011.593319
14. Mahringer, C.A., Dittrich, K., Renzl, B.E.: Interdependent routines and innovation processes – An ethnographic study of Scrum teams. In: Academy of Management Proceedings. Academy of Management (2019). https://doi.org/10.5465/AMBPP.2019.11891abstract
15. Mahringer, C.A.: Exploring routine ecologies – A characterization and integration of different perspectives on routines. University of Stuttgart, Dissertation, Stuttgart (2019)
16. Schwaber, K., Sutherland, J.: The Scrum guide. The definite guide to Scrum: The rules of the game (2020)
17. Feldman, M.S.: Organizational routines as a source of continuous change. Organ. Sci. **11**, 611–629 (2000). https://doi.org/10.1287/orsc.11.6.611.12529
18. Dittrich, K., Guérard, S., Seidl, D.: Talking about routines: the role of reflective talk in routine change. Organ. Sci. **27**, 678–697 (2016). https://doi.org/10.1287/orsc.2015.1024
19. Sele, K., Grand, S.: Unpacking the dynamics of ecologies of routines: mediators and their generative effects in routine interactions. Organ. Sci. **27**, 722–738 (2016). https://doi.org/10.1287/orsc.2015.1031
20. Pentland, B.T., Feldman, M.S.: Narrative networks: patterns of technology and organization. Organ. Sci. **18**, 781–795 (2007). https://doi.org/10.1287/orsc.1070.0283
21. Danner-Schröder, A., Geiger, D.: Unravelling the motor of patterning work: toward an understanding of the microlevel dynamics of standardization and flexibility. Organ. Sci. **27**, 633–658 (2016). https://doi.org/10.1287/orsc.2016.1055
22. Pentland, B.T., Kim, I.: Narrative networks in routine dynamics. In: Feldman, M.S., Pentland, B.T., D'Adderio, L., Dittrich, K., Rerup, C., Seidl, D. (eds.) Cambridge Handbook of Routine

Dynamics, pp. 184–195. Cambridge University Press, Cambridge (2021). https://doi.org/10.1017/9781108993340.016
23. Pentland, B.T., Recker, J., Wolf, J.R., Wyner, G.: Bringing context inside process research with digital trace data. J. Assoc. Inf. Syst. **21**, 5 (2020). https://doi.org/10.17705/1jais.00635
24. Pentland, B.T., Recker, J., Wyner, G.: Rediscovering handoffs. Acad. Manag. Discov. **3**, 284–301 (2017). https://doi.org/10.5465/amd.2016.0018
25. Goh, K., Pentland, B.T.: From actions to paths to patterning: toward a dynamic theory of patterning in routines. Acad. Manag. J. **62**, 1901–1929 (2019). https://doi.org/10.5465/amj.2018.0042
26. Pentland, B.T., Mahringer, C.A., Dittrich, K., Feldman, M.S., Wolf, J.R.: Process multiplicity and process dynamics: weaving the space of possible paths. Organ. Theory **1**, 1–21 (2020). https://doi.org/10.1177/2631787720963138
27. van den Aalst, W.M.P.: Process Mining: Discovery, Conformance and Enhancement of Business Processes. Springer, Heidelberg (2011). https://doi.org/10.1007/978-3-642-19345-3
28. Kummer, T.-F., Recker, J., Mendling, J.: Enhancing understandability of process models through cultural-dependent color adjustments. Decis. Support Syst. **87**, 1–12 (2016). https://doi.org/10.1016/j.dss.2016.04.004
29. Grisold, T., Wurm, B., Mendling, J., Vom Brocke, J.: Using process mining to support theorizing about change in organizations. In: Proceedings of the 53rd Hawaii International Conference on System Sciences (2020)
30. Hwang, S.-Y., Yang, W.-S.: On the discovery of process models from their instances. Decis. Support Syst. **34**, 41–57 (2002). https://doi.org/10.1016/S0167-9236(02)00008-8
31. Malinova, M., Mendling, J.: Cognitive diagram understanding and task performance in systems analysis and design. MIS Q. (In Press)
32. Aalst, W.: Making work flow: on the application of petri nets to Business Process Management. In: Esparza, J., Lakos, C. (eds.) ICATPN 2002. LNCS, vol. 2360, pp. 1–22. Springer, Heidelberg (2002). https://doi.org/10.1007/3-540-48068-4_1
33. Yeshchenko, A., Di Ciccio, C., Mendling, J., Polyvyanyy, A.: Visual drift detection for sequence data analysis of business processes. IEEE Trans. Visual. Comput. Graph. (2021). https://doi.org/10.1109/TVCG.2021.3050071
34. Melcher, J.: Process Measurement in Business Process Management: Theoretical Framework and Analysis of Several Aspects. KIT Scientific Publishing, Karlsruhe (2014). https://doi.org/10.5445/KSP/1000025397
35. Burattin, A., Kaiser, M., Neurauter, M., Weber, B.: Eye tracking meets the process of process modeling: a visual analytic approach. In: Dumas, M., Fantinato, M. (eds.) BPM 2016. LNBIP, vol. 281, pp. 461–473. Springer, Cham (2017). https://doi.org/10.1007/978-3-319-58457-7_34
36. Chapela-Campa, D., Mucientes, M., Lama, M.: Discovering infrequent behavioral patterns in process models. In: Carmona, J., Engels, G., Kumar, A. (eds.) BPM 2017. LNCS, vol. 10445, pp. 324–340. Springer, Cham (2017). https://doi.org/10.1007/978-3-319-65000-5_19
37. Keegan, B.C., Lev, S., Arazy, O.: Analyzing organizational routines in online knowledge collaborations: a case for sequence analysis in CSCW. In: Proceedings of the 19th ACM Conference on Computer-Supported Cooperative Work & Social Computing, pp. 1065–1079. ACM (2016). https://doi.org/10.1145/2818048.2819962
38. Gabadinho, A., Ritschard, G., Mueller, N.S., Studer, M.: Analyzing and visualizing state sequences in R with TraMineR. J. Stat. Softw. **40**, 1–37 (2011). https://doi.org/10.18637/jss.v040.i04
39. Gu, Z., Eils, R., Schlesner, M.: Complex heatmaps reveal patterns and correlations in multidimensional genomic data. Bioinformatics **32**, 2847–2849 (2016). https://doi.org/10.1093/bioinformatics/btw313

40. Pentland, B.T., Feldman, M.S., Becker, M.C., Liu, P.: Dynamics of organizational routines: a generative model. J. Manage. Stud. **49**, 1484–1508 (2012). https://doi.org/10.1111/j.1467-6486.2012.01064.x
41. Pentland, B.T., Hærem, T., Hillison, D.: Comparing organizational routines as recurrent patterns of action. Organ. Stud. **31**, 917–940 (2010). https://doi.org/10.1177/0170840610373200
42. Howard-Grenville, J.: The persistence of flexible organizational routines: the role of agency and organizational context. Organ. Sci. **16**, 618–636 (2005). https://doi.org/10.1287/orsc.1050.0150
43. D'Adderio, L.: The replication dilemma unravelled: how organizations enact multiple goals in routine transfer. Organ. Sci. **25**, 1325–1350 (2014). https://doi.org/10.1287/orsc.2014.0913
44. Boe-Lillegraven, S.: Transferring routines across multiple boundaries: a flexible approach. Res. Sociol. Organ. **61**, 31–53 (2019). https://doi.org/10.1108/S0733-558X20190000061003
45. vom Brocke, J., Zelt, S., Schmiedel, T.: On the role of context in business process management. Int. J. Inf. Manag. **36**, 486–495 (2016). https://doi.org/10.1016/j.ijinfomgt.2015.10.002
46. Mendling, J., Berente, N., Seidel, S., Grisold, T.: Pluralism and pragmatism in the information systems field: the case of research on business processes and organizational routines. Data Base Adv. Inf. Syst. **52**, 127–140 (2021). https://doi.org/10.1145/3462766.3462773
47. Mendling, J., Pentland, B.T., Recker, J.: Building a complementary agenda for business process management and digital innovation. Eur. J. Inf. Syst. **29**, 208–219 (2020). https://doi.org/10.1080/0960085X.2020.1755207
48. Ariouat, H., Hanachi, C., Andonoff, E., Benaben, F.: A conceptual framework for social business process management. Procedia Comput. Sci. **112**, 703–712 (2017). https://doi.org/10.1016/j.procs.2017.08.151
49. Suša Vugec, D., Tomičić-Pupek, K., Vukšić, V.B.: Social business process management in practice: overcoming the limitations of the traditional business process management. Int. J. Eng. Bus. Manag. **10**, 1–10 (2018). https://doi.org/10.1177/1847979017750927
50. Geiger, D., Danner-Schröder, A., Kremser, W.: Getting ahead of time—Performing temporal boundaries to coordinate routines under temporal uncertainty. Adm. Sci. Q. **66**, 220–264 (2021). https://doi.org/10.1177/0001839220941010
51. Feldman, M.S.: Making process visible: alternatives to boxes and arrows. In: The Sage Handbook of Process Organization Studies, pp. 625–635. Sage, London (2016). http://dx.doi.org/10.4135/9781473957954.n41

Predicting Process Structure After a Disruption: An Example from the Clinical Documentation Process

Inkyu Kim[1](✉), Kenneth A. Frank[1], Julie Ryan Wolf[2], and Brian T. Pentland[1]

[1] Michigan State University, East Lansing, MI, USA
kiminky1@broad.msu.edu
[2] University of Rochester, Rochester, NY, USA

Abstract. Using data from the audit trail of an electronic medical record system, we examine the effects of a disruption on the clinical documentation process. We use process mining to construct a network that describes the process and then we use a latent factor selection model to analyze changes to that network. Rather than attempting to discover a particular process model, our goal is to identify theory-based factors that explain change and stability in the overall pattern of actions. We conduct the analysis at two levels of granularity and we compare time periods with and without disruption. The paper contributes to current research on routine dynamics as network dynamics by demonstrating the use of network science to predict the structure of an organizational routine.

Keywords: Routine dynamics · Network dynamics · Latent factor models

1 Introduction

Organizational routines (defined as recognizable, repetitive patterns of interdependent action carried out by multiple actors) are a foundational element in the science of organization [1]. Over the last 20 years, the field of routine dynamics has focused on the mechanisms of endogenous change: change that occurs in the absence of external influences.

In this paper, we turn our attention to the effects of exogenous disruptions: When routines are disrupted, are how does the overall pattern of action change? Other papers have studied disruptions (e.g., [2]), but this is the first to apply concepts from network science to explain the dynamics of organizational routines. We model routines as directed

This research was supported by the National Science Foundation under Grant No. SES-1734237. Any opinions, findings, and conclusions or recommendations expressed in this material are those of the author(s) and do not necessarily reflect the views of the National Science Foundation. This research was also supported in part by University of Rochester CTSA (UL1 TR002001) from the National Center for Advancing Translational Sciences (NCATS) of the National Institutes of Health (NIH). The content is solely the responsibility of the author(s) and does not necessarily represent the official views of the National Institutes of Health.

A. Marrella and B. Weber (Eds.): BPM 2021 Workshops, LNBIP 436, pp. 221–231, 2022.
https://doi.org/10.1007/978-3-030-94343-1_17

graphs [3], known in the process management literature as "directly follows graphs" [4]. Using latent factor selection models [5], we study the effects of disruptions on the structure of the network. In social networks, mechanisms like reciprocity, homophily, and preferential attachment contribute to formation and dissolution of network ties [6], but analogous network-based mechanisms have never been defined or investigated in the context of organizational routines.

This paper offers a first step towards defining and testing mechanisms that drive routine dynamics. As a sophisticated tool to model and measure process change and theory to explain patterns of actions, process mining and routine dynamics have potential to be mutual complementary, but there has been disconnection between two areas [7]. In this study, we provide implications with respect to intersection of process management and routine dynamics by using a deductive method of process mining to examine dynamics of routines in organizations [8].

2 Background

Routine dynamics concerns understanding the mechanisms that influence stability or change in action patterns [9]. When action patterns persist over time, this persistence can be interpreted in several ways, such as inertia [10], resistance [11], persistence [12], regeneration [13], or resilience [14]. Schulz [15] offers an encyclopedic list of mechanisms that keeps routines "on track", ranging from very macro (institutional norms) to very micro (neuronal priming). Network structure does not appear in the list of mechanisms because at that time, nobody was thinking about modeling routines as directed graphs. But as routines are repeated related actions, they can be understood in terms of the evolution or stability of networks representing the relationships among actions. Theoretical explanations of routine persistence did not consider the structure of the routine itself.

2.1 Routine Dynamics as Network Dynamics

Organizational routines can be seen as patterns of action [1, 16]. These patterns can be represented as a valued, directed graph where the vertices represent categories of action and the edges represent sequential relations between those categories [3]. In process mining, this is called a "directly follows graph" (DFG) [4]. Where a conventional social network represents relations between actors (e.g., people), a DFG represents relations between categories (or clusters) of actions. In research on organizational routines, these graphs are most often referred to as "narrative networks" [17].

To model network dynamics, we need to explain edge formation/dissolution, which is the fundamental mechanism of network dynamics [6]. The structure of the network changes as edges are added or deleted. In social network research, models that predict edge formation or deletion are often referred to as selection models because they predict how people select other people as interaction partners [18]. There are some well-established selection mechanisms in social networks, such as homophily and preferential attachment, that drive network dynamics (edge selection). Our goal here is to begin, for the first time, to identify and test mechanisms that drive routine dynamics.

To address this issue, we use two different levels of granularity; 1) a fine-grained level consisting of screens and clicks and 2) a coarse-grained level composed of touchpoint and handoffs. In this study, we use the expanded concept of handoffs as a sequential relationship between not just people, but also between events [3]. As a repetitive patterns of sequentially related events, we can view organizational routines as a series of handoffs [19].

Our main theoretical concern is the extent to which network structure itself has an influence on routine dynamics. If it does, it would provide a new theoretical mechanism to explain and predict the inertia/persistence/resilience of routines, complementary to existing explanations.

2.2 Research Setting and Data

This analysis is based on data from one outpatient dermatology clinic at the University of Rochester Medical Center (URMC). We compare the pattern of action before and after the start of flu season (September 1, 2016). From prior research on this clinic, we know that flu season causes a measurable change in the EHR audit trail, as shown in Fig. 1 [7]. Figure 1 describes changes in the structure of the clinical documentation process over time compared to a fixed reference. In particular, it uses cosine distance to show how the repertoire of actions used in the process changes over time. Each point in Fig. 1 represents one day. It shows that the repertoire of actions changes abruptly at the beginning of flu season. This is because a new set of actions is added to the system to track and support seasonal vaccination. Later, at the end of flu season, the repertoire of actions changes back. However, Fig. 1 is purely descriptive. Our goal here is to move beyond description to provide a theory-based model that can help explain how the pattern of action responds to this exogenous disruption.

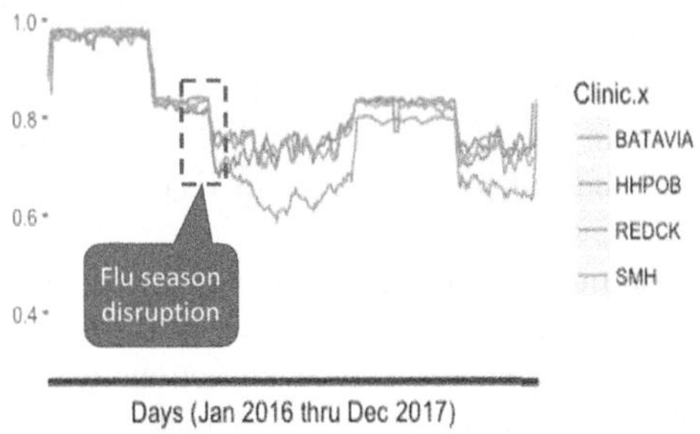

Fig. 1. Flu season disrupts the normal clinical routine (Adapted from Pentland, Vaast and Wolf [7])

In this paper, we focus on three weeks of data from one clinic. We examine the pattern of action two weeks prior to flu season and one week after. This is a small subset of the data collected for the larger project, but it is adequate to demonstrate our approach. Unlike a typical exercise in process mining, our goal is not to discover a process model. Rather, we are trying to identify factors that influence changes in the process.

2.3 Two Levels of Granularity

We analyze data extracted from the audit trail of the EPIC Electronic Medical Record (EMR) system. The subset of records used here includes detailed, time-stamped records of EMR utilization in 627 patient visits at one clinic from August 18, 2016 (two weeks before the start of flu season) to September 7, 2016 (one week after). The middle column of Fig. 2 includes a brief example of the audit trail data.

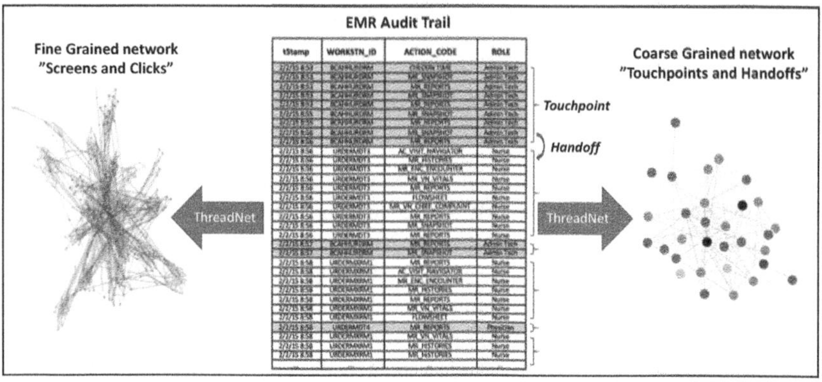

Fig. 2. EMR audit trail can be interpreted at two levels of granularity

Figure 2 shows how *ThreadNet* (Pentland et al. 2020) can be used to convert EMR audit trails into networks at two different levels of granularity. At the center of Fig. 2 is a small part of an audit trail where each row is a time-stamped action. On the left side of Fig. 2, *ThreadNet* can produce a very fine-grained network by considering each unique row as a node. At the fine-grained level, edges represent pairs of actions within the EMR system. On the right side of Fig. 2, *ThreadNet* can produce a coarse grained network where each node consists of a group of actions and each edge can be interpreted as a *handoff* [3]. For example, in Fig. 2, there is a handoff between the Admin Tech and the Nurse, and then another handoff back to the Admin Tech. We test our theory of network dynamics at both of these levels.

We examine this disruption at two levels of granularity because we expect the effects to be different. At the fine-grained level, flu season directly changes the repertoire of actions in the network. On September 1, a set of actions was added to the network. Office staff were prompted to ask all incoming patients about their vaccination status and patients were invited to schedule (or receive) a vaccination. This had a visible impact on the clinical documentation process, but the main work of the dermatology clinic was not otherwise affected. This would lead us to expect changes to the pattern of action at the fine-grained level, but not necessarily at the coarse-grained level. Pentland, Vaast and Wolf [7] report that the medical staff (physicians and nurses) were unaware of any particular changes in their work processes for the time period in question. From their point of view, there was no disruption at all.

3 Hypothesis Development

Network dynamics can be defined in terms of two basic processes: edge formation and edge dissolution [6]. As with models of social networks, we recognize that formation of new edges is a different process than dissolution of existing edges. In a network model of a routine, an edge represents the sequential relationship of two actions or events. New sequential relationships might form for a variety of reasons that seem difficult to predict and generalize (experimentation, workaround, error, etc.) Therefore, we focus on mechanisms that influence the persistence (or dissolution) of existing edges. For brevity, we discuss and phrase the hypotheses in terms of handoffs, in the general sense of the term used by Pentland, Recker and Wyner [3]. A handoff is simply an edge in a narrative network.

3.1 Frequency of Handoffs

Handoffs that occur frequently represent the stereotypical "ruts in the road" [13] that define routinized patterns of action. For example, in Fig. 2, the patient checks into the clinic with the Admin Tech, who hands off to the nurse to record vital signs, etc. We operationalize frequency of handoffs by counting the edges between touchpoints in the network. Some handoffs are reinforced more frequently and may also be enabled or constrained by technological, material or organizational structures. We expect frequent handoffs to persist after a disruption to the network.

H_1: More frequent handoffs are more likely to persist after a disruption.

3.2 Speed of Handoffs

Handoffs can also be weighted according to how long they take to perform, on average, using time-stamp data from the event log. Faster handoffs represent quicker ways of getting things done. The speed of handoffs is operationalized as how long it takes for an actor to perform handoffs. In other words, we compute mean duration of each handoff in the network. Clinics are busy places, so we hypothesize that faster handoffs (with shorter mean duration), are more likely to persist after a disruption.

H_2: Faster handoffs are more likely to persist after a disruption.

3.3 Paths and Betweenness

The edge betweenness is defined as the number of shortest paths going through an edge in the network [20]. In a narrative network, paths represent ways of doing things. An edge that has greater betweenness is on more paths [21]. If a disruption changes how things get done, handoffs with high betweenness are less likely to dissolve (and more likely to persist or "bounce back") than handoffs with low betweenness:

H_3: Handoffs with higher betweenness are more likely to persist after a disruption.

3.4 Coherence

Coherence represents the number of nodal attributes that remain the same across an edge [3]. For example, do both actions occur in the same place? Are they performed by the same actor? Do they require the same tools or technology? This can be easily operationalized in a narrative network, where each node is defined by a number of contextual factors, such as place, actor and technology. When more factors change, the context is less coherent. The logic of this hypothesis is similar to the logic for effects of homophily in social networks ("birds of a feather…"), but the mechanism is different.

H_4: More coherent handoffs are more likely to persist after a disruption.

4 Formal Model

Our goal is to predict the frequency of all edges in the DFG that represents the clinical documentation process. To do so, we use the previous state of the process (at time *t* – *1*) to predict the current state of the process (at time *t*). We can express our four hypotheses in terms of the formal model shown in Eq. (1):

$$\begin{aligned} w_{ijt} &= \beta_1\left(w_{ijt-1}\right) + \beta_2\left(\overline{speed}_{ijt-1}\right) \\ &\quad + \beta_3\left(betweenness_{ijt-1}\right) + \beta_4\left(coherence_{ijt-1}\right) + \theta_i + \theta_j + u_i v_j + e \end{aligned} \quad (1)$$

In this model, the time period *t* represents one week. This time scale makes sense because the pattern of action is generally very stable and the disruption from the start of flu season occurs on a specific day (September 1). The dependent variable in this model is w_{ijt}, which represents the frequency of each edge in the network (between actions *i* and *j*) during time period *t*.

The term w_{ijt-1} represents the frequency of edges from the previous time period, as in H_1. $\overline{speed}_{ijt-1}$ reflects the average speed or duration of the edge w_{ij}, as in H_2. $betweenness_{ijct-1}$ represents the number of network paths that pass through w_{ijc}, as in H_3. $coherence_{ijt-1}$ represents the extent to which actions i and j share a coherent material context, as in H_4.

θ_i and θ_j are random effects relating to the base rate of actions *i* and *j*. If *i* and *j* occur more or less often, that will directly influence how often w_{ij} occurs. In terms of this model, they function like control variables. As we apply the model here, θ_i and θ_j reflect the change in the repertoire of actions described in Fig. 1.

Lastly, $u_i v_j$ represents the similarity between pairs of nodes on each dimension (action *i* and *j*) of latent space and $e_{i,j}$ is the error term.

5 Results

We estimate this model using a latent factor model including random effects [22, 23] to control for unobserved network effects. This model has the advantage of estimating all of the edges in the network at once, rather than treating them as independent. This model is implemented in the R package *amen* (https://cran.r-project.org/web/packages/

amen/amen.pdf) which uses an MCMC (Markov Chain Monte Carlo) procedure for model estimation. This package provides regression coefficients, standard errors and significance levels, but does not provide an R^2 for the model.

We conduct four different versions of the analysis. First, as mentioned above, we examine two different levels of granularity: the fine-grained "screens and clicks" level and the coarse-grained "touchpoints and handoffs" level. At the fine-grained level, the nodes are defined by a single attribute – actions – so there is no way to compute coherence or test H_4. At the coarse-grained level, nodes are defined by two attributes – actor and workstation – so we can compute coherence and test H_4.

Second, we compare two different time periods: one where there is a disruption and one where there is not. While a more elaborate analysis could be conducted, this provides an indication of how the disruption compares to the normal, steady state operations. The results of these analyses are shown in Table 1.

5.1 Frequency of Handoffs (H_1)

It should be no surprise that the frequency of an edge (handoff) in one time period is a good predictor of its frequency in the next time period. At both levels of granularity, with or without disruption, $\mathbf{H_1}$ is supported. Metaphorically, the ruts in the road today predict where the traffic will go tomorrow.

However, at the fine-grained level, the disruption appears to change the strength of this effect. Even when we control for other factors (such as the base rate of actions, θ_i and θ_j), the influence of handoff frequency is decreased by more than 50% from the prior week. Perhaps we should say: the ruts in the road today predict where the traffic will go tomorrow *unless the road is blocked*.

5.2 Duration of Handoffs (H_2)

Contrary to expectations, the average duration of the handoff does not influence persistence, except perhaps at the coarse-grained level. Faster handoffs are slightly more likely to persist after disruption. Even in a very busy clinic, where time is presumably valuable, the waiting time from one action to the next appears to have a very minor influence on persistence. Conceptually, this could be interpreted as undermining the idea that routines enhance efficiency, but it would be worth testing this hypothesis on a larger, more diverse set of processes.

5.3 Betweenness of Handoffs (H_3)

Betweenness provides different results at different levels of granularity. It is significant at the fine-grained level, but not at the coarse-grained level. We do not have an explanation for this difference, so further investigation is needed. However, at the fine-grained level, we can see that the magnitude of this effect is more than doubled when there is a disruption. In a narrative network, betweenness means that an edge (or a node) is on more paths; it is a useful for performing a larger variety of tasks. When there is a disruption, the useful handoffs are more likely to persist.

Table 1. Results of analysis at two level of granularity[1]

	Not disrupted	Disrupted
	t − 1: Aug 18–Aug 24 t: Aug 25–Aug 31	t − 1: Aug 25–Aug 31 t: Sept 01–Sept 07
Fine-grained (clicks/screens)		
H_1: Frequency	1.912***	0.975***
	(0.060)	(0.039)
H_2: Duration	−0.044	0.014
	(0.024)	(0.025)
H_3: Betweenness	0.213***	0.469***
	(0.037)	(0.039)
Random effect: θ_i	0.405	0.507
	(0.065)	(0.082)
Random effect: θ_j	0.389	0.430
	(0.063)	(0.067)
# nodes (unique action)	146	137
# edges (pairs of actions)	1,848	1,800
Coarse-grained (touchpoints/handoffs)		
H_1: Frequency	1.052***	1.063***
	(0.086)	(0.088)
H_2: Duration	0.008	0.064*
	(0.027)	(0.032)
H_3: Betweenness	0.108	0.141
	(0.071)	(0.075)
H_4: Coherence	0.499***	0.608***
	(0.124)	(0.146)
Random effect: θ_i	0.684	0.575
	(0.162)	(0.145)
Random effect: θ_j	0.602	0.571
	(0.109)	(0.148)
# nodes (touchpoints)	86	80
# edges (handoffs)	919	838

5.4 Coherence of Handoffs (H_4)

Because of the way coherence is defined, we are only able to test its effect at the coarse-grained level. At that level, coherence is a strong predictor of edge persistence. When a pair of actions is more coherent (performed by the same actor, or performed in the

[1] The values for the random effects (θ_i, θ_j) indicate variances.

same location), it is more likely to persist (with or without a disruption). This finding is interesting because this effect is significant even when we control for the effect of frequency. Coherence appears to influence the persistence of network structure over and above the more familiar effect of repetition.

5.5 Changes in the Base Rate of Actions

In a sense, we are interpreting random effects in Eq. (1) (θ_i and θ_j) as control variables: Controlling for changes in base rates of the actions, what drives changes in the pairs of actions? The magnitude of these effects is far greater than their standard error in all four conditions (the *amen* package does not provide p-values for these parameters, so we rely on the eyeball test for significance). Thus, we can safely say that in all four conditions, there is considerable variation among actions in the tendency to be part of a chain of actions defining a routine. However, when we compare the coefficients in Table 1, with and without disruption, they are of similar magnitude and sign in all four conditions (with or without the disruption at both levels of granularity).

6 Discussion

This paper represents a first step towards a theory of routine dynamics as network dynamics. The empirical foundation for this theory is generated through process mining, which is usually used to discover a stationary model of a process. Here, we are using process mining to help build theory about stability and change in routines. The contribution here goes beyond the specific findings about a particular dermatology clinic. The main contribution concerns the general idea of using network models to develop new theory about routine dynamics.

This approach to routine dynamics requires the analysis of network time-series data, as discussed in [24]. Here, we have limited our analysis to three time periods because we wanted to focus on the disruption of flu season. In a more elaborate model, we could extend the analysis to a longer time series. While increasingly sophisticated network time series models are available [25], they need to be carefully adapted to narrative networks. As of this writing, we are working on how to apply the *amen* R-package to multi-period time series analysis.

Even with the limited analysis we present here, there still some substantive insights. For example, the strongest predictor of edge persistence from one time period to the next is the frequency (weight) of that edge. In other words, routines tend to be repetitive. It would be disingenuous to count this well-established fact as a contribution, but it does lend face validity to the approach. The more interesting finding is that the magnitude of this effect is reduced by roughly 50% after a disruption. Likewise, the influence of betweenness is more than doubled after a disruption.

While these effects are interesting, the first key contribution here is not the specific magnitude of these effects. It is the overall approach to estimating them in the first place. This approach depends on using process mining to construct narrative networks of the process. While this class of network (a kind of directly follows graph, or DFG) has severe limitations as a process model [4], it provides a sensitive indicator of process stability and change. If the DFG is stationary, so is the process. Given these graphs, we are able to begin testing hypotheses about the factors that drive their stability and change using analytical tools like latent factor models [23].

The second key contribution here is theoretical. Hypotheses 1–4 represent a first attempt at defining formal, generalizable mechanisms that govern the dynamics of narrative networks. The analysis presented here is the first time that the effects of frequency, duration, betweenness or coherence on routine dynamics have been investigated empirically. These mechanisms may seem simplistic, but so do some of the key mechanisms that drive the dynamics of social networks ("birds of feather…", "the rich get richer…", "the friend of my friend…"). Perhaps simplicity is a virtue.

The last contribution of the study concerns the possible implications of this line of inquiry for process management and the BPM life cycle [27]. Routines are hard to change, as they tend to resist intervention and snap back into their old shapes. This study suggests that it may be possible to identify what aspects of routines contribute most to inertia and resistance.

7 Conclusion

Findings from a single case should be regarded as preliminary. We cannot build a generalizable theory from three weeks of data in one dermatology clinic. However, the findings do provide a variety of encouraging mysteries. For example, why does betweenness influence persistence at the fine-grained level but not the coarse-grained level? Where there is mystery, there is an opportunity for learning, so it seems that there may be more to learn from network models of routines.

References

1. Feldman, M.S., Pentland, B.T.: Reconceptualizing organizational routines as a source of flexibility and change. Adm. Sci. Q. **48**, 94–118 (2003)
2. Edmondson, A.C., Bohmer, R.M., Pisano, G.P.: Disrupted routines: team learning and new technology implementation in hospitals. Adm. Sci. Q. **46**, 685–716 (2001)
3. Pentland, B.T., Recker, J., Wyner, G.: Rediscovering handoffs. Acad. Manag. Discov. **3**, 284–301 (2017)
4. van der Aalst, W.M.: A practitioner's guide to process mining: limitations of the directly-follows graph. Procedia Comput. Sci. **164**, 321–328 (2019)
5. Hoff, P.D.: Bilinear mixed-effects models for dyadic data. J. Am. Stat. Assoc. **100**, 286–295 (2005)
6. Snijders, T.A.: The statistical evaluation of social network dynamics. Sociol. Methodol. **31**, 361–395 (2001)
7. Pentland, B.T., Vaast, E., Wolf, J.R.: Theorizing process dynamics with directed graphs: a diachronic analysis of digital trace data. MIS Q. **45**, 967–984s (2021)
8. Grisold, T., Wurm, B., Mendling, J., Vom Brocke, J.: Using process mining to support theorizing about change in organizations. In: The 53rd Hawaii International Conference on System Sciences (2020)
9. Feldman, M.S., Pentland, B.T., D'Adderio, L., Lazaric, N.: Beyond routines as things: introduction to the special issue on routine dynamics. Organ. Sci. **27**, 505–513 (2016)

10. Gilbert, C.G.: Unbundling the structure of inertia: resource versus routine rigidity. Acad. Manag. J. **48**, 741–763 (2005)
11. Becker, M.C., Lazaric, N., Nelson, R.R., Winter, S.G.: Applying organizational routines in understanding organizational change. Ind. Corp. Chang. **14**, 775–791 (2005)
12. Howard-Grenville, J.A.: The persistence of flexible organizational routines: the role of agency and organizational context. Organ. Sci. **16**, 618–636 (2005)
13. Birnholtz, J.P., Cohen, M.D., Hoch, S.V.: Organizational character: on the regeneration of camp poplar grove. Organ. Sci. **18**, 315–332 (2007)
14. Grote, G., Weichbrodt, J.C., Günter, H., Zala-Mezö, E., Künzle, B.: Coordination in high-risk organizations: the need for flexible routines. Cogn. Technol. Work **11**, 17–27 (2009)
15. Schulz, M.: Staying on track: a voyage to the internal mechanisms of routine reproduction. In: Becker, M.C. (ed.) Handbook of Organizational Routines, pp. 228–257. Edward Elgar, Cheltenham (2008)
16. Cohen, M.D., et al.: Routines and other recurring action patterns of organizations: contemporary research issues. Ind. Corp. Change **5**, 653–698 (1996)
17. Pentland, B.T., Kim, I.: Narrative networks in routine dynamics. In: Feldman, M.S., Pentland, B.T., D'Adderio, L., Dittrich, D., Rerup, C., Seidl, D. (eds.) Cambridge Handbook of Routine Dynamics. Cambridge University Press (2021)
18. Steglich, C., Snijders, T.A., Pearson, M.: Dynamic networks and behavior: separating selection from influence. Sociol. Methodol. **40**, 329–393 (2010)
19. Kremser, W., Schreyögg, G.: The dynamics of interrelated routines: introducing the cluster level. Organ. Sci. **27**, 698–721 (2016)
20. Pentland, B.T., Rueter, H.H.: Organizational routines as grammars of action. Adm. Sci. Q. **39**, 484–510 (1994)
21. Girvan, M., Newman, M.E.: Community structure in social and biological net-works. Proc. Natl. Acad. Sci. **99**, 7821–7826 (2002)
22. White, D.R., Borgatti, S.P.: Betweenness centrality measures for directed graphs. Soc. Netw. **16**, 335–346 (1994)
23. Hoff, P.D., Raftery, A.E., Handcock, M.S.: Latent space approaches to social network analysis. J. Am. Stat. Assoc. **97**, 1090–1098 (2002)
24. Hoff, P.D.: Multiplicative latent factor models for description and prediction of social networks. Comput. Math. Organ. Theory **15**, 261–272 (2009)
25. Pentland, B.T., Recker, J., Wolf, J.R., Wyner, G.: Bringing context inside process research with digital trace data. J. Assoc. Inf. Syst. **21**, 5 (2020)
26. Rossetti, G., Cazabet, R.: Community discovery in dynamic networks: a survey. ACM Comput. Surv. (CSUR) **51**, 1–37 (2018)
27. Dumas, M., La Rosa, M., Mendling, J., Reijers, H.A.: Fundamentals of Business Process Management. Springer, Heidelberg (2018). https://doi.org/10.1007/978-3-662-56509-4

Capabilities for Digital Process Innovation: Results of an Ongoing Action Research Study

Ralf Plattfaut(✉) and Vincent Borghoff

South Westphalia University of Applied Sciences, Soest, Germany
{plattfaut.ralf,borghoff.vincent}@fh-swf.de

Abstract. Organizations around the globe are faced with digital transformation. However, many of these organizations, especially small- and medium-sized enterprises, lack the corresponding individual skills as well as the organizational capabilities to implement digital process innovations and change the corresponding organizational routines. This research-in-progress paper reports on preliminary results of an action research study. We conceptualize organizational digital process innovation capabilities as dynamic capabilities and create a theoretical understanding of these capabilities with their corresponding microfoundations, i.e., the individual employee's skills. Based on an in-depth action research study we will be able to show in how far outside consulting support will change individuals' skills as well as organizational capabilities. We thus contribute to both theory on organizational capability-building for digital process innovation as well as dynamic capability theory.

Keywords: Digitalization · Dynamic capabilities · Organizational capabilities · Digital innovation

1 Introduction

Virtually all organizations around the world are currently dealing with digital transformation. Albeit there is no established and agreed-upon definition for the term, it is clear that recent technological advancements enable and require drastic innovations, improvements, or adjustments of business model, decision making, and operational processes [1, 2]. Especially as operational processes can be understood as organizational routines, their digitalization and innovation is difficult. Digital process innovation, therefore, relies on the presence and characteristics of digital organizational capabilities, culture, and structures [3].

However, organizations often lack the capabilities to drive digital process innovation on their own. Academic literature, as well as general media, has reported on many cases where companies and organizations seek outside help to transform themselves and their processes. To give an example, Plattfaut reports on a traditional company using external consultants to drive their digital process innovation initiative and build corresponding capabilities [4].

Particularly, we often observe a lack in skills of individual employees with regards to digitalization. Exemplarily, Matt et al. report that current employees "may have a

A. Marrella and B. Weber (Eds.): BPM 2021 Workshops, LNBIP 436, pp. 232–242, 2022.
https://doi.org/10.1007/978-3-030-94343-1_18

[…] less tech-savvy mindset and may lack the required technological capabilities to cope with the upcoming changes" and call for corresponding research [1]. Other authors agree that the digital transformation requires different skill-sets: Domain experts need to be able to navigate the digitalized world and technical specialists need to have a digital mindset [5].

In our ongoing research project, we study the impact of external digital process innovation consulting on individual skills, and organizational capabilities. We aim to deepen our understanding in how far external digitalization impulses have an impact on organizational capabilities, individual skills to implement digital process innovations and, thus, to consciously change business processes (or organizational routines). During this research effort, we employ dynamic capability theory as a theoretical frame for digital process innovation capabilities. Here, we also aim at contributing to dynamic capability theory at large through a deeper understanding of microfoundations of dynamic capabilities on a more practice-oriented view [6].

As a case organization, we selected a small company with only 15 employees. A focus on a small organization enables us to study individual skills of a larger proportion of the workforce. As such, we can also gain a deeper understanding of the development of organizational capabilities. Moreover, in such a setting, the planned intervention will be more visible and thus the results appear to be magnified.

To achieve this research objective, we conduct a qualitative action research (AR) study [7] over the course of 13+ months. In early 2020, we collected in-depth qualitative data on the status quo of an SME. In a first intervention in summer 2020, we optimized three business processes in the organization with the help of lightweight digitalization measures such as Robotic Process Automation (RPA). In a second intervention in winter 2020/2021, we optimized further processes which resulted in additional digital process innovations. After both interventions, we evaluated the impact of our action on skills, and capabilities using a further round of qualitative data collection. In the next months, we will conduct another round of data collection to assess whether the interventions have led to lasting change.

In this research-in-progress paper, we report on the preliminary results of this ongoing AR study. For this, the article is structured as follows. In the next section, we present our theoretical background in terms of digital process innovation, capabilities, and skills. Afterward, we discuss our (planned and partially executed) research approach. Next, we present the preliminary results of our research. Last, we close with a discussion and conclusion.

2 Theoretical Background

2.1 Digital Process Innovation

We observe increasing importance of information technology (IT) in today's world, with regards to both existing IT solutions such as Enterprise Resource Planning, Customer Relationship Management, or Supply Chain Management systems as well as new technologies such as the internet of things, blockchain, or artificial intelligence. Through a combination of them together with innovation regarding business models, strategies, and processes they become transformative for the business at large [8].

Digital transformation, and here digital process innovation, enables organizations to both optimize existing business models, work routines, or capabilities and to radically reinvent themselves. New IT solutions are used to enhance the efficiency and effectiveness of decision-making and business processes across the whole organization [9]. Digital process innovation then depicts the use of these digital technologies to create new or adapt existing business processes [10]. The used technologies enable organizations to improve process quality with a simultaneous reduction in costs [3]. Exemplarily, multiple organizations use digital process automation techniques such as RPA to increase the efficiency of business processes by 30% [4, 11]. Pentland et al. argue that especially in digital environments, small changes in existing routines can lead to more extensive changes in the subsequent development [12]. These changes can be triggered by a change in either the human or the technical agency [13]. Therefore, introducing new technologies within an organization can lead to changes within the employees' agency to adapt routines.

2.2 Organizational Digital Process Innovation Capabilities

Dynamic capabilities can be seen as an extension of the Resource-Based View (RBV) to overcome its more or less static alignment [14, 15]. The concept was introduced by Teece et al. [16] as "the firm's ability to integrate, build, and reconfigure internal and external competencies to address rapidly changing environments". It extends the RBV by adding the structures of markets, as an external factor of changing environments, and processes, as internal adoption to those environments, to the concept [16]. In the following the theory of dynamic capabilities has been conceptualized by other authors, e.g. Teece [17], Zollo and Winter [18], Eisenhardt and Martin [19], or Helfat [20].

For this paper we will follow the definition of Barreto [21], who sees dynamic capabilities as "the firm's potential to systematically solve problems, formed by its propensity to sense opportunities and threats, to make timely and market-oriented decisions, and to change its resource base". As such, organizational dynamic capabilities can be disaggregated into sensing, seizing, and transforming [17]. Sensing represents the capability to recognize opportunities and threats, seizing the ability to address sensed opportunities in terms of selecting, and transforming the process of modifying the resource base.

Dynamic capabilities develop inside an organization out of specific routines and/or recombination and integration of resources [19]. This development does not take place spontaneously; it is rather to be understood as a recurring routine and develops out of an intentional effort in change [22]. These routines base on multiple factors, like positions and paths as an internal factor or the nature of the specific market environment an organization is operating in [19]. Blyler and Coff [23] argued that social capital is a necessary (though not sufficient) condition for the existence of a dynamic capability, as only through social interaction resources are connected and recombined. The importance of this social factor becomes more apparent when considering heterogeneous individual knowledge, and skills between individuals and organizational positions [24].

This emphasizes a more individual view on the foundations of dynamic capabilities, as routines and organizational attitudes arise out of individual perceptions, and skill [25]. Eisenhardt, Furr, and Bingham [26] and Helfat and Peterat [27] pay more attention to the individual level, with a focus on executives' cognitive capabilities. Salvato and Vassolo

[28] found evidence for an overemphasis in research on the managerial level when analyzing routine and organizational change. Research on individual employees can help to close this gap by drawing more attention to the operational level [29]. In addition, the lack of internal capabilities may also affect the success of innovation. Internal barriers in capabilities may hinder successful change. This approach combines institutional theory and dynamic capabilities. External interventions represent an opportunity to break existing routines and spur organizational and routine change [30].

2.3 Skills Needed for Digital Process Innovation

To achieve organizational change towards digitalization and implement digital process innovations, employees do not only need the right mindset but also corresponding skills. Here, one can distinguish between expertise and managerial IT skills [31]. Expertise includes technical skills, like knowledge of programming languages, operational systems, or databases. In addition, knowledge in technology management, as well as interpersonal skills and business functional knowledge is needed [32]. As such, organizations need skills for both technological and transformative change [33]. While the first addresses the adoption and operation of digital solutions, the second describes the transformational characteristics of digital solutions on an organizational level. As such, employees do not only need mere IT skills but profound knowledge in creating digital process innovations. This emphasizes the need for a good understanding of the current business to leverage the potential of digital technologies [34, 35].

3 Research Approach

3.1 Methodology and Case Background

In this study, we follow an AR approach. According to Susman [36], the cyclic AR method consists of five steps: (1) diagnosing, (2) action planning, (3) action taking, (4) evaluation, and (5) specifying learning. This cycle presupposes the establishment of a client system, as the research environment, where the interaction between the researcher and the included practitioners is located. This interaction is based on a collaborative spirit, which allows direct influence on the knowledge and skills of participants [37]. AR as a method is also ideal for creating and analyzing change in organizations, which is a main focus of this study [38].

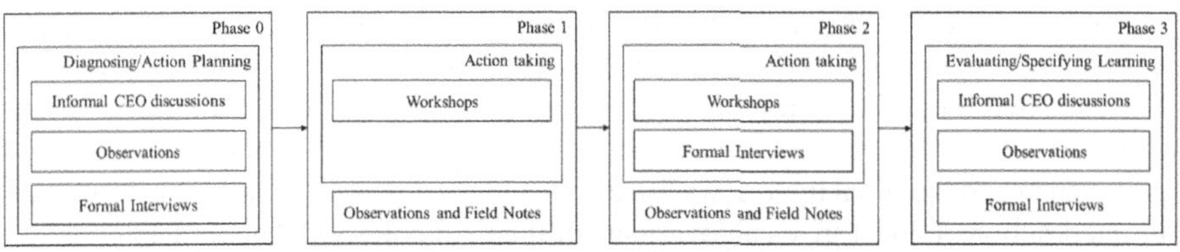

Fig. 1. Research and data collection approach

Four our study, we follow a linear, two-cyclic, AR approach [39]. Coad and Yourdon [40] used a similar approach when intervening in several organizations with a comparable intention to influence skill and behavior. We structured our AR approach into four higher-level phases, as shown in Fig. 1. Phase 0 covers the diagnosing and action planning AR steps. Here, we analyze the organizations status regarding digital capabilities prior to the intervention and identify the potential for digitalization. Phase 1 and 2 mainly consist of action taking. This includes the optimization and digital innovation of business processes as well as the conduction of topic-specific workshops. In phase 1, four student teams use lightweight IT to redesign and automate the administrative processes identified in the prior step, in coordination with the management of GROW. These automations are implemented and refined by a working student within phase 2. With this external intervention, we aim at disrupting routines and stimulating the employees reflection on further digital potential [30, 41]. As lightweight IT enables the development of digital solutions without deeper IT knowledge, it enables business employees to proactively digitalize processes and routines within their domain [42]. During the ongoing phase 3, we analyze the achieved change of the employees' digital capabilities, change in organizational routines, as well as respective changes within the organization towards digitalization in general. This covers the evaluation of the taken action by matching organizations and researcher's perspective.

For our research, we selected GROW as our case study organization. GROW is a regional business development agency.[1] Although GROW is organized as a private company, it is owned by the local and regional administrations from the corresponding region. The main goals for GROW are to attract new firms to the region and help the existing firms to prosper. For this, GROW provides services for regional companies in various aspects. These services are often embedded in publicly funded projects, e.g., to increase the energy efficiency of companies in the region, improve attractiveness for companies to hire families, or increase digital skills. GROW has 15 employees who work in the corresponding knowledge-intensive processes.

We use three types of data, formal interviews, as well as observations, and taken field notes from interactions with GROW, collected within all four phases of our study. In a primary phase (Phase 0), we collected data on the initial situation of the organization as well as the organizational routines concerning digitalization and digital transformation. Here, we were able to observe employees in their environment, gather information from informal discussions with the CEO, and conducted formal, semi-structured, interviews with employees of GROW.

In phases 1 and 2, we use applied workshops on the respective automation approaches to collect data in the form of direct observations of the interaction between the consulting teams and the GROW representatives, as well as field notes taken from interactions between the both mentioned outside the respective workshops. For phase 2, we additionally conducted interviews with employees, who are in direct touch with the developed solutions, setting a focus on the initial interaction with the artifacts. In the ongoing phase 3, we plan to conduct final interviews, here with a focus on the achieved changes through our intervention. Table 1 gives an overview of the total collected data to date.

[1] Name of the company pseudonymized for confidentiality reasons.

Table 1. Data collected to date

Formal Interviews		
Phase	Interviewee	Function
0	I1	Digital Industry Consultant
	I2	Project Manager
	I3	Digital Industry Consultant
	I4	Digital Health Consultant
2	I5	Project Manager
	I6	Digital Industry Consultant
	I7	Marketing Specialist
	I8	Executive Assistance
Observations and field notes		
Phase	Type	Collected Data
1–2	Workshops	Observations and Field Notes from 4 Workshops
0–3	Consulting Interactions	Field Notes from 7 interaction between the consulting teams and GROW

4 Preliminary Results

In the following, we will present the preliminary results of our study. In phase 0, we identified seven themes regarding the status quo ante of the organization with regards to digital process innovation, organizational capabilities, and employees' individual capabilities and skills.[2] Building on this, we describe our actions and observations with regards to these themes from both action phases 1 and 2. The results of phase 3 are expected in the second half of 2021 and will conclude this research in progress.

Results of Phase 0

Theme 1: Share of administrative routines. All interviewed employees describe a relatively big amount of administrative work. These activities are considered to be non-value adding and take time and focus away from the original task, i.e., to attract new firms to the region and help existing firms develop. The individually perceived share of the non-value adding activities ranges from 30 to 50%, or, as one employee put it: "In the area of administrative tasks, I would say, that it has a share of 40 to 50% of total" [I3].

These administrative activities originate either from internal administrative inefficiencies due to a lack of process digitalization or from complex interactions with external parties such as the shareholders (e.g., the local district authority) or funding organizations on state or federal level. "Some of our administrative tasks are GROW internal, e.g., travel expense accounting, administration of external invoices. But there are also

[2] An earlier version of these results of Phase 0 was presented at the International Conference Wirtschaftsinformatik 2021.

topics which [...] revolve around the communication with the responsible federal ministry" [I4]. In some cases, the administrative effort was doubled, i.e., the same report was prepared twice, once for internal purposes and once for external stakeholders.

Theme 2: Front runner among the public organizations in the region. Although the high share of administrative routines could be reduced through additional internal digitalization, GROW can be seen as a frontrunner in digitalization when compared to the main shareholder, i.e., the district authority. This becomes especially visible as for some processes the district authority acts as a service provider. "Everything that has to do with the district authority is, in fact, more likely to have paper-based interface" [I1]. In contrast to this, GROW has started to digitalize its internal processes to a greater extent. Exemplarily, the organization switched to a modern full-fledged Office 365 environment and is currently migrating to a cloud-based CRM solution. Local governments view GROW as being progressive in regards to digitalization.

Theme 3: Low internal technical digitalization capabilities. In general, it can be stated, that GROW has very limited internal capabilities when it comes to the field of digitalization. This is highlighted by two observations. Firstly, GROW has no internal IT department but depends on external providers. These IT providers are either part of the district authority, e.g., for telephone services, or are sourced on the external market, e.g., for the CRM system. This reliance on different external providers hinders further digitalization. Taking both examples, "because the [telephone] switchboard is currently still in the district authority, there is no call register in our CRM system" [I3].

Theme 4: Digitalization skills or capabilities is no recruiting criteria. In addition, GROW does not include digitalization skills in its recruitment criteria. Both interviews and analyzed job advertisements show that specific IT or digitalization knowledge is not required from potential applicants.

Theme 5: Mindset of Passenger and not of Driver. Employees see themselves as pure users of digital technologies. There is no impetus for self-driven digitalization. Exemplarily, one employee sees himself as "one of those who use digital solutions with pleasure when they are offered. But I do not start, conceptualize or develop something myself or make any innovations in that field" [I2]. There is a strong understanding that someone should digitalize GROW, as long as these digital solutions are easy to use.

Theme 6: Innovation is driven bottom-up as well as top-down. Building on the fifth theme, we could also observe that innovation is mainly driven by the top management. There are rare instances when ideas also are generated by employees, but only in specific domains. Digital innovations "which affect the entire GROW workforce are ultimately initiated by the management. So the individual ideas may come from different people, but the actual impulse comes from the management because they have to make the general decisions" [I4].

Theme 7: Low individual IT development skills. As already mentioned, the individual IT development skills are fairly low. Most interviewees see themselves as interested users but have no, or nearly no capability in terms of development (I1, I2, and I4). I3 has some basic IT background but is no real programmer. Guided, less code-heavy development is possible for him, "but in general, when it comes to programming skills, I would say that I am still scratching the surface in that area" [I4].

Phase 1: Changes Through the First Intervention

In phase 1, four groups of four to five information systems students (18 students in total) analyzed four processes, identified ideas for digitalization jointly with GROW employees, and implemented corresponding digital process innovations prototypically. As an example, one process analyzed was the travel expense management process. Before, employees of GROW filed their travel expenses using excel sheets that were printed and processed on paper. The paper documents were checked by the administrative support, signed off by the CEO, and finally paid by administrative support again. Here, the process is now supported using workflow management systems and rules-based routing. As such, new technology was incorporated into the process, i.e., a digital process innovation was implemented.

Still, during phase 1 some observations with regards to the seven themes identified above could be made. Firstly, through the two solutions that came into use, the amount of administrative work was reduced. It stayed, however, still on a high level (ad theme 1). Secondly, the observation that internal digitalization capabilities were low also hold true in phase 1 (ad theme 3). Employees of GROW were still in the passenger mindset (ad theme 5) and, thus, the collection of requirements for the process changes were difficult and mainly driven by the student consultants. Thirdly, the processes that were seen as candidates for innovation were mostly nominated by the top management, i.e., the CEO (ad theme 6).

Phase 2: Changes Through the Second Intervention

In the second phase, one student worked at GROW for four months as an external consultant. The student worked originally on five different processes. One example process is the appointment management. GROW as a regional business development agency has the main goal to attract new firms to the region and help the existing firms to prosper. As such, a lot of the work of the employees is done through meetings with customer representatives. Here, a new meeting scheduling system was introduced that was integrated into the existing calendar solution. Customers can now schedule meetings and appointments directly on the website of GROW which reduces administrative overhead for the employees. According to GROW employees, the "*solution is very user friendly*" with "*a lot of useful aspects*" (I7).

Again, during phase 2 we could directly observe some aspects with regards to the seven themes identified above. Firstly, the implemented digital process innovations reduced the administrative work within the respective processes to a great extent (ad theme 1). Moreover, with regards to the appointment management process, the first employee switched the role and started to drive the innovation. The process was suggested by an employee who was also involved in the selection of the solution (ad themes 5 and 6).

5 Concluding Discussion

Our preliminary results indicate that GROW's internal capabilities to change organizational routines and adopt digital process innovations were limited based on a lack of digital skills and capabilities of the individual employees. GROW heavily relies on external assistance for adopting new digital technologies. We could perceive a beginning change within this status. As the employees are introduced to lightweight solutions, enabling easy digitalization of business processes, they begin to reflect prior stable organizational routines and are moving towards a more digital mindset, embracing the potential of digital technology for process and business model innovations. However, these initial findings need to be reassessed in phase 3 of the research project. Here, we aim at identifying whether the external interventions really allow building internal process digitalization capabilities. Based on these findings we aspire to reach a deeper understanding of the development of organizational capabilities through the individual capabilities of employees on the operational level.

The preliminary and the aspired results of the overall research project will be limited to a certain extent. Firstly, there is the risk of residual researcher's bias. Through a study design with a delineation between researchers actively involved in the intervention phase and researchers with a focus on qualitative data analysis, we hope to overcome this bias to a large extent. Secondly, as the results are and will be generated in a single case organization, generalizability needs to be further discussed and analyzed.

References

1. Matt, C., Hess, T., Benlian, A.: Digital transformation strategies. Bus. Inf. Syst. Eng. **57**(5), 339–343 (2015). https://doi.org/10.1007/s12599-015-0401-5
2. Ciriello, R.F., Richter, A., Schwabe, G.: Digital innovation. Bus. Inf. Syst. Eng. **60**(6), 563–569 (2018). https://doi.org/10.1007/s12599-018-0559-8
3. Wiesböck, F., Hess, T.: Digital innovations. Electron. Mark. **30**(1), 75–86 (2019). https://doi.org/10.1007/s12525-019-00364-9
4. Plattfaut, R.: Robotic process automation - process optimization on steroids? In: Krcmar, H., Fedorowicz, J., Boh, W.F., Leimeister, J.M., Wattal, S. (eds.) Proceedings of the 40th International Conference on Information Systems, ICIS 2019. Association for Information Systems (2019)
5. Kane, G.C., Palmer, D., Nguyen Phillips, A., Kiron, D., Buckley, N.: Achieving digital maturity: adapting your company to a changing world. MIT Sloan Manag. Rev. **2017** (2017)
6. Wenzel, M., Danner-Schröder, A., Spee, A.P.: Dynamic capabilities? Unleashing their dynamics through a practice perspective on organizational routines. J. Manag. Inq. **105649262091654** (2020). https://doi.org/10.1177/1056492620916549
7. Baskerville, R.L., Wood-Harper, A.T.: A critical perspective on action research as a method for information systems research. J. Inf. Technol. **11**, 235–246 (1996). https://doi.org/10.1080/026839696345289
8. Urbach, N., et al.: The impact of digitalization on the IT department. Bus. Inf. Syst. Eng. **61**(1), 123–131 (2018). https://doi.org/10.1007/s12599-018-0570-0
9. Sebastian, I.M., Ross, J.W., Beath, C., Mocker, M., Moloney, K.G., Fonstad, N.O.: How big old companies navigate digital transformation. MIS Q. Exec. **16**, 197–213 (2017)
10. Fichman, R., Santos, B.D., Zheng, Z.: Digital innovation as a fundamental and powerful concept in the information systems curriculum. Manag. Inf. Syst. Q. **38**, 329–353 (2014)

11. Lacity, M.C., Willcocks, L.P.: Robotic process automation at telefónica O2. MIS Q. Exec. **15**, 21–35 (2016)
12. Pentland, B.T., Liu, P., Kremser, W., Haerem, T.: The dynamics of drift in digitized processes. MIS Q. **44**, 19–47 (2020). https://doi.org/10.25300/MISQ/2020/14458
13. Leonardi, P.M.: When flexible routines meet flexible technologies: affordance, constraint, and the imbrication of human and material agencies. MIS Q. **35**, 147 (2011). https://doi.org/10.2307/23043493
14. Priem, R.L., Butler, J.E.: Is the resource-based view a useful perspective for strategic management research? AMR **26**, 22–40 (2001). https://doi.org/10.5465/amr.2001.4011928
15. Helfat, C.E., Peteraf, M.A.: The dynamic resource-based view: capability lifecycles. Strat. Mgmt. J. **24**, 997–1010 (2003). https://doi.org/10.1002/smj.332
16. Teece, D.J., Pisano, G., Shuen, A.: Dynamic capabilities and strategic management. Strat. Mgmt. J. **18**, 509–533 (1997). https://doi.org/10.1002/(SICI)1097-0266(199708)18:7%3c509::AID-SMJ882%3e3.0.CO;2-Z
17. Teece, D.J.: Explicating dynamic capabilities: the nature and microfoundations of (sustainable) enterprise performance. Strat. Mgmt. J. **28**, 1319–1350 (2007). https://doi.org/10.1002/smj.640
18. Zollo, M., Winter, S.G.: Deliberate learning and the evolution of dynamic capabilities. Organ. Sci. **13**, 339–351 (2002). https://doi.org/10.1287/orsc.13.3.339.2780
19. Eisenhardt, K.M., Martin, J.A.: Dynamic capabilities: what are they? Strat. Mgmt. J. **21**, 1105–1121 (2000). https://doi.org/10.1002/1097-0266(200010/11)21:10/11%3c1105::AID-SMJ133%3e3.0.CO;2-E
20. Helfat, C.E.: Dynamic Capabilities. Understanding Strategic Change in Organizations. Blackwell, Oxford (2007)
21. Barreto, I.: Dynamic capabilities: a review of past research and an agenda for the future. J. Manag. **36**, 256–280 (2010). https://doi.org/10.1177/0149206309350776
22. Ambrosini, V., Bowman, C.: What are dynamic capabilities and are they a useful construct in strategic management? Int. J. Manag. Rev. **11**, 29–49 (2009). https://doi.org/10.1111/j.1468-2370.2008.00251.x
23. Blyler, M., Coff, R.W.: Dynamic capabilities, social capital, and rent appropriation: ties that split pies. Strat. Mgmt. J. **24**, 677–686 (2003). https://doi.org/10.1002/smj.327
24. Felin, T., Powell, T.C.: Designing organizations for dynamic capabilities. Calif. Manag. Rev. **58**, 78–96 (2016). https://doi.org/10.1525/cmr.2016.58.4.78
25. Felin, T., Foss, N.J.: Strategic organization: a field in search of micro-foundations. Strateg. Organ. **3**, 441–455 (2005). https://doi.org/10.1177/1476127005055796
26. Eisenhardt, K.M., Furr, N.R., Bingham, C.B.: CROSSROADS—microfoundations of performance: balancing efficiency and flexibility in dynamic environments. Organ. Sci. **21**, 1263–1273 (2010). https://doi.org/10.1287/orsc.1100.0564
27. Helfat, C.E., Peteraf, M.A.: Managerial cognitive capabilities and the microfoundations of dynamic capabilities. Strat. Mgmt. J. **36**, 831–850 (2015). https://doi.org/10.1002/smj.2247
28. Salvato, C., Vassolo, R.: The sources of dynamism in dynamic capabilities. Strat. Mgmt. J. **39**, 1728–1752 (2018). https://doi.org/10.1002/smj.2703
29. Wohlgemuth, V., Wenzel, M., Berger, E.S., Eisend, M.: Dynamic capabilities and employee participation: the role of trust and informal control. Eur. Manag. J. **37**, 760–771 (2019). https://doi.org/10.1016/j.emj.2019.02.005
30. Bapuji, H., Hora, M., Saeed, A., Turner, S.: How understanding-based redesign influences the pattern of actions and effectiveness of routines. J. Manag. **45**, 2132–2162 (2019). https://doi.org/10.1177/0149206317744251
31. Aral, S., Weill, P.: IT assets, organizational capabilities, and firm performance: how resource allocations and organizational differences explain performance variation. Organ. Sci. **18**, 763–780 (2007). https://doi.org/10.1287/orsc.1070.0306

32. Kim, G., Shin, B., Kim, K., Lee, H.: IT capabilities, process-oriented dynamic capabilities, and firm financial performance. JAIS **12**, 487–517 (2011). https://doi.org/10.17705/1jais.00270
33. Wiesboeck, F.: Thinking Outside of the IT Capability Box. AMCIS 2018 Proceedings, vol. (2018)
34. Lu, Y., Ramamurthy, K.: Understanding the link between information technology capability and organizational agility: an empirical examination. MIS Q. **35**, 931 (2011). https://doi.org/10.2307/41409967
35. Lyytinen, K., Yoo, Y., Boland, R.J., Jr.: Digital product innovation within four classes of innovation networks. Info Syst. J. **26**, 47–75 (2016). https://doi.org/10.1111/isj.12093
36. Susman, G.I.: Action research: a sociotechnical systems perspective. In: Morgan, G. (ed.) Beyond method. Strategies for social research. Sage, Beverly Hills, Calif. (1985)
37. Coghlan, D., Shani, A.B.: Creating action research quality in organization development: rigorous, reflective and relevant. Syst. Pract. Action Res. **27**(6), 523–536 (2013). https://doi.org/10.1007/s11213-013-9311-y
38. https://doi.org/10.2307/25148642
39. Baskerville, R., Wood-Harper, A.T.: Diversity in information systems action research methods. Eur. J. Inf. Syst. **7**, 90–107 (1998). https://doi.org/10.1057/palgrave.ejis.3000298
40. Coad, P., Yourdon, E.: Object-oriented Analysis. Yourdon Press, Englewood Cliffs, NJ (1991)
41. Glaser, V.L.: Design performances: how organizations inscribe artifacts to change routines. AMJ **60**, 2126–2154 (2017). https://doi.org/10.5465/amj.2014.0842
42. Aguirre, S., Rodriguez, A.: Automation of a business process using robotic process automation (RPA): a case study. In: Figueroa-García, J.C., López-Santana, E.R., Villa-Ramírez, J.L., Ferro-Escobar, R. (eds.) WEA 2017. CCIS, vol. 742, pp. 65–71. Springer, Cham (2017). https://doi.org/10.1007/978-3-319-66963-2_7

Digital Twin of an Organization: Are You Serious?

Markus C. Becker[1](✉) and Brian T. Pentland[2]

[1] University of Southern Denmark, Odense, Denmark
mab@sam.sdu.dk
[2] Michigan State University, East Lansing, MI, USA
pentland@broad.msu.edu

Abstract. Digital twins are becoming established tools for physical devices and systems. Their success has raised the promise and the "grand challenge" of digital twins of organizations. To the extent that organizations include networks of interdependent processes, human agency and conflict, and learning, we argue that building valid, reliable digital twins of organizations involves substantial challenges that might require new methods and new science as compared to current models of business processes. For instance, we need a richer, more complete vocabulary to describe interdependence and regulatory mechanisms that managers can use to design and intervene in networks of interdependent processes. Models from biological sciences might provide fruitful inspiration, given that regulatory networks play important roles in biological systems. If we are serious about digital twins of organizations, we need to push the boundaries of process methods and process science.

Keywords: Digital twins · Routine dynamics · Regulatory networks

1 Introduction

A digital twin is a computational model of a system that is sufficiently valid and reliable that it can be used for design, prediction, maintenance, and other valuable use cases. The Gartner Group [1] defines a digital twin as "a software design pattern that represents a physical object with the objective of understanding the asset's state, responding to changes, improving business operations and adding value." Digital twins have been created for a wide variety of physical systems, such as buildings and automobiles. Generally speaking, artifacts with digital twins are created through a digitalized design process; the digital twin arrives before its physical sibling is born [2].

Recently, the idea of a Digital Twin of an Organization (DTO) has grabbed the imagination of scholars and practitioners in the Business Process Management community (e.g., https://www.my-invenio.com/). In many respects, the basic idea of DTO is not new. Operations management is built on the idea of using mathematical models to design and predict organizational performance (e.g., Little's Law for queueing [3]). The general idea of modeling production processes is an established practice in industrial

A. Marrella and B. Weber (Eds.): BPM 2021 Workshops, LNBIP 436, pp. 243–254, 2022.
https://doi.org/10.1007/978-3-030-94343-1_19

engineering [4, 5] and tools for creating detailed simulations of "virtual factories" are commercially available (e.g., https://www.flexsim.com/factory-simulation/).

So, when Caporuscio et al. [6] refer to DTO as a "grand challenge," they are talking about taking this general idea to a whole new level: "representing all the elements and connections of an organizational system in virtual models, which can be perpetually simulated and analyzed to achieve continuous assessment and optimization of the organization." That is a remarkably broad vision. Our purpose in this paper is to contribute to the debate on DTOs and inform nascent efforts of developing DTOs by contributing clarifications and suggestions from an interdisciplinary perspective that, we hope, might be useful in this endeavor. For this paper, we therefore want to narrow and clarify what we mean by a *digital twin of an organization.* These ideas can be refined in many ways [7], but we focus here on four baseline criteria:

1. ***Organizations include multiple, interdependent processes.*** A single process is not an organization. For example, *order-to-cash* is always linked with *purchase-to-pay,* and each of these major processes typically consists of many specialized, interdependent sub-processes.
2. ***Organizations involve agents with the capacity to make choices, learn from experience, and pursue their own objectives.*** These agents are typically human, but not necessarily. Because agents do not necessarily have the same objectives, conflict is common [8].
3. ***A digital twin is a model that serves a purpose.*** A DTO should help with managing the organization: designing the organization, planning, decision-making on organizational policy in functional domains, etc. If the purpose is simple, the model might be simple. But even simple organizations include multiple, interdependent processes carried out by at least some human agents.
4. ***A digital twin is more than a dashboard.*** Dashboards provide visualization of how tasks are accomplished in an organization. Their contribution is to guide managerial attention to *where* intervention might be required. Dashboards provide visualization, but digital twins provide simulation and prediction [6]. DTOs can contribute not just to "continuous assessment" but also "optimization of the organization" [6] when they enable managers to predict and simulate how the organization might behave when it faces new or different situations.

It should be clear from these criteria that creating a DTO will extend the current state-of-the-art in process mining and modeling. Without question, great strides have been made in making processes visible and providing support for managing and improving processes. A key part of that success has been the ability to "open the black box" to reveal the structures and mechanisms that connect inputs to outputs. Technical innovations, such as neural network models [9, 10], object-centric approaches [11], and mining of simulations [12, 13] are constantly improving the state-of-the-art. Perhaps because of this on-going progress, the DTO idea has entered the conversation.

Our position is that a valid, reliable DTO is not a realistic goal unless we can offer convincing solutions to key issues concerning what are relevant key features of organizations and how to model them. If we are serious about digital twins of organizations, we need to push the boundaries. New methods and new science may be required. Towards this end, we raise four key questions:

1. ***What makes DTO so challenging?*** At a minimum, novel challenges in building digital twins of *organizations* arise from (a) the inclusion of human agents and (b) the inclusion of multiple interdependent processes. To highlight the challenges in creating a DTO, we draw on a recently published example from the literature on routine dynamics [14].
2. ***In light of these challenges, what kinds of models best reflect the true fabric of organization?*** While finite state machines (e.g., Petri nets) are a reasonable way to model deterministic business processes, modelling an "organizational system" is a bigger challenge. It may be useful to take inspiration from biological systems that are modelled as networks of interdependent regulatory processes that give rise to emergent, adaptive behavior.
3. ***How can DTO help managers design effective interventions?*** We envision the main use case for DTO to be helping managers understand *how* to intervene to solve organizational problems, design organizational structures, policies, and incentives. This might involve understanding emergent phenomena.
4. ***How can we evaluate a DTO?*** To the extent that managers need to address problems that extend beyond prior experience (e.g. how to integrate a novel technology?), fitting a model to historical data does not seem adequate. A DTO needs to be built and evaluated with respect to generalizable theory.

2 What Makes DTO so Challenging?

Scaling from simple physical systems to complex social systems involves moving along at least two dimensions: from physical to social and from simple to complex, as shown in Fig. 1. We can define each dimension separately, but in real systems, they interact.

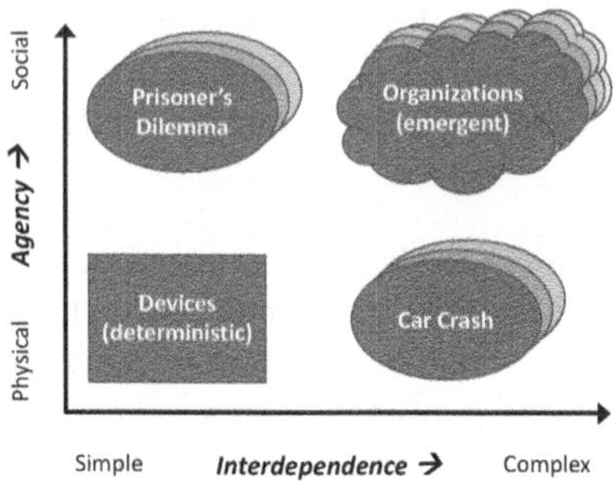

Fig. 1. Scaling from simple physical to complex social systems

The move from physical systems to social systems involves adding *agency*, which we can define simply as the ability to act otherwise [15]. Agency involves the capacity to reflect on the past and anticipate the future [16]. Agency is important even in very simple social systems. For example, in a classic prisoner's dilemma, there are two participants with two choices: to cooperate or defect [17]. This choice is not predetermined or mechanistic. Agents can behave strategically in their own interest so there is always the possibility of conflict [18].

The move from simple to complex involves adding interdependent sub-systems. Interdependence takes many forms, so a succinct definition remains elusive [19]. Basically, interdependence adds complexity. For example, Leonardi [20] studied the use of digital twins to model car crashes. Since modern vehicles are digitally designed to begin with, crash testing virtual cars is much cheaper and faster than crash testing real cars. Each part is relatively simple, but the safety of the vehicle depends on the ensemble of parts and their interconnections and interactions, which are extremely complex.

Now, imagine simulating a car crash where some of the individual parts have agency, learn from experience, and can choose how they will react to sudden deceleration. The analogy to organizations is imperfect, but instructive. The combination of agency, interdependence and learning adds to the challenge of creating valid, reliable models that can help managers to make design and policy decisions. If we take the DTO idea seriously, we see at least four major issues that need to be confronted: discovery, interdependence, agency and learning. The first is methodological, the others substantial.

2.1 Discovering the True Fabric of Organizing

While physical objects may be created through a digitalized design process, organizations generally are not. That means the digital twin has to be *discovered* from observation of an organization that is up-and-running. This has been the core challenge of process mining since its inception. van der Aalst [21] argues that process mining allows us to recognize what he calls the "true fabric of business processes." In that article, he is particularly concerned about modeling concurrency:

> Business processes tend to be highly concurrent and non-monolithic. ... The empirical nature of process mining helps managers, consultants, and process analysts to better understand the "fabric of real business processes" and, thus, also see the limitations of conventional process modeling languages. The challenge is to link elegant succinct formal models like Petri nets to behavior actually observed in reality [21: 690].

It is worth remembering that with the current state of the art in process mining, linking observed data to a specific Petri net can be challenging. Different ways of filtering an event log result in different Petri nets [22]. Applying different mining algorithms to the same event log generally results in different Petri nets [23]. Even for single, mechanistic processes, the "true fabric" remains elusive.

As we move beyond single, mechanistic processes, we expect the problem of discovery to become more difficult. Organizational systems that include humans have latent structures that are difficult to observe (e.g., truces, cultural assumptions and values). A classic problem in social science is that these structures go undetected until they are breached [24]. In other words, there are likely to be unobserved boundary conditions on a DTO that can only be discovered by trial and error, when the model fails.

2.2 Networks of Interdependent Processes

Processes in organizations are often concurrent, but they are also *interdependent* [17, 25]. The fabric of organizing is laced with interdependence. It is like the air: it surrounds everything, but it is hard to see. Within a process, various kinds of task interdependence (pooled, sequential, reciprocal) are common [26]. Across an organization, there is also role, reward, goal, knowledge and epistemic interdependence [17, 27]. For example, the reward (incentive) for one activity may depend on when, where or how another (physically and temporally remote) activity is performed. Like other latent structures, these forms of interdependence may be difficult to observe.

Unfortunately, interdependence between networks of processes is not well understood [28, 29]. Sometimes, there are clusters of interdependent routines with well-defined interfaces [28]. Other times, there are networks of interdependent processes that shape other processes, as described by [30]. The key point is that networks of interdependent processes can generate emergent behavior [31]. By emergent, we mean that the "existence and nature" of the behavior "depend upon entities at a lower level, but the behavior is neither reducible to, nor predictable from, properties of entities found at the lower level" [32: 103].

Interdependence of processes is, therefore, key to modeling the true fabric of organizing. Networks of interdependent processes appear to be important drivers of emergent behavior but modeling them is an area where new method and science seems to be needed [29].

2.3 Modeling Systems with Agency

In their analysis of digital twin technology, Gartner Group [1] refers to "physical object[s]." Yet the phrase *digital twin of an organization* implies aspirations that extend beyond physical objects. Business processes often include human agents. Rather than a digital twin of a car, a DTO would be a digital twin of the car *and its driver*.

We use the term "agent" rather than "actor" to signal that humans have *agency* [16]. Humans have "the ability to remember the past, imagine the future, and respond to present circumstances" [32: 100]. For instance, they can act on their intentions concerning future outcomes, their interpretations of the context, themselves, and others, incentives provided by the context, and their orientation towards past, present or future [16, 33–35]. Each of these mechanisms adds texture to the true fabric of organization.

Agency introduces the need for monitoring and control, which is a central issue in organizational design [7, 36]. The classic "principal-agent" problem in contracting and organizational design stems from the fact that agents can choose their own course of action to serve their own interests. Agency introduces a new challenge to modeling organizations because we cannot assume that all the "parts" are working towards the same goal [37]. There is always the possibility of conflict.

2.4 Modeling Systems with Learning

If a digital twin of an organization was possible today, it would most likely be invalid tomorrow. This is because organizations are open systems that change over time in response to their environment and the particular interests and capabilities of internal and external stakeholders [25, 27, 38]. Organizations undergo an on-going process of emergence [39]. Routines can change through mere repetition [32]. Further, the people in organizations learn from experience and from each other; they change roles; they enter and exit the organization [40–42]. And with learning, there is always forgetting [43]. These learning processes pose an on-going threat to the validity and reliability of a digital twin for purposes of prediction.

Together, these issues (discovery, interdependence, agency and conflict, and learning) underscore the enormous challenge involved in scaling from simple physical objects to complex organizations. In the next section, we describe an example that will help make these issues more concrete.

3 Example: An Agile Software Development Organization

Goh and Pentland [14] use a routine dynamics perspective to describe an organization that develops videogames using SCRUM, an agile development methodology. Goh and Pentland [14] wanted to explain changes in the complexity of the patterns of action enacted by game developers as the project progressed from sprint to sprint. This example could be regarded as an outlier because project work is likely to be less structured and less mechanistic than production work [44]. Nevertheless, it serves to make some useful points.

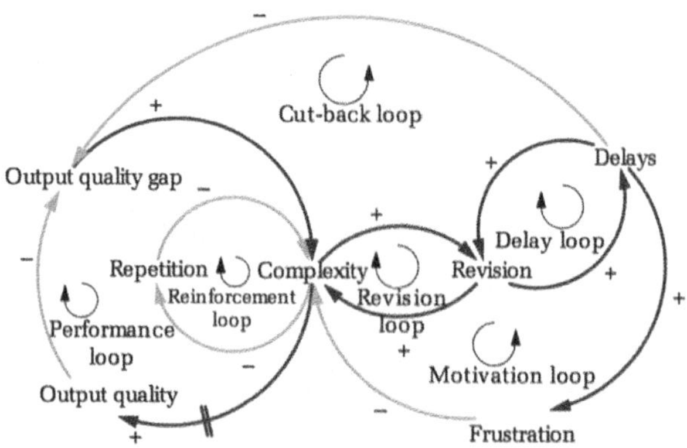

Fig. 2. Interdependent processes in a game development project

Figure 2 shows a causal diagram that includes a set of interdependent processes that influenced patterns of action in game development. This diagram is a qualitative description derived from participant observation and interviews with the game developers. The figure contains the following six causal loops:

- ***Reinforcement.*** Through repetition, the pattern of action becomes simpler and better defined.

- ***Output quality gaps.*** Dealing with quality problems increases the complexity in the pattern of actions.
- ***Revisions.*** Dealing with revisions increases the complexity in the pattern of actions.
- ***Delays.*** Revisions lead to delays, which lead to more revisions and frustration among the developers. Delays also lead to cutbacks.
- ***Motivation.*** When developers get more frustrated, they tend to revert to established work patterns.
- ***Cutbacks.*** Reducing the scope of the project helps reduce quality problems.

There are some key points to glean from this example. First, these processes are highly interdependent and not easily decomposable: delays, revisions, frustration, cutbacks and quality problems form a Gordian Knot that cannot be unraveled. To understand how they generated the behavior of the organization, they need to be considered together. This exemplifies the idea of a 'network of regulatory mechanisms', as described below. Second, these processes are quite generic. Schedule, scope, budget and quality are classic dimensions of any project. It is not surprising that they showed up in this project. Third, it shows that the human element is inextricably bound up in the process network. Frustration and motivation reflect the individual interpretations and emotions of the game developers. Fourth, it shows that exogenous contextual factors (such as scope and schedule) can directly influence the internal dynamics of the process network.

Finally, there is one key point is that not shown directly in Fig. 2 but is essential to its significance: this complex web of interacting processes had a dramatic influence on the execution of the game development work itself. From sprint to sprint, the way the developers worked (as captured in the event log based on their SCRUM sheets) was different. Of course, the extent of contextual influence on a particular process is an empirical question. But in a network of interdependent processes, each process creates context for the others [30]. This dynamic web of interdependence, colored by human agency and learning, is what gives rise to the true fabric of organizing.

4 Models of Organizations: Beyond Organizations as Mechanisms

What kinds of models are best suited to address these challenges and align the focus on the key elements described above (discovery, dynamic web of interdependence, agency and learning) and their implications for valid and reliable DTOs? To address this question, it is useful to consider the overall perspective taken in tackling the challenge of developing DTOs. While we need to avoid overly simplistic metaphors, organizations as *mechanisms* or as *organisms* are two common ways of thinking about organizations [38, 45]. Because the concept of a digital twin emerged from the realm of physical systems, it is natural to think in terms of organizations as mechanisms. This metaphor is consistent with the practice of modeling business processes as finite state machines (e.g., Petri nets). However, as we scale from modeling one or a small number of business processes to modeling larger organizational systems, this way of conceiving of organizations might become less suitable and productive.

Latching onto the challenges identified above, we note that it may be fruitful to take inspiration from how the biological sciences model complex systems. From epigenetics to ecologies, recent theorizing highlights networks of interdependent regulatory processes that give rise to emergent, adaptive systems [46, 47]. These may offer a potentially useful way to think of the dynamic web of interdependence that characterizes complex social organizations.

4.1 Networks of Regulatory Mechanisms

In scaling from simple to complex systems, networks of interdependent processes must play a central role. As mentioned above, process models are often limited to task interdependence (sequential flow). Other kinds of interdependence (e.g., involving rewards and incentives) may be equally consequential in an organizational system. For example, the interdependence between scope, schedule, budget and quality involves the rewards and incentives of various stakeholders, as well as physical constraints. For a valid and reliable digital twin of an organization, it seems important to recognize a more complete range of types of relations and interdependence. We need to understand what drives – and regulates – the dynamics of networks of interdependent processes. If these dynamics are emergent, DTOs need to capture the drivers of those emergent dynamics.

The causal loops in Fig. 2 are reminiscent of the regulatory pathways that are used to model emergent dynamics in the biological sciences, from epigenetics to ecological systems. While the connection is just metaphorical at this time, we expect that it could lead to theoretical and practical insights.

First, by highlighting networks of regulatory pathways as a focus for theory on process interdependence, the connection points to possible means of managerial intervention. In pharmacology, the search for therapeutic drugs is often guided by efforts to reinforce (or disrupt) regulatory pathways [48]. The theory helps guide the practice. Second, it points to a toolbox of analytical and modeling techniques with a track record of application to networks of regulatory pathways and emergent phenomena. We therefore propose regulatory networks as a focus for efforts of developing digital twins that also address social issues and ultimately, digital twins of organizations.

4.2 Discovering Regulatory Networks

As mentioned, one challenge in developing DTO is that the digital twin has to be discovered from observation of an ongoing operation. Given a running organization, it is not clear how accurately the regulatory networks can be recovered through process mining. Process mining for system dynamics [12], for instance, is in its infancy.

Rather than modeling these regulatory mechanisms from a dynamic systems perspective (as systems of differential equations), it may be helpful to think in terms of enabling and constraining relations. Towards that end, declarative mining [49, 50] might be a fruitful way to discover regulatory process networks. In a declarative model, the process is represented as a set of constraints and can be executed in all possible ways as long as these constraints are respected [51: 9238].

The issue is not whether declarative (vs. imperative) notation is better or more expressive. The issue is how to discover dependencies between processes that might not otherwise be apparent (e.g., patterns of behavior that reflect cultural norms). These dependencies may regulate patterns of action in the organization. To design effective interventions, managers may need models that embody a richer, more complete set of relations and regulatory mechanisms.

5 Evaluating a DTO

Fitting a model to observed behavior is a good first step, but the challenge of creating a useful DTO goes beyond linking "succinct formal models like Petri nets to behavior actually observed in reality" [21: 690]. To support managerial intervention, design and policy making, a DTO may need to simulate or predict behavior that has *not yet been observed*. For that reason, the DTO needs to incorporate a theoretical understanding of the underlying phenomena that generalizes beyond a particular set of historical observations.

For physical applications, like automotive crash testing, good theoretical foundations are available. Using finite-element representations for each part of a vehicle, along with knowledge of material properties and physics, engineers can simulate virtual crashes that have never occurred in reality [19].

Unfortunately, for organizational systems, the theoretical foundations are not as well understood and not as deterministic. We can discover models using state-of-the-art techniques, but we can only evaluate them against observed data. That may be sufficient for day-to-day adjustments, but for novel problems, it may not. We need to advance the science, not just the method.

6 Conclusion: A Challenge Worth Pursuing

Without question, a valid, reliable digital twin of a human organization is a challenging goal. Organizations are not deterministic mechanisms; they are woven from interdependence, agency and learning. They contain latent structures that are difficult to observe. While challenging, the goal is worth pursuing for at least two main reasons, one practical and one theoretical.

6.1 A Platform for Designing Interventions for Managers

The practical motivation for DTO is to help managers design effective interventions to solve organizational problems. Given that complex systems are difficult to comprehend for human actors, a DTO would be particularly valuable for managers who are boundedly rational and organizations that are highly complex [52]. Similarly, a DTO would be useful for boundedly rational organization designers [27] in their efforts to create an organization such that it has desired performance features, e.g., provides coordination where needed or handle cooperation problems [53]. For instance, it would enable prototyping and "crash testing" new organizational designs *in silico* [27]. If the costs of trial-and-error experimentation with different organization designs is high, or if the consequences of complex organization designs are difficult to anticipate without a model or simulation, then a DTO would be a particularly attractive tool.

To fulfil these roles, the DTO needs to provide valid, reliable predictions of what would happen if something in the organization were changed. By focusing on business processes, the DTO would already be focused centrally on what generates performance – if nothing happens in the organization, no performance is generated. The ability to model the mechanisms that regulate the system of interdependent processes seems to be central to providing such valid, reliable predictions. If valid and reliable regulatory networks can be discovered, it would thus be a valuable tool for management, offering the possibility to simulate how to tune the organization as a complex system. Thereby, it would help address questions such as how to intervene to increase an organization's performance, but also how to make changes 'stick'. Even intermediate, more limited models than the full digital twin envisioned by Caporuscio et al. [6] would therefore be of great use in practice, such as digital twins that also address social interaction and its consequences.

6.2 A Vehicle for Advancing Science

Research in process management and routine dynamics would benefit greatly from the conversations around DTO as a grand challenge. If our analysis of the challenge is even partly correct, then DTO provides an arena where research on routine dynamics can inform process management and vice versa. Currently, these research communities are quite separate [54]. The endeavor and practice of building DTO would provide a motivation for collaboration. It would provide opportunities to build the science, and may provide an excellent vehicle for discussion, controversy, and learning about questions such as limitations and the appropriate granularity, abstraction, and timescales. We expect that this might be a case of practice leading science, as the necessary data and experimentation would need to be grounded in practical managerial concerns. If we are serious about DTO, engaging in this dialogue seems a fruitful way forward.

References

1. Gartner Group (2019). https://www.gartner.com/en/newsroom/press-releases/2019-02-20-gartner-survey-reveals-digital-twins-are-entering-mai. Accessed 10 May 2021
2. Baskerville, R.L., Myers, M.D., Yoo, Y.: Digital first: the ontological reversal and new challenges for information systems research. MIS Q. **44**(2), 509–523 (2020)
3. Little, J.D.C.: A proof for the queuing formula: $L = \lambda W$. Oper. Res. **9**(3), 383–387 (1961). https://doi.org/10.1287/opre.9.3.383
4. Nåfors, D., Lindskog, E., Berglund, J., Gong, L., Johansson, B., Vallhagen, J.: Realistic virtual models for factory layout planning. In: 2017 Winter Simulation Conference (WSC), pp. 3976–3987. IEEE (2017)
5. Yildiz, E., Møller, C., Bilberg, A.: Demonstration and evaluation of a digital twin-based virtual factory. Int. J. Adv. Manuf. Technol. **114**(1–2), 185–203 (2021). https://doi.org/10.1007/s00170-021-06825-w
6. Caporuscio, M., Edrisi, F., Hallberg, M., Johannesson, A., Kopf, C., Perez-Palacin, D.: Architectural concerns for digital twin of the organization. In: Jansen, A., Malavolta, I., Muccini, H., Ozkaya, I., Zimmermann, O. (eds.) ECSA 2020. LNCS, vol. 12292, pp. 265–280. Springer, Cham (2020). https://doi.org/10.1007/978-3-030-58923-3_18
7. Puranam, P.: A future for the science of organization design. J. Organ. Des. **1**(1), 18–19 (2012)
8. Cyert, R.M., March, J.G.: A Behavioral Theory of the Firm. Blackwell, Oxford (1963)

9. Nolle, T., Seeliger, A., Thoma, N., Mühlhäuser, M.: DeepAlign: alignment-based process anomaly correction using recurrent neural networks. In: Dustdar, S., Yu, E., Salinesi, C., Rieu, D., Pant, V. (eds.) Advanced Information Systems Engineering. CAiSE 2020. LNCS, vol. 12127, pp. 319–333. Springer, Cham (2020). https://doi.org/10.1007/978-3-030-49435-3_20
10. Park, G., Song, M.: Predicting performances in business processes using deep neural networks. Decis. Support Syst. **129**, 113191 (2020). https://doi.org/10.1016/j.dss.2019.113191
11. van der Aalst, W.M., Berti, A.: Discovering object-centric petri nets. Fundam. Inform. **175**(1–4), 1–40 (2020)
12. Pourbafrani, M., van der Aalst, W.M.: PMSD: data-driven simulation using system dynamics and process mining. arXiv preprint arXiv:2010.00943 (2020)
13. Pourbafrani, M., van Zelst, S.J., van der Aalst, W.M.P.: Supporting automatic system dynamics model generation for simulation in the context of process mining. In: Abramowicz, W., Klein, G. (eds.) Business Information Systems. BIS 2020. Lecture Notes in Business Information Processing, vol. 389, pp. 249–263. Springer, Cham (2020). https://doi.org/10.1007/978-3-030-53337-3_19
14. Goh, K., Pentland, B.T.: From actions to paths to patterning: toward a dynamic theory of patterning in routines. Acad. Manag. J. **62**(6), 1901–1929 (2019)
15. Giddens, A.: The Constitution of Society: Outline of the Theory of Structuration. Univ of California Press, Oakland (1984)
16. Emirbayer, M., Mische, A.: What is agency? Am. J. Sociol. **103**(4), 962–1023 (1998)
17. Axelrod, R.: The Evolution of Cooperation. Basic Books, New York (1984)
18. Jensen, M.C., Meckling, W.H.: Theory of the firm: managerial behavior, agency costs and ownership structure. J. Financ. Econ. **3**(4), 305–360 (1976)
19. Raveendran, M., Silvestri, L., Gulati, R.: The role of interdependence in the micro-foundations of organization design: task, goal, and knowledge interdependence. Acad. Manag. Ann. **14**(2), 828–868 (2020)
20. Leonardi, P.M.: Car Crashes Without Cars: Lessons about Simulation Technology and Organizational Change from Automotive Design. MIT Press, Cambridge (2012)
21. van der Aalst, W.M.: Business process management as the Killer App for Petri nets. Softw. Syst. Model. **14**(2), 685–691 (2014). https://doi.org/10.1007/s10270-014-0424-2
22. Tax, N., Sidorova, N., van der Aalst, W.M.P.: Discovering more precise process models from event logs by filtering out chaotic activities. J. Intell. Inf. Syst. **52**(1), 107–139 (2018). https://doi.org/10.1007/s10844-018-0507-6
23. Janssenswillen, G.: Unearthing the Real Process Behind the Event Data: The Case for Increased Process Realism (Doctoral dissertation, Universiteit Hasselt) (2019)
24. Garfinkel, H.: Studies in Ethnomethodology. Prentice-Hall, Englewood Cliffs, NJ (1967)
25. Mintzberg, H.: The Structuring of Organizations. Prentice-Hall, London (1979)
26. Thompson, J.D.: Organizations in Action – Social Science Bases of Administrative Theory. Transaction Publishers, New York (1967)
27. Puranam, P.: The Microstructure of Organizations. Oxford University Press, Oxford (2018)
28. Kremser, W., Schreyögg, G.: The dynamics of interrelated routines: introducing the cluster level. Organ. Sci. **27**(3), 698–721 (2016)
29. Rosa, R.A, Kremser, W., Bulgacov, S.: Routine interdependence: intersections, clusters, ecologies and bundles. In: Feldman, M.S., Pentland, B.T., D'Adderio, L., Dittrich, D., Rerup, C., Seidl, D. (eds.) Cambridge Handbook of Routine Dynamics, Cambridge University Press (in press)
30. Kremser, W., Pentland, B.T., Brunswicker, S.: Interdependence within and between routines: a performative perspective. Res. Sociol. Organ. **61**, 79–98 (2019)
31. Kauffman, S.: The Origins of Order. Oxford University Press, New York (1993)

32. Hodgson, G.M.: Institutions and individuals: interaction and evolution. Organ. Stud. **28**(1), 95–116 (2007)
33. Feldman, M.S., Pentland, B.T.: Reconceptualizing organizational routines as a source of flexibility and change. Adm. Sci. Q. **48**(1), 94–118 (2003)
34. Howard-Grenville, J.A.: The persistence of flexible organizational routines: the role of agency and organizational context. Organ. Sci. **16**(6), 618–636 (2005)
35. Dionysiou, D.D.: Pragmatism & routine dynamics. Feldman, M.S., Pentland, B.T., D'Adderio, L., Dittrich, K., Rerup, C., Seidl, D. (eds.): Cambridge Handbook of Routine Dynamics. Cambridge University Press, Cambridge (2021)
36. Eisenhardt, K.M.: Agency theory: an assessment and review. Acad. Manag. Rev. **14**(1), 57–74 (1989)
37. Milgrom, P., Roberts, J.: Economics, Organization and Management. Prentice-Hall, Englewood Cliffs, NJ (1992)
38. Burns, T., Stalker, G.M.: The Management of Innovation. Tavistock, London (1961)
39. Tsoukas, H., Chia, R.: On organizational becoming: rethinking organizational change. Organ. Sci. **13**(5), 567–582 (2002)
40. Argote, L., Levine, J.M. (eds.): The Oxford Handbook of Group and Organizational Learning. Oxford University Press, New York, NY (2020)
41. Christensen, M., Knudsen, T.: Division of roles and endogenous specialization. Ind. Corp. Chang. **29**(1), 105–124 (2020)
42. Huber, G.P.: Organizational learning: the contributing processes and the literatures. Organ. Sci. **2**(1), 88–115 (1991)
43. Casey, A.J., Olivera, F.: Reflections on organizational memory and forgetting. J. Manag. Inq. **20**(3), 305–310 (2011)
44. Obstfeld, D.: Creative projects: a less routine approach toward getting new things done. Organ. Sci. **23**(6), 1571–1592 (2012)
45. Morgan, G.: Images of Organization. Sage, Thousand Oaks, CA (1979)
46. Mercer, T.R., Mattick, J.S.: Structure and function of long noncoding RNAs in epigenetic regulation. Nat. Struct. Mol. Biol. **20**(3), 300–307 (2013)
47. Montoya, J.M., Pimm, S.L., Solé, R.V.: Ecological networks and their fragility. Nature **442**(7100), 259–264 (2006)
48. Cui, W., Aouidate, A., Wang, S., Yu, Q., Li, Y., Yuan, S.: Discovering anti-cancer drugs via computational methods. Front. Pharmacol. **11**, 733 (2020)
49. Maggi, F.M., Bose, R.P.J.C., van der Aalst, W.M.P.: Efficient discovery of understandable declarative process models from event logs. In: Ralyté, J., Franch, X., Brinkkemper, S., Wrycza, S. (eds.) Advanced Information Systems Engineering. CAiSE 2012. LNCS, vol. 7328, pp 270–285. Springer, Berlin, Heidelberg (2012). https://doi.org/10.1007/978-3-642-31095-9_18
50. Maggi F.M.: Declarative process mining. In: Sakr, S., Zomaya, A. (eds.) Encyclopedia of Big Data Technologies. Springer, Cham (2018). https://doi.org/10.1007/978-3-319-63962-8_92-1
51. Rovani, M., Maggi, F.M., de Leoni, M., van der Aalst, W.M.P.: Declarative process mining in healthcare. Expert Syst. Appl. **42**(2015), 9236–9251 (2015)
52. Simon, H.A.: Administrative Behavior, 4th edn. The Free Press, New York (1997)
53. Gulati, R., Wohlgezogen, F., Zhelyazkov, P.: The two facets of collaboration: cooperation and coordination in strategic alliances. Acad. Manag. Ann. **6**(1), 531–583 (2012)
54. Pentland, B.T., Vaast, E., Ryan Wolf, J.: Theorizing process dynamics with directed graphs: a diachronic analysis of digital trace data. MIS Q. **45**(2) 967–984 (2021)

14th International Workshop on Social and Human Aspects of Business Process Management (BPMS2 2021)

14th International Workshop on Social and Human Aspects of Business Process Management (BPMS2 2021)

The involvement of human aspects in business process management takes place both on a social and individual level. Social information systems [1] such as social media, Enterprise 2.0, and social platforms are spreading quickly in society, organizations, and economics. The integration of business process management and social information systems is becoming more and more widespread [2, 3]. New approaches for using social information systems in combination with business process management appear frequently.

Social information systems are used both in external and internal business processes. Companies can co-create products and services, e.g., companies integrate customers into product development to capture ideas and features. Thus, communication with the customer is increasingly bi-directional. Social information systems differ from traditional information systems by enabling emergent interactions [4]. Emergent interactions are interactions that are defined during run-time by two or more stakeholders. No plan or approval from a supervisor or management is necessary. Emergent interactions enable the articulation of personal into collective knowledge thus representing mechanisms for harnessing collective intelligence in the digital age.

The integration of business process management and social information systems enables the creation of new business models using social platforms. Prominent examples are TripAdvisor, Uber, and AirBnB. By using the value-creating mechanisms of social information systems, business models became possible which were not realizable before; e.g., AirBnB uses a crowdsourcing model for quality control by using users' reviews of apartments. In this way, a quality assessment of products and services became possible that had previously been too costly.

Human aspects complement the social perspective on business process management. The fact that more and more enterprises are using business process management implies that the human individual is involved in a multitude of business processes. Individuals must cope with multiple process contexts and thus must administer data appropriately. Digital assistants such as Alexa [5] integrate individuals in processes that could not interact with conventional computers. Human aspects of business process management relate to the individual who creates a process model, to the communication among people, during and after the process execution, and to the social process of collaborative modeling. They also relate to the interaction/collaboration/coordination/cooperation that should be implemented in the business process or to specific human-related aspects of the business process itself and their representations in models.

Given this background, the goal of the BPMS2 workshop [6] is to explore how social information systems integrate with business process management [7], and how business process management may profit from this integration [8]. Furthermore, the workshop investigates the human aspects introduced into business process management by involving human actors. Examples are the use of crowdsourced knowledge and tasks and the need for new user interfaces, e.g., augmented reality and voice bots.

Six papers were accepted for presentation at the BPMS2 2021 workshop. In their paper "Pair Modeling: Does One Plus One Add Up?", Vinicius Stein Dani, Mahendrawathi Er, Jelmer Jan Koorn, Jan Martijn van der Werf, Henrik Leopold, and Hajo A. Reijers focus on the effect of conducting modeling tasks in pairs on process model quality and participants' satisfaction. In an experiment, the authors let novice modelers perform modeling tasks individually and in pairs. The experiment did reveal that the participants are highly satisfied when performing modeling tasks in pairs.

Fernanda Gonzalez-Lopez, Jorge Munoz-Gama, Paolo Fabia, and Alberto Jara propose in their paper "On the Systematic Analysis of Process Landscape Method Implementation using Grounded Theory" an evaluation procedure for the systematic analysis of a landscape method implementation through grounded theory.

In the paper titled "Business Process Improvement in Industry 4.0: An Interorganizational Perspective", by Vítor Ribeiro, João Barata, and Paulo Rupino Da Cunha, design science research was chosen to propose an approach for interorganizational business processes improvement in decentralized contexts of Industry 4.0. The authors also suggest a draft of a BPMN extension mechanism for IOBP 4.0.

Lars Drewes and Volker Nissen conduct an experimental study in their paper titled "The Effect of Self-Healing on Process Acceptance" to investigate the influence of self-healing in the case of process error. The authors measured acceptance of a generic purchasing process by means of a questionnaire in the three dimensions of attitude (cognitive, affective, and conative). The study shows that there is a significant difference in acceptance between processes with and without self-healing in the cognitive dimension.

In their paper "PACAs: Process-Aware Conversational Agents", Luis Fernando Lins, Glaucia Melo, Toacy Oliveira, Paulo Alencar, and Donald Cowan present an approach for building a novel type of conversational agent – called PACAs. They leverage the Business Process Model and Notation (BPMN) standard to assist in runtime process execution. The authors also describe a use case in which PACAs are implemented based on Rasa – a chatbot framework – and Camunda Engine – a business process or workflow management system.

In their paper "Intelligent Robots in Business Processes: A Students' Perspective", Amy Van Looy, Gilles Bauwens, and Krzysztof Zaniewski address robotics with more advanced levels of intelligence. They show that gender and academic discipline only have a limited impact in predicting the acceptance of intelligent robots at work, but different adoption profiles seem to matter most. Based on these findings, the authors discuss the role of students, universities, and employees to better prepare for the future digitalization of the workplace.

We wish to thank all the people who submitted papers to BPMS2 2021 for having shared their work with us, the many participants creating fruitful discussion, and the members of the BPMS2 2021 Program Committee, who made a remarkable effort in reviewing the submissions. We also thank the organizers of BPM 2021 for their help with the organization of the event.

Selmin Nurcan
Rainer Schmidt

References

1. Schmidt, R., Alt, R., Nurcan, S.: Social information systems. In: Proceedings of the 52nd Hawaii International Conference on System Sciences, pp. 2642–2646. Hawaii (2019)
2. Schmidt, R., Nurcan, S.: BPM and social software. In: Ardagna, D., et al., (eds.) Business Process Management Workshops, pp. 649–658. Springer Berlin Heidelberg (2009)
3. Bruno, G., et al.: Key challenges for enabling agile BPM with social software. J. Softw. Maintenance Evol. Res. Pract. **23**, 297–326 (2011). https://doi.org/10.1002/smr.523
4. Schmidt, R., Kirchner, K., Razmerita, L.: Understanding the business value of social information systems – a research agenda. In: 53rd Hawaii International Conference on System Sciences (HICSS), pp. 2639–2648 (2020). https://doi.org/10.24251/HICSS.2020.321
5. Schmidt, R., Alt, R., Zimmermann, A.: A conceptual model for assistant platforms. In: 54th Hawaii International Conference on System Sciences (HICSS), pp. 4024–4033. Wailea (2021). https://doi.org/10.24251/HICSS.2021.490
6. Nurcan, S., Schmidt, R.: Introduction to the first international workshop on business process management and social software (BPMS2 2008). In: Business Process Management Workshops, pp. 647–648 (2009)
7. Schmidt, R., Nurcan, S.: Augmenting BPM with social software. In: Business Process Management Workshops, pp. 201–206. Ulm (2010)
8. Erol, S., et al.: Combining BPM and social software: contradiction or chance? J. Softw. Maintenance Evol. Res. Pract. **22**, 449–476 (2010). https://doi.org/10.1002/smr.460

Organization

Workshop Organizers

Selmin Nurcan	University of Paris 1 Panthéon-Sorbonne, France
Rainer Schmidt	Munich University of Applied Sciences, Germany

Program Committee

Jan Bosch	Chalmers University of Technology, Sweden
Marco Brambilla	Politecnico di Milano, Italy
Lars Brehm	Munich University of Applied Science, Germany
Norbert Gronau	University of Potsdam, Germany
Holger Günzel	Munich University of Applied Sciences, Germany
Barbara Keller	Munich University of Applied Sciences, Germany
Kathrin Kirchner	Technical University of Denmark, Denmark
Ralf Klamma	RWTH Aachen University, Germany
Andrea Marrella	Sapienza Università di Roma, Italy
Massimo Mecella	Sapienza University of Rome, Italy
Michael Möhring	Munich University of Applied Sciences, Germany
Selmin Nurcan	Université Paris 1 Panthéon-Sorbonne, France
Mohammad Ehson Rangiha	City, Universityof London, UK
Gustavo Rossi	Universidad Nacional de La Plata, Argentina
Flavia Santoro	Universidade do Estado do Rio de Janeiro, Brazil
Rainer Schmidt	Munich University of Applied Sciences, Germany
Miguel Angel Sicilia	University of Alcala, Spain
Pnina Soffer	University of Haifa, Israel
Johannes Tenschert	FAU Erlangen-Nürnberg, Germany
Barbara Weber	University of St. Gallen, Switzerland
Moe Thandar Wynn	Queensland University of Technology, Australia
Alfred Zimmermann	Reutlingen University, Germany

Pair Modeling: Does One Plus One Add Up?

Vinicius Stein Dani[1(✉)], Mahendrawathi ER[2], Jelmer J. Koorn[1], Jan Martijn E. M. van der Werf[1], Henrik Leopold[3], and Hajo A. Reijers[1]

[1] Utrecht University, Princetonplein 5, 3584 CC Utrecht, The Netherlands
{v.steindani,j.j.koorn,j.m.e.m.vanderwerf,h.a.reijers}@uu.nl
[2] Institut Teknologi Sepuluh Nopember, Surabaya, Indonesia
mahendra_w@is.its.ac.id
[3] Kühne Logistics University, Großer Grasbrook 17, 20457 Hamburg, Germany
henrik.leopold@the-klu.org

Abstract. Pair programming is a technique where two programmers work together. This technique offers benefits such as improved quality of the code, faster task completion, and increased participant satisfaction. Existing research in Business Process Management has not studied the advantages of pair process modeling in comparison to individual modeling. In this paper, we focus on the effect of conducting modeling tasks in pairs on process model quality and participants' satisfaction. In an experiment, we let novice modelers perform modeling tasks individually and in pairs. Contrary to pair programming findings, we could not statistically support that pair modeling improves the process model's quality. Similar to pair programming, the experiment did reveal that the participants are highly satisfied when performing modeling tasks in pairs. Considering this result, instructors or managers may want to use the pair setup for training purposes.

Keywords: Process modeling · Pair modeling · BPMN

1 Introduction

Business process models play an important role in supporting different stages of the Business Process Management (BPM) lifecycle [6]. The process modeling task often requires collaboration among different process modelers [8,11]. Collaborative modeling is a process where multiple stakeholders (e.g., domain experts, process analysts) actively and collaboratively contribute to creating a model [8,20]. In this paper, we refer to *pair modeling* as a specific form of collaborative process modeling where people model processes in pairs.

Many studies in collaborative process modeling have broken away from the early focus on output or quality of process modeling to investigate the process of conducting collaborative process modeling, i.e., the process of process modeling [8]. Previous research reported that collaborative process modeling resembles

A. Marrella and B. Weber (Eds.): BPM 2021 Workshops, LNBIP 436, pp. 261–272, 2022.
https://doi.org/10.1007/978-3-030-94343-1_20

pair programming [9]. The benefits of pair programming compared to individual programming are, among others: (1) improved quality of the produced code, supported by significantly lower end defect content, (2) faster task completion, and (3) increased participant satisfaction [5]. Given the benefits of pair programming, and the similarities between coding and modeling, it seems reasonable to expect similar benefits from pair modeling. Existing research in BPM has not studied the potential advantages of pair modeling as opposed to individual modeling in an experimental setting.

In this paper, we attempt to fill the gap in the existing process modeling literature. We conduct a pair modeling experiment, and introduce a new way of assessing the quality of process models. Previous studies assessed the quality of process models based on syntactic, semantic, and pragmatic aspects of the process models. We acknowledge that the quality of a process model will depend on the process modeling guidelines taught during the participant's training. Thus, instead of measuring the quality of process models as in previous studies, we use a proxy measure related to the learning process. We use a new quality metric called *infringement* to identify when the participants violate the modeling guidelines taught during their training. In addition to that, we assess the participants' satisfaction with the pair modeling. Therefore, we focus on two benefits from pair programming. We conduct an experiment followed by the application of the System Usability Scale (SUS) questionnaire [2] to, respectively, investigate whether pair modeling results in fewer infringements, and higher participants' satisfaction.

The experiment participants are Information Systems students who have completed a BPM course. We believe that students are good representatives of novice process modelers. The experiment results show no statistical support that pair modeling reduces infringements compared to individual modeling. However, we found that process modelers are highly satisfied when performing modeling tasks in pairs. The findings may have implications for instructors or managers in designing the process modeling course delivery or deciding on the use of a pair setup for training purposes.

This paper is structured as follows. In the following section, we discuss related work. Section 3 describes the quality metric of process modeling used in this research, followed by descriptions of the research model and hypotheses in Sect. 4, and by the experiment design and execution in Sect. 5. Section 6 provides the results of the experiment. A discussion and a conclusion are provided in Sects. 7, and 8, respectively.

2 Related Work

Several works investigated the process of conducting collaborative process modeling [9,19,20]. In [19], the authors looked into the process of process modeling via a controlled environment. In [20], the author investigated collaborative process modeling in a workshop-based environment, and discovered that the more the workshop participants engage, the higher is the model's quality. In [9], the authors reported that collaborative process modeling resembles pair programming. Pair programming is a well-established technique with a driver and a controller role. Several authors have investigated pair programming as reported in [5,15,18,22]. In [5], a set of benefits of pair programming is reported. The first benefit is the ability for the programmer in the controller role to spot typos made by the programmer in the driver role, which otherwise would not be observed until a later stage in the programming task. Other benefits are improved quality, supported by significantly lower defect content, faster task completion, and increased participant satisfaction. In [22], the authors identify that pair programming enhanced learning for computer science students. Similarly, in [15,18], the authors identified that pair programming supports students while learning programming notions and improves their interest in learning this skill. In addition, they observed that pair programming contributes to reducing dropout rates [18] and increasing retention of the material being studied [15].

Different aspects, e.g., syntactic, semantic, and pragmatic quality, are typically used to assess a process model's quality [1,13,16,21]. A set of guidelines regarding process modeling quality is proposed in [16], while in [1] the authors investigate the effect of using process modeling guidelines during the process modeling task. In [21], the authors identified a set of anti-patterns based on the analysis of over two thousand process models produced by novice process modelers. In [13], the authors analyzed over five hundred process models, and identified different inconsistencies from a set of pre-selected process modeling guidelines. These inconsistencies are related to different aspects of process modeling (e.g., labeling, structure, layout). We are interested in learning process modeling and how working individually or in pairs influences the results of the process modeling task. We acknowledge that the quality of the learning process would depend on the delivery of the course, particularly process modeling guidelines taught by the lecturers. So, instead of measuring the quality of the process model based exclusively on syntactic, semantic, or pragmatic quality, *we use infringements as a new quality metric*, described in more detail in the next section.

Previous studies focused on different aspects related to process modeling, e.g., the process of process modeling and the quality of the process models. These studies did not compare the outcomes from pair modeling in terms of quality, and the participants' satisfaction with individual process modeling. This is what we are interested in, and outline in this study. Different from previous studies, we conducted an experiment to emulate the pair programming setup for the process modeling task.

3 The Quality of Process Models

To assess the quality of process models, we selected six guidelines as quality metrics. They relate to activity labeling, the use of start and end events, joining multiple end events into one (whenever possible), and outgoing and incoming flow. We defined the expected representation for each quality metric. We consider an *infringement* to be any representation that differs from the ones expected.

To obtain the expected representations, we applied a questionnaire to a group of BPM lecturers. We asked them to select, from different types of possible representations for each of the six quality metrics, which one they used to teach process modeling. For example, for the activity labeling, the representation could be the "use of verb + object" (e.g., "Create report") instead of "action + noun" (e.g., "Creation of report"). Based on the lecturers' responses, we obtained the expected representation of the six quality metrics as summarized in Table 1.

Table 1. The six quality metrics (QM) used for the experiment's process model assessment. Each representation different from the expected, detected within a process model, is considered to be an *infringement*

QM	Title	Expected representation
1	Activity labelling	Use of "verb + object"
2	Start event	Explicit use of Start event
3	End event	Explicit use of End event
4	Join of multiple End events	Join multiple End events using join-gateway
5	Outgoing flow from activity	Use of gateway for outgoing flow from activity
6	Incoming flow to activity	Use of gateway for incoming flow to activity

4 Research Model and Hypotheses

Our research model is presented in Fig. 1. We expect the process modeling setup, whether conducted individually or in pairs, to influence the quality of the process models as measured in terms of the sum of infringements, and the user satisfaction as measured in terms of the SUS score. The factor (i.e., independent variable) considered is the setup in which the process modelers perform the modeling task; thus, we have a single factor with two levels: individual and pair. The response variables (i.e., dependent variables) are the sum of infringements made by the participants in their process models, and the SUS Score. On a secondary level, we are also interested in knowing which of the quality metrics the students can apply well, and which quality metrics they tend to violate.

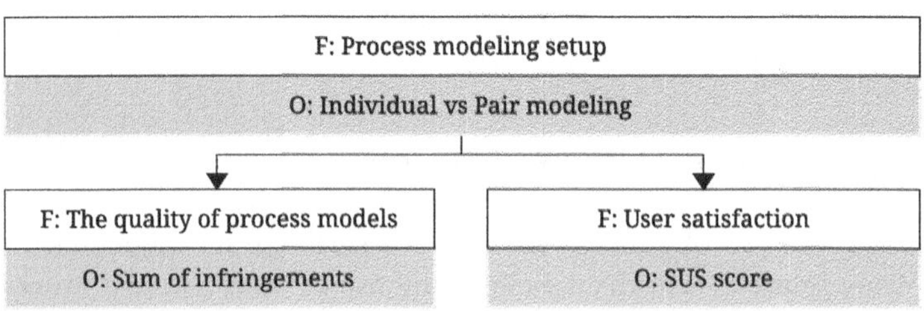

Fig. 1. Research model, where *F* represents the theoretical factor, and *O* the operationalization of the factor in our study

We expect that conducting process modeling in pairs will help the participants to remind each other of what guidelines they have been taught in the course to apply these guidelines in their model. We hypothesize that: *pair modeling will reduce the sum of infringements compared to individual modeling* (H1). Similar to previous pair programming studies, we also expect participants' satisfaction to be higher during pair modeling. We hypothesize that: *participants are more satisfied during pair modeling than during individual modeling* (H2).

5 Research Method

We conducted an experiment involving randomly selected bachelor's degree students from a higher education institution in Indonesia. All students had completed a BPM course. Overall, there were 28 participants involved in the experiment. Before the experiment, we sent a pre-experiment questionnaire to gather generic information about the participants, e.g., age, and level of exposure to process modeling. The participants' age varied between 19 and 22 years old. Most of the participants (61%) have known Business Process Model and Notation (BPMN) for less than a year. Most of the participants rated their experience with BPMN as moderate (75%), or low (18%). Based on this information, we could assume that the participants were novice process modelers.

5.1 Materials and Procedures

The materials used in our experiment were two process descriptions. We selected the process descriptions considering that both should result in process models with similar control-flow complexity [3,10] and of small size (i.e., less than 16 activities). Process description 1 is the ZooClub process as presented in [4], and Process description 2 is the FixComp process as presented in [6]. The participants used the Camunda Modeler[1] to create the process models. To conduct the experiment in a completely online manner, we used Zoom Video Communications[2]. We simultaneously shared the respective process description to the participants on each run of the experiment via email. Finally, we created a script to split participants into specific breakout rooms in Zoom meetings automatically. We analysed the models delivered by the participants on the six quality metrics as explained in Sect. 3.

5.2 Experiment Design

Figure 2 depicts our experiment design, where participants were divided into two groups. Each group performed two runs of the modeling task. For the first run, the participants in each group worked individually. A different process description was assigned to each group. For the second run, the participants worked in

[1] https://camunda.com/download/modeler/.
[2] https://zoom.us/.

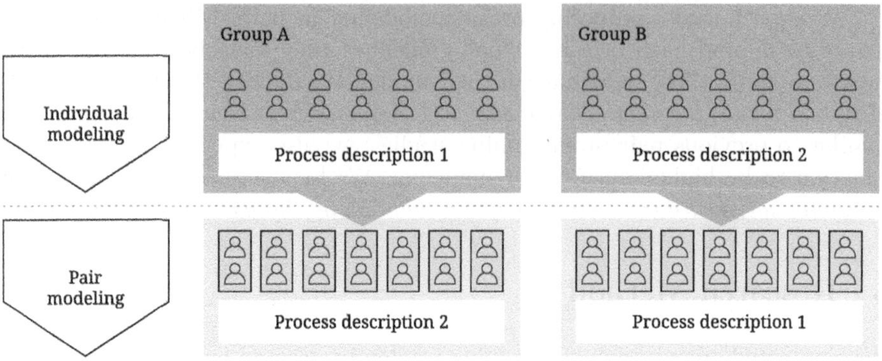

Fig. 2. Experiment design overview

pairs, and the process descriptions were switched from the first run. The first and second run results were process models created, respectively, individually and in pairs.

5.3 Pilot Test and Dry-Run

Before the experiment, we conducted a pilot test with 8 participants. Based on the pilot test, we improved the experiment design, and achieved the one presented in Sect. 5.2. We also learned from the pilot test that it would be essential to prepare the participants, and the researchers, for the experiment environment. Thus, we conducted a dry-run before the actual experiment. The dry-run was conducted with the actual participants. We simulated the general steps taken during the actual experiment without making the participants work on any process description. Overall, with the dry-run, we could test experiment technicalities and flow.

5.4 Experiment Execution

Figure 3 presents a detailed overview of the experiment execution. First, we automatically, and randomly, divided participants into two groups with a splitting script. A 10-min introduction and preparation session followed. Then, as the first run began, the participants opened the process modeling tool, and started sharing and recording their screen on the tool. The participants modeled the process description for 30 min. After the given time, we stopped the meeting tool, and the recordings were saved. All participants submitted their individual process models, and had a 5-min break to rest. After the break, we ran the pairings script again to divide the participants into as many pairs as could be formed into each group. The result of the script was used to automatically generate the "breakout rooms" for the second run of the experiment. Before the second run, there was a 5-min warm-up preparation session. The participants performed the process modeling in pairs for another 30 min during the second run. Finally, we

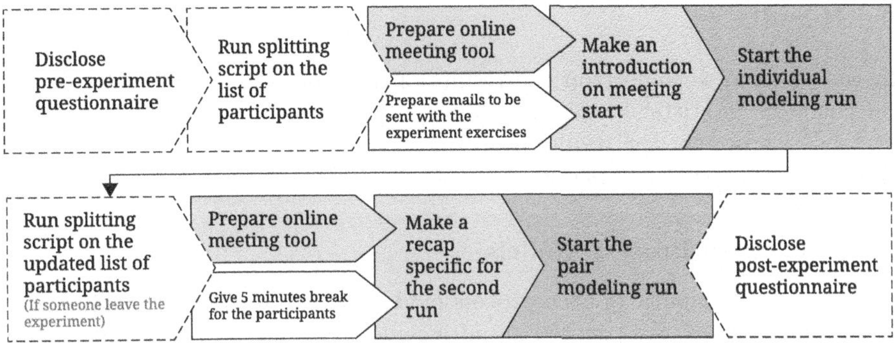

Fig. 3. Experiment execution overview

asked the participants to answer a post-experiment questionnaire at the end of the experiment. In total, the whole experiment took roughly 80 min.

5.5 Countering Threats to Validity

Several factors could influence the experiment and pose threats to validity. Thus, we worked to control and minimize the impacts of these factors. As suggested by Figl et al. [7], the quality of a process model often depends on the process analyst's expertise. We used the blocking technique [23] when splitting the participants into groups to diminish the threat of different expertise levels. The participants' grades during the BPM course were used in the splitting process, so we obtained two groups with an equivalent average grading.

Another threat that could affect our experiment result is if one of the process descriptions used as input of the process modeling task was much easier than the other. We used two process descriptions with similar textual lengths to avoid this. This resulted in process models of equivalent control-flow complexity. We designed our experiment to measure the output from the same participant for both process descriptions in a crossover design. In this way no participant models the same process description more than once, which also works to diminish the learning problem.

The modeling tool used by the participants may also play a role in the outcome of the modeling. There are many modeling tools available with different features, e.g., Signavio, Camunda Modeler, etc. Because we wanted the participants to use their knowledge as much as possible, we chose Camunda Modeler, which was the tool they used during the course. Finally, we avoided mentioning to the participants how the experiment's output would be assessed. We did this to avoid participant bias and demand characteristics [17], where the participants change their behavior according to what is expected from them during the experiment. Also, we reinforced that the participants' involvement and results would be kept anonymous.

6 Results

In this section, we report on the results of our work. In Sect. 6.1, we present the conducted statistical analysis to investigate the quality of process models. In Sect. 6.2, we present the results from the application of the SUS questionnaire to assess participants satisfaction.

6.1 Quality of Process Models

We assessed the quality of each process model created individually and in pairs by the participants. We counted an infringement whenever a quality metric did not meet the expected representation (see Sect. 3). We did not count the number of times a certain quality metric is violated. When we saw that the participant's process model violated the quality metric standard of a specific metric at least once, we counted it as one infringement of the quality metric. Therefore, a process model may have up to six infringements.

Figure 4 display the number of infringements per participant throughout both sessions. On average, the participants in groups A and B made two infringements per process model. Interestingly, the spread in the sum of infringements is wider when the modeling task is performed individually. When they performed the modeling task in pairs, the range of infringements was between 1 and 3. These results show that modeling in pairs allows the transfer of knowledge between the individuals with a better recollection of the standard guidelines with the ones with a lower recollection of the standard guidelines.

Figure 5 depicts the resulting infringements per participant per modeling setup. Each column represents one infringement (from QM1 up to QM6), and

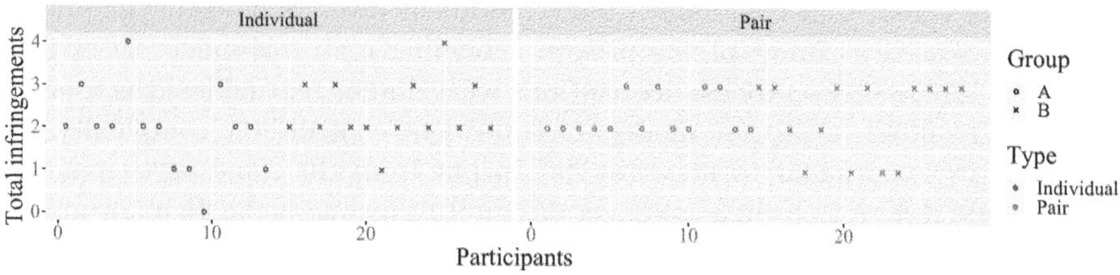

Fig. 4. Number of infringements per participant

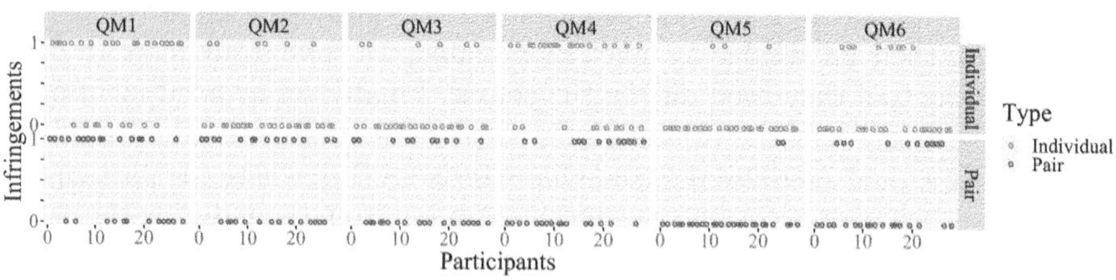

Fig. 5. Resulting infringements per participant per modeling setup

shows how each participant performed both individually and in a pair. The number 1 indicates the participant violated the quality metric (i.e., made an infringement) for the process modeling setup; and, 0 otherwise. Several findings are evident from Fig. 5. Overall, the participants:

- appear to remember clearly the quality metric "the use of gateway for outgoing flow from an activity" (QM5). This quality metric had the least infringements both when the participants worked individually or in pairs. The infringement of QM5 was even less when participants worked in pairs;
- produced fewer infringements related to "activity labeling" (QM1), and "join multiple end events" (QM4) when working in pairs;
- produced more infringements on the quality metrics "explicit use of start event" (QM2), "explicit use of end event" (QM3), and "use of gateway for incoming flow to an activity" (QM6) when working in pairs. This may be due to the rush to start and end the process modeling task when performing the pair modeling, so the participants suppress the start and end events from their models. It may also be easier for a person to spot a mistake in labeling and correct the mistake just in time than to do the same for more complex structures.

To investigate if conducting process modeling tasks in pairs will reduce the sum of infringements compared to conducting process modeling tasks individually, we performed statistical analysis over the results from the process modeling tasks. Because our data was not normally distributed, we applied the Mann-Whitney U test [12]. We did not find a statistically significant difference in the sum of infringements made by the participants when working individually or in pairs ($U = 77$, $p = 0.46$). Hence, we cannot statistically support the hypothesis that pair modeling will reduce the sum of infringements compared to individual modeling.

To check for possible confounding variables, we investigated two other perspectives: the groups and the process descriptions. Figure 6 shows the boxplots in terms of the sum of infringements across different modeling setups (Fig. 6a), groups (Fig. 6b), and process descriptions (Fig. 6c). First, we checked whether or not groups A and B scored similarly. Second, we tested if process descriptions

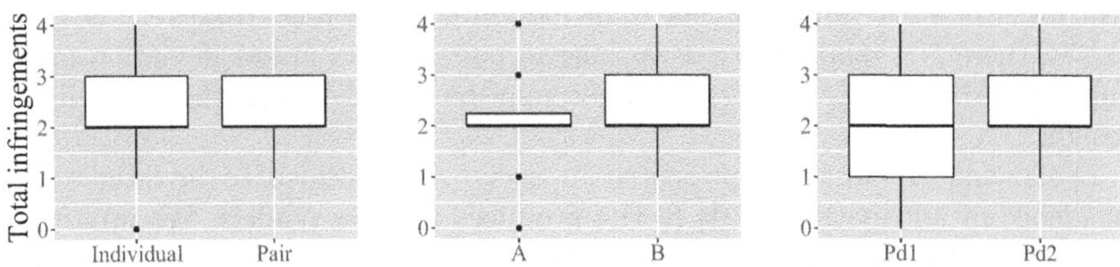

(a) Process modeling setup (b) Group (c) Process description

Fig. 6. Data distribution

1 and 2 resulted in equivalent scores. This analysis was conducted because if there were a statistically significant difference in any of these two perspectives, this could indicate that something hampered the results. We learned that there was no statistically significant difference between the groups ($U = 69$, $p = 0.27$) and process descriptions ($U = 54$, $p = 0.28$).

6.2 User Satisfaction

In addition to the sum of infringements, we also wanted to know the participants' satisfaction with pair modeling compared to individual modeling. To achieve this, we used the responses from the post-experiment questionnaire. Among others, this questionnaire consisted of ten questions that compose the SUS questionnaire. From the responses, we calculated the final SUS-score being 72.25, which is a score between "Good" and "Excellent" [14].

Next to these questions, we presented participants the following statements, which they should rate on a five-point Likert-scale: "If I could choose between performing the process modeling task individually or in pairs, I would prefer to do it in pairs", and 71% of the participants answered that they "Agree" or "Totally agree". Furthermore, 86% of the participants answered they "Agree" or "Totally agree" to the statement "I would model in pairs again". Concerning the learning process: 64% answered they "Agree" or "Totally agree" to "I felt stimulated by pair modeling to learn more", while 21% were "Neutral". Finally, 89% answered they "Agree" or "Totally agree" to "Pair modeling helps in learning", while 11% were "Neutral".

7 Discussion

Compared to the pair programming findings benefits [5], our findings are mixed. On the one hand, we found analogous results to pair programming in terms of participants' satisfaction: the participants enjoy working in pairs and prefer to do it again in the future. On the other hand, our findings differ from pair programming in terms of the effect on end defects. The statistical test shows no significant difference in the sum of infringements between pair modeling and individual modeling. Thus, we cannot conclude that conducting pair modeling will lead to fewer infringements than individual modeling. However, as Fig. 4 shows, there is a tendency that working in pairs results in less deviation in the sum of infringements. A bigger sample may be useful to test this preliminary insight.

Following the analysis presented in Sect. 6, we decided to investigate the incidence of anti-patterns [21] in the produced process models. We wanted to see how the participants performed according to common mistakes reported in the BPM literature. When working in pairs, we noticed that participants make less infringements in terms of (a) hanging elements (i.e., the number of elements within a pool that do not contain incoming sequence flows) and (b) message flow misuse across swim lanes. We also observed that a recurrent pattern during the individual process modeling did not occur during the pair setup: the pattern observed was the use of control-flow direction from right to left in some parts of the model. We infer that this kind of pattern may also be easiest to improve when modelers work together. Further study is needed to investigate these tentative findings more properly.

8 Conclusion

In this study, we conducted an experiment to investigate whether pair modeling results in higher quality process models when compared to individual modeling. We found no statistically significant difference between the two process modeling setups. We did discover that novice process modelers are highly satisfied when performing the process modeling task in pairs. They clearly prefer pair modeling over individual modeling.

Our findings have implications for both the educational and industrial fields. Novice process modelers prefer pair modeling over individual modeling when performing a process modeling task. This suggests that instructors and managers should consider using a pair modeling setup for training purposes.

In our experiment, we focused on introducing pair modeling in a classroom setting. For future work, we want to further investigate why novices prefer pair modeling, and how collaborative modeling methods can improve their modeling skills. We think pair modeling helps novices in improving their modeling skills. However, more research is needed to study the effects of such methods on modeling skills.

References

1. Avila, D.T., Cigana, R.P., Fantinato, M., Reijers, H.A., Mendling, J., Thom, L.H.: An experiment to analyze the use of process modeling guidelines to create high-quality process models. In: Hartmann, S., Küng, J., Chakravarthy, S., Anderst-Kotsis, G., Tjoa, A.M., Khalil, I. (eds.) DEXA 2019. LNCS, vol. 11707, pp. 129–139. Springer, Cham (2019). https://doi.org/10.1007/978-3-030-27618-8_10
2. Brooke, J.: SUS: A Quick and Dirty Usability Scale, pp. 189–194. Taylor & Francis (1996)
3. Cardoso, J.: How to measure the control-flow complexity of web processes and workflows, pp. 199–212. Future Strategies Inc., January 2005
4. Carmona, J.: The alignment of formal, structured and unstructured process descriptions. In: van der Aalst, W., Best, E. (eds.) PETRI NETS 2017. LNCS, vol. 10258, pp. 3–11. Springer, Cham (2017). https://doi.org/10.1007/978-3-319-57861-3_1
5. Cockburn, A., Williams, L.: The Costs and Benefits of Pair Programming, pp. 223–243. Addison-Wesley Longman Publishing Co., Inc. (2001)
6. Dumas, M., la Rosa, M., Mendling, J., Reijers, H.A.: Fundamentals of Business Process Management. Springer, Heidelberg (2013). https://doi.org/10.1007/978-3-642-33143-5
7. Figl, K.: Comprehension of procedural visual business process models: a literature review. Bus. Inf. Syst. Eng. **59**(1), 41–67 (2017)

8. Forster, S., Pinggera, J., Weber, B.: Collaborative business process modeling. In: Rinderle-Ma, S., Weske, M. (eds.) Proceedings of EMISA 2012 - Der Mensch im Zentrum der Modellierung, Vienna, Austria, 13–14 September 2012. LNI, vol. P-206, pp. 81–94. GI (2012). https://dl.gi.de/20.500.12116/17744
9. Forster, S., Pinggera, J., Weber, B.: Toward an understanding of the collaborative process of process modeling. In: Deneckère, R., Proper, H.A. (eds.) Proceedings of the CAiSE'13 Forum at the 25th International Conference on Advanced Information Systems Engineering (CAiSE), Valencia, Spain, 20 June 2013. CEUR Workshop Proceedings, vol. 998, pp. 98–105. CEUR-WS.org (2013). http://ceur-ws.org/Vol-998/Paper13.pdf
10. Gruhn, V., Laue, R.: Approaches for business process model complexity metrics. In: Abramowicz, W., Mayr, H.C. (eds.) Technologies for Business Information Systems, pp. 13–24. Springer, Dordrecht (2007). https://doi.org/10.1007/1-4020-5634-6_2
11. Hahn, C., Recker, J., Mendling, J.: An exploratory study of IT-enabled collaborative process modeling. In: zur Muehlen, M., Su, J. (eds.) BPM 2010. LNBIP, vol. 66, pp. 61–72. Springer, Heidelberg (2011). https://doi.org/10.1007/978-3-642-20511-8_6
12. Jonathan, L., Feng, J.H., Hochheiser, H.: Research Methods in Human-Computer Interaction, 2nd edn. Morgan Kaufmann (2017)
13. Leopold, H., Mendling, J., Günther, O.: Learning from quality issues of bpmn models from industry (extended abstract). In: CEUR Workshop, vol. 1701, no. 4, pp. 36–39 (2016)
14. Lewis, J.R.: The system usability scale: past, present, and future. Int. J. Hum.-Comput. Interact. **34**(7), 1–2 (2018)
15. McDowell, C., Werner, L., Bullock, H.E., Fernald, J.: Pair programming improves student retention, confidence, and program quality. Commun. ACM **49**(8), 90–95 (2006)
16. Mendling, J., Reijers, H.A., van der Aalst, W.M.P.: Seven process modeling guidelines (7PMG). Inf. Softw. Technol. **52**(2), 127–136 (2010)
17. Nichols, A.L., Maner, J.K.: The good-subject effect: investigating participant demand characteristics. J. Gener. Psychol. **135**(2), 151–166 (2008)
18. Papadakis, S.: Is pair programming more effective than solo programming for secondary education novice programmers? A case study. Int. J. Web-Based Learn. Teach. Technol. **13**(1), 1–16 (2018)
19. Pinggera, J., et al.: Tracing the process of process modeling with modeling phase diagrams. In: Daniel, F., Barkaoui, K., Dustdar, S. (eds.) BPM 2011. LNBIP, vol. 99, pp. 370–382. Springer, Heidelberg (2012). https://doi.org/10.1007/978-3-642-28108-2_36
20. Rittgen, P.: End-user involvement and team factors in business process modeling. In: Proceedings of the HICSS, pp. 180–189 (2012)
21. Rozman, T., Polančič, G.: Analysis of most common process modelling mistakes in BPMN process models. In: BPM and Workflow Handbook, vol. 121, pp. 233–246. Elsevier BV, October 2008
22. Slaten, K.M., Droujkova, M., Berenson, S.B., Williams, L., Layman, L.: Undergraduate student perceptions of pair programming and agile software methodologies: verifying a model of social interaction. In: Agile Development Conference (ADC 2005), pp. 323–330. IEEE (2005)
23. Voss, D., Dean, A., Draguljic, D.: Design and Analysis of Experiments Springer Texts in Statistics. Springer, Heidelberg (2017). https://doi.org/10.1007/978-3-319-52250-0

On the Systematic Analysis of Process Landscape Method Implementation Using Grounded Theory

Fernanda Gonzalez-Lopez(✉), Jorge Munoz-Gama, Paolo Fabia, and Alberto Jara

Pontificia Universidad Católica de Chile, Santiago, Chile
fgonlop@uc.cl

Abstract. Process Landscapes are critical elements towards the systematic management of a process portfolio, representing its process architecture from a high-level perspective. Consequently, a wide range of methods to generate Process Landscapes is available in the literature, whose effectiveness has been extensively validated, mostly through case studies. However, little is known or reported about the implementation of landscape methods with the human team, e.g. facilitators, common pitfalls, or lessons learned. To fill this research gap, this article proposes the Landscape Method Implementation Grounded Theory based evaluation (LanMI-GT) procedure for the systematic analysis of a landscape method implementation through Grounded Theory. This was generated using a Design Science Research approach. The contribution of this work is twofold. First, the proposed procedure would enable practitioners to systematically analyze the implementation with a human team of a specific landscape method. Second, the procedure is demonstrated by analyzing a well-known Process Landscape Method in a real context, and the findings thereof are discussed in the paper.

Keywords: Process Landscape Modeling · Grounded Theory · Business Process Management · Socio-technical Context

1 Introduction

Process Landscapes are graphical models of the architecture of a given process portfolio from a high-level of abstraction [7]. These representations – sometimes also called process maps [23] or process architectures [6] – depict coarse-grained processes and their relations and additionally may include other types of information, e.g. grouping of processes. A Process Landscape provides an enterprise view of processes [28]. Thus, it yields valuable information for prioritizing process-related projects: it is a key asset for the systematic management of a given process portfolio [17]. Moreover, the existence of a Process Landscape should be the starting point of the process lifecycle [7].

A. Marrella and B. Weber (Eds.): BPM 2021 Workshops, LNBIP 436, pp. 273–285, 2022.
https://doi.org/10.1007/978-3-030-94343-1_21

In line with this, the need for methods for generating Process Landscapes has gained increasing attention among process initiatives over the last years [16]. A wide variety of Process Landscape Methods have been proposed based on different notations (i.e. standard [28] or dedicated [3]), approaches (e.g. systems thinking [4] or clustering techniques [8]), origins (i.e. academia [18] or practice [17]), and scopes (i.e. general [14] or more specific [15]). The effectiveness of these proposals has been extensively validated by demonstrating that they allow generating a valid Process Landscape, mostly through a particular case study.

Despite its relevance, Process Landscape generation is still an under-researched topic [19] with a number of open challenges. In particular, there is limited understanding about the Implementation of Process Landscape Methods in practice, i.e. the application of these types of methods in a socio-technical context [10]. Little is known and reported about the main challenges faced during implementation, the facilitators that ease the task, or which are the most common pitfalls, and lessons learned. This type of knowledge is crucial for understanding and improving the Implementation of Process Landscape Methods.

The goal of the present work is to fill this research gap by proposing a novel procedure for the systematic analysis of Process Landscape Method Implementation. The proposal is termed Landscape Method Implementation Grounded Theory based evaluation (LanMI-GT). LanMI-GT is based on Grounded Theory [9], i.e. a research approach for building knowledge based on the systematic collection and analysis of qualitative data about a phenomenon that is largely unknown [9]. Additionally, the procedure integrates Grounded Theory with Process Mining [1], i.e. a set of techniques for process analysis based on their execution data. The research leading to proposing LanMI-GT followed the Design Science Research (DSR) paradigm [21].

The contribution of our work is twofold. First, LanMI-GT would enable practitioners to systematically analyze the implementation of a particular Process Landscape Method under a specific socio-technical context. Second, this article presents an instantiation of LanMI-GT using a well-known Process Landscape Method in a common context, and the findings thereof are discussed. By doing this, we both illustrate the types of insights about a Process Landscape Method Implementation that can be generated with LanMI-GT, and provide the lessons learned in a particular instantiation that may apply in a similar scenario.

The remainder of the paper is organized as follows. Section 2 discusses related work. The methodology of our work is described in Sect. 3 (including problem identification, objectives for the solution, development, and demonstration). Section 4 presents the LanMI-GT procedure and its instantiation is described in Sect. 5. The conclusions of the work are provided in Sect. 6.

2 Related Work

This section presents the related work for our proposed procedure for the systematic analysis of a Process Landscape Method Implementation, i.e. LanMI-GT. This section is structured according to the conceptual model shown in

Fig. 1. This conceptual model represents that a given `Context` is subject to the `Implementation` of a specific `Method` that generates a `Process Landscape`. These elements are described in the following.

Context. The Context defines the organizational setting in which a Method is implemented. Such a context is characterized by both social (e.g. people, conceptual structures, values and goals, desires and fears, and norms and budgets) and technical (e.g. software, hardware, and techniques) aspects.

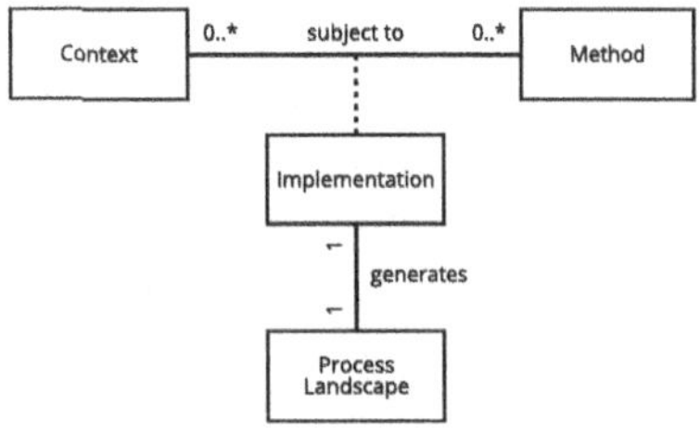

Fig. 1. Conceptual model for Process Landscape Method analysis.

Method. A Process Landscape Method is composed of a set of concepts used in the method, a procedure defining the steps of the method, and a notation for producing a Process Landscape [26]. Examples of Process Landscape Methods include the Fractal Enterprise Model (FEM) [3], the Riva method [14], and the BPTrends Associates (BPTA) method [17]. These Methods can be assessed in terms of their design aspects, e.g. goals, stakeholders, and theoretical foundations. A pioneering work in this line is the framework by Green and Ould [13] that allows describing a Method in terms of the form of the resulting Process Landscapes, contents of the method in terms of its constructs and theoretical foundations, purpose of the method and its resulting Process Landscapes, and lifecycle of the generated Process Landscapes. Similarly, new frameworks have been proposed [10,22] by considering additional aspects of design evaluation, such as the organization size for which it is suitable. Despite their usefulness for understanding aspects (such as the scope and underlying rationale of a Method), these approaches do not focus on uncovering insights about a Method's Implementation in a socio-technical Context.

Implementation. The Implementation consists of the application of a Process Landscape Method in a given Context. To evaluate the implementation, naturalistic research approaches of the phenomenon are mostly used. Typically, this consists of a more or less standardized case study design, e.g. [3,14]. The reported results for these case studies are rather aggregated without providing details about human-related aspects. Diverging from case studies, an experiment for evaluating Process Landscape Method Implementation has been proposed [11] to assess user perceptions after they have applied the Method themselves. This kind of evaluation, though suitable for comparison and for better understanding method implementation issues, has a number of limitations such as being a post-task evaluation, and not being specifically designed to observe socio-technical aspects of a Method's Implementation.

Process Landscape. The Process Landscape is the output of the Process Landscape Method that represents a high-level process architecture. Process Landscapes are conceptual models and, as such, subject to evaluation in terms of their quality by using dedicated frameworks, e.g. [20]. In the cited case studies assessing Implementation, Process Landscapes are evaluated by their final users in terms of their effectiveness for representing the high-level process architecture in a particular Context.

Against the previously described background, the present work focuses on the limitations of currently available approaches for assessing the Implementation of a Process Landscape Method in a given socio-technical Context. In this work, we propose an analysis procedure based on Grounded Theory [9]. Grounded Theory is an inductive research approach based on the systematic collection and analysis of qualitative data about a phenomenon [9]. This theory can be used as a data coding technique for structured analysis [24]. In this scenario, a code is understood as a word or short phrase assigning an essence-capturing attribute for a portion of language-based or textual data [25]. As a data coding technique, Grounded Theory offers several methods, e.g. initial coding, focused coding, and axial coding.

Despite the novelty of our proposal in using Grounded Theory as the basis for studying the phenomenon of Process Landscape Method Implementation, the use of Grounded Theory in Business Process Management (BPM), though incipient, is not new. For instance, Grounded Theory has been used to explore the goals and motivations of organizations when designing their Process Landscapes [19], and how expert modelers understand quality in declarative models [2]. These works prove Grounded Theory as a suitable approach for exploring a phenomenon that is largely unknown [24], as is the case of implementing a given Process Landscape Method in a given context.

3 Methodology

The LanMI-GT was developed according to the Design Science Research Methodology (DSRM) by [21]. The steps of this DSRM are: identify the problem and motivate, define objectives for the solution, design and development, demonstration, evaluation, and communication. The DSRM steps are discussed in the following as applied to our work, leaving the evaluation step for future research. As discussed in Sect. 1 and 2, the **problem** targeted by the work is the limited support for systematically analyzing the implementation of Process Landscape Methods in a way that considers socio-technical aspects. This is an open challenge for better understanding Process Landscape Methods as Implemented in specific contexts. Accordingly, the LanMI-GT procedure was designed with the following design **objectives** (DO) and respective design decisions:

- *DO1. Provide a means for capturing the socio-technical aspects of a method's implementation.* This objective was addressed by using a qualitative approach, particularly Grounded Theory coding techniques, as a basis for the structured analysis of the method's implementation phenomenon.
- *DO2. Provide a means for comparing the theoretical method's procedure with how the method's procedure was executed in the implementation.* This objective was tackled by creating an event log from the qualitative data to which

Process Mining techniques could be applied for understanding the executed procedure and comparing it to the theoretical method's procedure.

The **developed** LanMI-GT procedure is presented in Sect. 4. The design was used for analyzing a particular Process Landscape Method in a real Context. Such **demonstration** is presented in Sect. 5.

4 Procedure

This section presents the Landscape Method Implementation Grounded Theory based evaluation (LanMI-GT) procedure. LanMI-GT is composed of the eight steps (S) described in the following.

S1. Verbal data transcript. Generate a verbal data transcript of each interaction, i.e. meeting, taking place within the Implementation of the Method. An example of a portion of verbal data transcript is shown in the *transcript* column of Table 1.
S2. Initial coding. Code the transcripts generated in the previous step using initial coding, i.e. assign codes that break the data up in a way that the codes stick close to the data [5]. In our case, interventions of the participants of the meeting were assigned (a set of) codes describing the activity taking place and the involved attitudes, strategies, and opinions, as shown in the *content-based code* column of Table 1.
S3. Focused and axial coding. Explore the issues that emerge during the implementation: activities, attitudes, strategies, and opinions. Towards this end, further stages of Grounded Theory are applied to coded data from the initial coding stage, namely focused and axial coding. Focused coding requires deciding which are the most relevant initial codes – in terms of significance and/or frequency – and constituting them into categories [5]. Axial coding relates categories and specifies their properties and dimensions to provide coherence and structure to the collected data [5]. For instance, codes may be organized via a mind map.
S4. Content analysis. Summarize the main findings of the content analysis. For example, this may be done by building a conceptual framework.
S5. Initial coding categorization. Assign a step of the Method's procedure to each piece of coded data. For instance, the code with id B.2.99 in Table 1, was assigned to a step named "Process" as shown in the *category-based code* column.
S6. Descriptive model discovery. Discover a descriptive process model for exploring the Process Landscape Method in terms of how it was executed in practice. This model is discovered using Process Mining techniques [1] on the data generated in the previous step.
S7. Descriptive model analysis. Analyze the discovered descriptive model.
S8. Analysis integration and reporting. Integrate content analysis with the descriptive model analysis and report the findings.

Table 1. Portion of a coded transcript from the second meeting of a department.

Transcript	Id	Content-based code	Category-based code
Interviewee 1: Right, we should name that process "Regulatory compliance control"	B.2.99	Suggest process name	Process
Consultant 1: To which stakeholder does it relate to?	B.2.100	Ask for related stakeholders	Stakeholder

5 Demonstration

This section describes the demonstration of the LanMI-GT procedure (presented in Sect. 4) for analyzing the Implementation of a particular Process Landscape Method in a real Context. This provides the reader with both: (i) an example of the application of the procedure and the types of findings that it conveys, and (ii) findings of the Implementation of the Method [17] in a local government organization that can be generalized to similar cases. To provide the reader with the information needed to understand the demonstration of LanMI-GT, this section – following the conceptual model in Fig. 1 – not only analyzes the Implementation of the Method, but it also describes the Method itself, the Context in which it is implemented, and the resulting models. A more complete description of this demonstration is available in [12].

5.1 Context

The organization was the public administration of Puente Alto, Santiago, Chile. Over the last years, the different municipal departments have been documenting and analyzing their key processes, as part of a municipal management improvement plan. Most recently, the municipal departments have been assigned the task of generating their Process Landscapes. We supported 13 departments in this endeavor, from which we report the work with 4 of them, namely Community Development, Environment and Sanitation, Social Care, and Information Technology. The cited work was executed between August 2020 and January 2021, during the COVID-19 pandemic, using video-conferences (via Google Meets) and a modeling software (i.e. Archi).

5.2 Method

The method was based on the BPTA method [17] and adapted according to the project at hand. The concepts used in the method – value chain, stakeholder, process, value stream, and classification – are presented in Table 2 as defined in the method. The procedure of the Method is represented by the BPMN process model in Fig. 2. The initial step is to select the value chain that will be analyzed. The second step consists of the identification of the stakeholders of the selected value chain. The third step is defining processes: at least one for each stakeholder. In the fourth step, each value stream is labeled. The output of this step is the Value Chain Model that shows processes within the value chain of the organization and how they relate to stakeholders via value streams. Finally, processes identified in the previous steps are organized into three process categories: primary, support, and management. The output of this step is the Process Landscape that shows the processes classified into the named process categories. The notation used for the Value Chain Model and Process Landscape is ArchiMate [27].

Table 2. Definition of the concepts as used in the BPTA method [17].

Concept	Definition
Value chain	Comprehensive set of work carried out in an organization to provide a product/service to an external customer
Stakeholder	Groups that are internal/external to the organization and that have some interest in the success/failure of a given value chain. They can be related by generalization relations
Process	Processes depict the high-level interaction between a stakeholder and a value chain based on the concept of value stream. They can be related by resource flow and sequence relations
Value stream	Set of activities that are triggered by a requirement of and that are completed by the fulfillment of the named requirement
Classification	Categories in which a process can be classified in the context of a given value chain. Primary processes are involved in core activities, support processes provide resources that allow executing primary processes, and management processes provide rules to which process need to conform (such as strategies and budgets)

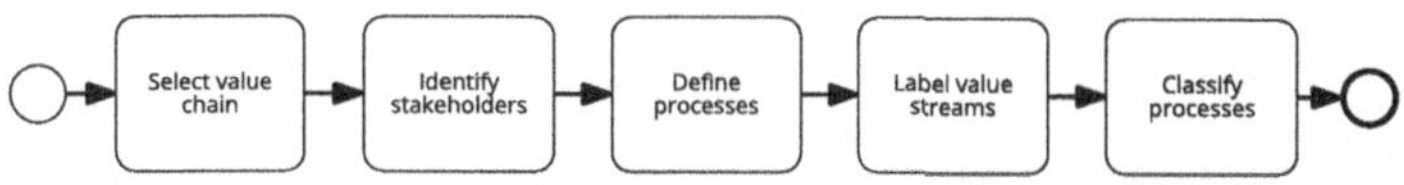

Fig. 2. Procedure of the Process Landscape Method, based on [17].

5.3 Implementation

For each department, the method was applied along a three-meeting cycle. The article reports on the work conducted with 4 departments, i.e. 12 meetings. Each meeting had 4 participants: 2 of the paper's authors in the role of "consultants" (same for all departments) and 2 domain experts of the organization in the role of "interviewees" (different for each department). For each meeting, one consultant was in charge of conducting the meeting and the other was responsible for using the modeling software while sharing his screen with the other participants of the meeting. Models were built in real-time during the meetings and, in some cases, were edited for readability prior to their final delivery. The LanMI-GT procedure, as applied for analyzing this Implementation, is described in the following.

S1. Verbal data transcript. Meetings were recorded (using OBS Studio). One author transcribed the verbal data from the recordings of the meetings (using Google docs). A portion of a coded transcript is provided in Table 1, where the column labeled *transcript* shows how verbal data was transcribed and the column labeled *id* provides an identifier for the coded portions of data.
S2. Initial coding. Each transcript was coded by one of the authors and when in doubt, a discussion was held until consensus was reached. The column labeled *content-based code* in Table 1 shows the initial coding for the given portion of the transcript.
S3. Focused and axial coding. This was an iterative task of selecting and organizing the coded data into more general categories and relating them. For this, we used a mind map where the nodes represented codes or more general conceptualizations that describe a set of codes/conceptualizations, and the arcs represented relations between the aforementioned elements. For the sake of space, the complete mind map (all data) is available in the technical report in [12].
S4. Content analysis. We synthesized the mind map generated in the previous step into the conceptual framework shown in Fig. 3. It allowed us to identify key issues and lessons in regard to the following implementation aspects: preparation, tool support, meetings, and concepts of the method. In parallel to this, we also observed two other main issues that emerged: diagnosis opportunities using the Process Landscape, and also a diversity of attitudes of the interviewees. For the sake of space, we refer the reader to [12] for a detailed description of the conceptual framework.

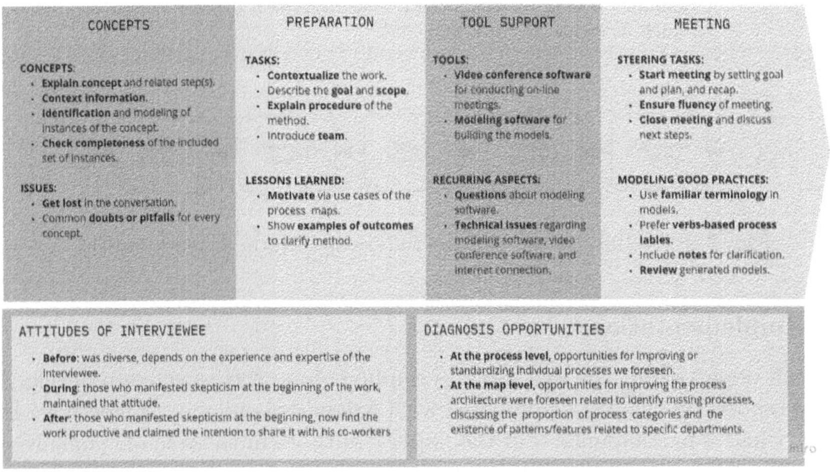

Fig. 3. Conceptual framework resulting from using LanMI-GT in the project.

S5. Initial coding categorization. The *category-based code* column in Table 1 shows a classification of content-based codes (i.e. initial coding) into the following categories: "Stakeholder", "Value Chain", "Process", "Value Stream", "Classification", and "Other". The first five categories correspond to the steps of the procedure of the method (see Sect. 5.2), and the latter is applied to whatever does not fit into the first five categories.

S6. Descriptive model discovery. Before applying Process Mining [1], we first generated logs based on data in the *category-based code* column of Table 1. For each meeting, a `csv` file was generated in which each row corresponded to the consecutive values in the category-based code column of the meeting's coding. A meeting-based case id and the sequence information were added to each row. Though not mandatory for applying Process Mining, timestamps could be added to the log instead of sequence information for conducting other types of analysis, e.g. performance analysis (not covered in this paper). After each log was generated, we imported them to Disco, a Process Mining software that generates (via a discovery algorithm) diagrams representing a descriptive process model that fits the provided data. Alternatively, this step could be performed using Network Analysis instead of Process Mining techniques.

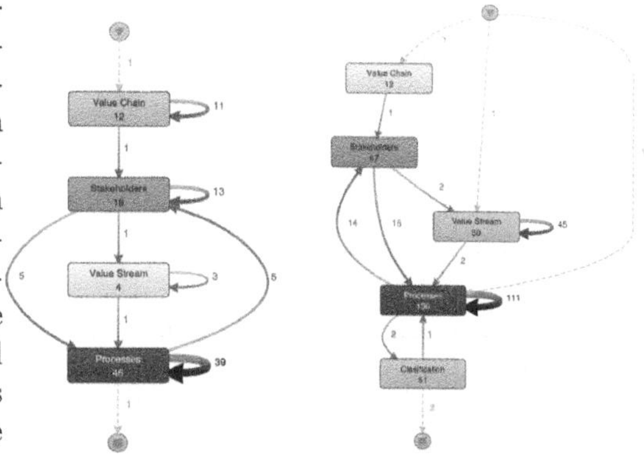

(a) First meeting (b) All meetings

Fig. 4. Descriptive model of the Implementation of the Process Landscape Method in one of the departments of the project (after filtering "Other").

Descriptive models were generated at the meeting level (i.e. 12 models, e.g. Fig. 4a) as well as at the department level (i.e. 4 models, e.g. Fig. 4b) using Disco. The discovered models are depicted as Directly Follows Graphs. In this notation, nodes represent steps of the procedure of the method (i.e. categories "Stakeholder", "Value Chain", "Process", "Value Stream", and "Classification") and directed arcs indicate the flow. Start and end nodes are shown as circles. Numbers within node and arc labels enrich the model by providing frequency information.

S7. Descriptive model analysis. The actual execution of the Method's procedure was analyzed mainly in terms of tasks and control flow of the discovered models (e.g. Fig. 4). It was observed that the category "Other" accounted for approximately a quarter of the codes registered (per meeting and per department), indicating that the Implementation of the Method involved some additional tasks than the ones defined by its theoretical procedure. The interested reader is referred to [12] for details of the analysis.

S8. Analysis integration and reporting. Each content-based code is linked to a department and a meeting via an *id* (see Table 1). Using this information allows for integrating the content and the descriptive model analyses. Findings of this analysis are reported in Sect. 5.5.

5.4 Process Landscape

A Process Landscape generated for a department is provided in Fig. 5. The full example, including the corresponding Value Chain Model is available in [12]. Process Landscapes were validated with each department organization in terms of representing the process architecture of the respective department. This was done at the end of the last meeting, as well as after sending the final versions of the generated models to the department, by asking them to either manifest their approval or their observations.

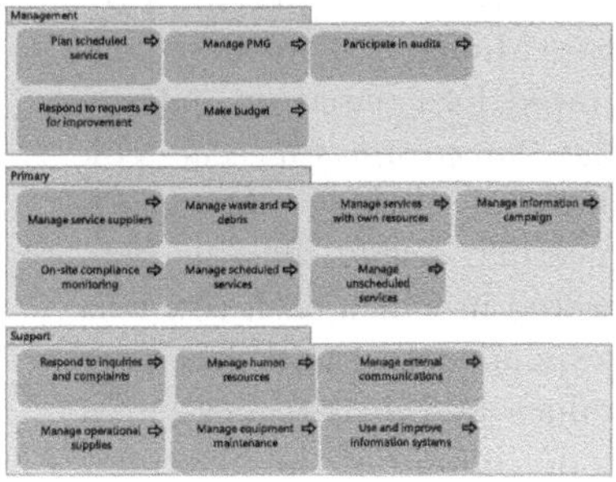

Fig. 5. Process Landscape example.

5.5 Findings

In the following, we summarize key insights of the Implementation of the BPTA method in the local government of Puente Alto, integrating the content and the descriptive model analysis. For the sake of space, we refer the reader to [12] for a complete and detailed description of our findings.

Execution of Method's Procedure. The sequence of steps of the Method's theoretical procedure was followed to a large extent during its implementation in all four departments, though regularly going back to previous steps. What is more, we observed additional tasks that supported the Method's theoretical procedure: (i) preparation tasks, i.e. contextualize the work within the organization, describe the goal of the method and scope of the work, explain the procedure of the method, and introduce the team of consultants; and (ii) coordination tasks, i.e. start meeting, ensure fluency of meeting, and close meeting. These discovered tasks would explain, to a large extent, the codes in the "Other" category.

Technical Aspects. The method was implemented using two software tools: a video-conference software for running the on-line meetings (i.e. Google Meet), and a modeling software (i.e. Archi). Even though interviewees were familiar with the former, they did not know the latter. Consequently, curiosity questions regarding the modeling software were raised throughout the different meetings. Regarding the notation used in the models (i.e., ArchiMate), it was rapidly understood by the interviewees though it was new to them. Along different meetings, a number of technical issues emerged related either to software or internet connection. However, these issues were infrequent and easily solved. Overall, the technical setting proved adequate for the method's implementation.

Social Aspects. All interviewees were familiar with the concept of process, as it was part of the growing organizational process culture. However, the remaining concepts (i.e. value chain, stakeholder, value stream, and classification) were not understood by them all and it was necessary to invest a little time in explaining them. It was also possible to observe differences between the four different departments in terms of their process training, expectations, and motivation. When these factors were lower, it was useful to spend more time in preparation tasks. In this regard, the following lessons were learned: (i) discussing use cases of a Process Landscape enhances motivation, (ii) showing examples of the outcomes (Process Landscape and Value Chain Model) eases the understanding of the method. It was also observed that, once the Value Chain Model or the Process Landscape were at an advanced stage, interviewees spontaneously made observations on how these models would aid them in improving their operations. In particular, the following opportunities were foreseen: standardization of individual processes, identification of missing processes, balancing the proportion of process categories, and pattern discovery within specific departments.

6 Conclusions

The article presents LanMI-GT, a novel Grounded Theory-based procedure for the systematic evaluation of the socio-technical aspects of a Process Landscape Method Implementation. Moreover, the proposal uses Process Mining techniques to enrich the socio-technical analysis.

To demonstrate our proposal, we described using LanMI-GT for assessing the Implementation of a well-known Process Landscape Method (i.e. BPTA [17]) in four departments of a public administration organization. We found that the sequence of steps prescribed by the method was observed, but that this was complemented by preparation and coordination tasks in attendance to social aspects of the context. Also, we observed that the on-line technical setting consisting of video-conference meetings with real-time modeling was successful for the work. We also identified lessons learned, issues that may arise in a similar context, and diagnosis opportunities that emerged as a by-product of the work.

As discussed in the conceptual model for Process Landscape Method analysis (see Fig. 1), there is an interplay between the context, the method, the implementation, and the resulting Process Landscapes. A limitation of our work is that, though it acknowledges this interplay, LanMI-GT does not include specific steps or guidelines for establishing relations between the evaluation of all these elements. We deem this as the main limitation of our proposal and expect to address it in future work. In terms of design, LanMI-GT is in an early developmental stage. Accordingly, our research agenda considers conducting an evaluation of LanMI-GT for closing a DSR design cycle reported in this paper.

Acknowledgments. This work was supported by ANID FONDECYT project 3210147.

References

1. van der Aalst, W.M.P.: Process Mining - Data Science in Action. Springer, Heidelberg (2016). https://doi.org/10.1007/978-3-662-49851-4
2. Abbad Andaloussi, A., Davis, C.J., Burattin, A., López, H.A., Slaats, T., Weber, B.: Understanding quality in declarative process modeling through the mental models of experts. In: Fahland, D., Ghidini, C., Becker, J., Dumas, M. (eds.) BPM 2020. LNCS, vol. 12168, pp. 417–434. Springer, Cham (2020). https://doi.org/10.1007/978-3-030-58666-9_24
3. Bider, I., Perjons, E., Elias, M., Johannesson, P.: A fractal enterprise model and its application for business development. Softw. Syst. Model. **16**(3), 663–689 (2016)
4. Burton, H.O., Pennotti, M.C.: The enterprise map: a system for implementing strategy and achieving operational excellence. Eng. Manag. J. **15**(3), 15–20 (2003)
5. Charmaz, K.: Constructing Grounded Theory. Introducing Qualitative Methods Series. SAGE (2014)
6. Dijkman, R., Vanderfeesten, I., Reijers, H.A.: Business process architectures: overview, comparison and framework. Enterp. Inf. Syst. **12**(2), 129–158 (2016)
7. Dumas, M., La Rosa, M., Mendling, J., Reijers, H.A.: Business Process Management. Springer, Heidelberg (2013)
8. Dumont, T., Fettke, P., Loos, P.: Towards multi-dimensional clustering of business process models using latent Dirichlet allocation. In: MKWI 2016, pp. 69–80 (2016)
9. Glaser, B., Strauss, A.: The Discovery of Grounded Theory: Strategies for Qualitative Research. Aldine Publishing Company (1967)
10. Gonzalez-Lopez, F., Bustos, G.: Business process architecture design methodologies - a literature review. Bus. Process. Manag. J. **25**(6), 1317–1334 (2019)
11. Gonzalez-Lopez, F., Bustos, G.: Evaluating methodologies for business process architecture design - a pilot study. In: ZEUS 2019 (2019)
12. Gonzalez-Lopez, F., Munoz-Gama, J., Fabia, P., Jara, A.: Technical report: on the systematic analysis of process landscape method implementation using grounded theory (2021). trlanmigt.haplab.org
13. Green, S., Ould, M.: A framework for classifying and evaluating process architecture methods. SPIP **10**(4), 415–425 (2005)
14. Green, S., Ould, M.A.: The primacy of process architecture. In: CAiSE Workshops, pp. 154–159 (2004)
15. Gruhn, V., Köhler, A.: Modelling, analysis and improvement of mobile business processes with the MPL method. In: ICEIS 2009, pp. 89–94 (2009)
16. Harmon, P., García, J.: The state of business process management 2020 - a BPTrends report (2020). https://www.bptrends.com
17. Harmon, P.: Business Process Change: A Business Process Management Guide for Managers and Process Professionals. Elsevier, Amsterdam (2014)
18. Kudryavtsev, D., Grigoriev, L.: Systemic approach towards enterprise functional decomposition. In: CEC 2011, pp. 310–317 (2011)
19. Malinova, M., Leopold, H., Mendling, J.: An empirical investigation on the design of process architectures. In: Alt, R., Franczyk, B. (eds.) Proceedings of the 11th International Conference on Wirtschaftsinformatik, vol. 2, pp. 1197–1211 (2013)

20. Moody, D., Sindre, G., Brasethvik, T., Solvberg, A.: Evaluating the quality of information models: empirical testing of a conceptual model quality framework. In: ICSE 2003, pp. 295–305 (2003)
21. Peffers, K., Tuunanen, T., Rothenberger, M., Chatterjee, S.: A design science research methodology for information systems research. J. Manag. Inf. Syst. **24**(3), 45–77 (2007)
22. Pesantes, M., Mitre, H.A., Lemus, C.: Evaluation of process architecture design methods. In: Murgante, B., et al. (eds.) ICCSA 2013. LNCS, vol. 7973, pp. 144–159. Springer, Heidelberg (2013). https://doi.org/10.1007/978-3-642-39646-5_11
23. Poels, G., García, F., Ruiz, F., Piattini, M.: Architecting business process maps. Comput. Sci. Inf. Syst. **17**(1), 117–139 (2020)
24. Recker, J.: Scientific Research in Information Systems. Springer, Heidelberg (2012). https://doi.org/10.1007/978-3-642-30048-6
25. Saldaña, J.: The Coding Manual for Qualitative Researchers. Sage (2009)
26. Sandkuhl, K., Seigerroth, U.: Method engineering in information systems analysis and design. Softw. Syst. Model. **18**, 1833–1857 (2019)
27. The Open Group: ArchiMate 3.0.1 Specification (2017)
28. Van Nuffel, D., De Backer, M.: Multi-abstraction layered business process modeling. Comput. Ind. **63**(2), 131–147 (2012)

Business Process Improvement in Industry 4.0: An Interorganizational Perspective

Vítor Hugo Ribeiro(✉), João Barata, and Paulo Rupino Cunha

University of Coimbra, CISUC, DEI, Coimbra, Portugal
{vhribeiro,barata,rupino}@dei.uc.pt

Abstract. Industry 4.0 calls for end-to-end digital integration of supply chains and a new boundary-spanning logic of process design. The shift is from shared operation to shared transformation. Design science research was chosen to (1) propose an approach for interorganizational business processes improvement in decentralized contexts of Industry 4.0 (IOBP 4.0) and (2) draft a BPMN extension for IOBP 4.0. The results are relevant to guide the fourth industrial revolution with increasingly shared and digitalized business processes. For theory, our work contributes to the emerging business process management logic of digital transformation: support for coordinated touchpoints, flexible infrastructure, and empowered participants. For practice, we propose a continuous improvement approach for IOBP 4.0 that ensures manufacturing visibility in collaboration networks. Managing the punctuated equilibrium of boundary-spanning business processes will be a priority for this decade.

Keywords: Interorganizational business process · Industry 4.0 · Business process improvement · BPMN extension

1 Introduction

Business process management (BPM) has enabled organizations to move beyond functional boundaries. Much has changed since the pioneer contributions of BPM, but the boundaryless nature of business processes is more evident than ever. Furthermore, in the digital transformation era of industry (alias Industry 4.0 or I4.0), cooperation, communication, and integration within and between organizations become priorities. Therefore, process models representing "how work is done" must support downstream planning of operations, upstream assessment, and decentralized continuous improvement.

Industry 4.0 is leveraged by multiple technologies such as the Internet of Things (IoT), cyber-physical systems, cloud computing, mobile systems, or artificial intelligence shaping the smart factory infrastructure. The overall aim is to integrate and digitalize distributed business processes and redesign supply chains. For example, a company may be manufacturing final products with 3D printers, while, at the same time, their partners produce accessories and raw materials needed to satisfy the customer's order. It is now clear that a new agenda is necessary to promote synergies between BPM and digital innovation in the industry [1, 2].

A. Marrella and B. Weber (Eds.): BPM 2021 Workshops, LNBIP 436, pp. 286–298, 2022.
https://doi.org/10.1007/978-3-030-94343-1_22

The collaborative nature of Industry 4.0 highlights the need to manage interorganizational business processes (IOBP) [3]. The study presented by [4] is an example of this trend. The authors present an approach to merge different process models collaborating in the production of artifacts. However, the resulting process models are often incomplete (e.g., some parts may be private) and challenging to share in organizations that need to compete in collaborative production networks. BPMN (OMG's Business Process Model and Notation - BPMN 2.0) is one of the primary standards in process modeling, including elements like tasks, events, and data objects [5]. However, BPMN cannot represent all the details of interorganizational practices [3]. This shortcoming is particularly relevant in two aspects of decentralized manufacturing: (1) details on the process interfaces with third party entities and (2) the specification of the enabling technologies of digital transformation in Industry 4.0. Therefore, BPMN extensions emerge as a promising solution to extend the vocabulary of the notation [6].

Contacts with industry managers revealed that rudimentary practices are still the norm, with process models (1) created independently by each organization in the supply chain, (2) supported by separate documentation (e.g., procedures and requirement lists), and (3) lacking a boundaryless approach to the design, improvement, and audit of IOBP. Moreover, despite the ISO 9001 requirements to adopt a process approach [7], the traditional focus of quality audits tends to be the internal documentation, missing crucial details in distributed environments. Two research objectives are formulated to address this gap: (1) create the foundations for a BPMN extension, more precisely a Conceptual Domain Model of the Extension (CDME) and (2) devise a set of design principles to continuously improve interorganizational business processes in companies adopting Industry 4.0.

The remainder of this paper is structured as follows. Section 2 presents background literature on Industry 4.0 and business processes. Next, the research approach is introduced. The results of the DSR cycle follow. Afterward, the discussion enumerates design guidelines for IOBP 4.0 design and continuous improvement. Finally, Sect. 6 presents the conclusions, limitations, and an outlook for the future.

2 Background

2.1 Decentralized Manufacturing Networks and Interorganizational Business Processes in the Industry 4.0 Era

Shifting from single-site to multi-site manufacturing comes with the need for decentralized decisions and more complex flows of data and activities. Collaborative networks also call for autonomous teams of humans and machines equipped with advanced computing power. Therefore, new process modeling languages and methods are necessary for the Industry 4.0 era [8]. However, when "parts" of manufacturing processes are enacted in different locations/settings, it is necessary to deal with moments of disruption (e.g., when a new system implemented) and stability [9], exploiting manufacturing capabilities not restricted to a single organization.

Modeling and improvement in BPM are two sides of a single coin, and popular quality standards like ISO 9001:2015 suggest a process approach to management. Following this standard, companies can adopt the PDCA cycle [7] and, for each step in Plan (P) – Do (D) – Check (C) – Act (A), continuously improve their business processes. BPM is "*the art and science of overseeing how work is performed in an organization to ensure consistent outcomes and to take advantage of improvement opportunities*" [10]. However, "*shifting from strategic interactions (driven by reduction of transaction costs) to transformational interaction (driven by collaborative transorganizational development) appears to be difficult to achieve in practice in a network setting*" [11].

Process, infrastructure, and people are fundamental building blocks of BPM culture [12] and quality culture. First, organizations should focus on the lifecycle of process identification (1), discovery (2), analysis (3), redesign (4), implementation (5), monitoring, and controlling (6), in which the process models assume a crucial role [10]. Second, BPM promotes the alignment between the process goals and the organizational infrastructure. Third, actors are expected to follow the processes as documented and modeled [1]. Nevertheless, the complexity of BPM in the digital transformation era needs to balance the traditional stability and predictability of work practices with the uncertainty and dynamic nature of change [2, 9]. Moreover, the emerging cyber-physical infrastructure must maximize process exploitation and leverage exploration capabilities to foster continuous improvement in decentralized manufacturing.

Recent research points to the necessity to move beyond the organization borders in modeling process details, incorporating process deviations and the constraints/opportunities for sociotechnical change [13] while keeping the process compliant and traceable. Representing social, technical, and transformational elements in process models is one of the challenges for research in this area.

Interorganizational business processes are interrelated activities shared and executed by two or more entities to achieve an objective of value to the partners [14]. Globalization and technological advances increase the need for collaboration within supply chains [15]. Therefore, entities involved in IOBP 4.0 development need to establish a relationship supported by technical, behavioral, legal, and strategic mechanisms [16].

However, balancing the needs of real-time control and compliance with decentralized decision-making and flexibility can be challenging [17]. As stated by [18], this type of collaboration arrangement offers "*significant opportunities at strategic level, as well as significant challenges at tactical level, in order to properly combine flexible and effective inter-organization collaborations with traditional internally managed processes*". Examples include the need for transparency between internal business processes and the "external part" [19], coordination and management of process interdependencies [3], and a clear definition of responsibilities [20]. In addition, companies must address the semantic gap caused by diverse internal process language/specifications [21] and the autonomy that each organization requires to implement their strategies at a different pace. Therefore, mechanisms to reduce the degree of coupling between the internal and external interfaces must be put in place [22].

The investments required by partnering across organizations in the digital transformation era require agility and joint innovation mechanisms to support continuous improvement [14]. However, when business process management is geographically dispersed [23] and transversal to different power structures, it is crucial to deploy innovative policies to allow traceability metrics for each activity [24].

The IOBP modelling activities involve the identification of the behavioural and functional aspects, for the execution of activities (what is done), occurring events and their visibility (e.g., private, public) [15]. Additionally, organizational aspects must be considered to identify the several business partners, their roles and responsibilities, and the resources used and exchanged [15]. Despite the essential contributions recently proposed to synthesize IOBP in a unified visualization [4], we could not find an approach in the literature to assist the entire lifecycle of IOBP 4.0 transformations at both design-time and run-time (operation).

2.2 Business Process Modelling and Extensions for Industry 4.0

The main goal of BPMN is to support BPM activities with a straightforward notation comprehensible by different domain experts. BPMN can be used to represent complex processes, for example, in manufacturing [25]. Another advantage is that BPMN has a well-defined language meta-model that facilitates model exchangeability [6]. Moreover, the BPMN meta-model contains a specification of elements for the definition of language extensions [6], which is particularly useful for adapting to new contexts.

Diagrams can be shared across organizations and partners using an XML-based interchange format. Therefore, our research gathers inspiration in:

- BPMN extensions for industry: PyBPMN extension [26] for cyber-physical systems is the most cited. Additional studies in this area include modeling industrial internet-of-things scenarios [27], business process fragments for manufacturing [28], requirements of process synchronization [25] and ubiquitous business process modeling [29]. Nevertheless, "*business process modelling remains unproven for all the processes encountered in manufacturing enterprises*" [28].
- BPMN extensions for interorganizational contexts: Some studies focus on time-aware business process modeling. For example, processes must "*adhere to a wide range of temporal requirements which rise from legal, regulatory, and managerial rules*" [30]. Notably, the first contribution with an approach for IOBP model design was presented by [31], using messages and pools for each organization. [32] presents a comprehensive BPMN extension for collaborative business processes, focusing on concepts related to the execution of collaborative tasks, privacy, confidentiality, state of execution of tasks, data management, and activity monitoring. The authors propose a set of new elements and illustrate them with examples.

It is now possible to extend these important contributions to the field of manufacturing, IOBP, and Industry 4.0 adoption. The following section describes the research approach towards IOBP4.0: interorganizational business processes for Industry 4.0 that balances compliance and change by design, adhering to the needs of multiple manufacturing organizations sharing a common production aim.

3 Research Approach

Design science research (DSR) is a problem-solving paradigm that relies on kernel theories to produce innovative artifacts [33]. The authors of [33] suggest an iterative process starting with the problem identification and motivation, define objectives of a solution, design and development, demonstration, evaluation, and communication [33]. Complementarily, the FEDS framework [34] was proposed to evaluate DSR projects, which considers the possibility of a "quick & simple" summative evaluation.

The DSR cycle reported in this paper includes a review of synergies between Industry 4.0 and IOBP – summarized in Sect. 2. First, we obtained 80 hits in Google Scholar using the keyword combination "BPMN extension" AND ("industry 4.0" OR "digital transformation"), excluding patents and citations. Only ten results were found in using "BPMN extension" AND ("inter-organizational business process" OR "interorganizational business process". Then, we searched for recent papers focusing on Industry 4.0 and digital transformation in BPM to understand the trends in these fields.

The methodology to design the artifacts [33] was adapted from [35], using UML profiles, and later improved by [36], with the analysis and conceptualization of the domain [6]. First, we conceptualized what continuous improvement means in the context of IOBP. Second, we identified key attributes in the literature to represent IOBP 4.0 and support (decentralized) digital transformation. Third, we created a CDME as a UML class diagram. Finally, we conducted a summative evaluation [34] of the results with two companies adopting Industry 4.0.

Company CC1 is a major European paper pulp production company, and CC2 is a small technical metal coatings provider. CC1 had an ongoing digital transformation project for the forest management process (integrating companies in production, logistics, inspection, transformation). CC2 created a new product line partially executed by external partners. Both companies are ISO 9001-certified and interested in continuously improving their processes in collaborative environments.

4 Modeling and Improving IOBP 4.0

The team created three foundational artifacts. We were first aggregating the necessity of "change" in digital transformation and gathered inspiration in the PDCA to describe how interorganizational business processes can continuously improve (Table 1).

After describing the lifecycle of IOBP 4.0 improvement, we extracted attributes to create the IOBP 4.0 extension from the literature (Table 2). These attributes highlight specific concepts that adhere to Industry 4.0 decentralized scenarios, such as the need to trace resources and activities, share information, execute distributed decisions and promote transformation in the technological infrastructure.

Table 1. Continuous improvement of IOBP 4.0: a Plan-Do-Check-Act approach.

Lifecycle	Description	Ref.
Shared planning (P)	IOBP 4.0 requires preparation and commitment from the different parties. Companies may compete for the same resources (e.g., machines) that must be scalable and optimized. Each "part" of the process must ensure flexibility by design. The organization involved in collaborative improvement must specify goals to achieve (e.g., IT investments and expected results for the overall shared goal)	[1, 14, 24, 27]
Shared execution (D)	IOBP 4.0 can be described by core BPMN elements (e.g., processes, tasks, events, resources, and data objects). Messages are important but insufficient to detail (1) interorganizational execution (e.g., who decides to cancel the process, quality criteria, performance indicators) and (2) particularities of Industry 4.0 (e.g., new technologies adopted in decentralized process parts)	[15, 25, 30, 37]
Shared monitoring (C)	IOBP 4.0 requires monitoring the performance of shared elements (e.g., process execution-level agreements). In addition, new challenges emerge from decentralized manufacturing (e.g., real-time data sharing) and protected logs for auditability purposes	[24, 37]
Shared digital transformation (A)	IOBP 4.0 improvements can be implemented by each actor independently or in cooperation. Thus, mindful actors and digitalization are inseparable	[1, 2, 9]

Table 2. Key attributes of IOBP 4.0.

Attribute	Description	Ref.
Confidentiality	Organizations may have restrictions on sharing information or managing customer-owned data. As a result, decisions may occur under incomplete information	[15, 19, 32, 38]
Responsibility	Shared processes require shared responsibility for innovation, execution, and monitoring	[15, 20, 37]
Authority	Global and local actors must be defined, and their decisional capacity specified in different scenarios	[3, 37]
Touchpoint	It is necessary to define when a message is required and the impact on all the stakeholders of the main process (e.g., customers may interact with the process at specific points, assessors' touchpoints, or interaction between cyber and physical elements of the process)	[28, 31]
Transparency	Partner organizations should embrace transparency to improve trust and process activities synchornization	[19, 38]
Compliance	Multiple regulations (voluntary and enforced) may compete in different geographical locations	[3, 23]
Traceability	Activities, resources, data, and decisions must be traceable within the entire process lifecycle	[24]
Interface	Shared elements (e.g., task, data) must have an interface to enable actors' intervention (e.g., app)	[22, 32]
Collaborative	Collaborative BPMN elements are critical. Parallel or sequential execution may be in collaboration	[32, 39]
Autonomy	Autonomous tasks and decisions (e.g., single-organization process improvement) must be identified	[37]
Digital infrastructure	Digitalized activities require technological devices to retrieve data (e.g., sensors), interact (mobile devices) and produce value with data	[8, 26]
Digital transformation phase	BPMN elements (e.g., task) have specific transformation stages (planned, development, deployed)	[1, 2, 11]

(*continued*)

Table 2. (*continued*)

Attribute	Description	Ref.
Target innovation	BPMN elements can be classified in terms of innovation status (state-of-the-art, outdated, actual, stable)	[1, 2, 11]

Finally, we produced the CDME for IOBP 4.0 (Fig. 1).

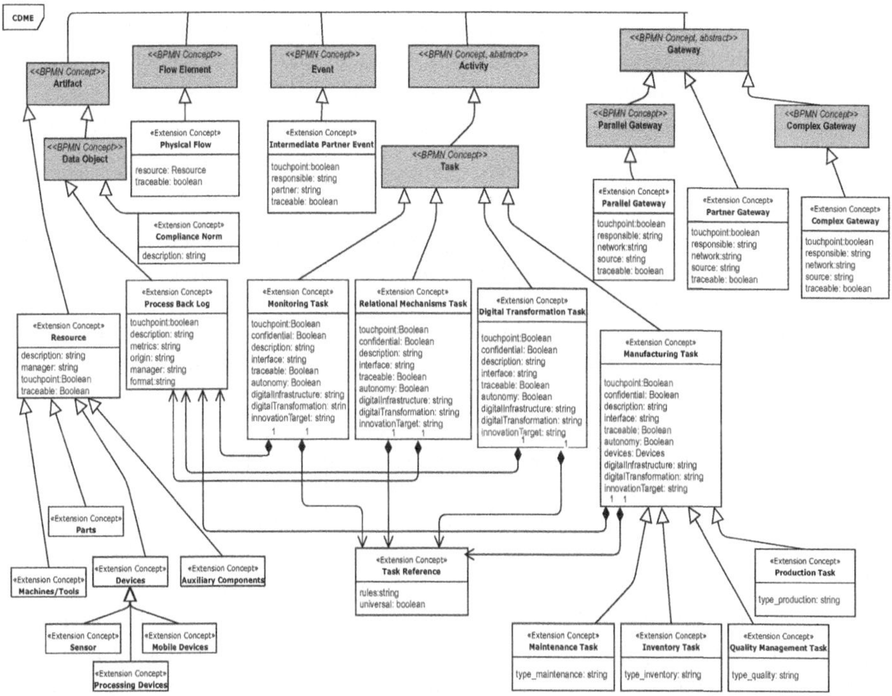

Fig. 1. Conceptual domain model of the extension for IOBP 4.0.

Four types of resources (on the bottom-left of Fig. 1) are essential in the context of Industry 4.0 [40]: machines/tools parts; devices (e.g., sensors, mobile devices); and auxiliary components that may be used and shared by the business partners in the manufacturing activities, each with a visibility classification (e.g., touchpoint, traceable). In addition, the task concept was extended with (1) manufacturing particularities and supplemented with (2) IOBP tasks for monitoring, (3) managing relationships, and (4) digital transformation. The latter three concepts are aimed at creating synergies among process partners. Moreover, the manufacturing-related tasks can be quality control, inventory control, production, and maintenance [28]. The goals is to cover the scope of operational and management activities of IOBP 4.0. Additionally, each task has a classification regarding visibility (e.g., private, touchpoint, traceable) and technological strategy (e.g., digital infrastructure, innovation target).

Finally, the data object was extended to represent (1) the several compliance regulations that each actor must follow while executing their activities [23]. Moreover, this object can also be broadened to represent the process backlog: information related to the monitorization of the business process [24] and analysis. The "Partner Gateway" extends the gateway concept, representing the actor involved in the gateway path decision. The event concept was extended with the intermediate partner event (event raised by a partner's decision in specific moments of the business process) [3, 37]. The flow element extension represents the physical exchange of resources across business processes [28]. The following section discusses the main findings of this DSR cycle and suggested guidelines for the continuous improvement of IOBP 4.0.

5 Discussion

Process activities need to be monitored and controlled across the collaboration network involved in Industry 4.0 investments. For example, some activities may need to comply with specific regulations (affecting one or multiple partners). The manufacturing stages may also require transport/sharing resources, represented by the physical flow. At the same time, partners' (independent/agreed) decisions raise the necessity to include the partner gateway and the partner event. Moreover, Industry 4.0 adds new challenges to traditional interorganizational process management because companies are changing their digital infrastructure in cycles of stabilization (exploitation) and destabilization (digital exploration), affecting each partner's BPMN element in particular ways.

PDCA cycle was considered suitable by the project participants familiar with ISO 9001, suggesting simple steps for continuous improvement in distributed environments. However, Table 1 also reveals issues when operating in collaborative networks. For example, governance, risk, and compliance (GRC) are more complex and involve interdependencies between partners [31], which is challenging to represent in traditional BPMN models. Nevertheless, we agree with [23] that GRC management is an opportunity to improve business processes, achieve genuine cost savings, and improve their competitive positions.

Due to the complex and dynamic nature of organizations, markets, and technologies in Industry 4.0, more complete models are necessary to represent work practices and the stage of digital transformation to design new systems or improve the operation of existing ones. According to the domain experts contacted during our DSR, a standard notation can assist the global process actors to manage activities and coordination of tasks (e.g., similarly to how Gantt-charts are usually adopted in project management to share information between partners). Furthermore, those models can be included in a common repository, shared by all actors, and integrate into their contractual agreements. Thus, the models can be helpful for the "top-down" communication of the global process owner and to collaboratively design, change, and promote innovation and improvement in boundary-spanning processes of Industry 4.0. However, despite the popularity of

BPMN (as happens in ISO 9001 certified industries), we cannot confirm the acceptance by the industry at this stage.

The artifacts developed in this cycle and the discussion with practitioners allowed us to derive the following design principle on how to develop IOBP 4.0 models:

- Adopt a top down IOBP 4.0 modeling approach for BPMN elements. Then, choose a bottom-up description of digital transformation attributes. While the former address the common (shared) business objective, the latter emerges from the negotiated contribution of all partners in the network and a trade-off between individual strategies and overall collaboration value;

Additionally, this cycle allowed the identification of process improvement activities that may be supported by the IOBP 4.0 models:

- Use business process models to negotiate continuous improvement initiatives among the partner organizations and establish an integrated digital transformation program;
- Continuously update IOBP 4.0 models. Industry 4.0 investments must be communicated to all interested parties and its performance monitored over time;
- Identify priorities for shared innovation in specific parts of the process. Industry 4.0 is enabled by end-to-end digital integration of supply chains, local weak points (e.g., partners not producing as expected) may need adjustments;
- Explore business process simulation techniques to evaluate the impact of changes.

6 Conclusion

This paper presents the results of a design science research project aiming to create (1) a shared PDCA approach to continuously improve interorganizational business processes in Industry 4.0 contexts and (2) the grounds for a BPMN extension for IOBP 4.0. Five main design guidelines are suggested to create IOBP 4.0 models that portray how industries collaborate and support shared continuous improvement planning, execution, and evaluation.

There are also limitations that need to be stated and opportunities for the next DSR cycle. First, although we have identified a lifecycle for the digital transformation of IOBP 4.0 and an extension, we have used a specific combination of keywords in our literature review. Other attributes may be included via search improvements and insights from the practitioners. Second, the artifacts produced in this cycle are essential to change the traditional (separate) process models. However, we do not yet have evidence of its benefits in the entire collaboration network. Our contribution includes the proposal of design guidelines for the creation/transformation of boundary-spanning IOBP 4.0, balancing the needs of digital transformation, which is essential, but also challenging when we evaluate change "over time" [2]. Third, the companies that agreed to participate in our work sharing their models are not representative of the entire industry. Other companies adopting Industry 4.0 can be added to the study. Fourth, the focus of this cycle was on manufacturing-related IOBP 4.0, but the model can be extended to other interorganizational business processes, for example, purchasing, marketing, or services.

Fifth, further evaluation will need an external ISO 9001 process audit. This limitation was already considered in preparation for the next cycle. We have included ISO 9001-certified companies adopting industry 4.0 with processes that need to be shared by at least another organization with an independent decision hierarchy. Moreover, it will be essential to evaluate synergies of BPMN extensions for Industry 4.0 and enterprise architecture approaches for digital transformation [41]. For example, Archimate [42] supports high granular modeling, viewpoints relevant for interorganizational contexts (e.g., business process cooperation viewpoint), and representation of both physical and digital layers of digital transformation.

The next DSR cycles will focus on developing the graphical representation for the extension and evaluating the organizational (e.g., synergies in identifying process improvements) and social (e.g., the usability of the IOBP 4.0 models) implications of its adoption by the case companies.

Acknowledgments. This work is co-funded by national funds through the FCT - Foundation for Science and Technology, I.P., within the scope of the project CISUC - UID/CEC/00326/2020 and by European Social Fund, through the Regional Operational Program Centro 2020.

References

1. Baiyere, A., Salmela, H., Tapanainen, T.: Digital transformation and the new logics of business process management. Eur. J. Inf. Syst. **29**, 238–259 (2020)
2. Mendling, J., Pentland, B.T., Recker, J.: Building a complementary agenda for business process management and digital innovation. Eur. J. Inf. Syst. **29**, 208–219 (2020)
3. Legner, C., Wende, K.: The challenges of inter-organizational business process design - a research agenda. In: ECIS (2007)
4. Kunchala, J., Yu, J., Yongchareon, S., Liu, C.: An approach to merge collaborating processes of an inter-organizational business process for artifact lifecycle synthesis. Computing **102**(4), 951–976 (2019). https://doi.org/10.1007/s00607-019-00770-z
5. Object Management Group, Inc.: Business Process Model and Notation (BPMN) v2.0.2 (2014)
6. Braun, R.: BPMN extension profiles - adapting the profile mechanism for integrated BPMN extensibility. In: 17th IEEE Conference on Business Informatics, vol. 1, pp. 133–142 (2015)
7. ISO: ISO 9001:2015 Quality management system – Requirements. International Organization for Standardization, Geneva (2015)
8. Petrasch, R., Hentschke, R.: Process modeling for industry 4.0 applications: towards an industry 4.0 process modeling language and method. In: 2016 13th International Joint Conference on Computer Science and Software Engineering, JCSSE 2016, pp. 1–5 (2016)
9. Lyytinen, K., Newman, M.: Punctuated Equilibrium, process models and information system development and change: towards a socio-technical process analysis. Sprouts Working Papers on Information Systems, vol. 6, Paper 1 (2006)
10. Dumas, M., La Rosa, M., Mendling, J., Reijers, H.A.: Fundamentals of Business Process Management, 2nd edn. Springer, Heidelberg (2018). https://doi.org/10.1007/978-3-662-56509-4
11. Yström, A., Ollila, S., Agogué, M., Coghlan, D.: The role of a learning approach in building an interorganizational network aiming for collaborative innovation. J. Appl. Behav. Sci. **55**, 27–49 (2019)

12. Vom Brocke, J., Schmiedel, T., Recker, J., Trkman, P., Mertens, W., Viaene, S.: Ten principles of good business process management. Bus. Process Manag. J. **20**, 530–548 (2014)
13. Queiroz, M.M., Fosso Wamba, S., Machado, M.C., Telles, R.: Smart production systems drivers for business process management improvement: an integrative framework. Bus. Process Manag. J. **26**, 1075–1092 (2020)
14. Bala, H., Venkatesh, V.: Assimilation of interorganizational business process standards. Inf. Syst. Res. **18**, 340–362 (2007)
15. Bouchbout, K., Alimazighi, Z.: Inter-organizational business processes modelling framework. In: CEUR Workshop Proceedings, vol. 789, pp. 45–54 (2011)
16. Martins, C.T., Soares, A.L.: Dissecting inter-organizational business process modeling: a linguistic and conceptual approach. In: Camarinha-Matos, L.M., Afsarmanesh, H., Ollus, M. (eds.) Network-Centric Collaboration and Supporting Frameworks. PRO-VE 2006. IFIP International Federation for Information Processing, vol. 224, pp. 221–228. Springer, Boston (2006). https://doi.org/10.1007/978-0-387-38269-2_23
17. Giaglis, G.M., Paul, R.J., Doukidis, G.I.: Simulation for intra- and inter-organisational business process modelling. In: WSC90 Proceedings, USA, pp. 1297–1304 (1996)
18. Bocciarelli, P., D'Ambrogio, A., Paglia, E., Giglio, A.: An HLA-based BPMN extension for the specification of business process collaborations. In: 2017 IEEE/ACM DS-RT Proceedings, pp. 1–8 (2017)
19. Norta, A.H.: Exploring dynamic inter-organizational business process collaboration (2007)
20. Van der Aalst, W.M.P.: Loosely coupled interorganizational workflows: modeling and analyzing workflows crossing organizational boundaries. Inf. Manag. **37**, 67–75 (2000)
21. Zhang, D.: Web services composition for process management in e-business. J. Comput. Inf. Syst. **45**, 83–91 (2005)
22. van der Aalst, W.M.P., Weske, M.: The P2P approach to interorganizational workflows. In: Dittrich, K.R., Geppert, A., Norrie, M.C. (eds.) CAiSE 2001. LNCS, vol. 2068, pp. 140–156. Springer, Heidelberg (2001). https://doi.org/10.1007/3-540-45341-5_10
23. Schoenthaler, F., Augenstein, D., Karle, T., Draghici, A., Popescu, A.-D., Gogan, L.M.: Design and governance of collaborative business processes in industry 4.0. Proc. - Soc. Behav. Sci. **i**, 544–551 (2015)
24. Breu, R., et al.: Towards living inter-organizational processes. In: 2013 IEEE International Conference on Business Informatics, pp. 363–366 (2013)
25. Traganos, K., Spijkers, D., Grefen, P., Vanderfeesten, I.: Dynamic process synchronization using BPMN 2.0 to support buffering and (un)bundling in manufacturing. In: Fahland, D., Ghidini, C., Becker, J., Dumas, M. (eds.) BPM 2020. LNBIP, vol. 392, pp. 18–34. Springer, Cham (2020). https://doi.org/10.1007/978-3-030-58638-6_2
26. Bocciarelli, P., D'Ambrogio, A., Giglio, A., Paglia, E.: A BPMN extension for modeling cyber-physical-production-systems in the context of industry 4.0. IEEE 14th International Conference on Networking, Sensing and Control 599–604 (2017)
27. Engels, G., Strothmann, T., Teetz, A.: Adapt cases 4 BPM - a modeling framework for process flexibility in IIoT. In: Proceedings - IEEE International Enterprise Distributed Object Computing Workshop, EDOCW, pp. 59–68 (2018)
28. Erasmus, J., Vanderfeesten, I., Traganos, K., Grefen, P.: Using business process models for the specification of manufacturing operations. Comput. Ind. **123**, 103297 (2020)
29. Yousfi, A., Hewelt, M., Bauer, C., Weske, M.: Toward uBPMN-based patterns for modeling ubiquitous business processes. IEEE Trans. Ind. Inform. **14**, 3358–3367 (2018)
30. Cheikhrouhou, S., Kallel, S., Guermouche, N., Jmaiel, M.: Enhancing formal specification and verification of temporal constraints in business processes. In: Proceedings - 2014 IEEE International Conference on Services Computing, SCC 2014, pp. 701–708 (2014)
31. Fedorowicz, J., et al.: Business process modeling for successful implementation of interorganizational systems. In: AMCIS 2005 (2005)

32. Amdah, L., Anwar, A.: A DSL for collaborative business process. In: 2020 International Conference on Intelligent Systems and Computer Vision, ISCV 2020, pp. 1–6 (2020)
33. Peffers, K., Tuunanen, T., Rothenberger, M.A., Chatterjee, S.: A design science research methodology for information systems research. J. Manag. Inf. Syst. **24**, 45–77 (2007)
34. Venable, J., Pries-Heje, J., Baskerville, R.: FEDS: a framework for evaluation in design science research. Eur. J. Inf. Syst. **25**, 77–89 (2016)
35. Stroppi, L.J.R., Chiotti, O., Villarreal, P.D.: Extending BPMN 2.0: method and tool support. In: Dijkman, R., Hofstetter, J., Koehler, J. (eds.) BPMN 2011. Lecture Notes in Business Information Processing, vol. 95, pp. 59–73. Springer, Heidelberg (2011). https://doi.org/10.1007/978-3-642-25160-3_5
36. Braun, R., Schlieter, H.: Requirements-based development of BPMN extensions: the case of clinical pathways. In: 2014 IEEE 1st International Workshop on the Interrelations Between Requirements Engineering and Business Process Management, REBPM 2014 – Proceedings, pp. 39–44 (2014)
37. Mircea, M., Ghilic-Micu, B., Stoica, M., Sinioros, P.: Inter-organizational performance and business process management in collaborative networks. Econ. Comput. Econ. Cybern. Stud. Res. **50**, 107–122 (2016)
38. Liu, D.-R., Shen, M.: Workflow modeling for virtual processes: an order-preserving process-view approach. Inf. Syst. **28**, 505–532 (2003)
39. Luís Osório, A., Camarinha-Matos, L.M.: Distributed process execution in collaborative networks. Robot. Comput. Integr. Manuf. **24**, 647–655 (2008)
40. Zor, S., Schumm, D., Leymann, F.: A proposal of BPMN extensions for the manufacturing domain. In: ICMS Proceedings, pp. 1–6 (2011)
41. Zimmermann, A., Schmidt, R., Jugel, D., Möhring, M.: Evolution of enterprise architecture for intelligent digital systems. In: Dalpiaz, F., Zdravkovic, J., Loucopoulos, P. (eds.) RCIS 2020. LNBIP, vol. 385, pp. 145–153. Springer, Cham (2020). https://doi.org/10.1007/978-3-030-50316-1_9
42. The Open Group: ArchiMate® 3.1 Specification (2018)

The Effect of Self-healing on Process Acceptance

Lars Drewes(✉) and Volker Nissen

Technische Universität Ilmenau, 98693 Ilmenau, Germany
{lars.drewes,volker.nissen}@tu-ilmenau.de

Abstract. According to the process acceptance theory, the acceptance of processes can have an influence on their correct execution. If deviations and manipulations of processes of any kind are to be excluded, it is necessary to understand which factors lead to a change in acceptance. To this end an experimental study is conducted to investigate the influence of self-healing in case of process error. A generic purchasing process is implemented and made available online. The acquisition of participants takes place via the crowdsourcing platform Amazon's Mechanical Turk. Process acceptance is measured by means of a questionnaire in the three dimensions of attitude (cognitive, affective and conative). The study shows that there is a significant difference in acceptance between processes with and without self-healing in the cognitive dimension. The results of the tests indicate that self-healing is not automatically more accepted than repeated manual effort. A tentative explanation is that a loss of participants' control of the process leads to this apparently counterintuitive result.

Keywords: BPM · Process acceptance · Process experiment · Amazon's mechnical turk · Social influence

1 Introduction

The functional orientation that prevailed in companies for decades has meanwhile often been replaced by process orientation to improve enterprise performance and competitiveness. However, good process design and active process management are essential prerequisites for achieving process innovation, cost reduction, and customer satisfaction [1–4]. In this regard, it is crucial for the processes to be carried out in accordance with their underlying design. The problem is that the individual behavior of the participants in the process is often disregarded when designing, controlling, and optimizing processes. However, in the end these people decide whether a process is initiated and executed as required. According to the process acceptance theory, this decision is based on the acceptance that the process executor (acceptance subject) assigns to the process (acceptance object) within a certain (acceptance) context. If this acceptance is too low, modifications of that process occur or even a refusal of the execution takes place. It is also possible that unofficial processes, so-called shadow processes, arise as a result. Similarly, existing substitutional processes may experience a much higher acceptance and for this reason are preferred [3–5].

A. Marrella and B. Weber (Eds.): BPM 2021 Workshops, LNBIP 436, pp. 299–311, 2022.
https://doi.org/10.1007/978-3-030-94343-1_23

Such process deviations cannot only result in business disadvantages, such as loss of customer satisfaction and thus loss of sales but can also pose a potential risk to people and the environment. An example is the persistent outbreak of MRSA bacteria in hospitals, which is due to the lack of attention paid to hygiene processes. A deviation into illegality is also possible if, for example, embargo processes are circumvented, which may result in severe penalties [3, 4, 6].

Consequently, it is of great relevance to investigate and understand influencing factors of process acceptance. Müllerleile [5] established a basic theoretical concept on the subject of process acceptance along with its measurement. Furthermore, he identified influencing factors on acceptance by means of a qualitative empirical study, in the form of expert interviews. Within this study 28 possible influencing factors on process acceptance were identified [5].

However, to understand the mechanisms of process acceptance more precisely and to be able to derive optimization potentials and direct actions in business practice, further investigations are necessary that determine the effect strength of individual factors as well as potential interactions between such factors.

In this contribution, the factor self-healing in case of a process error, as defined in Sect. 3.2, will be examined. This factor was chosen because nowadays automated checks and corrective actions often play an important role in the execution of business processes. To this end, we pursue the following research questions:

1. RQ1: How can the factor process self-healing in the context of process error be operationalized?
2. RQ2: Is a process with self-healing capabilities more accepted than a process without such capabilities?

2 Method

To investigate the effects of the factor self-healing on process acceptance, an online experiment is conducted. For this purpose, a virtual procurement system with the associated purchasing process is implemented and made available online. This enables the participants of the experiment to experience the process actively and directly. The crowdsourcing platform Amazon's Mechanical Turk (MTurk) is used to acquire participants, as it allows for a cost-efficient realization with a large sample size [7]. The comparability of such studies with conventional studies conducted in a more classical experimental setting has been proven several times in different scientific disciplines. See for example Horton et al. [7], Berinsky et al. [8] and Paolacci et al. [9]. Mturk allows to define a profile for the inclusion of workers, so that individuals can only participate if they match certain characteristics. This option was used to exclude individuals who had already participated in one of our process acceptance studies. The intention was to exclude the possibility of being influenced by information from previous studies. All other persons could participate in the study.

Potential interfering variables, such as age, gender, origin, level of education, etc., are controlled by means of a block randomization of the participants. For this purpose, the necessary sample size is divided into equally sized blocks, to which each expression of the independent variable is assigned once. This allows both, random assignment of the expressions and increases the chance of equal distribution [10, 11]. Randomness is implemented in our case using the pseudorandom number generator of the Java class Random() [12]. To eliminate expectancy effects and other psychological confounders, a double-blind study is conducted. The participants do not know which group they are in or how many and which groups exist. On the other hand, the experimenter is represented by a server and thus an artificial entity that cannot transmit expectations and emotions [10, 13].

In addition to such mechanisms, further controls must be implemented in online-based experiments to prevent unintentional behavior during the study. These include conscious manipulation as well as unconscious behavior resulting in biased results. For example, crowdsourcing platforms can be exploited by individuals to monetarily enrich themselves quickly and without much effort by performing the available tasks not personally, but through software bots. These perform the tasks in an automated manner and in a way that generates quick rewards. From the researcher's point of view, bots generate useless data, as for instance questionnaires are filled with random answers. To prevent such automated approaches, a CAPTCHA is used to differentiate between humans and machines [14], as recommended by as Mason and Suri [15]. Within this study, the CAPTCHA takes the form of selecting one of three possible process models that matches a textual description. Each failed attempt is documented in the meta-data.

Furthermore, there is always the tendency that participants do not conscientiously fulfill the set task. This leads to unconsidered answers, which in turn appear as outliers in the statistical evaluation. In addition, some participants only skim the instructions or answer randomly. Overall, this leads to a reduction in the test strength of the experimental study. As a countermeasure, an instructional manipulation check (IMC) is used [16]. This methodological tool can be used to identify whether participants in a study fully read and understand the instructions, questions, and response options given. The IMC consists of a longer introductory text and a question with given answer options. The question itself is designed to be simple and based on the process previously performed, so that each participant can solve it without further problems. The actual test and thus the test for attention is realized by an instruction embedded in the text. Only following the instruction, in this case clicking on the title of the test, leads to the successful completion of the IMC. Participants who do not follow this instruction and instead simply answer the question fail the test and can repeat it. This forces participants to fully comprehend the information given [16]. Again, each failed attempt is documented in the meta-data of the experiment.

3 Setup of the Online Experiment

3.1 Operationalization of the Dependent Variable Process Acceptance

Within the theory of process acceptance developed by Müllerleile, the term process acceptance is defined as follows, considering the general definition of acceptance according to Lucke [17]: Process acceptance is the affirmative positive basic attitude of the process participants towards a process. This attitude results from an inner, rational or motivational-emotional conviction, by which the process is approved and recognized in its entirety. This conviction leads to a behavior which corresponds to an unchanged execution of the given process [5]. The measurement construct associated with the theory was developed using Rossiter's C-OAR-SE methodology [18] and grouped into the three dimensions of attitude acceptance: Cognitive, Affective, and Conative [19]. The first dimension includes the attitude towards an object based on available knowledge and reason. The affective dimension captures motivational-emotional feelings and the conative refers to the inner willingness to perform an action even if it ultimately does not have to be performed [5]. Each of the three dimensions comprises a set of question-answer pairs that has been operationalized using the DLF IIST binary approach [18]. Accordingly, the participant can only choose between the answer options "Yes" and "No". He/she is therefore forced to give an answer. Table 1 shows an example of the cognitive dimension of the questionnaire used here[1]. The answers are effect coded according to their loading [21]. In case of a positive loading, "Yes" is coded with 1 and "No" with −1. For a negative loading, vice versa accordingly. In addition to these questions, each dimension contains a five-point bipolar Likert scale coded in the interval [−2; 2]. In the case of our experiment, for the cognitive dimension accordingly:

[Overall I reject this way of purchasing material] −2 −1 0 1 2 [Overall I approve this way of purchasing material]

To form the three acceptance indicators, the coded responses per dimension are aggregated, which are then included in the statistical evaluations.

Table 1. Cognitive dimension of the measurement construct (following Müllerleile [5])

Question	Scale	Loading
I think this way of managing purchases is useful	Yes/No	+
I think this way of managing purchases is important	Yes/No	+
I think this way of managing purchases is complicated	Yes/No	−
I think this way of managing purchases is laborious	Yes/No	−
I think this way of managing purchases takes a long time	Yes/No	−
I think I'm happy with the result	Yes/No	+
I think this way of managing purchases is decent	Yes/No	+

3.2 Definition and Operationalization of the Independent Variable

As a result of his qualitative research, Müllerleile hypothesized that process participants feel better when a process is able to heal itself when an error occurs [5]. From this,

[1] Due to space limitations, the other two dimensions are not shown. These can be found in [20].

the following hypothesis can be derived: If a process implements self-healing mechanisms that handle a process error, the acceptance of the process increases in the cognitive/affective/conative dimension, compared to a process without such an implementation.[2] Before an operationalization of the independent variables can be performed, the two theoretical constructs process error and self-healing must be defined.

In the literature, the term process error is often associated with process quality. This is justified by the measurement of quality, among other things, on the basis of a faultless process execution and output [22, 23]. According to the DIN EN ISO 9000:2015-11 standard, quality measures the extent to which a set of inherent characteristics of an object fulfills previously defined requirements. Conversely, a failure to meet such requirements is considered an error [24].

The term self-healing originates from the field of biology and describes the ability of a biological system to carry out a healing autonomously [25]. For example, fractures of bones are automatically repaired by the system, the living organism [26]. This paradigm is now to be transferred to the process acceptance domain. In this case, the object to be healed would be the acceptance object and thus the process with all its elements. In self-healing it is assumed that the initiation and the healing must be carried out by the process itself. Since acceptance originates from the process participants, they are classified within the theory as acceptance subjects [4]. Thus, they are not directly involved in the healing.

In summary, the process must have entities that are capable of self-healing. This includes, for instance, IT systems utilized to execute parts of the process. In the IT domain, systems are considered self-healing if they can rescue themselves from an unintended state and return to the intended normal state [25]. What is actually meant by a deviation from the normal state depends on the considered IT system and is not always clearly definable. The transition from "normal" to "not normal" is often fuzzy, but nevertheless the system must be able to detect it. Accordingly, it is also necessary to define thresholds, such as an error tolerance, at which the system should intervene [23, 25].

This is summarized in the following working definition: In the context of process acceptance research, self-healing is understood to mean that an artificial entity returns a process instance to the state prior to the occurrence of a process error, so that the desired end result is achieved with the desired quality, whereby a previously defined error tolerance determines the intervention of the entity. A process error is understood as any deviation from the requirements to be met by the process, whereby the quality of the process is measured.

After defining all theoretical constructs, it is necessary to operationalize the independent variable self-healing in case of a process error. Since the effect of a potential self-healing on acceptance is to be investigated, it is necessary to first define the error occurring in the process. For this purpose, the objective of the process and, derived from it, its requirements must be defined in accordance with the working definition. A generic purchasing process (cf. Sect. 3.3) serves as the basis for the experiment. The objective of

[2] Since the difference between the hypotheses concerns only the dimensions of attitude, they are combined here in one hypothesis for simplicity. Each dimension is statistically evaluated individually.

the process is to generate a purchase order for materials and send it to the supplier. Thus, the process closes with the event of sending the order. One requirement resulting from it is it that the used procurement system must send the order correctly. Not fulfilling this requirement corresponds to a process error that results in missing the central objective of the process. A corresponding self-healing of the process takes place by the underlying IT system eliminating the error of the corrupted communication link and accordingly restoring the previous process state. This means the order form including all entries and approvals is reconstructed automatically. The participant only has to resend the order. This manual action is necessary so that the participant notices that self-healing has taken place. If the sending would also be included into the healing, he would not notice it. In the control group of the experiment, in which no self-healing happens, the process instance terminates in fail-mode after the error occurs. In this case the participant is forced to start a new process instance.

The chosen type of self-healing or the algorithmic component of the form reconstruction seems very banal and ubiquitous in systems at the first moment. However, it is sufficient to test the influence of self-healing on process acceptance. For the respective participant, self-healing means that the communication problem has been solved, so that the objective of sending the purchase order could be achieved. A point of additional interest here for us was to investigate whether a standard procedure for self-healing, common in many IT systems today, really raises the acceptance of the respective process.

3.3 Process Design

A generic purchasing process as is common in enterprises will serve as the basis for the experiment (cf. Fig. 1). This workflow is comparable to a standard purchase-to-pay process in an SAP ERP system. To enable participants without work experience in purchasing departments to get along with this process, its basic activities and visualization are similar to classic e-commerce systems like Amazon. The purpose behind this is to achieve the highest possible generalizability of the results [13].

Before the actual process starts, the participant is introduced to the study. Participants are in the role of an employee of the purchase department and have accordingly the task to procure materials. They receive all necessary information to be able to carry out the process (in its respective variant) correctly. After this introduction, the participant must pass the mentioned CAPTCHA. If this is passed successfully, the actual process starts, which runs as follows:

First, a specialized department creates a so-called purchase requisition (PR), which contains all materials to be ordered. In the experiment, this function is performed by the server. It generates a PR and sends it to the participant so that it appears in his cockpit. Each line there represents a PR and contains information about the requester, the required material, the quantity, the recommended supplier and the recommended payment method.

Each of these requirements must now be converted by the participant into a purchase order. For this purpose, the requested materials are to be selected from a catalog of possible materials, the corresponding quantities are to be determined and the supplier and payment method are to be chosen. Subsequently, it is necessary to request a release from the purchasing manager, who in the experiment is again represented by the server and thus by an algorithm. The release takes place after a pre-specified time, which is identical for each participant, to avoid any variance. After release the functionality of the send button is enabled and the order can be sent by the participant.

Then, to apply and examine the self-healing factor, a sending error is issued. In group 1, where no self-healing is available in the process, the participant is led back to the purchasing cockpit, must create the order again from scratch and have it released. In group 2, participants are directly returned to the order, which is still filled with the previous entries and is still released. The process can therefore automatically repair the error and restore its status before the error occurred. The second attempt of sending out the purchase order is confirmed successfully in both groups.

Finally, the mentioned IMC is presented to the participant before the questionnaire for measuring acceptance is displayed.

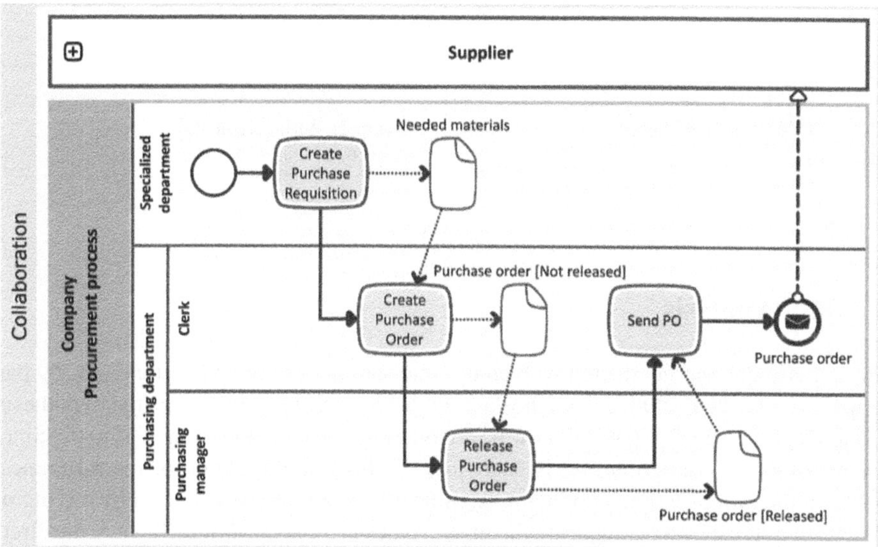

Fig. 1. Purchasing process used in the experiment

3.4 Experimental Design and Calculation of the Sample

Based on the variation of a single independent variable within the experiment, we rely on a single factor univariate design [13]. Before the experimental design can be set up, it must be determined how the independent variable is to be expressed. The basic purpose of this experiment is to determine whether self-healing capabilities of a process after a process failure lead to a change in process acceptance. As was already mentioned above, two groups are formed to fit the difference hypothesis. One group receives a process with these capabilities while the other group does not. This leads to the formation of a simple two-group plan (cf. Table 2).

Within the two-group experimental plan, a two-sided t-test can be used to identify a significant difference between the groups [21]. To calculate the optimal sample size, the use of this test and the fulfilment of all its requirements is now postulated. To optimize sampling, further parameters must be accounted for, in particular the required effect strength. Since this field of research is still relatively unexplored and it is not possible to determine the probable effect strength from the results of previous studies, an effect strength of $d = 0.2$ (low effect) according to Cohen [27] is deemed sufficient here [28]. In planning the selectivity of the t-test, established standard values are employed. Alpha is set to 0.05 and power to 0.80. This leads to an optimal sample size of 788 participants. Ideally, these should be divided equally among the groups (cf. Table 2). With this information the necessary server architecture and the experiment could be planned.

Table 2. Experimental plan of this study

Self-healing process capabilities	
Group 1: Without self-healing $n_1 = 394$	Group 2: With self-healing $n_2 = 394$

4 Data Analysis

The experiment was conducted with an optimal sample size of 788 participants. 69 participants had to be excluded from the data set due to missing data. To test the hypothesis, a statistical significance test will now be conducted for each dimension of attitude. Since the hypothesis is directional, it is necessary to check the direction of the differences between the groups before conducting significance tests. This is achieved by relying on descriptive statistics by comparing the mean values [13]. As shown in Table 3, the direction contradicts the hypothesis since the group without self-healing has a higher mean acceptance for the process than the other group. In practice this means that participants, on the occurrence of the process error, preferred to cancel the process and start again from scratch, because they accept manual intervention more than automated process self-healing. Usually, when designing or optimizing a process, it is intuitively assumed that automation is a better way to correct a process failure than manual intervention, especially because it is quicker. But according to the data collected and based on the selected process, the direction of action assumed from the theory is false. However, the results can still be useful for improvements in process management or process optimization. Consequently, further statistical analysis was carried out to check whether these differences are significant and what effect strength they have.

Since there are two groups within the experiment and the samples are independent from each other, an independent t-test shall be used for significance testing. For this purpose, the corresponding prerequisites must be checked first. The normal distribution prerequisite can be skipped due to a sample size >30 and the use of a two-sided test,

Table 3. Mean effects on the process acceptance per dimension

Group	Cognitive	Affective	Conative
Without self-healing	5.5763	6.3503	3.9124
With self-healing	4.7808	5.9808	3.4877

because then the t-test is considered robust against violations [29–31]. The required variance homogeneity can be checked with a Levene test. Table 4 shows the results. In the cognitive dimension the difference is significant. This means that the null hypothesis is rejected and thus the variances are classified as not homogeneous. In the other two dimensions, however, variance homogeneity can be assumed [30].

Table 4. Levene test, calculated in IBM SPSS v27

Dimension	F	Sig.
Cognitive	5.629	0.018
Affective	1.405	0.236
Conative	0.534	0.465

In case of missing homogeneity, a Welch test can be used instead of the t-test [32]. In consequence, a Welch test is calculated for the cognitive dimension and a t-test for the affective and conative dimensions.

Table 5. Significance tests per dimension, calculated in IBM SPSS v27

Dimension	Test	T	Df	Sig.
Cognitive	Welch-test	2.644	704.064	.008
Affective	t-test	0.902	717	.367
Conative	t-test	1.268	717	.205

As can be seen in Table 5, a statistically significant difference can only be found in the cognitive dimension ($p < .05$). For additional validation, a Mann-Whitney-U test is applied to the data as a non-parametric test. There the significance value is .028 and thus also leads to the rejection of the null hypothesis. It can be stated that the participants who experience a process without self-healing abilities have a significantly higher value of process acceptance within the cognitive dimension than the participants of the other group. The calculated effect strength is $d = .197$ according to Cohen [27] and can therefore be classified as a small effect. For the final evaluation of the converse hypothesis 'If a process implements self-healing mechanisms that handle a process error, the acceptance of the process decreases in the cognitive dimension, compared to a process without such an implementation', we use the paradigm of Westermann and Hager [33]. Therefore, the effect sizes achieved must be related to the average effect size of such studies and evaluated against this. Since there are no comparable studies in this field of research, a medium effect (d = .5) according to Cohen [27] is used to evaluate the hypothesis. This effect strength occurs most frequently in a broad cross-section of psychological studies [10]. Overall, the hypothesis of the cognitive dimension can be assumed to be limitedly proven [33].

5 Conclusion and Discussion

5.1 Discussion of the Results

As the statistical analysis shows, a significant difference in process acceptance can only be found in the cognitive dimension between the two test groups. This dimension measures acceptance with questions that capture the participant's attitude based on individual knowledge or ideas about the object to be judged. The object is a process that can or cannot compensate for the occurrence of an emerging error by restoring all entered data and triggered actions, which includes the release of the order already issued. The results of the tests indicate that self-healing is not automatically more accepted than repeated manual effort. This contradicts previous assumptions based on theory, where we assumed that participants would prefer to have less effort or to achieve the lowest possible process throughput time. The reasons for this apparently counterintuitive participant behavior in the case of self-healing are still to be analyzed in detail. First investigations suggest that the participants either do not trust an algorithm that performs reconstructions or that they believe their individual input caused the error. It should be noted that the failure was basically indicated (failure in sending procedure), but no exact description of the failure or an indication of the possible reasons. Therefore, it is possible that the participants search for the cause of the failure in their own actions. In addition, after reconstruction in the process variant with self-healing, the inputs could no longer be changed due to the order release that had already been issued. If participants viewed their data input as the source of error without an option to change this data after the error occurred, this may be the reason for the counterintuitive acceptance results in the cognitive dimension. Thus, this study may have identified a new influencing factor of process acceptance, the individually felt loss of control over the process.

5.2 Limitations of the Study

Within this study a laboratory experiment has been conducted that exposes the participants to an artificially created environment or situation. This is associated with limitations that must be considered. To improve the generalizability of the results, a generic purchasing process was chosen, which was derived from a similar process in the broadly used SAP ERP system. Furthermore, the process and the visualization are comparable to purchasing processes of classical e-commerce systems in the private sector. There are also some advantages of this type of investigation compared to more classical field studies in organizations. These include a large sample size, good traceability of the results, and appropriate mechanisms to avoid manipulations and control interfering variables [13].

5.3 Implications for Further Research

The results presented here refer to a special process (purchasing). While this is indeed a very common process in enterprises, the investigations carried out here should be extended to other types of business processes in order to better generalize results. Furthermore, the results should be validated by field experiments.

Based on the assumption that the participants' loss of control had an influence on the acceptance of process self-healing, another experiment is currently implemented, again resorting to the platform MTurk. Here, the participants will be given full control on the data after order reconstruction (self-healing).

Of further interest is a closer examination of socio-demographic variables, which were considered as interfering in this study and whose influence was eliminated by block randomization. Nevertheless, these were recorded in the questionnaire as well as the acceptance data. Thus, an analysis is possible and planned. For this purpose, hierarchical regression will be used. The analysis should provide information on whether, for example, age, gender, or the current country of residence had an influence on the process acceptance of a particular person. This will also help to fine-tune future experiments using digital process implementations and MTurk.

References

1. Becker, J., Kahn, D.: Der Prozess im Fokus. In: Becker, J., Kugeler, M., Rosemann, M. (eds.) Prozessmanagement, pp. 3–16. Springer, Heidelberg (2005). https://doi.org/10.1007/3-540-27153-8_1
2. Hammer, M., Champy, J.: Reengineering the Corporation. A Manifesto for Business Revolution. Collins Business Essentials, New York (2006)
3. Müllerleile, T., Nissen, V.: When processes alienate customers. Towards a theory of process acceptance. In: van der Aalst, W., et al. (eds.) S-BPM ONE - Scientific Research. Lecture Notes in Business Information Processing, vol. 170, pp. 171–180. Springer, Cham (2014). https://doi.org/10.1007/978-3-319-06065-1_13
4. Müllerleile, T., Ritter, S., Englisch, L., Nissen, V., Joenssen, D.W.: The influence of process acceptance on bpm. an empirical investigation. In: Aveiro, D., Caetano, A. (eds.) 2015 IEEE 17th Conference on Business Informatics (CBI), pp. 125–132. IEEE, Piscataway (2015)
5. Müllerleile, T.: Prozessakzeptanz. Theoretische und empirische Untersuchung der Akzeptanz und Ablehnung betrieblicher Prozesse. 1st edn. Springer Gabler, Wiesbaden (2019). https://doi.org/10.1007/978-3-658-27103-9
6. Militz, M.: Strategien zur Eingrenzung multiresistenter Erreger im Krankenhaus. Trauma Berufskrankh **10**, 140–145 (2008)
7. Horton, J.J., Rand, D.G., Zeckhauser, R.J.: The online laboratory: conducting experiments in a real labor market. Exp. Econ. **14**, 399–425 (2011)
8. Berinsky, A.J., Huber, G.A., Lenz, G.S.: Evaluating online labor markets for experimental research Amazon.com's mechanical turk. Polit. Anal. **20**, 351–368 (2012)
9. Paolacci, G., Chandler, J., Panagiotis, I.G.: Running experiments on Amazon mechanical turk. Judgm. Decis. Mak. **2010**, 411–419 (2010)
10. Sedlmeier, P., Renkewitz, F.: Forschungsmethoden und Statistik für Psychologen und Sozialwissenschaftler. Für Psychologen und Sozialwissenschaftler. 3rd edn. Pearson, Hallbergmoos (2018)

11. Muche, R., Rohlmann, F., Büchele, G., Gaus, W.: Randomisierung in klinischen Studien in der Rehabilitationsforschung: Grundlagen und praktische Aspekte. Die Rehabil. **41**, 311–319 (2002)
12. Oracle: Random (Java Platform SE 8). https://docs.oracle.com/javase/8/docs/api/java/util/Random.html#nextInt. Accessed 24 Mar 2021
13. Döring, N., Bortz, J.: Forschungsmethoden und Evaluation in den Sozial - und Humanwissenschaften, 5th edn. Springer, Heidelberg (2016). https://doi.org/10.1007/978-3-642-41089-5
14. von Ahn, L., Blum, M., Langford, J.: Telling humans and computers apart automatically. Commun. ACM **47**, 57–60 (2004)
15. Mason, W., Suri, S.: Conducting behavioral research on Amazon's mechanical turk. Behav. Res. Methods **44**, 1–23 (2012)
16. Oppenheimer, D.M., Meyvis, T., Davidenko, N.: Instructional manipulation checks. Detecting satisficing to increase statistical power. J. Exp. Soc. Psychol. **45**(4), 867–872 (2009)
17. Lucke, D.: Akzeptanz. Legitimität in der "Abstimmungsgesellschaft". VS Verlag für Sozialwissenschaften, Wiesbaden, s.l. (1995)
18. Rossiter, J.R.: Measurement for the Social Sciences. The C-OAR-SE Method and Why It Must Replace Psychometrics. Springer, New York (2011). https://doi.org/10.1007/978-1-4419-7158-6
19. Müller-Böling, D., Müller, M.: Akzeptanzfaktoren der Bürokommunikation. Oldenbourg, München (1986)
20. Drewes, L., Nissen, V.: The effect of process length on process acceptance. In: Reussner, R.H., Koziolek, A., Heinrich, R. (eds.) 50. Jahrestagung der Gesellschaft für Informatik INFORMATIK 2020. Back to the Future. Lecture Notes in Informatics (LNI), vol. 307, pp. 635–648 (2021)
21. Bortz, J., Schuster, C.: Statistik für Human- und Sozialwissenschaftler. Extras online, 7th edn. Springer, Heidelberg (2016)
22. Atzert, S.: Strategisches Prozesscontrolling. Koordinationsorientierte Konzeption auf der Basis von Beiträgen zur theoretischen Fundierung von strategischem Prozessmanagement. Gabler Verlag/Springer Fachmedien Wiesbaden GmbH Wiesbaden, Gabler, Wiesbaden (2011)
23. Christ, J.P.: Intelligentes Prozessmanagement. Marktanteile ausbauen, Qualität steigern, Kosten reduzieren. Springer Gabler, Wiesbaden (2015)
24. Deutsches Institut für Normung e. V. (DIN): Quality management systems. Fundamentals and vocabulary, vol. 01.040.03; 03.120.10. Beuth Verlag, Berlin (2015)
25. Ghosh, D., Sharman, R., Raghav Rao, H., Upadhyaya, S.: Self-healing systems—survey and synthesis. Decis. Support Syst. **42**, 2164–2185 (2007)
26. Hager, M.D., Greil, P., Leyens, C., van der Zwaag, S., Schubert, U.S.: Self-healing materials. Adv. Mater. **22**, 5424–5430 (2010)
27. Cohen, J.: Statistical Power Analysis for the Behavioral Sciences, 2nd edn. Erlbaum, Hillsdale (1988)
28. Hussy, W., Jain, A.: Experimentelle Hypothesenprüfung in der Psychologie. Hogrefe Verlag für Psychologie, Göttingen (2002)
29. Boneau, C.A.: The effects of violations of assumptions underlying the test. Psychol. Bull. **57**, 49–64 (1960)
30. Coolican, H.: Research Methods and Statistics in Psychology, 6th edn. Psychology Press, London (2014)
31. Sawilowsky, S.S., Blair, R.C.: A more realistic look at the robustness and type II error properties of the t test to departures from population normality. Psychol. Bull. **111**, 352–360 (1992)

32. Ruxton, G.D.: The unequal variance t-test is an underused alternative to student's t-test and the Mann-Whitney U test. Behav. Ecol. **17**, 688–690 (2006)
33. Westermann, R., Hager, W.: Entscheidung über statistische und wissenschaftliche Hypothesen: Zur Differenzierung und Systematisierung der Beziehungen. Zeitschrift für Sozialpsychol. **13**, 13–21 (1982)

PACAs: Process-Aware Conversational Agents

Luis Fernando Lins[1(✉)], Glaucia Melo[2], Toacy Oliveira[1], Paulo Alencar[2], and Donald Cowan[2]

[1] Federal University of Rio de Janeiro, Rio de Janeiro, Brazil
luisfernandolins@poli.ufrj.br, toacy@cos.ufrj.br
[2] Computer Systems Group, University of Waterloo, Waterloo, Canada
{gmelo,palencar,dcowan}@uwaterloo.ca

Abstract. Processes are an essential concept in modern society, promoting the standardization, documentation, and control of complex interactions between businesses, governments, and individuals. Therefore, it is essential to provide approaches and tools that support the participants during the execution of these processes. However, current solutions for process guidance are usually proprietary, difficult to follow, and not effective in supporting these participants in their daily processes. In this paper, we present an approach for building a novel type of conversational agents – which we call PACAs – that can communicate with workflow management systems, using the BPMN standard, to support the execution of a process. We also describe a use case in which a PACA is implemented by using Rasa – a chatbot framework – and Camunda Engine – a workflow management system. We believe that promoting the integration of processes and conversational agents is key to simplifying process execution and, thus, making business processes more widespread and prevalent in the academy and industry.

Keywords: Business process · Conversational agent · Workflow · Chatbot · Camunda · Rasa

1 Introduction

Processes are pervasive in personal and organizational settings. They can support a wide variety of tasks, ranging from scheduling a trip [8] to hiring new employees or admitting students into a university program [4]. For this reason, organizations are increasingly relying on automated processes to improve efficiency and accuracy in the execution of their tasks.

In many cases, processes require people to perform tasks manually. These processes are known as human workflows [2]. Although mechanisms have been proposed to visualize, control, and manage processes, performing tasks in a process involving numerous human decisions can be more complicated than it seems. First, tasks may have either a strict or a loose order, and this might leave participants confused in complex scenarios. Second, dependencies between tasks can

A. Marrella and B. Weber (Eds.): BPM 2021 Workshops, LNBIP 436, pp. 312–318, 2022.
https://doi.org/10.1007/978-3-030-94343-1_24

be challenging to track, in which case, not being aware of the information that is required *a priori*, people may start executing tasks they are not able to finish properly. Third, processes that rely heavily on humans may differ based on their context or require access to ancillary documentation so that errors are avoided [3]. Owing to these challenges, process participants should have a solution to guide them through process instances.

In recent years, a new technology called "conversational agents" (CAs) [5] has been gaining increased attention. CAs are programs that assist in human-computer interactions (HCI) [6] through spoken and written language. Although CAs are flexible enough to support several types of HCI scenarios, they typically lack business process concepts, i.e., they do not consider the context of a process instance when interacting with users.

In this paper, we investigate an approach for integrating process concepts into conversational agents. This integration leads to agents of a novel type, which we call process-aware conversational agents (PACAs). The paper is structured as follows. Section 2 presents related work on process management in conversational agents. Section 3 provides an overview of our proposed solution and introduces a guide that can help with the development of PACAs. Section 4 describes how a PACA can be used in a real-life scenario involving trip planning. Finally, Sect. 5 presents concluding remarks and future work.

2 Related Work

A few conversational agents that assist with tasks have been developed by the academic community, and several process management tools are available in the industry. In 2017, Cranshaw et al. built and evaluated an app called Calendar.help [1] that uses the concept of process workflows for assisting with scheduling and rescheduling meetings. In 2018, Toxtli et al. created a tool called TaskBot [7] with the goal of helping teams to complete assignments. The authors intended to reduce the onus of context switching by using the chatbot to assign tasks to team members and remind them to start or finish their tasks. These solutions, although valuable, use an empiric process notation, lacking the compatibility that a universal modeling standard, such as BPMN, can provide. Furthermore, these related works do not provide an easy integration with any business process management (BPM) tools.

In recent years, there has been a rise in the number of BPM tools available in the industry. Tools such as Kissflow[1], Process Street[2], and Camunda[3] were developed to help businesses and their employees track the execution of essential and recurring processes. Nevertheless, when business processes become too large, employees can make mistakes during their execution [3], and these process management tools, owing to their complexity, may not adequately help

[1] https://kissflow.com/.
[2] https://www.process.st/.
[3] https://www.camunda.com/.

these workers. Therefore, the tool that we present in this paper aims to complement BPM tools – in this case, Camunda – to make them more user-friendly by leveraging natural language understanding via chatbots.

3 Proposed Solution

This section describes our approach and its implementation, offering an overview of how PACAs are configured and used, and presenting the outline of a guide for generating PACAs.

PACAs are developed by integrating chatbots with a Workflow Management System (WFMS) to control the execution of a BPMN process with natural language. Our proposed solution is based on Rasa, an open-source chatbot framework, and a WFMS called Camunda Engine. We have decided to use Camunda Engine because it is an open-source framework that is well documented and widely used both in academia and industry. In our proposed solution, Rasa will perform Natural Language Understanding (NLU) to determine what the user is saying. Then, Rasa Core, the framework's back end, will decide which action the chatbot should take next. The Core will call Camunda Engine's API to know which tasks are available, present them to the user in natural language, and, once the user tells the chatbot that a task is complete, the Core will complete that task in Camunda.

Once the chatbot is running, a user can communicate with it through the command line or API requests. It can also be run on Rasa X, a default GUI provided by the Rasa framework, or connected to Facebook Messenger, Slack, Telegram, and other messaging services.

We have split the life cycle of a PACA into two phases: configuration and usage. The former is shown in Fig. 1, while the latter is represented in Fig. 2. During the configuration phase, the process analyst creates a process model in Camunda Modeler[4], makes necessary modifications to the model (Fig. 1.1), and deploys the process model to Camunda Engine (Fig. 1.2). Then, the developer will examine this BPMN model (Fig. 1.3) and convert process modeling elements from the BPMN diagram into Rasa files (Fig. 1.4). Finally, as soon as the Rasa Framework and Camunda Engine are running (Fig. 1.5), Rasa will be able to communicate with Camunda Engine, giving rise to a fully functioning Process-Aware Conversational Agent (Fig. 1.6).

With an operational PACA, the user can then use it to navigate the process. Whenever a user says they want to start a process (Fig. 2.1), the PACA will start a process instance in Camunda Engine (Fig. 2.2) and then provide a list of all the available tasks for the user (Fig. 2.3). The user will then execute one of the available tasks (Fig. 2.4) and tell PACA when they have finished that task (Fig. 2.5). PACAs can also request task-related information if necessary. If that is the case, after the task is completed, the chatbot can request those pieces of information (Fig. 2.6), which will then be provided by the user (Fig. 2.7). Finally,

[4] https://camunda.com/products/camunda-platform/modeler/.

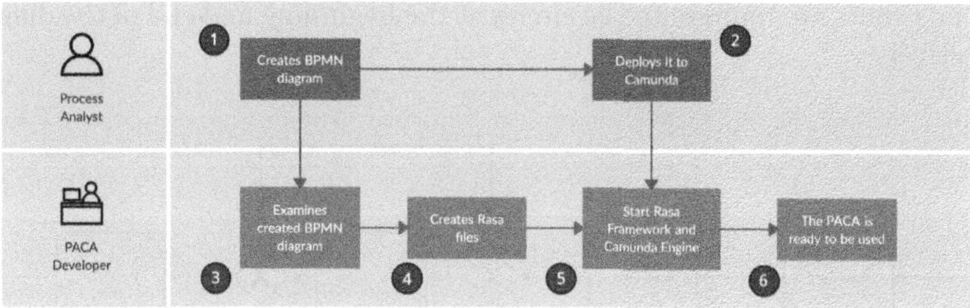

Fig. 1. Overview of the necessary steps to configure a PACA.

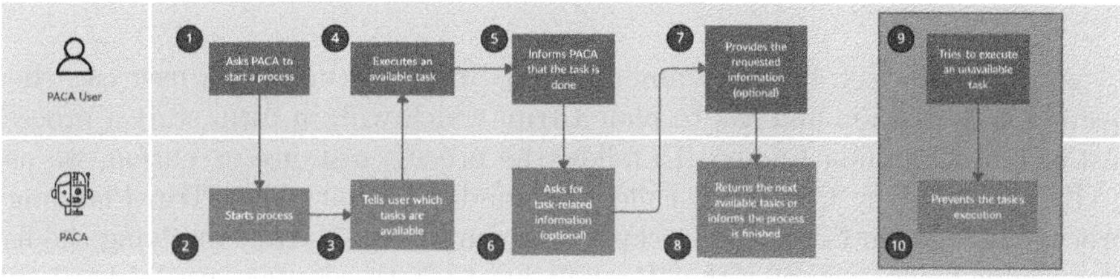

Fig. 2. Overview of the necessary steps to use a PACA.

the PACA will reply with the next available task(s) (Fig. 2.8) – in which case, the flow would go back to step 5 – or with a message indicating that the process is finished if the execution has reached the end event. In an exception flow, if the user tries to execute an unavailable task (Fig. 2.9), PACA will prevent that task's execution (Fig. 2.10).

To assist in the development of PACAs, we have created a PACA Generation Guide, which should be used by the developers and process analysts who are working to create PACAs. The guide consists of multiple steps and substeps, but in this paper, we will only mention the main ones. They are: (i) Preparing the BPMN; (ii) Building the default Rasa chatbot; (iii) Adding process-specific pieces to the Rasa chatbot; and (iv) Running Camunda and the chatbot. The complete description of each of these steps is available at http://www.repositorio.poli.ufrj.br/monografias/projpoli10033716.pdf.

4 Use Case: The Trip Planning PACA

We will now explore a PACA that was built for a "Trip Planning" scenario. This scenario serves as an excellent example because the process is well known and has dependencies between tasks. The code for the "Trip Planning" PACA that was developed is available on GitHub[5]. Figure 3 presents the diagram for this process. It includes four user tasks (represented as rectangles), process gateways (represented as diamonds), and sequence flows (represented as arrows). The start

[5] https://github.com/luis-f-lins/process-aware-conversational-agent.

and end events are represented as circles at the beginning and end of the diagram, respectively.

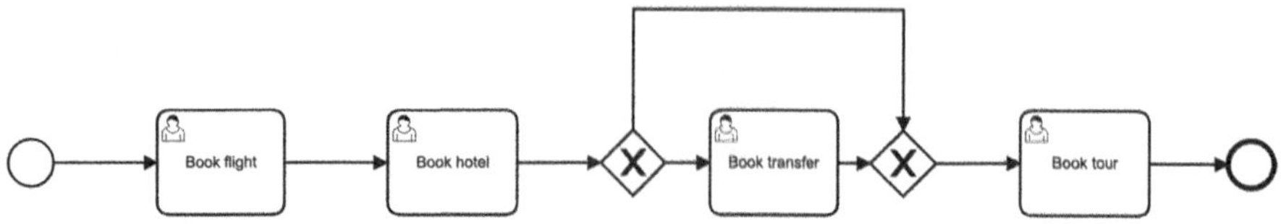

Fig. 3. BPMN diagram for the "Trip Planning" process.

As shown in Fig. 4, after Camunda and Rasa are running, a user can then begin a conversation and ask to plan a trip, which will, in turn, start a process instance in Camunda Engine. To follow the process instance execution, we use a GUI provided by Camunda, called Camunda Cockpit. The "Trip Planning" process instance in Camunda Cockpit is shown in Fig. 5. After receiving the list of available tasks, including the "Book flight" task, the user can proceed to book the flight and inform the chatbot that the task is finished. In this case, before replying with the list of the next available tasks, the chatbot will ask the user for a required piece of information, which is the flight date. After the user provides the flight date, this piece of information will be stored in Camunda Engine, as shown on the list at the bottom of Fig. 5. After all the required information is provided and all tasks are executed, the PACA informs the user that the process is finished.

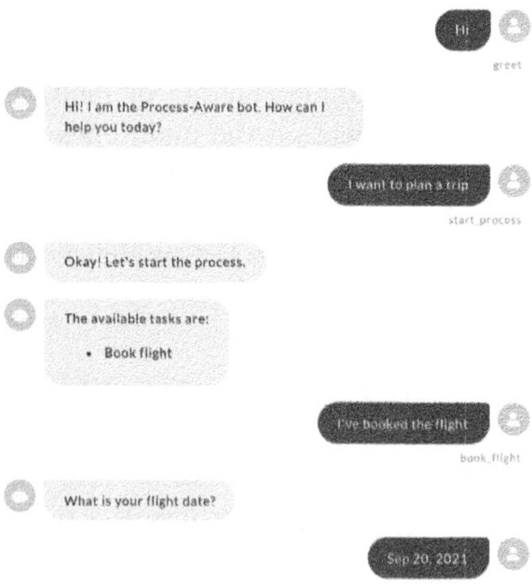

Fig. 4. Conversation with a PACA in a "Trip Planning" scenario.

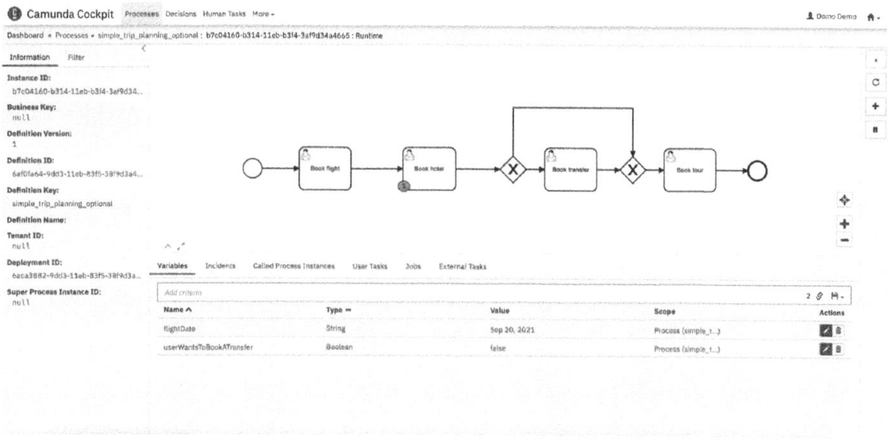

Fig. 5. Process instance shown in Camunda Cockpit after the "Book flight" task.

5 Conclusion and Future Work

In this paper, we have presented an approach and tool that embeds business process models into conversational agents. Besides having the potential to help users navigate their day-to-day processes, we believe this approach is a first step towards establishing a more conversational, chatbot-based strategy for supporting process execution. Future work might involve implementing more process examples as chatbots and a formal evaluation of the proposal, which might be based on user studies [1]. Other possible future endeavors include creating PACAs that support different processes and automatically recognize the process that the user wants to start, training RASA models automatically after parsing a BPMN process, and associating process mining methods with the PACAs.

References

1. Cranshaw, J., et al.: Calendar.help: designing a workflow-based scheduling agent with humans in the loop. In: Proceedings of the 2017 CHI Conference on Human Factors in Computing Systems, pp. 2382–2393 (2017)
2. Dang, J., Toklu, C., Hampel, K., Enke, U.: Human workflows via document-driven process choreography. In: 2008 International MCETECH Conference on e-Technologies (MCETECH 2008), pp. 25–33. IEEE (2008)
3. Dijkman, R., La Rosa, M., Reijers, H.: Managing large collections of business process models-current techniques and challenges. Comput. Ind. **63**(2), 91–97 (2012)
4. Dumas, M.: Fundamentals of Business Process Management. Springer, Berlin (2018). https://doi.org/10.1007/978-3-662-56509-4
5. Gnewuch, U., Morana, S., Maedche, A.: Towards designing cooperative and social conversational agents for customer service. In: ICIS (2017)
6. Helander, M.G.: Handbook of Human-Computer Interaction. Elsevier, Amsterdam (2014)

7. Toxtli, C., Monroy-Hernández, A., Cranshaw, J.: Understanding chatbot-mediated task management. In: Proceedings of the 2018 CHI Conference on Human Factors in Computing Systems, pp. 1–6 (2018)
8. Weske, M.: Business Process Management: Concepts, Languages, Architectures. Springer, Berlin, New York (2012). https://doi.org/10.1007/978-3-642-28616-2

Intelligent Robots in Business Processes: A Students' Perspective

Amy Van Looy(✉), Gilles Bauwens, and Krzysztof Zaniewski

Faculty of Economics and Business Administration, Department of Business Informatics and Operations Management, Ghent University, Tweekerkenstraat 2, 9000 Ghent, Belgium
Amy.VanLooy@UGent.be

Abstract. The fourth industrial revolution urges companies to incorporate technologies such as artificial intelligence and robotics in their business processes. The success of emerging technologies to innovate business processes, however, largely depends on the acceptance by employees, e.g., as being the end users for human-robot collaboration. The current paper addresses a yet under investigated technology, namely robotics with more advanced levels of intelligence. The existing literature about intelligent robots is still scarce and mainly limited to applications in particular sectors. Since intelligent robot applications will increase in the medium term, our purpose is to add an interdisciplinary perspective by examining the degree to which acceptance differs among master students as highly-skilled future employees. Our survey shows that gender and academic discipline only have a limited impact to predict the acceptance of intelligent robots at work, but different adoption profiles seem to matter most for robotic process changes. Finally, we discuss the role of students, universities and employees to better prepare for the future digitalization of the workplace.

Keywords: Intelligent robots · Acceptance · Students · Future employees

1 Introduction

Due to the continuing digitalization, professional life is changing tremendously [11]. The COVID-19 pandemic only accelerated this evolution [2, 16]. The fourth industrial revolution has started, which is characterized by an advanced use of artificial intelligence (AI), big data and robotics [10]. Since engineers and programmers are making constant progress in the capabilities of robots, future employees will have to deal with a totally different workplace. This also accounts for higher-skilled work.

The body of knowledge that focusses on the implications of robotics for business processes is mostly limited to machine-based robots [5], whereas literature on intelligent robotics primarily focuses on social robots [4]. We take a broader and multidisciplinary perspective by looking at intelligent robots, which are machinery systems with comprehensive improvements in perception, decision-making and performance, and which can simulate human behaviors, emotions and thinking [19]. We focus on the acceptance of those robotic process changes by future higher-skilled employees because of employees'

A. Marrella and B. Weber (Eds.): BPM 2021 Workshops, LNBIP 436, pp. 319–331, 2022.
https://doi.org/10.1007/978-3-030-94343-1_25

crucial role for a successful adoption of human-robot cooperation [20]. We narrow down to master students, who most likely have to collaborate with intelligent robots and are considered as higher-skilled based on the latest trends.

Our research question is: **"To which degree do differences exist among master students to accept intelligent robots for business processes?"** We performed a quantitative research in one West-European university, where we intend to find out how master students (i.e., as future higher-skilled employees) feel about intelligent robotics in their work practices. The choice for a single case is motivated by controlling some contextual variables that are outside our scope (e.g., culture). As master students will soon enter the professional market, it is important for employers to understand their attitudes and behavior towards using intelligent robots, and how this will affect the performance of business processes. Additionally, it is relevant for educational institutions to gain more insight into the factors relevant for students to accept intelligent robots in a digital economy. We include students from different backgrounds (i.e., alpha, beta and gamma) for obtaining a wide spectrum of opinions.

In Sect. 2, we analyze existing literature [8]. Section 3 explains our survey design. We follow an established theory on IT acceptance to provide clear and repeatable results [18]. We investigate sub groups among master students to uncover differences in acceptance with ANOVA-based tests (Sect. 4). Section 5 shows implications for companies, educational institutions and students. We conclude in Sect. 6.

2 Research Background

2.1 Defining Intelligent Robots

The literature generally distinguishes between five robot types. Each type comes with its own level of intelligence, purpose and set of prominent features [7]. The first type are *industrial robots*, which are automatically controlled, (re)programmable robots with most applications in assembly, order picking, welding, etc. Only some industrial robots have intelligent capabilities to adjust their operations. The key benefits of these robots are robustness, speed and precision. Secondly, *service robots* are mostly used in private and public services. They are characterized by a higher degree of intelligence, human-robot interaction and clever mobility. Different service robots exist, with humanoid robots being one of the most commonly researched robots (Sect. 2.2). An example of humanoid robots is SAR (Socially Assistive Robot) which can complete complex physical tasks with a social interface for convincing a user [1]. Thirdly, *ubiquitous robots* [9] rely on small networked portable computer devices. In contrast to service robots, their interface and service do not have spatial limitations so they provide seamless interfaces. The fourth type are *genetic robots*, which are genome-based evolutionary robots characterized by a progressive growth mechanism to provide continuous optimization to cope with changing conditions. The last type are *bio robots*, which are often perceived as typical sci-fi robots for offering a more extensive set of applications (e.g., prosthetic limbs or artificial hearts) [7]. For our study, all robots types are considered as long as there is a sufficient level of intelligence for conducting cognitive tasks.

2.2 Existing Studies About Students' Acceptance of Intelligent Robots

We started with a Systematic Literature Review [8] to get a comprehensive overview of the related gaps. We therefore set up inclusion and exclusion criteria to filter the search results, as explained in Table 1.

Table 1. SLR protocol.

Research question guiding this review	What is the state of research on students' acceptance for using intelligent robots in business processes (i.e., for their future work practices)?
Searched sources	Web of Science, Scopus, Springer, IEEE
Search terms	1/ "adoption", "acceptance"
	2/ "intellig* robot*", "AI", "artificial intelligence", "machine learning"
	3/ "student*", "graduate*", "adolescen*", "millennial*"
Search strategy	Peer reviewed journals and conference papers Search terms contained in articles' title, abstract or keywords
Inclusion criteria	Written in English, until September 2020 Main focus on robot acceptance and use for professional use Should deal with intelligent robots
Exclusion criteria	Should not deal with AI variants in other disciplines Should not use of robots for educational, teaching purposes Should not use machine-based robots (e.g., warehouse, surgery tools)
Quality criteria	Only peer-reviewed studies

Based on our SLR filtering, we started with over 3,404 papers to end with 25 papers about students' acceptance of intelligent robots in business processes. Overall, the literature is still relatively scarce, with four research gaps. A first gap is related to the sector distribution. The found literature mostly focuses on education, human health and social work (72% of the papers), while other sampled studies have no concrete business application involved (4 papers). Other papers are only weakly linked to students, although this target group will certainly have to work with intelligent robotics. Consequently, considering students from various backgrounds can add new information to the body of knowledge concerning robotic process changes.

The second gap deals with the research methods. It turns out that experiments are the most popular analysis method, with 14 out of 25 papers in our SLR. Survey information can broaden the scope towards multiple scientific disciplines.

Furthermore, starting from a well-known IT acceptance theory will make our results easier to understand and simpler to reproduce by others. Our SLR sample covers eight papers using the UTAUT theory. However, most of those are not strictly focusing on students or checking the influence of other context variables than ours. The present study, therefore, aims to discover the differences with respect to gender, academic discipline and innovation adoption profile (Sect. 3.2).

Finally, the SLR reveals that most research focusses on humanoid robotics (11 papers), although intelligent robotics can come in various forms (Sect. 2.1). We thus include a wider scope of robots. Given these four gaps, our study will be a valuable addition to extend the body of knowledge.

3 Research Method

Our survey was based on the unified theory of acceptance and use of technology (UTAUT). We combined UTAUT1 [17] and UTAUT2 [18], which we translated to the acceptance of intelligent robots by master students (i.e., for human-robot collaboration in their future work). The UTAUT theory is widespread, which was also confirmed in our SLR (i.e., mentioned by eight out of 25 papers), and which makes our research easier to understand. Moreover, UTAUT makes our research reproducible to confirm our findings in other regions and/or a larger student sample.

3.1 Selection of Respondents

Data was gathered between December 2020 and mid-March 2021. We targeted master students of all faculties belonging to one West-European university in order to be able to control some contextual variables (e.g., culture, government policies and national labor market) that could bias a comparative basis between the academic disciplines of alpha, beta and gamma faculties [12]. Due to the COVID-19 pandemic, almost all teaching activities happened online and we were unable to collect the data physically [16]. Instead, the student representatives were contacted and asked to distribute the survey within their faculty. This resulted in 207 respondents, of which 55 were deleted because they completed less than half of the survey (i.e., resulting in a profound quality check). Although having some gender bias ($n = 152$, male = 53, female = 99), the master students were spread among all faculties and disciplines (Table 2), covering almost all sectors. The average survey duration was 17 min.

3.2 Variables

In addition to the UTAUT acceptance factors, three context factors were included for our comparative basis: *gender, academic discipline* and *innovation adoption experience* [13]. Gender was directly proposed by UTAUT. We extended UTAUT by looking at differences among academic disciplines, as an alternative to sector. Instead of differentiating between voluntary or mandatory use, we predicted attitude and behavioral intention by the respondents' personal innovation adoption profiles [13]. Table 3 shows that all innovation profiles were covered along the disciplines.

Table 2. Profile based on the academic discipline and faculty (N = 152).

Discipline	Frequency	%	Faculty	Frequency	%
Alpha	63	41.4%	Arts and philosophy	5	3.3%
			Law and criminology	21	13.8%
			Economics and business administration	22	14.5%
			Psychology and educational science	1	0.7%
			Political and social science	14	9.2%
Beta	5	34.2%	Science	37	24.3%
			Engineering and architecture	9	5.9%
			Bioscience engineering	6	3.9%
Gamma	37	24.3%	Medicine and health science	33	21.7%
			Veterinary medicine	0	0%
			Pharmaceutical sciences	4	2.6%

Table 3. Cross tabulation of innovation adoption by academic discipline (N = 152).

Diffusion levels	Alpha	Beta	Gamma	Total
Innovators	3	2	1	6
Early adopters	9	11	4	24
Early majority	19	22	13	54
Late majority	25	13	16	54
Laggards	7	4	3	14

The two dependent variables were *attitude towards using intelligent robots* and *behavioral intention to use intelligent robots* in business processes. The acceptance factors were measured by 47 statements on a 5-point Likert scale (1 = strongly disagree; 5 = strongly agree), as literally taken from UTAUT. We conducted factor analyses to create the latent factors. After re-arranging the statement sequence per factor and reversing two statements (i.e., from a negative to a positive formulation), we decided to exclude *price value* ($\alpha = 0.53$) and *human facilitation* ($\alpha = 0.63$). We initially expected to exclude price because it is difficult to predict future technology prices. Regarding human versus non-human facilitation, students preferred consulting FAQs and fora rather than human interactions. By excluding these two factors, we obtained our research model for analysis with a total of 10 acceptance factors (Fig. 1).

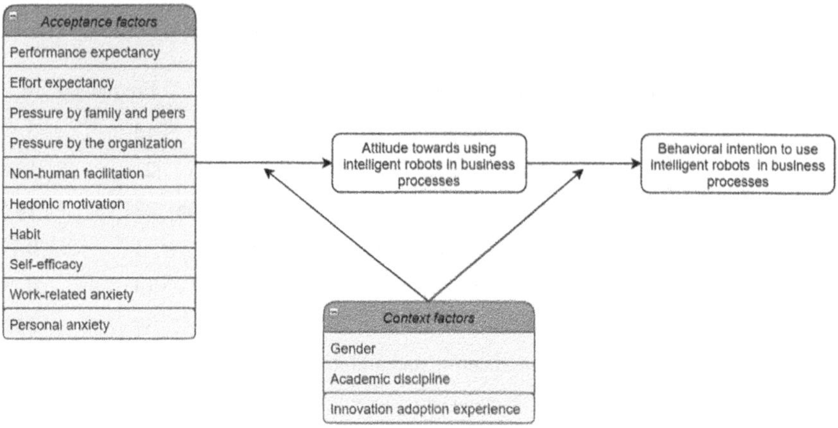

Fig. 1. Initial research model.

3.3 ANOVA-Based Statistical Tests

The tests for statistical differences were based on null- and alternative hypotheses.

- H_0: no difference exists among the categories of a contextual factor regarding attitude towards using intelligent robots and behavioral intention, respectively.
- H_a: at least one significant difference exists among the categories regarding attitude towards using intelligent robots and behavioral intention, respectively.

To select the appropriate ANOVA-based tests, we first assessed the ANOVA assumptions along with our relatively small sample size. Each sub sample was independent from the others as master students belonged to one academic discipline. The same goes about gender and adoption profile. Finally, the outcome and acceptance variables were continuous, obtained after factor analyses. We started with checking the normality assumption in our outcome and acceptance variables. Following the Kolmogorov-Smirnov and the Shapiro-Wilk tests, we found that our outcome variables were not normally distributed. Similarly, we found that only two acceptance variables were normally distributed (i.e., habit and personal anxiety). The eight other acceptance variables were not normally distributed. Due to a small sample size, we could not use the central limit theorem. Furthermore, we used the Levene's test to check the homogeneity of variances of the outcome variables. When considering gender and academic discipline, we could not reject the null-hypothesis of equal variances. However, when considering innovation adoption profiles, we found different variances for the outcome variable 'behavioral intention to use intelligent robots' on a 5% significance level. We used the same method to check homogeneity of variance for the acceptance variables. Again, when considering gender and academic discipline, we did not find different variances. When considering innovation adoption profiles, we found one acceptance variable that did not have equal variances on a 5% significance level, namely effort expectancy.

Consequently, we conducted Welch's ANOVA tests [3], which are parametric tests but known for being reliable when scale variables are non-normally distributed under the condition that each group has more than 15 observations. When considering academic disciplines, Welch's was accompanied by the parametric Games-Howell post-hoc test, which does not assume equal sample sizes or normal distributions, and can be performed for equal and unequal variances [6, 15]. When considering gender (i.e., two categories), post-hoc testing was not required because differences were directly derived from ANOVA. Given the small number of innovation adoption profiles, we turned to the Kruskal-Wallis H-test because some subsamples had less than 15 observations. For pairwise comparisons, we adjusted the significance values by a Bonferroni correction.

3.4 Evaluation Criteria

External validity remained limited because all respondents were master students from a single university. This deliberate choice gave us control for a number of contextual variables (e.g., culture, government policies and national labor market). Although our sample size was relatively limited, credibility was addressed by including the established theoretical underpinnings of UTAUT 1 and UTAUT 2, with validated statements. As all Cronbach alphas of the constructs were higher than or equal to 0.7, we achieved high levels of internal consistency. Moreover, the UTAUT theory [17, 18] contributed to content validity. Furthermore, the results from our factor analyses provided us with a high level of construct validity. Another check for consistency was implemented by a control question to check the outcome variables (i.e., attitude and behavioral intention to use). Correlations with the control variable were highly significant for both outcome variables ($P = 0.000$), showing internal consistency and data reliability. For the attitude variable, the correlation coefficients indicated a moderate-to-strong linear relationship with the control variable (Pearson's $r = 0.798$; Spearman's rho $= 0.752$; Kendall's tau_b $= 0.628$; $P = 0.000$). We observed a moderate linear relationship for the control variable and the behavioral intention variable (Pearson's $r = 0.623$; Spearman's rho $=$ 625; Kendall's tau_b $= 0.508$; $P = 0.000$).

4 Results

We report on IT familiarity, professional expectations and ANOVA-based tests.

4.1 IT Familiarity

We asked our respondents about their awareness of certain IT-related terms and behaviors. All respondents turned out to have a smartphone during several years already, except for seven respondents. Almost all master students were also well acquainted with using websites or apps in daily life, especially social media, streaming of music and TV and online newspapers. Average experience was present for using online traffic data, sharing documents in the cloud, and online games.

In Table 4, we present students' IT awareness of new IT terms related to Industry 4.0. Interestingly, we found low awareness of the more complicated terms (e.g., Industry

Table 4. Awareness of new IT terms (N = 152).

Have you ever heard this item?	No	Via education	Via news, magazines	Via hobbies	Via friends, family	Other
Industry 4.0	127	21	5	0	2	2
Business intelligence	100	36	19	3	11	2
Process mining	60	47	34	11	27	5
Machine learning	35	81	46	14	36	7
Smart technology	25	56	90	9	34	11
Robotics	13	55	91	11	48	9
Digitalization	4	108	118	29	77	6
AI	4	100	111	25	78	10
Algorithms	1	120	74	20	53	13
Automation	1	111	106	25	79	7

4.0, business intelligence and process mining). Easier keywords related to IT (e.g., automation and digitalization) were known via education, news and magazines. For the more technical terms (e.g., algorithms and artificial intelligence,), education was key to spread awareness of new IT developments. We will look more closely into the related implications for education in the discussion, Sect. 5.

4.2 Professional Expectations

We also asked our respondents about their perceived meaningfulness of future work in order to gain insight into their professional expectations (Table 5). We observed that future employees especially appreciate exercising skills and self-development as the most important aspects of a job, closely followed by pursuing a purpose. On the other spectrum, we found that autonomy and money were perceived as the least important aspects for their future workplace. Social relationships were only ranked with average importance, which is a remarkable observation from the perspective of intelligent robotics and human-robot collaboration as well. Similarly, the aspect of self-esteem and recognition was almost evenly spread between the importance levels, with a light skew to being more important. Such professional expectations are valuable to consider by employers for employees to feel comfortable with their job. More reflections will follow in the discussion section.

4.3 ANOVA-Based Results for Academic Discipline

Regarding the acceptance of intelligent human-robot collaboration, our tests showed no differences in the outcome variables between students from different academic disciplines ($p = 0.606$ and $p = 0.295$). However, we observed differences for effort expectancy

Table 5. Ranking job expectations to do meaningful work (N = 124).

What aspects do you find most important in your future job?	Most or very important	Rather (less) important	Very less or least important
Pursuing a purpose	52	38	34
Social relationships	27	57	40
Exercising skills and self-development	68	41	15
Self-esteem and recognition	48	42	34
Autonomy	18	33	73
Earning money (salary)	35	37	52

($p = 0.024$), non-human facilitation ($p = 000$), self-efficacy ($p = 0.042$) and work-related anxiety ($p = 0.000$). Via a Games-Howell post-hoc, we found differences between the alpha and gamma students for non-human facilitation, between alpha and beta students for work-related anxiety and between beta and gamma students for all post-hoc tested variables (Table 6).

Table 6. Games-Howell post-hoc testing for acceptance factors by discipline (N = 127).

Games-Howell	Alpha versus Beta	Alpha versus Gamma	Beta versus Gamma
Effort expectancy	NS	NS	0.592**
Non-human facilitation	NS	0.432**	0.725***
Self-efficacy	NS	NS	0.461**
Work-related anxiety	0.661***	NS	−0.711***

(NS $P > 0.100$; * $P < 0.100$; ** $P < 0.050$; *** $P < 0.001$)

4.4 ANOVA-Based Results for Gender

Based on a Welch's ANOVA, we did not find gender differences in attitude ($p = 0.907$) or in behavioral intention ($p = 0.466$). We also tested for gender differences in the acceptance variables, resulting in four statistically significant differences: effort expectancy ($p = 0.008$), non-human facilitation ($p = 0.016$), self-efficacy ($p = 0.000$) and work-related anxiety ($p = 0.000$). A post-hoc analysis was not needed due to the fact that there were only two gender groups.

4.5 ANOVA-Based Results for Innovation Adoption Profile

Based on the respondents' innovation adoption profiles [13], we conducted a Kruskal-Wallis ANOVA test, and observed major differences between the subsamples. Regarding the outcome variables, the statistical differences were between the most and least innovative groups (i.e., which are typically the first to use a new technology and those at the other side of this spectrum) (Table 7).

Table 7. Pairs statistically different for the outcome variables in the post-hoc analysis.

Attitude towards using intelligent robots	Behavioral intention to use intelligent robots
Laggards versus Early majority	Laggards versus Early majority
Laggards versus Early adopters	Laggards versus Early adopters
Laggards versus Innovators	Laggards versus Innovators
Late majority versus Early majority	Late majority versus Early adopters
Late majority versus Early adopters	Late majority versus Innovators
Late majority versus Innovators	

The Kruskal-Wallis test indicated that only one acceptance variable was not statistically different, namely pressure by family and peers ($p = 0.705$). Similarly, for all other acceptance variables, there was a gap between the most and least innovative users in technology. The precise pairs can be found in Table 8.

Table 8. Significant different pairs after post-hoc analysis of acceptance variables.

Performance expectancy • Laggards versus Innovators • Late majority versus Innovators **Effort expectancy** • Late majority versus Early adopters **Habit** • Late majority versus Innovators **Work-related anxiety** • Innovators versus Late majority • Innovators versus Laggards **Personal anxiety** • Early adopters versus Late majority • Early adopters versus Laggards	**Pressure by the organization** • Laggards versus Innovators • Late majority versus Innovators • Early adopters versus Innovators **Non-human facilitation** • Late majority versus Early adopters **Hedonic motivation** • Laggards versus Early adopters • Laggards versus Innovators • Late majority versus Early adopters • Late majority versus Innovators • Early majority versus Innovators **Self-efficacy** • Late majority versus Innovators

5 Discussion

This study has shown that the acceptance of intelligent robots in business processes depends on innovation adoption profiles rather than gender or discipline (Fig. 2).

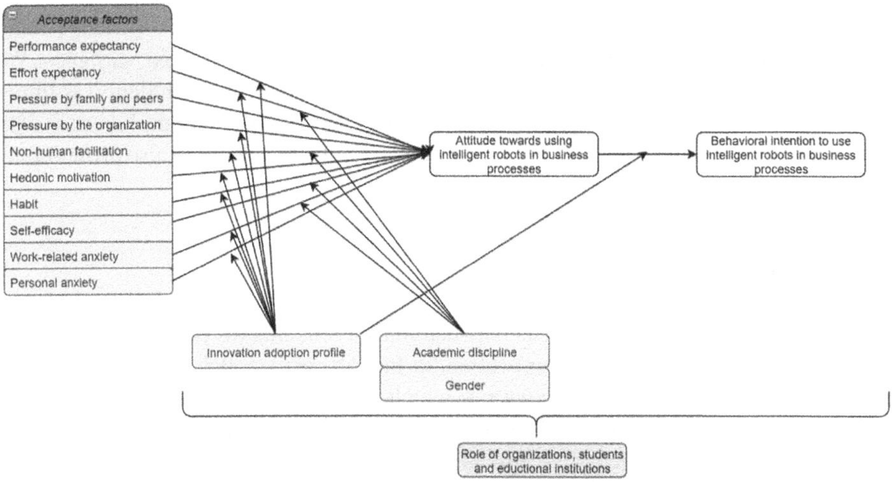

Fig. 2. Revised research model.

5.1 Implications Concerning the Role of Universities, Students and Employers

As stated in Sect. 4.1, the role of education is important in spreading awareness around new IT topics. As robotics advance, the curricula of universities and secondary schools should be reconsidered, and AI subjects should be more prominently presented to students of all disciplines to help them prepare for human-robot collaboration. Given the importance of skills and self-development for meaningful work (Sect. 4.2), the future work generation wishes to improve knowledge during studies and at work. Additionally, a mindset of lifelong learning will become key to keep up with developments of emerging IT. Finally, the Covid-19 crisis has urged companies to adopt new forms of working [16], and governments are encouraging employees to telework. Consequently, most companies are expanding their digital infrastructure to accommodate to such more advanced digital working environments, for which it is of utmost importance to provide sufficient information and guidance. Furthermore, the Covid-19 crisis has encouraged companies to rethink and redesign business processes [14, 16]. From our study, we can derive recommendations for employers such as the importance of supporting lifelong learning. Efficiently using technological resources means improving human capital via trainings. Next, we have uncovered that non-human facilitation plays a central role in the acceptance of intelligent robots in business processes. Especially helpdesks, manuals and FAQs help to improve efficacy.

5.2 Research Limitations

Our findings face two data collection limitations, starting with the relatively small sample size. First, the switch from offline to online classrooms due to COVID-19 has avoided us to physically attend all targeted courses, and we became dependent on the goodwill of students. However, using a profound theoretical background and following objective test parameters make our research reproducible. We are also working on a follow-up research with semi-structured interviews to gain deeper insights. Secondly, we acknowledge being limited to one West-European university for controlling certain context variables. We mainly intend to open the debate about the future role of universities, students (as future employees) and employers. Alternatively, the scope can include older employees as well to differentiate between digital native backgrounds. Future research can also differentiate between types of intelligent robots.

6 Conclusion

We have investigated the differences between master students (as highly-skilled future employees) for accepting intelligent robots in human-robot collaboration. This study is situated in the Industry 4.0 revolution, which is introducing advanced intelligence levels into robotics. A literature screening has provided scare evidence, being focused on specific sectors (e.g., education, human health, social work) and specific robots (e.g., humanoid and social robots). By adding context factors (i.e., gender, academic disciplines and innovation adoption profiles) and a novel target group (i.e., master students), our research has uncovered new knowledge based on a survey design. Master students still lack familiarity with relevant IT terms characterizing Industry 4.0, showing the crucial role that schools and universities have on IT adoption. Future employees mostly care about exercising skills for self-development and about pursuing a purpose to perceive meaningful work. Autonomy and money are typically experienced as less important for the future work generation. Such considerations can help employers for keeping their workforce satisfied. Although we have observed limited differences between alpha and other disciplines, more differences exist between beta and gamma students. Also the gender differences have been limited to four factors. Given that a person's innovation adoption profile is key, we advise to stimulate robot implementations into business processes by means of different stimuli.

Acknowledgement. With support of the faculty's student excellence program.

References

1. Abdi, J., Al-Hindawi, A., Ng, T., Vizcaychipi, M.P.: Scoping review on the use of socially assistive robot technology in elderly care. BMJ Open **8**, e018815 (2018)
2. Almeida, F., Santos, J.D., Monteiro, J.A.: The challenges and opportunities in the digitalization of companies in a post-COVID-19 world. IEEE Eng. Manage. Rev. **48**(3), 97–103 (2020)
3. Box, G.E.: Non-normality and tests on variances. Biometrika **40**(3–4), 318–335 (1953)
4. Gonnot, A., Michel, C., Marty, J.-C., Cordier, A.: Social robots in collaborative learning: consequences of the design on students' perception. In: 11th International Conference on Computer Supported Education (2019)
5. Kaan, H.L., Ho, K.Y.: Clinical adoption of robotics in endoscopy: challenges and solutions. JGH Open **4**(5), 790–794 (2020)
6. Keselman, H.J., Games, P.A., Rogan, J.C.: Protecting the overall rate of type I errors for pairwise comparisons with an omnibus test statistic. Psychol. Bull. **86**(4), 884–888 (1979)

7. Kim, J.-H., Choi, S.-H., Park, I.-W., Zaheer, S.A.: Intelligence technology for robots that think [application notes]. IEEE Comput. Intell. Mag. **8**(3), 70–84 (2013)
8. Kitchenham, B.: Guidelines for performing systematic literature reviews in software engineering (version 2.3). Keele University and University of Durham (2007)
9. Krumm, J.: Ubiquitous Computing Fundamentals. CRC Press, London (2009)
10. Lasi, H., Fettke, P., Kemper, H.-G., Feld, T., Hoffmann, M.: Industry 4.0. Bus. Inf. Syst. Eng. **6**(4), 239–242 (2014)
11. Parviainen, P., Kääriäinen, J., Tihinen, M., Teppola, S.: Tackling the digitalization challenge: how to benefit from. Int. J. Inf. Syst. Proj. Manag. **5**, 63–77 (2017)
12. Recker, J.: Scientific Research in Information Systems. Springer, Heidelberg (2013)
13. Rogers, E.M.: Diffusion of Innovations. Simon and Schuster, New York (2003)
14. Seetharaman, P.: Business models shifts: impact of Covid-19. Int. J. Inf. Manag. **54**, 102173 (2020)
15. Shingala, M.C., Rajyaguru, A.: Comparison of post hoc tests for unequal variance. Int. J. New Technol. Sci. Eng. **2**(5), 22–33 (2015)
16. Van Looy, A.: How the COVID-19 pandemic can stimulate more radical business process improvements: using the metaphor of a tree. Knowl. Process Manag. (2021). https://doi.org/10.1002/kpm.1659
17. Venkatesh, V., Morris, M.G., Davis, G.B., Davis, F.D.: User acceptance of information technology. MIS Q. **27**(3), 425–478 (2003)
18. Venkatesh, V., Thong, J.Y., Xu, X.: Consumer acceptance and use of information technology. MIS Q. **36**(1), 157–178 (2012)
19. Wang, T.-M., Tao, Y., Liu, H.: Current researches and future development trend of intelligent robot: a review. Int. J. Autom. Comput. **15**(5), 525–546 (2018). https://doi.org/10.1007/s11633-018-1115-1
20. Van Looy, A.: Adding intelligent robots to business processes: a dilemma analysis of employees' attitudes. In: Fahland, D., Ghidini, C., Becker, J., Dumas, M. (eds.) BPM 2020. LNCS, vol. 12168, pp. 435–452. Springer, Cham (2020). https://doi.org/10.1007/978-3-030-58666-9_25

9th International Workshop on DEClarative, DECision and Hybrid Approaches to Processes (DEC2H 2021)

9th International Workshop on DEClarative, DECision and Hybrid Approaches to Processes (DEC2H 2021)

The rules, contextual conditions, and expected outcomes that drive the conduction of processes are often implicit in process flows, process activities, or even tacit knowledge in the head of employees. However, the analysis and automation of decision processes should be made explicit in order to let automated systems analyse, support, or execute business process models. The declarative modeling paradigm aims to directly express the business rules or constraints underlying a given process. The paradigm has gained momentum, and in recent years several declarative notations have emerged, including a.o., Declare, Dynamic Condition Response (DCR) Graphs, Decision Modeling and Notation (DMN), and fragment-based Case Management (fCM). Recently, there has been a surge of interest in hybrid approaches, combining the strengths of declarative and procedural modeling paradigms.

In the workshop on DEClarative, DECision and Hybrid approaches to processes (DEC2H), we are interested in the application and challenges of decision- and rule-based modeling in all phases of the business process management lifecycle: identification, discovery, analysis, redesign, implementation, and monitoring.

DEC2H 2021 is the ninth edition of the workshop – formerly known as DeHMiMoP (Workshop on Declarative/Decision/Hybrid Mining and Modelling for Business Processes) till its sixth edition. DEC2H 2021 attracted five high-quality international submissions. Each paper was reviewed by at least three members of the Program Committee. Of all the submitted manuscripts, the top three were accepted for presentation.

Haarmann et al. present a novel model-driven approach for the management of knowledge-intensive processes with fCM. In particular, it is designed to let knowledge workers specify objectives in the form of first order logic statements expressing expected states, that is, conditions over data and activities. Goossens et al. describe an approach that resorts to deep-learning techniques to extract information about decision models from natural-language texts. In particular, it focuses on the inference of decision dependencies in the form of decision requirement diagrams and decision logic clauses synthesised as decision tables. Hildebrandt et al. formally define Timed DCR graphs with Data, a novel modeling language and notation for the declarative specification of processes. It extends classical DCR graphs by adding decision and input events, binding the execution to DMN-like conditional expressions and contextual data. The workshop concluded with the invited talk of Thomas T. Hildebrandt, entitled "Transparent, Explainable and Maintainable Digital Models of Decisions and Processes Regulated by Law". He illustrated the evolution of Timed DCR Graphs with Data through the real-world case studies offered by the EcoKnow project. The objective of the EcoKnow project is to provide technologies and methods for effective, co-created, and compliant adaptive case management for knowledge workers. During his talk, he showed also the process engine-as-a-service and tools to assist the mapping between textual procedures and laws to models.

We thank the authors for their noteworthy contributions and the members of the Program Committee for their invaluable help in the reviewing and discussion phase of the manuscripts. We hope that the reader will benefit from the reading of these papers to learn more on the latest advances in research about declarative, decision, and hybrid approaches to business process management.

September 2021

Søren Debois
Claudio Di
Ciccio Tijs
Slaats Jan Vanthienen

Organization

Workshop Chairs

Søren Debois	IT University of Copenhagen, Denmark
Claudio Di Ciccio	Sapienza University of Rome, Italy
Tijs Slaats	University of Copenhagen, Denmark
Jan Vanthienen	KU Leuven, Belgium

Program Committee

Rafael Accorsi	Accenture, Switzerland
Andrea Burattin	Technical University of Denmark, Denmark
Alessio Cecconi	WU Vienna, Austria
Carl Corea	University of Koblenz-Landau, Germany
João Costa Seco	Universidade Nova de Lisboa, Portugal
Massimiliano de Leoni	University of Padua, Italy
Riccardo De Masellis	Uppsala University, Sweden
Chiara Di Francescomarino	Fondazione Bruno Kessler, Italy
Rik Eshuis	Eindhoven University of Technology, The Netherlands
Robert Golan	DBmind Technologies Inc., USA
María Teresa Gómez-López	Universidad de Sevilla, Spain
Xunhua Guo	Tsinghua University, China
Amin Jalali	Stockholm University, Sweden
Dimitris Karagiannis	University of Vienna, Austria
Krzysztof Kluza	AGH University of Science and Technology, Poland
Fabrizio M. Maggi	Free University of Bozen-Bolzano, Italy
Artem Polyvyanyy	University of Melbourne, Australia
Flavia M. Santoro	Universidade do Estado do Rio de Janeiro, Brazil
Stefan Schönig	Universität Regensburg, Germany
Lucinéia H. Thom	Universidade Federal do Rio Grande do Sul, Brazil
Han van der Aa	Humboldt University of Berlin, Germany
Wil M. P. van der Aalst	RWTH Aachen University, Germany
Mathias Weske	Hasso Plattner Institute, University of Potsdam, Germany

Modeling Objectives of Knowledge Workers

Stephan Haarmann(✉), Anjo Seidel, and Mathias Weske

Hasso Plattner Institute, University of Potsdam,
Prof.-Dr.-Helmert-Str. 2-3, 14482 Potsdam, Germany
{stephan.haarmann,mathias.weske}@hpi.de, anjo.seidel@student.hpi.de

Abstract. Traditionally, business process models are imperative, limiting process behavior to a small set of variants. Purely imperative modeling languages are therefore unsuited for knowledge-intensive processes, which differ from case to case and are driven by knowledge workers. For this reason, declarative, data-centric, and hybrid approaches have been developed. However, in existing approaches, knowledge workers receive little support for late-goal modeling and subsequent planning of their actions. To address this issue, we present a model-driven approach that allows knowledge workers to specify objectives. Objectives are first order logic statements about a desired future execution state of a model's case, regarding data and activities. Those logic expressions are a prerequisite for automated planning.

Keywords: Case management · Goal modeling · Planning

1 Introduction

Knowledge-intensive processes are unrepeatable and unpredictable [3]. They are multi-variant and primarily driven by informed decisions of knowledge workers. It is completely up to the knowledge workers to make decisions and to align their actions with goals and constraints. While more structured processes can be specified in imperative process models, traditional modeling languages are unsuited for knowledge-intensive processes [3,21]. However, novel languages, such as declarative [9,17] and data-centric [4,10,12,20] approaches, address the needs of knowledge-intensive processes.

Once a model has been created, it can be implemented, e.g., by using process engines [23]. The engine tells users which activities they may perform in a given state. Due to the flexibility of knowledge-intensive processes, there can be many possible next activities, though. Furthermore, the relationship between a particular activity and a process objective may not be apparent. Little support is provided for modeling goals/objectives at run-time and planning ahead, albeit planning being an common knowledge-intensive task [18] and late goal modeling being an important prerequisite [3]. Therefore, we address the following research questions in this paper:

A. Marrella and B. Weber (Eds.): BPM 2021 Workshops, LNBIP 436, pp. 337–348, 2022.
https://doi.org/10.1007/978-3-030-94343-1_26

RQ1 What kind of objectives are within the scope of a process model/case?
RQ2 How can objectives be specified according to a model's scope?

Based on the fragment-based case management approach [8], this paper presents a framework for specifying objectives. The specification of objectives is the first step toward planning actions accordingly. Our approach is demonstrated by a proof of concept implementation.

The remainder of this paper is structured as follows: In Sect. 2, we briefly present the fragment-based case management approach and an example. Section 3 describes the framework for specifying objectives followed by an implementation (Sect. 4). We present related work in Sect. 5 before concluding our paper (Sect. 6).

2 Fragment-Based Case Management

The fragment-based case management approach (fCM) [8] divides a knowledge-intensive process into multiple fragments. Each fragment is highly structured and models imperative control flow and declarative data flow. The fragments of one process can be dynamically instantiated and combined at run-time constrained only by data requirements of activities. In case management, processes evolve around data [10]. In fCM, the data of a case is detailed in a data model, which includes classes, associations, and multiplicity constraints, and a state transition system for each class. Finally, each case model specifies a data-based termination condition, which specifies the goal of the case. fCM is a hybrid approach that combines data modeling with imperative as well as declarative process modeling.

In the remainder of this section, we present each element of fCM case models in more detail using an illustrative example. The data model describes the classes of data objects as well as binary associations among them. Associations must be existential and one-to-one or one-to-many. Consider the submission and reviewing phase at an academic conference. The conference is the central case class, which is instantiated once for each case. According to the data model (Fig. 1), each conference has up to 100 paper submissions. Each paper has up to five reviews and up to one decision, which is based on three to five reviews.

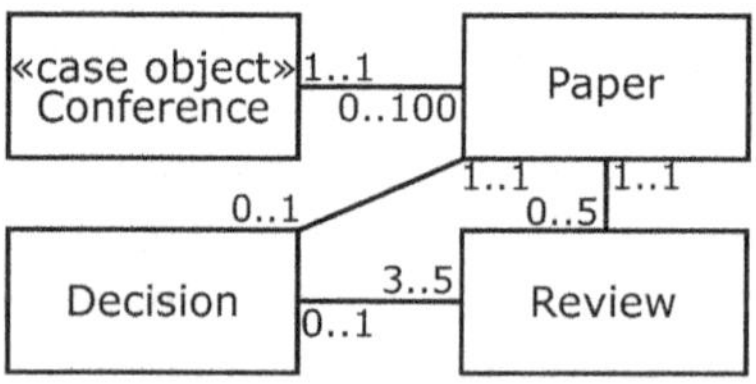

Fig. 1. Data model for submitting and reviewing papers at an academic conference.

For each class, we define a state transition system (see Fig. 2). Each data object belongs to a class and has a state. The state transition system defines

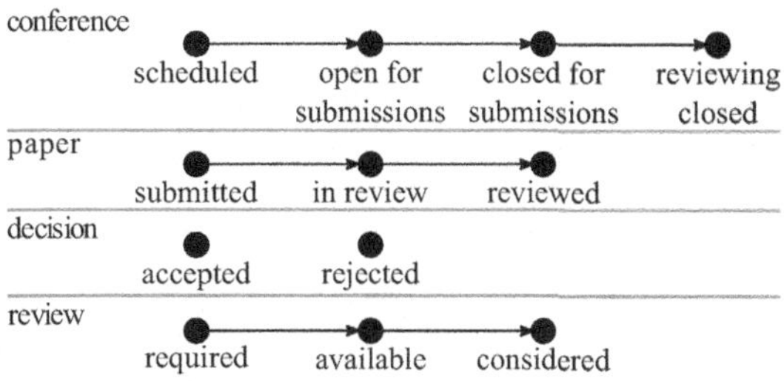

Fig. 2. State transition systems for the submission and reviewing example.

how this state can change. A paper, for example, can be in one of the states submitted, in review, and reviewed. The state may be changed from submitted to in review and from in review to reviewed. A decision, on the other hand, is either in state accepted or rejected and cannot be changed.

Fragments resemble small BPMN [15] process models. They are acyclic control flow graphs, which can contain start-events, activities, and data-based XOR-gateways. Furthermore, an activity has a set of input sets representing alternative data inputs and a set of output sets representing alternative data outputs. Fragments can begin with a start event that triggers a new case. They can also begin with an activity. In the latter case, they can be started by knowledge workers if the case is already running and if the data requirements of the activity are satisfied. Fragments can, in general, run repeatedly and concurrently. Within the scope of the domain model and the object behavior, new fragments can be added at run-time, e.g., to adapt to unforeseen situations.

The example consists of the fragments depicted in Fig. 3. Fragment f1 is the only one with a start event. It starts a new case and runs exactly once. Fragment f2 handles the submissions and is executed once for each submission. Fragment f3 assigns reviewers and fragment f4 creates reviews. Multiple instances of f3 and f4 can run concurrently. Finally, fragment f5 handles the decision.

The termination condition, i.e., the data condition specifying the goal of the case, is a predicate logic clause. The predicates are valid data object states. The termination condition of the example is conference[reviewing closed] and requires an object of class conference in state reviewing closed.

The semantics of a case model involve information from all the model's components. Here, we limit the discussion to a short walk through. The case begins once the conference is scheduled (f1). The conference is opened for submissions. Before the submission is closed, papers can be submitted (f2). Eventually, the submission is closed (f1). All papers are updated to state in review. Reviews are assigned (f3) and papers consequently reviewed (f4). As soon as all reviews that have been assigned to a paper have been created, the activity decide on paper is enabled (f5). If there are less than three reviews, decide on paper assigns an additional one. Otherwise, it may also reject or accept the paper and create a decision object in the respective state. If there are already five reviews, no additional review is allowed. Once all papers have been reviewed, the reviewing can be closed (f1). More details on fCM's semantics can be found in [5].

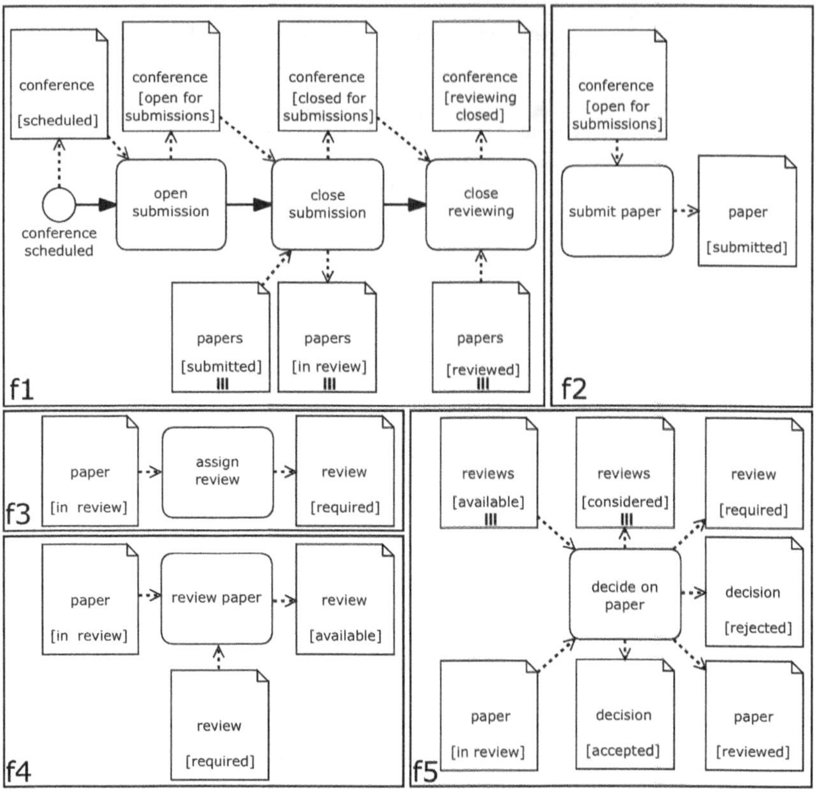

Fig. 3. Fragments defining the case behavior for submitting and reviewing papers.

Due to space limitation, we limit the example to a simple case model. A more detailed version of the example is provided with the implementation[1].

3 A Framework for Specifying Objectives

According to the Merriam-Webster dictionary, an objective is "something toward which effort is directed"[2]. We define an objective as a (partial) goal, which knowledge workers want to achieve during process execution. An objective can be case-specific and temporary. It is, therefore, often defined at run-time (cf. late goal modeling [3]).

Objectives can be very different in nature. Users may want to reach a certain state, execute a specific action, or satisfy timing constraints. Objectives can be linked to certain artifacts, e.g., a particular paper whose reviewing should be completed. In our model-driven approach, we are interested in objectives

[1] https://github.com/bptlab/fCM-query-generator (2021/05/17).
[2] https://www.merriam-webster.com/dictionary/objective (2021/04/27).

within the scope of a case model. The model contains classes, including states, associations, and multiplicity constraints, as well as activities that operate on data objects. The objective "*Make a decision for all papers*" is within the scope of the example case model. "*Set the score of paper 13 to 5*" is not because data operations on attributes are not defined by the model. In this section, we present a framework for defining objectives based on a case model. Therefore, we define a set of objectives $G_{M,I} = \{g_1, g_2, g_3, \dots\}$ for the sample case model M and a case I.

3.1 Data Objectives

Data objects in fCM belong to classes, they have states, and are linked to one another. Objects also have unique IDs, which allows distinguishing objects. This information can be used in objectives to make statements about the number of objects and their states, links, and identities.

Let ID be a set of object identifiers, S a set of data object states, and C a set of classes defined by a case model, then we define the data objects of a case by sets O and L as follows.

- O is a finite set of data objects, where each object $o \in O$ is a tuple $o = (id, c, s)$, such that $id \in ID, c \in C$, and $s \in S$.
- $L \subseteq \{\{o_1, o_2\} | o_1, o_2 \in O\}$ is a finite set of links. A link is an unordered pair of data objects.

The sets O and L change over time: When knowledge workers execute activities, they may change the state of objects, create new ones, or link them.

We use the set of objects and the set of links to specify objectives. In principle, all first order logic statements using these sets are possible. Here, we limit the presentation to a few generic and case specific examples that represent a range of different objectives. For the following objectives, knowledge workers may specify a class c, a state s, an ID id, and/or a multiplicity constraint with lower bound l and upper bound u.

- There exists an object of class c, $\exists(id', c', s') \in O : c' = c, id' \in ID, s' \in S$
- All objects of class c are in state s, $\forall(id', c', s') \in O : (c' = c) \implies (s' = s), id' \in ID$
- There exist at least l and at most u objects of class c, $l \leq |\{(id', c, s') \in O\}| \leq u$, $id' \in ID, s' \in S$
- There exists an object with the identity id in state s, $\exists(id', c', s') \in O : (id' = id) \wedge (s' = s), c' \in C$

Using these patterns, we can express the objective g_1 for the example: "*All papers should be in state reviewed*":

$$g_1 \equiv \forall(id', c', s') \in O : (c' = Paper) \Rightarrow (s' = reviewed), id' \in ID$$

So far, we did not consider the set of links. Using the information from set L, we can extend the already defined objectives and furthermore constrain the links of the objects. Therefore, let $c_l \in C$, $s_l \in S$, and $[l_l..u_l]$ be an additional class, state, and multiplicity constraint defined by the knowledge worker.

- There exist an object of class c linked to an object of class c_l, $\exists\{(id', c', s'), (id'', c'', s'')\} \in L : (c' = c) \wedge (c'' = c_l), id', id'' \in ID, s', s'' \in S$
- All objects of class c are linked to an object of class c_l, $\forall o_1 = (id', c, s') \in O, \exists o_2 = (id'', c_l, s'') \in O : \{o_1, o_2\} \in L, id', id'' \in ID, s', s'' \in S$
- There is an object of class c, and all linked objects of class c_l have state s_l, $\exists o_1 = (id', c, s') \in O, \forall\{o_1, (id'', c_l, s'')\} \in L : s'' = s_l, id', id'' \in ID, s' \in S$
- All objects of type c have at least l_l and at most u_l links to objects of class c_l, $\forall o_1 = (id', c, s') \in O : l_l \leq |\{\{o_1, (id'', c_l, s'')\} \in L\}| \leq u_l, id', id'' \in ID, s', s'' \in S$

Using these patterns, we can formulate two objectives g_2 and g_3 for the example fCM model to express the following goal. "*All papers should be linked to one decision and three to four reviews*":

$$g_2 \equiv \forall o_p = (id_p, Paper, s_p) \in O, \exists o_d(id_d, Decision, s_d) \in O : \{o_p, o_d\} \in L,\\ id_p, id_d \in ID, s_p, s_d \in S$$
$$g_3 \equiv \forall o_p = (id_p, Paper, s_p) \in O : 3 \leq |\{\{o_p, (id_r, Review, s_r)\} \in L\}| \leq 4,\\ id_p, id_r \in ID, s_p, s_r \in S$$

3.2 Activity Objectives

Objectives can involve activities. An activity instance can be in one of the states initial, control flow enabled, data flow enabled, enabled, running, and terminated (Fig. 4) [8]. Activities read and write data objects. Let $\mathcal{A}$ be the set of activity names in the case model and $S_\mathcal{A}$ be the set of activity states, we define the activity instances by a set A and the functions *read*, *write*, and *io*:

- $A \subseteq \mathcal{A} \times S_\mathcal{A}$ is a finite set describing activity instances. Since fragments can run repeatedly and concurrently, a single tuple may represent multiple activity instances, and multiple tuples with the same name may be in A.
- $read : A \rightarrow \mathcal{P}(\mathcal{P}(O))$ is a function assigning each tuple in A a set of input sets containing data objects. A single activity instance reads one input set when it is started. For example, if a conference has five papers p_1, p_2, p_3, p_4, and p_5 in state in review, then each of these papers is a possible input to assign review:

$$read(assign\ review, enabled) = \{\{p_1\}, \{p_2\}, \{p_3\}, \{p_4\}, \{p_5\}\}$$

The activity can be executed once for each paper.

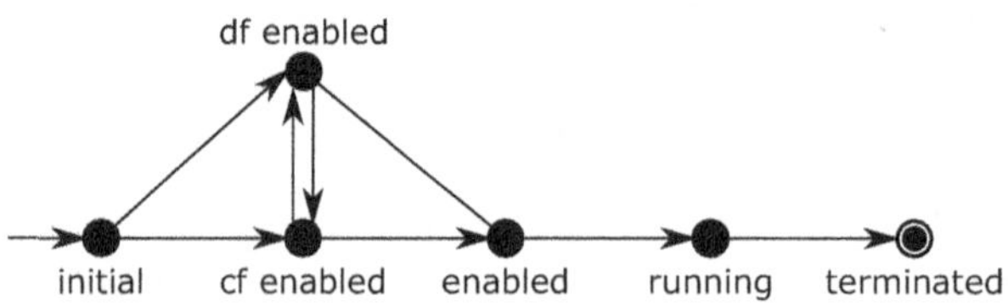

Fig. 4. Life cycle of an activity. The state of an activity can be used in an objective.

- $write : A \rightarrow \mathcal{P}(\mathcal{P}(O))$ is a function assigning each tuple in A a set of output sets containing data objects and is defined analogously to *read*. An activity instance writes one output set when it termiantes.
- $io : A \rightarrow \mathcal{P}(\mathcal{P}(O)) \times \mathcal{P}(\mathcal{P}(O))$ is a function that assigns each tuple in A a set of input-output combinations. For each $a \in A$ holds that $io(a) \subseteq read(a) \times write(a)$.

This information can also be used to define objectives. An activity may be required to be in a particular state. It may further be required to read or write certain objects or to instantiate a certain class. Let $\alpha \in \mathcal{A}$ be an activity name, $s_\alpha \in S_\mathcal{A}$ a state in the activity life cycle, $c \in C$ a class, $id \in ID$ an object ID, and $s \in S$ a data state.

- An instance of activity α must be in state s_α, $\exists(\alpha, s_\alpha) \in A$
- An instance of activity α that reads an object of class c must be in state s_α, $\exists a = (\alpha, s_\alpha) \in A : \exists R \in read(a) \wedge \exists(id', c, s') \in R, id' \in ID, s' \in S$
- An instance of activity α that writes an object with ID id in state s is in state s_α, $\exists a = (\alpha, s_\alpha) \in A : \exists W \in write(a) \wedge \exists(id, c', s) \in W, c' \in C$
- An instance of α in state s_α that creates a new instance of class c exists, $\exists a = (\alpha, s_\alpha) \in A, \exists(R, W) \in io(a) : \exists(id', c, s') \in W \wedge \nexists(id', c, s'') \in R, id' \in ID, s', s'' \in S$

Consider the objective g_4, "*Decide on paper should be enabled for paper 13*":

$$q_4 \equiv \exists a = (decide\ on\ paper, enabled) \in A, \exists R \in read(a) : (13, Paper, s') \in R, s' \in S$$

Objects that are read and written may also be defined more precisely using links.

3.3 Composed Objectives

Objectives can be combined using logical connections (e.g., $\wedge$, $\vee$, and $\neg$). Since the sets O, L, A and the functions $read, write, io$ change over time, objectives can also be temporally ordered. We allow defining a partial temporal order between objectives. If two objectives are ordered, one has to be achieved before the other.

Consider the following objective for the conference example (see Sec. 2). "*A decision for all papers should be made based on 3 to 4 reviews. Furthermore, at least 25 papers should be accepted, and activity close reviewing should be enabled.*"

$$g_5 \equiv (25 \leq |\{o_p = (id_p, Paper, s_p) \in O | \exists\{p, (id_d, Decision, accepted)\} \in L\}),$$
$$id_p, id_d \in ID, s_p \in S$$
$$g_6 \equiv ((close\ reviewing, enabled) \in A)$$
$$g_7 \equiv g_5 \wedge g_6$$

Objectives can be used to plan the execution of processes. Consider the following scenario (g_{10}): during decide on paper, the program committee decides that a meta-review is necessary for paper 13. They therefore want to create a fourth review, and make the decision as quickly as possible:

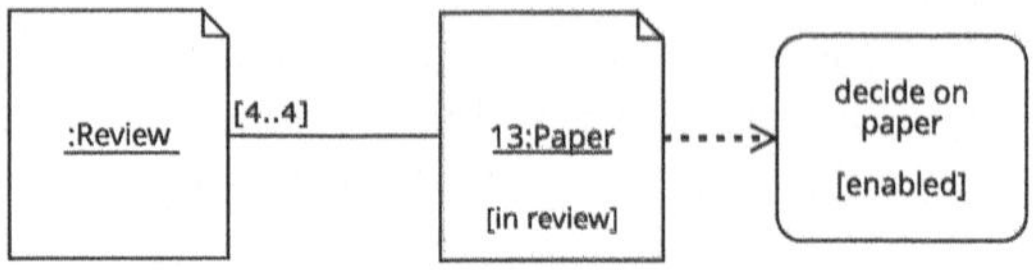

Fig. 5. Visualization of the objective g_{10}. Activity "decide on paper" reads paper with ID 13, which must be linked to four reviews.

$$g_8 \equiv p = (13, \mathit{Paper}, \mathit{in\ review}) \in O : |\{id' | \{p, (id', \mathit{Review}, \mathit{available})\} \in L, id' \in ID\}| = 4$$
$$g_9 \equiv p = (13, \mathit{Paper}, \mathit{in\ review}) \in O : \exists R \in \mathit{read}(\mathit{decide\ on\ paper}, \mathit{enabled}) : p \in R$$
$$g_{10} \equiv g_8 \wedge g_9$$

The objective is also depicted in Fig. 5. From right to left, the figure defines an instance of activity "decide on paper" in state enabled that reads the paper object with ID 13, which must be associated to four review objects. The notation is inspired by UML's object diagrams and BPMN-Q [2]. The shortest plan for the described scenario and objective g_{10} consists of two activities:

1. Complete decide on paper by creating a new review in state required.
2. Execute create review for the newly created review object.

Not every objective can be reached.

- An objective can be inconsistent, for example, when an objective requires all objects of a specific class to be in two states at the same time.
- An objective may contradict the case, e.g., if it requires fewer links or objects than exist. Objects and links cannot be deleted. The objective is unsatisfiable.
- An objective may contradict the case model, e.g., it may require more links than allowed by the multiplicity constraints. Some conflicts between objectives and case models can be resolved by adding new fragments at run-time [8].

Planning is often performed manually by the knowledge worker but can also be automated based on the process' state space [13]. We plan to present automated planning of knowledge-intensive processes with fCM models and respective objectives in a future publication.

4 Proof of Concept Implementation

Objectives are run-time goals specified by knowledge workers. An implementation should provide the means for specifying objectives, which can be used for planning. We present an implementation[3] that consists of a user interface for specifying objectives according to the framework presented in Sect. 3, and a compiler formalizing the objective for further planning.

To specify an objective, knowledge workers select an fCM case model. The model is parsed, and forms for specifying objectives are generated (see Fig. 6).

[3] Source and documentation: https://github.com/bptlab/fCM-query-generator.

They are based on the data objects and activities in the case model. These forms allow defining first order logic statements about data objects including their states, and about tasks that should be enabled. Data objectives can use existential and universal qualifications as well as multiplicity constraints. All conditions can be concatenated with the logic operators AND, OR, and NOT. That way model-driven objectives as presented in Sect. 3 can be specified.

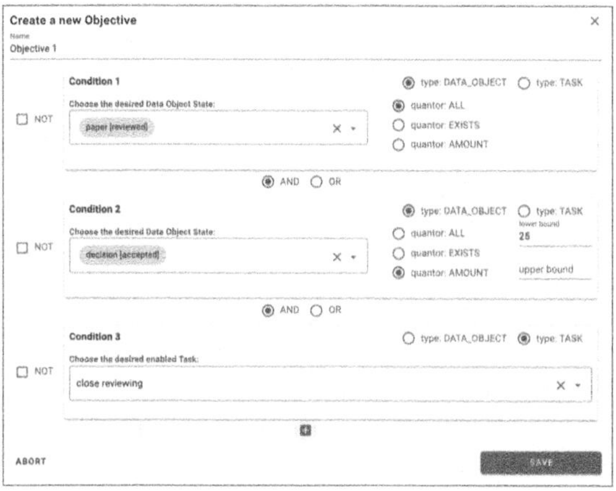

Fig. 6. The user interface for specifying an objective.

The specified objective is then compiled into a state space query, which can be used together with the colored Petri net formalization of the case model [5] for planning using CPNTools [11].

The current implementation has limitations. First, the objectives supported by the user interface contain neither links nor object identities. Objects read or written by an activity cannot be specified, and only the state enabled of activities is supported. Furthermore, objectives cannot be temporally related. The provided implementation can be improved by supporting all types of objectives as presented in the framework. Though, this may require the input of run-time information (i.e., the object identities and links).

5 Related Work

We structure related work in three parts: first, we discuss modeling approaches for knowledge-intensive processes. Next, we provide an overview on fCM. Finally, we look at goal modeling and planning for knowledge-intensive processes.

Knowledge-intensive processes have flexible behavior primarily driven by informed decisions of knowledge workers [3,21]. To specify such processes, various approaches have been proposed. Declarative approach define constraints between activities that must be satisfied but not an order of activities satisfying them. The most well-researched declarative approaches are DECLARE [17] and Dynamic Condition Response Graphs [9], which are based on linear temporal logic over finite traces. Declarative models can represent processes that have little structure concisely. However, they provide limited data support and can be difficult to comprehend—hybrid approaches, such as fCM, aim to improve this [1].

Artifact-centric process models focus on data objects and how they evolve. Processes are defined based on life cycles of artifacts. Among these approaches are the Guard-Stage-Milestone approach [10], PHILharmonicFlows [12], and BAUML [4]. Furthermore, the Guard-Stage-Milestone approach inspired the Case Management Model and Notation (CMMN) standard [16]. Artifact-centric approaches can emphasize the data-driven nature of knowledge-intensive processes. However, most constraints must be expressed on the level of data artifacts and the actual activities may be split among multiple artifacts. For an overview on knowledge-intensive and data-centric approaches, see the surveys [3] and [20].

The fCM approach is a hybrid approach [8]. At its core are highly structured activity-centric process fragments. The approach leverages both imperative control flow and declarative data flow. Furthermore, data models and class-specific behavioral models specify additional constraints. Recently, fCM has been extended with associations [6] and multiplicity constraints [5,7], including formal semantics based on colored Petri nets. While the approach combines the strengths of different approaches, fCM models have hidden dependencies, which harm comprehensibility. Comprehensibility, however, is important for manual planning.

One requirement of knowledge-intensive processes is late goal modeling [3]. Planning actions according to a goal is common in knowledge work [18]. Some approaches, such as CMMN [16], support planning, i.e., by specifying the respective scope. However, only few approaches support late goal modeling [3] and automated planning. Marrella et al. define SmartPM [14], which allows automatic adaptation of knowledge-intensive processes using planning to handle exceptions. If data deviates from the expectation, planning is used to find a sequence of actions that lead to a consistent state. Venero et al. [22] present a method for planning knowledge-intensive processes using Markov decision processes. Similarly, Sprovieri and Vogler [19] use planning to compose processes from partly structured models. Both approaches define goal values for data attributes. In comparison, our approach specifies objectives based on the abstract state of data objects, links between objects, as well as activities and their I/O-behavior.

6 Conclusion

Models of knowledge-intensive processes are often abstract and/or difficult to comprehend. It is, therefore, hard to plan knowledge workers' actions. Yet, goal modeling and planning are important aspects of knowledge work. To overcome the challenges, we proposed an approach based on fragment-based case management. It allows knowledge workers to specify objectives during run-time on the basis of a case model and a case. Data objects, their links, and activities can be used to define logic expressions about a desired state of the model. This formalization of objectives is crucial for automated planning.

However, there are some limitations to the approach. Objectives are limited to the information in the case model. More precisely, data objects, their states and links, and the state of activities. Other aspects, such as timing constraints, are not supported. In the future, the approach may use data from past executions (i.e., event logs) or enriched process models to allow a wider range of objectives.

Nevertheless, the approach shows that it is possible to formalize model-driven objectives. This is the basis of planning actions automatically during the execution of a case model. As the actual planning is a complex topic in itself, we plan to address it in a future publication. We believe that this work is an important step towards more end-user friendly case management.

References

1. Andaloussi, A.A., Burattin, A., Slaats, T., Kindler, E., Weber, B.: On the declarative paradigm in hybrid business process representations: a conceptual framework and a systematic literature study. Inf. Syst. **91**, 101505 (2020)
2. Awad, A.: BPMN-Q: a language to query business processes. In: Enterprise Modelling and Information Systems Architectures - Concepts and Applications, Proceedings of the 2nd International Workshop on Enterprise Modelling and Information Systems Architectures (EMISA 2007), St. Goar, Germany, 8–9 October 2007, pp. 115–128 (2007)
3. Di Ciccio, C., Marrella, A., Russo, A.: Knowledge-intensive processes: characteristics, requirements and analysis of contemporary approaches. J. Data Semant. **4**(1), 29–57 (2015)
4. Estañol, M., Queralt, A., Sancho, M.R., Teniente, E.: Artifact-centric business process models in UML. In: La Rosa, M., Soffer, P. (eds.) BPM 2012. LNBIP, vol. 132, pp. 292–303. Springer, Heidelberg (2013). https://doi.org/10.1007/978-3-642-36285-9_34
5. Haarmann, S., Montali, M., Weske, M.: Refining case models using cardinality constraints. In: Proceedings of Advanced Information Systems Engineering - 33rd International Conference, CAiSE 2021, Melbourne, VIC, Australia, 28 June–2 July 2021, pp. 296–310 (2021)
6. Haarmann, S., Weske, M.: Correlating data objects in fragment-based case management. In: Abramowicz, W., Klein, G. (eds.) BIS 2020. LNBIP, vol. 389, pp. 197–209. Springer, Cham (2020). https://doi.org/10.1007/978-3-030-53337-3_15
7. Haarmann, S., Weske, M.: Data object cardinalities in flexible business processes. In: Del Río Ortega, A., Leopold, H., Santoro, F.M. (eds.) BPM 2020. LNBIP, vol. 397, pp. 380–391. Springer, Cham (2020). https://doi.org/10.1007/978-3-030-66498-5_28
8. Hewelt, M., Weske, M.: A hybrid approach for flexible case modeling and execution. In: La Rosa, M., Loos, P., Pastor, O. (eds.) BPM 2016. LNBIP, vol. 260, pp. 38–54. Springer, Cham (2016). https://doi.org/10.1007/978-3-319-45468-9_3
9. Hildebrandt, T.T., Mukkamala, R.R.: Declarative event-based workflow as distributed dynamic condition response graphs. In: Honda, K., Mycroft, A. (eds.) Proceedings Third Workshop on Programming Language Approaches to Concurrency and communication-cEntric Software, PLACES, Paphos, Cyprus, 21 March 2010. EPTCS, vol. 69, pp. 59–73 (2010)

10. Hull, R., et al.: Introducing the guard-stage-milestone approach for specifying business entity lifecycles. In: Bravetti, M., Bultan, T. (eds.) WS-FM 2010. LNCS, vol. 6551, pp. 1–24. Springer, Heidelberg (2011). https://doi.org/10.1007/978-3-642-19589-1_1
11. Jensen, K., Kristensen, L.M., Wells, L.: Coloured Petri Nets and CPN tools for modelling and validation of concurrent systems. Int. J. Softw. Tools Technol. Transf. **9**(3), 213–254 (2007)
12. Künzle, V., Reichert, M.: PHILharmonicFlows: towards a framework for object-aware process management. J. Softw. Maint. Res. Pract. **23**(4), 205–244 (2011)
13. Marrella, A.: Automated planning for business process management. J. Data Semant. **8**(2), 79–98 (2019)
14. Marrella, A., Mecella, M., Sardiña, S.: SmartPM: an adaptive process management system through situation calculus, indigolog, and classical planning. In: Baral, C., Giacomo, G.D., Eiter, T. (eds.) Principles of Knowledge Representation and Reasoning: Proceedings of the Fourteenth International Conference, KR, Vienna, Austria, 20–24 July 2014. AAAI Press (2014)
15. Object Management Group: Business Process Model and Notation (BPMN), January 2014. https://www.omg.org/spec/BPMN
16. Object Management Group: Case Management Model and Notation (CMMN), December 2016. https://www.omg.org/spec/CMMN
17. Pesic, M., Schonenberg, H., van der Aalst, W.M.P.: DECLARE: full support for loosely-structured processes. In: 11th IEEE International Enterprise Distributed Object Computing Conference (EDOC), 15–19 October 2007, Annapolis, Maryland, USA, pp. 287–300. IEEE Computer Society (2007)
18. Pyöriä, P.: The concept of knowledge work revisited. J. Knowl. Manag.(2005)
19. Sprovieri, D., Vogler, S.: Run-time composition of partly structured business processes using heuristic planning. In: International Conference on Enterprise Systems, ES, Basel, Switzerland, 14–15 October 2015, pp. 225–232. IEEE (2015)
20. Steinau, S., Marrella, A., Andrews, K., Leotta, F., Mecella, M., Reichert, M.: DALEC: a framework for the systematic evaluation of data-centric approaches to process management software. Softw. Syst. Model. **18**(4), 2679–2716 (2018). https://doi.org/10.1007/s10270-018-0695-0
21. Swenson, K.D.: Position: BPMN is incompatible with ACM. In: La Rosa, M., Soffer, P. (eds.) BPM 2012. LNBIP, vol. 132, pp. 55–58. Springer, Heidelberg (2013). https://doi.org/10.1007/978-3-642-36285-9_7
22. Venero, S.K., Schmerl, B.R., Montecchi, L., dos Reis, J.C., Rubira, C.M.F.: Automated planning for supporting knowledge-intensive processes. In: Proceedings of Enterprise, Business-Process and Information Systems Modeling - 21st International Conference, BPMDS 2020, 5th International Conference, EMMSAD 2020, Held at CAiSE 2020, Grenoble, France, 8-9 June 2020, pp. 101–116 (2020)
23. Weske, M.: Business Process Management - Concepts, Languages, Architectures, 3rd edn. Springer, Heidelberg (2019)

Extracting Decision Dependencies and Decision Logic from Text Using Deep Learning Techniques

Alexandre Goossens(✉), Michelle Claessens, Charlotte Parthoens, and Jan Vanthienen

Leuven Institute for Research on Information Systems (LIRIS), KU Leuven, Leuven, Belgium
alexandre.goossens@kuleuven.be

Abstract. Decision models are increasingly being used in modeling business processes. Hence, extracting decision models automatically from texts would help decision modellers by reducing modeling time and supporting them in their analysis. In this paper, deep learning techniques are investigated to extract decision dependencies and conditional clauses directly from text. By using a large dataset of labeled and tagged sentences and NLP, deep learning techniques (BERT and BI-LSTM-CRF) are trained and tested on the identification of these items. The results show that the performance is sufficiently high to extract decision dependency and logic (semi)-automatically from text which provides a big step towards automatic decision modelling.

Keywords: Deep learning · Decision Model and Notation (DMN) · Decision model extraction

1 Introduction

Decision modelling allows to model a decision with its structure and logic, alongside process models such as Business Process Model and Notation (BPMN). As decision and process descriptions are often stored in textual documents (and often together), modelling can be a time-consuming task. Several directions have been explored to make decision modelling faster, such as decision model extraction from logs [6,9] or from process models [7]. Recently, more attention has been given to the extraction of decision dependencies [13] or decision logic [5] directly from text using patterns. Automatically extracting decision models from text would benefit all stakeholders as modelling time would greatly be reduced.

In this paper, deep learning techniques for the creation of decision models will be investigated, (i.e. the creation of Decision Model & Notation (DMN) models) with a dataset consisting of approximately 600 tagged and labeled sentences. To the best of our knowledge, decision model construction using deep learning techniques has not been researched before.

A. Marrella and B. Weber (Eds.): BPM 2021 Workshops, LNBIP 436, pp. 349–361, 2022.
https://doi.org/10.1007/978-3-030-94343-1_27

The paper is structured as follows: Sect. 2 explains the problem with an example; Sect. 3 deals with the related work; Sect. 4 describes the methodology; Sect. 5 discusses the experiments and the results; in Sect. 6 the limitations and future work are discussed and Sect. 7 concludes this paper.

2 Problem Statement

The creation of a decision or process model requires to identify all the components and relationships of the model, making it a knowledge intensive task. By using deep learning techniques and Natural Language Processing (NLP), the **main goal** of this paper is to **extract decision dependencies from text, construct a Decision Requirements Diagram (DRD) automatically** and **extract decision logic clauses from text**. Decision dependencies can be found both in sentences describing dependencies or logic, and decision logic can be found in logic sentences. Dependency and logic sentences can be classified manually or automatically (which is being researched). Next, DMN will be explained with an example.

Example 1. Eligibility for smallpox vaccine is dependent on risk and outbreak. If you are at risk or there is an outbreak then you are eligible for a smallpox vaccine. Risk is determined from the contact with a similar virus, labworker or smallpox virus exposure. You are at risk if you are in contact with a similar virus and you are a labworker. You are also at risk if there is smallpox virus exposure.

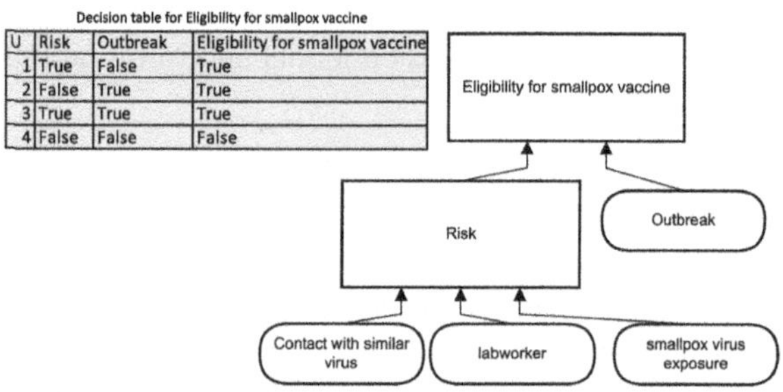

Fig. 1. DMN: DRD and one decision table of the example

Decision models represent decisions, their structure and logic [27]. In 2015, the Object Management Group (OMG) introduced a decision modelling standard called DMN [2], next to BPMN [1]. A DMN model represents the structure and the logic of a decision. The decision requirements level visualizes the structure in a Decision Requirements Diagram (DRD), containing decisions (rectangles), input data (rounded rectangles) and information requirements (solid arrows). Information requirements can refer to input data or the result of another decision. In the DMN model in Fig. 1, Eligibility for smallpox vaccine is the final decision and needs both outbreak information and the output of decision Risk as inputs. The decision logic level constitutes the other part of a DMN model and uses decision tables to represent the decision logic.

3 Related Work

Decision Model Extraction. Facilitating the construction of decision models is not a novel research direction. In [7] decision models are derived from process models and in [9] from process execution logs. Decision model extraction from text however is a more novel research direction. In [18], DMN models extraction with a pattern-based approach from structured natural language text is researched. By using pattern recognition, the automatic extraction of decision dependencies from paragraphs and generation of DRDs is investigated in [13]. In [5] decision rules and decision tables are derived automatically from a single sentence, not including decision dependencies.

Rule Extraction. Rule extraction from text is a complex research field with several investigated approaches. In [23,26] an emphasis on syntactic analysis is examined, resulting in a real application [20]. The previous approaches had difficulties dealing with sentence variations and in [19] dependency trees are used to deal with sentences with a similar meaning but different constructions to solve this issue. Other approaches using NLP have been investigated such as in [8] where Unified Modeling Language (UML) use cases were used for the extraction of business rules and Semantics Of Business Vocabulary And Business Rules (SBVR) vocabularies. In [11] rule extraction from legal documents is investigated.

Process Model Extraction. In the somewhat related field of process modelling, the extraction of small BPMN process model parts from text has been researched in [15] and the generation of BPMN models from group stories was examined in [4]. In [25] BPMN models generation from use cases semi-automatically is investigated and in [12] text-based process traces are investigated for the derivation of BPMN models. In [28], a BPMN model is extracted if a business policy is given as input following their definition and requires the identification of several BPMN elements. The approach described in [14] derives BPMN models from text if it is in a sequential format and no questions or process irrelevant sentences are present. In [3] the extraction of declarative process models from text was investigated. The approach described in [24] uses textual annotations where Annotated Textual Description of a Process(ATDP) is developed, providing structure in an unstructured text and allowing for formal reasoning.

4 Methodology

4.1 General Approach

A decision description mainly contains two kinds of sentences: sentences with decision dependencies (e.g.: Your job and age define your exposure.) or with decision logic (e.g.: When age is higher than 50 then vaccination period is May).

In the following sections, the general approach to (automatically) construct a decision model will be described. Next, the specific approach to extract decision dependencies and decision logic are described. The deep learning model is given the appropriate sentences to extract dependencies or logic. Irrelevant sentences are assumed to have been identified and removed. After the NLP steps (preprocessing and coreference resolution), the decision dependencies and logic will be extracted. This results in a list of concepts and their relationships and a list of conditions and consequences. From this a DRD can be automatically constructed and a decision table manually derived. The steps are visualized in Fig. 2.

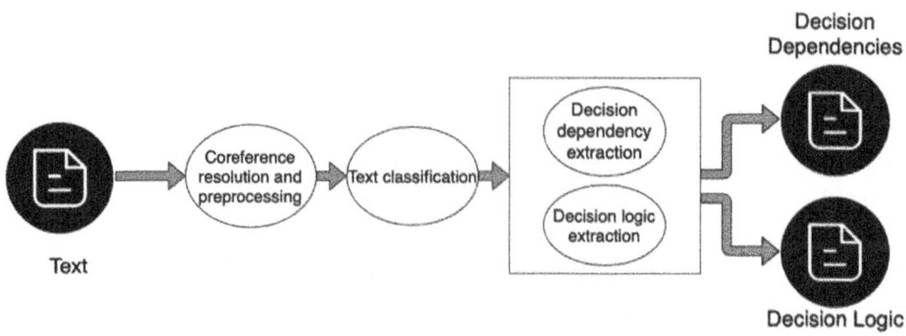

Fig. 2. From text to decision model

1. **Coreference Resolution & Preprocessing.** Coreference resolution resolves inconsistencies so that each concept is described consistently with the same words so that a constructed decision model uses the same words describing the same concepts in a consistent manner when build automatically. The open-source NeuralCoref4 package[1] was used to perform coreference resolution automatically. Next, some preprocessing needs to be performed. Firstly, all the words are lowercased and lemmatized, meaning that a word is replaced by a simpler common form of that word (e.g. Eligibility becomes eligible). Next, all stopwords and punctuation are removed. According to the previous steps the sentence *When the dog goes out for a walk, it is happy.* will become *when dog go for walk dog be happy.* Afterwards, the text is split into distinct parts or tokens. A token can be a word, a part of a word or a

[1] https://github.com/huggingface/neuralcoref.

punctuation symbol. Lastly, each token is converted into a number which will correspond to a tokenID which will then be used to find the matching word in the glossary of the corresponding pre-trained deep learning model.
2. **Text Classification.** Afterwards, the sentences are classified manually or automatically (under research) into dependency, conditional or irrelevant (to the decision model) sentences to ensure the right sentence is given to the right model.
3. **Decision Dependency and Logic Extraction.** As this paper mainly deals with this step, the topic is more elaborated upon in Sects. 4.2 and 4.3. But in general, a list of concepts and dependencies is identified and extracted. These are used for the construction of a DRD. In this step parts of business rules (condition and consequence clauses) are also identified and extracted. Note that the influence of capitals letters and stop word removal is assessed for both logic and dependency extraction as these can be part of the concepts.
4. **Result.** Finally the models will return a list of decision dependencies and a list with sentence clauses. From the decision dependency list, a DRD can be constructed and from the conditional sentences clauses list it is possible to derive a decision table resulting in a complete DMN model.

4.2 Decision Dependency Extraction

The extraction of decision dependencies requires extracting base concepts, derived concepts and the potentially present action verb. As such, both (explicit) dependency and conditional (or logic) sentences can be used for the extraction of decision dependencies (Fig. 5). Hence, deep learning models need to be trained on both types of sentences. The following definitions are derived from [13]:

Definition 1. *A concept is an attribute of a business object that can take on a value and plays a central role in a decision (e.g., weight of a person).*

Definition 2. *A base concept is considered as the input concept of a decision, this can be data or the output of another decision.*

Definition 3. *A derived concept is the output concept of a decision. The value of this concept depends on the value of its base concepts.*

Definition 4. *A dependency is the relation between a base and a derived concept. This can be represented as a tuple (ACT, BAS, DER) with ACT the action verb representing the kind of relation, BAS the base concept and DER the derived concept.*

Definition 5. *A DMN DRD is represented as a tuple of the form = (I, D, R). With I a set of input information items (i.e., concepts that are not the output of a decision), D a set of decisions and R a set of information requirements (i.e., dependencies).*

First, the sentences are tagged meaning that chunks of texts are given a label. Next, the classifiers are trained to identify these tags. Finally, the decision dependencies are derived and a DRD can be constructed. These are the steps:

Decision Dependency Tagging. To capture and classify the chunks of texts, the inside-outside-beginning (IOB) tagging format, also used for Named Entity Recognition (NER), is used [22]. If a chunk of text is irrelevant, an O(utside) tag is given. Tags are created for base concepts (BAS), derived concepts (DER) and action verbs (ACT). The first word of a concept is indicated with a B(eginning)-XXX tag and the other words are indicated with an I(nside)-XXX tag.

Decision Dependency Tag Extraction. Two techniques were investigated to extract the previously mentioned IOB-tags from a sentence.

BERT: By using a python package called NERDA[2] it is possible to fine-tune pre-trained transformer models for NER tasks. Both the Bidirectional Encoder Representations from Transformer (BERT)-base-cased and BERT-base-uncased models [10] were tested. Both models consist of 12 layers and 12 attention heads.

Bi-directional LSTM-CRF Models: In a second approach, a BiDirectional-Long Short-Term Memory- Conditional Random Field (Bi-LSTM-CRF) model is trained from scratch. The model architectures, and more specifically the Bi-LSTM component, enables the use of both past and future input features [17]. The CRF layer is put on top of the Bi-LSTM layer to optimize the prediction of current tags based on the previous tags. Moreover, the pre-trained word embedding GloVe [21] was implemented to generate word vector representations.

Deriving Decision Dependencies from the Extracted Tags

Concatenation of Tags. All tokens referring to the same concept are concatenated together i.e. all the I-XXX tags are concatenated to the corresponding B-XXX tag, resulting in a list of base concepts and derived concepts for each sentence.

Construction of Dependency Tuples. A decision tuple of the form (ACT,BAS,DER) is formed on sentence level. In conditional sentences no action verb is present.

DRD Construction. Based on the final dependencies, the DRD tuple (I, D, R) is composed and can be visualized.

4.3 Decision Logic Extraction

Logical structures can only be extracted from conditional sentences. With decision logic extraction, the splitting of a sentence into conditional, consequence and, if applicable, an exception clause is meant. The actual value extraction of concepts is future research. The following definitions are derived from [16]:

[2] https://pypi.org/project/NERDA/.

Definition 6. *A conditional clause is a subordinate clause that usually begins with the conjunction 'if' and states (multiple) possible results. The if-conjunction can also be replaced by a negative conjunction such as "except" or "unless", this will be referred to as the exception clause.*

Definition 7. *A consequence clause is dependent on the conditional clause and, if present, the exception clause. A special case is the else consequence clause, which holds if the conditional clause is false.*

Similarly to decision dependency extraction (Sect. 4.2), the same IOB-tag methodology is employed. The following tags are identified: COND for conditional clauses, EXCE for exception clauses, CONS for consequence clauses, ELSE for else consequence clauses and O for outside. The following steps apply:

Decision Logic Tag Extraction. Similar to the task of dependency tag extraction in Sect. 4.2, a finetuned BERT model is compared with a Bi-LSTM-CRF model trained from scratch. The models are trained on the problem specific dataset for logic extraction, and therefore aim to extract conditional clauses, exception clauses and (else) consequence clauses.

Tabular Representation of Conditional Sentence Structure. After the extraction of the IOB-tags, the following steps are performed:

Concatenation of Tags: The same technique as in Sect. 4.2 is used to concatenate the tokens. The result is a list per type of clause (i.e., conditional, exception, consequence, else consequence) for each sentence.

Tabular Representation of Conditional Sentence Structure: Finally, a table is constructed where each sentence is dedicated to a row, whereas the columns represent the type of subclause (see Fig. 5).

5 Experiments

In total 15 classifiers were constructed, each one of them trained and tested on the same training and test set. For Bi-LSTM-CRF experiments were conducted with and without preprocessing, for BERT-base-uncased experiments were conducted with and without the removal of stopwords and for BERT-base-cased experiments were conducted only with the removal of stopwords. The deep learning models were trained to predict IOB tags on a training set and evaluated on a separate testset for each of these models.

5.1 Collected Data

A representative, hybrid dataset consisting of approximately 600 sentences was collected and tagged manually (Table 1). The dataset consists of sentences retrieved from various sources such as websearches, Github sources, original or derived DMN descriptions. To further increase the diversity of the dataset, variants of sentences were constructed with synonyms or different constructions. Each one of these sentences was then classified as either explicit dependency sentence or conditional sentence. After collecting the dependency sentences, several patterns were identified ranging from the classical ones such as "X defines Y", "Y is defined by X" to more complex cases such as "X defines Y and V defines Z". Similar patterns are present in the conditional sentences.

For the decision dependency extraction dataset, the training set consists of 195 sentences for explicit dependency sentences and of 245 sentences for the conditional sentences. The test sets consist of 60 and 82 sentences respectively. For the decision logic extraction dataset, the dataset consists of 264 and 84 sentences for the training and test set respectively (see Table 1 for an overview). Note that the training size of conditional sentences differs between the two problems because sometimes in conditional sentences there are no dependencies to extract, e.g. sentences describing a formula. Moreover, the test set consists of sentences that were not present in the training set. No variations of sentences that were present in the training set were included into the test set.

Table 1. Training and test size of the problem specific datasets

Total dataset	Training size	Test size
All sentences	459	142
Decision dependency extraction dataset	Training size	Test size
Explicit dependency sentences	195	60
Conditional sentences	245	82
Decision logic extraction dataset	Training size	Test size
Conditional sentences	264	82

5.2 Results

Dependency Tag Extraction. The results for BERT-base uncased without stopword removal and for BI-LSTM-CRF without preprocessing are reported as these results were the best for each of the techniques. For each of the tags, Precision, Recall and F1-score are reported. Moreover, macro and micro averages are also reported, the difference between the two being that macro average assumes equal class frequency and micro is biased by the class frequency. As finding all the concepts is important, macro average results are used to decide on the performance of a technique. From the results, it can be seen that the recall values are generally above 0.8 for BERT. This indicates the model is able to identify the whole concept quite well. Lastly, it is possible that not all the inside and base tags are identified across sentences. In that case, the user can modify the extracted concept so that the program identifies it correctly, or the user can directly delete the duplicate concept or dependency. In Fig. 3, the results for both explicit dependency and conditional sentences are given.

Dependency Extraction		BERT base-uncased without stopword removal			BI-LSTM-CRF without preprocessing		
Explicit Dependency sentences	Level	Precision	Recall	F1-score	Precision	Recall	F1-score
	B-DER	0.79 ± 0.02	0.87 ± 0.03	0.83 ± 0.02	0.61 ± 0.06	0.62 ± 0.03	0.62 ± 0.02
	I-DER	0.84 ± 0.05	0.86 ± 0.03	0.85 ± 0.02	0.78 ± 0.02	0.63± 0.06	0.70 ± 0.03
	B-BAS	0.79 ± 0.02	0.88 ± 0.01	0.83 ± 0.01	0.58 ± 0.03	0.83± 0.02	0.68 ± 0.02
	I-BAS	0.87 ± 0.01	0.86 ± 0.05	0.86 ± 0.02	0.68 ± 0.02	0.71± 0.02	0.70 ± 0.01
	B-ACT	0.80 ± 0.02	0.93 ± 0.01	0.86 ± 0.01	0.82 ± 0.01	0.79 ± 0.00	0.80 ± 0.01
	AVG_MICRO	0.83 ± 0.02	0.87 ± 0.02	0.85 ± 0.02	0.68 ± 0.01	0.71 ± 0.01	0.70 ± 0.01
	AVG_MACRO	**0.82 ± 0.01**	**0.88 ± 0.02**	**0.85 ± 0.01**	0.70 ± 0.01	0.72 ± 0.01	0.70 ± 0.01
Conditional sentences	B-DER	0.89 ± 0.02	0.96 ± 0.02	0.92 ± 0.01	0.70 ± 0.08	0.78 ± 0.05	0.73 ± 0.03
	I-DER	0.90 ± 0.03	0.90 ± 0.03	0.90 ± 0.01	0.67 ± 0.11	0.72 ± 0.04	0.69 ± 0.05
	B-BAS	0.76 ± 0.01	0.85 ± 0.03	0.80 ± 0.01	0.55 ± 0.06	0.62 ± 0.07	0.57 ± 0.04
	I-BAS	0.93 ± 0.02	0.70 ± 0.02	0.80 ± 0.01	0.80 ± 0.04	0.33 ± 0.10	0.47 ± 0.10
	AVG_MICRO	0.88 ± 0.01	0.83 ± 0.01	0.85 ± 0.01	0.66 ± 0.07	0.58 ± 0.03	0.61 ± 0.05
	AVG_MACRO	**0.87 ± 0.01**	**0.85 ± 0.01**	**0.86 ± 0.01**	0.68 ± 0.05	0.62 ± 0.02	0.61 ± 0.05

Fig. 3. Results for dependency tag extraction

At the end of the decision dependency tag extraction decision dependency tuples are derived from sentences, e.g. if *Your job and age define your exposure* is given to the model, it will identify job and age as base concepts, exposure as derived concept and define as action verb. Finally, the following dependencies are derived (define, job, exposure), (define, age, exposure).

Logic Tag Extraction. The results for BERT-base uncased without stopword removal and for BI-LSTM-CRF without preprocessing are reported in Fig. 4 as these results were the best for each of the techniques.

Logic Extraction		BERT base-uncased without stopword removal			BI-LSTM-CRF without preprocessing		
Conditional sentences	Level	Precision	Recall	F1-Score	Precision	Recall	F1-Score
	B-CONS	0.88 ± 0.01	0.89 ± 0.02	0.88 ± 0.00	0.67 ± 0.07	0.53 ± 0.04	0.59 ± 0.04
	I-CONS	0.91 ± 0.01	0.94 ± 0.02	0.93 ± 0.02	0.82 ± 0.02	0.62 ± 0.09	0.70 ± 0.06
	B-COND	0.87 ± 0.00	0.94 ± 0.01	0.90 ± 0.00	0.71 ± 0.04	0.73 ± 0.06	0.71 ± 0.03
	I-COND	0.93 ± 0.03	0.89 ± 0.01	0.91 ± 0.02	0.73 ± 0.05	0.77 ± 0.03	0.75 ± 0.02
	B-ELSE	0.91 ± 0.00	0.87 ± 0.05	0.89 ± 0.03	0.36 ± 0.17	0.71 ± 0.08	0.47 ± 0.15
	I-ELSE	0.95 ± 0.05	0.97 ± 0.02	0.96 ± 0.01	0.31 ± 0.11	0.93 ± 0.03	0.46 ± 0.11
	B-EXCE	1.00 ± 0.00	0.80 ± 0.00	0.89 ± 0.00	1.00 ± 0.00	0.40 ± 0.14	0.56 ± 0.15
	I-EXCE	1.00 ± 0.00	0.65 ± 0.14	0.78 ± 0.10	1.00 ± 0.00	0.40 ± 0.12	0.56 ± 0.13
	AVG_MICRO	0.92 ± 0.02	0.92 ± 0.02	0.92 ± 0.02	0.69 ± 0.03	0.69 ± 0.03	0.69 ± 0.03
	AVG_MACRO	**0.93 ± 0.00**	**0.87 ± 0.00**	**0.89 ± 0.00**	0.70 ± 0.03	0.66 ± 0.05	0.60 ± 0.05

Fig. 4. Results for logic tag extraction

At the end of the decision logic tag extraction logic clauses are derived from sentences, e.g. if *When your job is situated in the first line health care, your exposure is high, otherwise your exposure is low.* is given to the model, the following clauses are extracted: Conditional clause: your job is situated in the first line health care; Consequence clause: your exposure is high; Else clause: your exposure is low. With these extracted clauses, it is now possible to derive a logic table. The conditional clauses will contain the inputs of a decision, the consequence clause the corresponding output and the else clause will contain the other output value in case the conditional clause does not hold.

5.3 Achievability of Automatic DRD Building

With these classifiers it now possible to (semi)-automatically extract decision dependencies and logic and construct a DMN model. Figure 5 shows a screenshot of the prototype to extract dependencies and logic from text. Note that the text is classified automatically using a custom build prototype classifier.

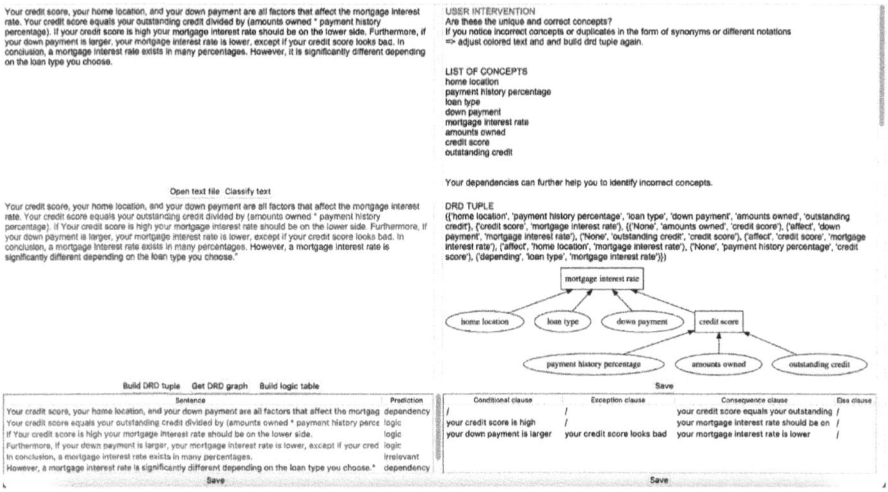

Fig. 5. Automatically deriving decision dependencies and logic and construction of a DRD

6 Discussion

Overall, BERT base-uncased without stopwords removal is the best performing model. BI-LSTM-CRF is performing the best without preprocessing but still is not performing as well as BERT. These models are assumed to perform better without stopword removal because some concepts do consist of stopwords so their removal would not allow for their complete identification. It is also important to note that the BI-LSTM-CRF model was trained from scratch and only with 5 epochs whilst BERT was a pretrained model. This could mean that if a bigger dataset was used and more epochs were used to train it on, better results could be achieved with BI-LSTM-CRF, this is considered future work.

The models also seem to have more difficulties identifying the beginning tags whilst the inside tags are a bit easier to identify. The identification of beginning tags is still accurate enough for a proof of concept. For decision dependency extraction, it can be concluded that using explicit dependency or conditional sentences yield more or less the same results and logic extraction from conditional sentences also yields good results.

Limitations and Future Work. There are some limitations that need to be pointed out. Firstly, coreference resolution struggles with the identification of synonyms and more complex cases. Secondly, currently the program is able to identify dependencies up until 2 levels within one sentence. This however covers a wide amount of sentences as rarely more than two levels are described within one sentence. Also, as this paper deals with decision model extraction from natural text, currently only the data type text is supported as decision input. Lastly, this dataset does not contain all subtleties of the English language. It is, however, still good enough to show a proof of concept.

As future work, the dataset will be increased with more sentences and patterns. In order to perform complete logic extraction, an ontology needs to be constructed so that the difference between a concept and a value can be learned. Together with a custom-built sentence classifier, a complete pipeline of text to decision model is available and comparing this with the rules-based pipeline [13] is also future work. A formal evaluation of the pipeline on full texts is also planned and will be integrated with the work of [14] to also work with textual descriptions of processes.

7 Conclusion

In this paper, a proof of concept for the extraction of decision dependencies and decision logic from text using deep learning is developed. With these classifiers, it is possible for decision modellers to (semi)-automatically extract decision dependencies and decision logic once the appropriate sentences are given as input.

References

1. OMG: Business process model and notation 1.0 (2010). https://www.omg.org/spec/BPMN/1.0
2. OMG: Decision model and notation 1.0 (2015). https://www.omg.org/spec/DMN/1.0/
3. van der Aa, H., Di Ciccio, C., Leopold, H., Reijers, H.A.: Extracting declarative process models from natural language. In: Giorgini, P., Weber, B. (eds.) CAiSE 2019. LNCS, vol. 11483, pp. 365–382. Springer, Cham (2019). https://doi.org/10.1007/978-3-030-21290-2_23
4. de AR Goncalves, J.C., Santoro, F.M., Baiao, F.A.: Business process mining from group stories. In: 2009 13th International Conference on Computer Supported Cooperative Work in Design, pp. 161–166. IEEE (2009)
5. Arco, L., Nápoles, G., Vanhoenshoven, F., Lara, A.L., Casas, G., Vanhoof, K.: Natural language techniques supporting decision modelers. Data Min. Knowl. Disc. **35**(1), 290–320 (2020). https://doi.org/10.1007/s10618-020-00718-4

6. Bazhenova, E., Buelow, S., Weske, M.: Discovering decision models from event logs. In: Abramowicz, W., Alt, R., Franczyk, B. (eds.) BIS 2016. LNBIP, vol. 255, pp. 237–251. Springer, Cham (2016). https://doi.org/10.1007/978-3-319-39426-8_19
7. Bazhenova, E., Weske, M.: Deriving decision models from process models by enhanced decision mining. In: Reichert, M., Reijers, H.A. (eds.) BPM 2015. LNBIP, vol. 256, pp. 444–457. Springer, Cham (2016). https://doi.org/10.1007/978-3-319-42887-1_36
8. Danenas, P., Skersys, T., Butleris, R.: Natural language processing-enhanced extraction of SBVR business vocabularies and business rules from UML use case diagrams. Data Knowl. Eng. **128**, 101822 (2020)
9. De Smedt, J., Hasić, F., vanden Broucke, S.K., Vanthienen, J.: Holistic discovery of decision models from process execution data. Knowl.-Based Syst. **183**, 104866 (2019)
10. Devlin, J., Chang, M.W., Lee, K., Toutanova, K.: BERT: pre-training of deep bidirectional transformers for language understanding. arXiv preprint arXiv:1810.04805 (2018)
11. Dragoni, M., Villata, S., Rizzi, W., Governatori, G.: Combining NLP approaches for rule extraction from legal documents. In: 1st Workshop on MIning and REasoning with Legal texts (MIREL 2016) (2016)
12. Epure, E.V., Martín-Rodilla, P., Hug, C., Deneckère, R., Salinesi, C.: Automatic process model discovery from textual methodologies. In: 2015 IEEE 9th International Conference on Research Challenges in Information Science (RCIS), pp. 19–30. IEEE (2015)
13. Etikala, V., Van Veldhoven, Z., Vanthienen, J.: Text2Dec: extracting decision dependencies from natural language text for automated DMN decision modelling. In: Del Río Ortega, A., Leopold, H., Santoro, F.M. (eds.) BPM 2020. LNBIP, vol. 397, pp. 367–379. Springer, Cham (2020). https://doi.org/10.1007/978-3-030-66498-5_27
14. Friedrich, F., Mendling, J., Puhlmann, F.: Process model generation from natural language text. In: Mouratidis, H., Rolland, C. (eds.) CAiSE 2011. LNCS, vol. 6741, pp. 482–496. Springer, Heidelberg (2011). https://doi.org/10.1007/978-3-642-21640-4_36
15. Ghose, A., Koliadis, G., Chueng, A.: Process discovery from model and text artefacts. In: 2007 IEEE Congress on Services (Services 2007), pp. 167–174. IEEE (2007)
16. Haspelmath, M., König, E., Oesterreicher, W., Raible, W.: Language typology and language universals: an international handbook, vol. 20. Walter de Gruyter (2001)
17. Huang, Z., Xu, W., Yu, K.: Bidirectional LSTM-CRF models for sequence tagging. arXiv preprint arXiv:1508.01991 (2015)
18. Kluza, K., Honkisz, K.: From SBVR to BPMN and DMN models. Proposal of translation from rules to process and decision models. In: Rutkowski, L., Korytkowski, M., Scherer, R., Tadeusiewicz, R., Zadeh, L.A., Zurada, J.M. (eds.) ICAISC 2016. LNCS (LNAI), vol. 9693, pp. 453–462. Springer, Cham (2016). https://doi.org/10.1007/978-3-319-39384-1_39
19. Lin, D., Pantel, P.: Discovery of inference rules for question-answering. Nat. Lang. Eng. **7**(4), 343–360 (2001)
20. Ono, T., Hishigaki, H., Tanigami, A., Takagi, T.: Automated extraction of information on protein-protein interactions from the biological literature. Bioinformatics **17**(2), 155–161 (2001)

21. Pennington, J., Socher, R., Manning, C.D.: Glove: global vectors for word representation. In: Proceedings of the 2014 Conference on Empirical Methods in Natural Language Processing (EMNLP), pp. 1532–1543 (2014)
22. Ramshaw, L., Marcus, M.: Text chunking using transformation-based learning. In: Third Workshop on Very Large Corpora (1995). https://www.aclweb.org/anthology/W95-0107
23. Riloff, E.: Automatically generating extraction patterns from untagged text. In: Proceedings of the National Conference on Artificial Intelligence, pp. 1044–1049 (1996)
24. Sànchez-Ferreres, J., Burattin, A., Carmona, J., Montali, M., Padró, L.: Formal reasoning on natural language descriptions of processes. In: Hildebrandt, T., van Dongen, B.F., Röglinger, M., Mendling, J. (eds.) BPM 2019. LNCS, vol. 11675, pp. 86–101. Springer, Cham (2019). https://doi.org/10.1007/978-3-030-26619-6_8
25. Sinha, A., Paradkar, A.: Use cases to process specifications in business process modeling notation. In: 2010 IEEE International Conference on Web Services, pp. 473–480. IEEE (2010)
26. Soderland, S.: Learning information extraction rules for semi-structured and free text. Mach. Learn. **34**(1), 233–272 (1999)
27. Vanthienen, J.: Decisions, advice and explanation: an overview and research agenda. A Research Agenda for Knowledge Management and Analytics (2021)
28. Wang, H.J., Zhao, J.L., Zhang, L.J.: Policy-driven process mapping (PDPM): discovering process models from business policies. Decis. Support Syst. **48**(1), 267–281 (2009)

Decision Modelling in Timed Dynamic Condition Response Graphs with Data

Thomas T. Hildebrandt[1(✉)], Håkon Normann[1,2], Morten Marquard[2], Søren Debois[2,3], and Tijs Slaats[1]

[1] Department of Computer Science, Copenhagen University, Copenhagen, Denmark
{hilde,normann,slaats}@di.ku.dk
[2] DCR Solutions, Njalsgade 76, Copenhagen, Denmark
{normann,mm,debois}@dcrsolutions.net
[3] Department of Computer Science, IT University of Copenhagen, Copenhagen, Denmark

Abstract. We provide a formal semantics for timed DCR graphs with decision events and data, and show how the model supports merging declarative decision modelling and process modelling in one notation. We exemplify the model by an expense report process and discuss how decisions may be presented using DMN jointly with the DCR process notation. If data domains are bounded, models are finite state and allow formal verification of both safety and liveness properties. The formal model is fully implemented in commercial process design and workflow management systems provided by DCR Solutions and also freely available for academic use.

Keywords: Declarative modelling · Data · Time · DCR Graphs · DMN

1 Introduction

Decision modelling and process modelling are two important tools used in the description of computations and not least the computations involved in business and case management processes. Traditionally, processes have often been described as procedures, e.g. using an imperative programming language or process modelling notation such as Business Process Model and Notation (BPMN) [8].

As running example in the paper we consider an expense report process. In the process, and employee can submit an expense report containing the type of expense (either cash payout or payment of bill) and the amount. Expenses below 200 Euro can be approved no matter the type. Payments of bills above 200, or cash payouts between 200 and 5000 can be payed out if a manager approves. Cash payouts above 5000 Euro are rejected no matter what.

While it is possible to model decision logic in a procedural way, e.g. as if, then, else rules or a BPMN diagram modelling a decision tree, the logical rules

A. Marrella and B. Weber (Eds.): BPM 2021 Workshops, LNBIP 436, pp. 362–374, 2022.
https://doi.org/10.1007/978-3-030-94343-1_28

behind a decision do often not prescribe an ordering of the computation. This suggests the use of a declarative notation such as a Decision Model and Notation (DMN) [9] decision table as exemplified for the expense report in Fig. 1. The table abstracts from the order in which the input is processed and can be associated to a decision activity [2] in the process model that computes the decision.

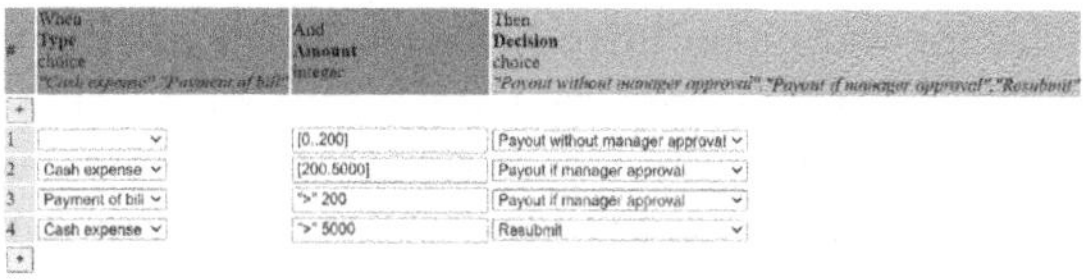

Fig. 1. DMN decision table for Decision event

Within the last decade there has been a growing awareness that imperative process notations such as BPMN are often too rigid for knowledge work processes. For instance, it gets complicated to model the expense report handling process if one wants the flexibility that expense forms can be partly updated, e.g. only the type or amount and resubmitted at any time before payout, and even more complicated if one wants to model time constraints, e.g. that a payout, if allowed, should happen between 7 and 14 days after the submission of the expense form and the approval of the manager (if required) should happen within 7 days from the submission.

This has led to the proposal of several variants of declarative process modelling languages, of which DECLARE [19] and DCR Graphs [11] are some of the most widely known. In the present paper we extend timed DCR graphs as introduced in [12] with decisions and data as implemented in the commercial design tool available at dcrsolutions.net (freely available for academic use) and illustrate how our running example of an expense approval process can be elegantly modelled.

Structure of Paper: The paper is structured as follows. In Sect. 2 we define timed DCR graphs with decisions and data and the running example. In Sect. 3 we provide pointers to related work and in Sect. 4 we briefly conclude and discuss future work.

2 Timed DCR Graphs with Decisions and Data

DCR graphs were introduced in [11] as part of the PhD thesis of Mukkamala [17] introducing a formal, declarative workflow process model providing a foundation for case management systems developed by our industrial partners. The work on the DCR notation and associated tool support was followed up in the PhD thesis of Slaats [22]. The notation is by now supported by mature modelling and simulation tools [3,14] and has been embedded in commercial case management systems widely used in Denmark by both central government institutions such as universities and local government institutions such as municipalities [4].

Since the seminal work [11], the DCR Graph modelling language has been extended in several ways, including modelling of time dependencies [12], nesting [10], dynamically spawned multi-instance subprocesses [5], and a theory of process refinement [5,18]. Several variants of DCR graphs and data have been presented in the past: A notion of data parameters and constraints guarded by Boolean constraints were also introduced early on in [17]. Parameters were instantiated when a process was initiated and could model that constraints may depend on data. Later, parameters where replaced by input events, receiving a value as part of their execution and holding the most recent value as part of the state [3]. Input events allow the setting of parameters to happen as part of the process execution and is used commercially to support modelling flexible web forms as DCR graphs [13].

In the present paper, we formalise the combination of timed DCR graphs from [12] with data as it is used in the commercial tool, and demonstrate how it provides a very flexible combination of process and decision modelling in one notation. We first define the syntax for expressions. To keep the presentation simple, we assume base values for data to be only of type Int (signed integer values) and Bool (Boolean), the extension to more data types is trivial and the commercial tool implements more types, including strings, enumerates and durations. We also have a void type for events without data. We let $\mathsf{Types} = \{\mathsf{Int}, \mathsf{Bool}, \mathsf{void}\}$ and V be the set of values.

Definition 1. *Assume a set E (of events) and $T : E \rightarrow Types$ a type function assigning at type to a subset of events. Let e^b and e^i range over events of type Boolean and Int respectively and let n range over integers and b over Booleans. The expressions over E,* $\mathsf{Exp_E}$ *ranging over Integer* $\mathsf{IExp_E}$*, Boolean expressions* $\mathsf{BExp_E}$ *and the void expression* void *are then given by the grammar below.*

$$\begin{aligned}
\mathsf{Exp_E} &:= \mathsf{BExp_E} \mid \mathsf{IExp_E} \mid \mathsf{void}\\
\mathsf{BExp_E} &:= e^b \mid b \mid \mathsf{IExp_E}\,\mathsf{IBOp}\,\mathsf{IExp_E} \mid \mathsf{BExp_E}\,\mathsf{BOp}\,\mathsf{BExp_E} \mid\\
&\quad\ \neg\mathsf{BExp_E} \mid \mathsf{if}\ \mathsf{BExp_E}\ \mathsf{then}\ \mathsf{Bexp_E}\ \mathsf{else}\ \mathsf{BExp_E}\\
\mathsf{IExp_E} &:= e^i \mid n \mid \mathsf{IExp_E}\,\mathsf{IOp}\,\mathsf{IExp_E} \mid \mathsf{if}\ \mathsf{BExp_E}\ \mathsf{then}\ \mathsf{Iexp_E}\ \mathsf{else}\ \mathsf{IExp_E}\\
\mathsf{IBop} &:= \ = \mid < \mid > \mid \leq \mid \geq\\
\mathsf{IOp} &:= + \mid - \mid *\\
\mathsf{Bop} &:= \mathit{and} \mid \mathit{or}
\end{aligned}$$

We can now define DCR graphs with decisions and data formally.

Definition 2. *A* timed DCR Graph with decisions and data *G is given by a tuple $(E, T, D, M, \rightarrow\!\bullet, \bullet\!\rightarrow, \bullet\!\rightarrow\!\times, \rightarrow\!\diamond, \rightarrow\!+, \rightarrow\!\%, L, l)$ where*

(i) E is a finite set of events,
(ii) $T : E \rightarrow$ Types *is the* type function

(*iii*) $D : E \to \mathsf{Exp_E} \uplus \{?\}$ *is the* decision function
(*iv*) $M = (Ex, Re, In) \in \big((E \rightharpoonup (\omega \times V_\bot) \times (E \rightharpoonup \infty) \times E \rightharpoonup \{\top\}\big)$ *is the* timed marking with data,
(*v*) $\rightarrow\!\!\bullet \subseteq E \times \omega \times \mathsf{BExp_E} \times E$, *is the* guarded timed condition relation,
(*vi*) $\bullet\!\!\rightarrow \subseteq E \times \infty \times \mathsf{BExp_E} \times E$, *is the guarded timed response relation,*
(*vii*) $\bullet\!\!\rightarrow\!\times, \rightarrow\!\diamond, \rightarrow\!+, \rightarrow\!\% \subseteq E \times \mathsf{BExp_E} \times E$ *are the guarded no-response, milestone, include and exclude relations respectively,*
(*viii*) L *is the set of labels,*
(*ix*) l *is the labelling function between events and labels.*

We require that decisions respect the type of events, i.e. if $D(e) \in \mathsf{BExp_E}$ *then* $T(e) = \mathsf{Bool}$, *if* $D(e) \in \mathsf{IExp_E}$ *then* $T(e) = \mathsf{Int}$ *and if* $D(e) = \mathsf{void}$ *then* $T(e) = \mathsf{Void}$. *If* $D(e) = ?$, *the event is an* input event, $T(e) \in \{\mathsf{Int}, \mathsf{Bool}\}$ *and data (of the right type) is provided by the environment when the event is executed. Finally, we require that the marking respect the type of events, i.e. if* $Ex(e) = (n, b)$ *then* $T(e) = \mathsf{Bool}$, *if* $Ex(e) = (n, z)$ *then* $T(e) = \mathsf{Int}$ *and if* $Ex(e) = (n, \bot)$ *then* $T(e) = \mathsf{Void}$. *For* $d \in \mathsf{Exp_E}$ *let* $[[d]]_M$ *be the inductively defined evaluation relation on events relative to the marking* M *where* $[[e]]_M = v$ *if* $Ex(e) = (n, v)$ *and defined in the obvious way for the rest of the cases.*

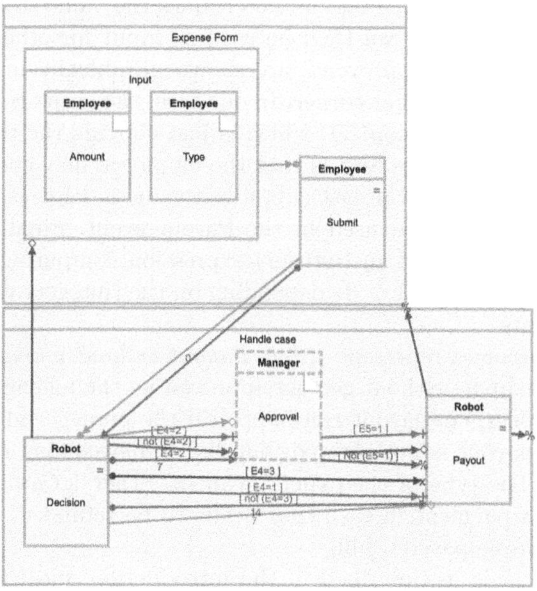

Fig. 2. Expense report with Decision

We explain the definition by introducing our running example of a timed DCR graph with decisions and data shown in Fig. 2 as it appears in the DCR

Solutions process design tool. The graph represents the rules of a process for handling an expense report. The expense report has a type and an amount and a decision resulting in either direct approval, requirement of manager approval or rejection (with the possibility of resubmission at any time before payout).

2.1 Events and Decisions

The nodes of a DCR graph is a set of labelled events E. This graph has six events E with id's $E1$, $E2$, $E3$, $E4$, $E5$, $E6$ and labelled respectively Amount, Type, Submit, Decision, Approve and Payout, and depicted as boxes with the label in the middle (the graphical notation does not show the identities of the events).

Compared to classical timed DCR graphs [12], the graph contains decision and input events, formally defined by the decision function $D : E \rightarrow \mathsf{Exp}_{\mathsf{E}} \uplus \{?\}$. The events $E1$, $E2$, and $E5$ are input events, i.e. $D(E) = \{?\}$ for $i \in \{E1, E2, E5\}$. The events labelled Submit, Decision,Payout are decision events.

Input events with a defined type, in the example the events labelled Amount, Type and Approve, are shown with a flipped paper in the upper right corner to indicate that data is provided as part of their execution. Input events with no defined type gets the null value and correspond to events in classical DCR graphs without data.

Decision events are assigned an expression by the function D, which is computed when the event is executed and determines the value stored for the event in the marking. This value can then be used as input for other decisions or as guards on relations. Decision events are shown graphically in the tool with a small = sign in the upper right corner. In our case, the event $E3$ labelled Submit is assigned a simple expression $E1$, which is just copying the value of the event $E1$ holding the amount. The idea is that the employee may change the amount after submission, but then the last submitted value is kept as the value of the submit event, which is then used in the Payout event. Finally, the event $E4$ labelled Decision is assigned an (integer) expression computing a decision with one out of three outcomes $\{1, 2, 3\}$ depending on the (integer) values of the Type and Amount events.

The three outcomes represent 1: has *Payout without manager approval*, i.e. the payout can happen without getting approved by the manager, and 2:*Payout if manager approves*, i.e. payout can happen if the manager also approves, and 3:*Resubmit*, i.e. payout is not allowed, but the employee may resubmit and get a new decision. The type of the expense can be either 1:*Cash expense*, i.e. the employee made a payment in cash that needs to be refunded, or 2:*Payment of bill*, i.e. the employee payed a bill.

The decision logic as described in the introduction and the DMN decision table in Fig. 1 can be expressed as the convoluted if, then, else statements,

$$\begin{aligned}&\mathsf{if(E0} < \mathsf{200)\ then\ 1\ else(if\ ((E0} >= \mathsf{200)\ and\ E1} = \mathsf{2)\ then\ 2\ else}\\&\quad\mathsf{(if\ (E0} >= \mathsf{5000)\ and\ E1} = \mathsf{1)\ then\ 3\ else\ 2))}\end{aligned}$$

Finally, the example in the figure also has three nesting events, Expense Form, Input and Handle case, which are not part of the formalisation given above, but implemented in the tool and formalised in [10]. They are just short hands for having the same relation from or to a group of events, making the diagram more readable. We see examples of this when we describe the meaning of relations below.

The marking M captures the current state of the DCR graph. A marking comprises three elements (Ex, Re, In) where $Ex : E \rightharpoonup (\omega \times V_\perp)$ is partial function of which the domain defines the set of executed events and the value of the function, when defined, provides the time since the last execution of the event and the value provided as result of a decision or as input, $Re : E \rightharpoonup \infty$ is partial function of which the domain defines the set of pending response events and the value of the function, when defined, provides the deadline for when the event must be executed in the future, and $In : E \rightharpoonup \{\top\}$ is a partial function defining the set of included events. Given an event $e \in E$ of a DCR graph with marking $M = (Ex, Re, In)$, we let $e \in Ex$ be short for $Ex(e)$ being defined, $e \in Re$ be short for $Re(e)$ being defined and $e \in In$ be short for $In(e)$ being defined.

In the graphical notation we indicate that an event is excluded by giving it a dashed border. In the example, the event $E5$ labelled Approve is excluded and the rest of the events are included. We indicate by a green checkmark that the event has been executed in the past and with a blue exclamation mark that the event is pending, i.e. required to be executed if it is included. There are no executed nor pending events in Fig. 2.

2.2 Relations

The relations $\rightarrow\!\!\bullet$ (in orange) and $\bullet\!\!\rightarrow$ (in blue) are the guarded, timed condition and response relations. We write $e \overset{[g]}{\rightarrow\!\!\bullet}_k e'$ for $(e, k, g, e') \in \rightarrow\!\!\bullet$, representing that e is a condition for e' with delay $k \in \omega$ and guard g, meaning that if g evaluates to true in the present marking, then the target event e' can not happen before k time steps[1] after the last execution of the source event e. We simply omit the guard if it is constantly true and omit the delay if it is 0, which correspond to the standard un-timed and un-guarded condition relation. In the example in Fig. 2 we have a standard condition relation from the (nesting of) input events Type and Amount to the event Submit, meaning that the type and amount needs to be filled out before the form can be submitted. Similarly write $e \overset{[g]}{\bullet\!\!\rightarrow}_k e'$ for $(e, k, g, e') \in \bullet\!\!\rightarrow$, representing that if g evaluates to true after execution of e in the current marking, then e' is a response to e with deadline k. Deadlines can be in the range $\infty = \omega \cup \{\omega\}$, i.e. a finite number or infinite ω, the latter meaning that the target event should happen eventually after the source event happens (or be excluded), allowing the specification of liveness properties. Again, if the guard is true we omit it and we omit the deadline if it is ω, corresponding to

[1] A time step is a discrete duration of time orthogonal to the execution of events. A time step can represent any duration, examples being a computers clock cycle, a day or a year. For the examples in this paper a time step is a day.

the standard untimed and unguarded response relation. In the example in Fig. 2 we have an unguarded response relation with delay 0 from Submit to Decision, meaning that the decision is required to be taken in the same timestep, i.e. the same day, as the form is submitted. We have a response with deadline 7 and guard $E4 = 2$ from Decision to Approval. This means that if the event $E4$, i.e. the Decision event, evaluates to 2 then Approval should happen within 7 days, unless it is excluded. Similarly, we have a response with deadline 14 and guard $not(E4 = 3)$ from Decision to Payout. This means that if the event $E4$, i.e. the Decision event, evaluates to anything but 3 (reject), then Payout should happen within 14 days, unless it is excluded or the obligation is removed again. We write $e \overset{g}{\bullet\!\!\to\!\times} e'$ for $(e, g, e') \in \bullet\!\!\to\!\times$, which is the guarded *no-response* relation shown in brown in the example. The meaning is that if the guard is true after execution of e then the event e' is set to not pending, i.e. any current obligation of executing e' is removed. We have a no-response relation with guard $E4 = 3$ from Decision to Payout. This means that if the event $E4$, i.e. the Decision event, evaluates to 3 (reject), then Payout is no longer required to happen within 14 days (unless another decision is made).

We write $e \overset{g}{\to\!\diamond} e'$ for $(e, g, e') \in \to\!\diamond$, which is the guarded *milestone* relation, shown in purple in the example. The meaning is that if the guard is true in the current marking and e is included and pending, then e' can not happen. In the example we have an unguarded milestone relation from Approval to Payout. This means that Payout can not happen if Approval is included and pending. Milestones are often used together with conditions, so the tool allows a shorthand for combining these two relations in one orange arrow looking like the milestone relation. We have such a combined milestone and condition with delay 7 from Decision to Payout, giving the manager 7 days to approve if necessary and not allowing payout if a new decision is required, which can happen if the Employee resubmit the expense form.

The relation $\overset{g}{\to\%}$ is the guarded exclude relation and the relation $\overset{[g]}{\to +}$ is the include relation. An exclusion (resp. inclusion) relation means that the targeted events is excluded (resp. included) from the graph when the source event happens. While excluded, an event cannot be executed, and any deadline on it and conditions from it are ignored. In the example we have an inclusion relation from Decision to Approve with the guard E4 = 2, and a guarded exclusion relation from Decision to Approve with the guard not(E4 = 2). This means that the event Approve is included if the outcome of decision is 2 (requires manager approval) and otherwise excluded. Similarly, we have an inclusion relation from Approve to Payout with the guard E5 = 1, and a guarded exclusion relation with the guard not(E5 = 1). This means that the event Payout is included if the outcome of manager approval is 1 (approve) and otherwise excluded.

2.3 Semantics

The dynamic behaviour of a DCR graph is defined by advancing time or execution of activities.

First we define when events are enabled and time can be advanced.

Definition 3 (Event and time step enabling). *Let* $(E, T, D, M, \rightarrow\!\bullet, \bullet\!\rightarrow, \bullet\!\rightarrow\!\times, \rightarrow\!\diamond, \rightarrow\!+, \rightarrow\!\%, L, l)$ *be a timed DCR graph with decisions and data. An event* $e \in E$ *is* enabled *for the marking* M*, writing* $\mathsf{enabled}(M, e)$ *if and only if:*

1. $e \in In$
2. $\forall e' \in In.\ e' \xrightarrow{[g]}\!\bullet_k e \wedge [[g]]_M = \mathtt{true} \implies (e' \in Ex \wedge k \leq Ex(e'))$
3. $\forall e' \in In.\ e' \xrightarrow{[g]}\!\diamond\, e \wedge [[g]]_M = \mathtt{true} \implies Re(e') = \bot$

For $n \in \omega$ *we say that the* time step n is enabled*, written* $\mathsf{enabled}(M, n)$*, if* $miner_G \geq n$*, where* $miner_G = min\{Re(e) \mid \exists e \in Re \cap In\}$ *is the minimal response deadline on an effectively included and effectively pending event.*

The conditions state that for an event e to be enabled, (1) it must be included. (2) Whenever e has a condition relation with guard that evaluates to true in the current marking from an included event e' with delay k, then this e' was executed at least k time steps ago. (3) No included e', with a milestone relation to e with guard that evaluates to true in the current marking, is pending. A time step n denotes that time advances n steps and can only happen if there are no included pending events with deadline less than n.

The definition of enabledness conservatively extends the definition of enabledness in timed DCR graphs without guards, data and decisions.

Lemma 1. *Given a graph G with all guards on relations being the constant* `true` *and all events of type* void*, i.e. for all* $e \in E$*,* $T(e) = \mathsf{void}$*), an event e is enabled according to Definition 3 iff it is enabled under the definition of enabledness for the corresponding timed DCR graphs* [12] *with no guards, decisions and data.*

In Fig. 2, time step events of any length and the Type and Amount events are enabled, the other events are blocked by conditions.

Below we define the immediate effect of executing an event e for a given marking M. First we introduce some notation for updating markings.

Notation. *When* $f : X \to Y$ *is a (possibly partial) function, we write* $f[x \mapsto y]$ *for the function* $f' : X \to Y$ *which is identical to* f*, except* $f'(x) = y$*. We apply this notation also to sets, taking* $f[x \mapsto y \mid P(x, y)]$ *to be the function* f' *which is identical to* f *except that* $f'(x) = y$ *for all* x, y *satisfying the given predicate* $P(x, y)$.

We now define the effect of executing an event. Compared to classical DCR graphs, the execution of an event e in a DCR graph with data and decisions have two kinds of executions: Executions of input events, where a value of the right type is to be provided as input during the execution and executions of decision events, where the value is computed by the expression associated to the event.

Definition 4. *Let G be a timed DCR graph with data and decisions and marking $M = (Ex, Re, In)$. The immediate effect of executing an enabled event e to be the marking* $\mathsf{effect}_G(M, e) = (Ex', Re', In')$ *where*

$$Ex' = Ex[e \mapsto (0, v)]$$

$$Re' = Re[e \mapsto \bot][e' \mapsto \bot \mid \exists e \overset{[g]}{\bullet\!\!\rightarrow\!\times} e'.[[g]]_M = \mathtt{true}\}]$$

$$[e' \mapsto k \mid \exists k'.\, e \overset{[g]}{\bullet\!\!\rightarrow}_{k'} e' \wedge [[g]]_M = \mathtt{true}$$

$$\wedge\, k = min\{k' \mid e \overset{g}{\bullet\!\!\rightarrow}_{k'} e' \wedge [[g]]_M = \mathtt{true}\}]$$

$$In' = In[e' \mapsto \bot \mid \exists\, e \overset{[g]}{\rightarrow\!\%} e'.[[g]]_M = \mathtt{true}]$$

$$[e' \mapsto \top \mid \exists\, e \overset{[g]}{\rightarrow\!+} e'.[[g]]_M = \mathtt{true}]$$

where v is a value of type $T(e)$ provided as input by the environment if $D(e) = ?$ and $v = [[D(e)]]_M$, if $D(e) \neq ?$.

The result of executing an event e is a new marking where the time since the last execution of the event e is set to 0 and the value the event e is first set to be non pending, and then all events e' with a no-response relation $e \overset{[g]}{\bullet\!\!\rightarrow\!\times} e'$ where the guard g evaluates to true is also set to non pending, and last all events e' with a response relation $e \overset{[g]}{\bullet\!\!\rightarrow} e'$ where the guard g evaluates to true is set to pending with the deadlines being the minimal deadline of such a response relation. Finally, all events e' with an exclude relation $e \overset{[g]}{\rightarrow\!\%} e'$ where the guard g evaluates to true is set to excluded, and then all events e' with an include relation $e \overset{[g]}{\rightarrow\!+} e'$ where the guard g evaluates to true is set to included.

The definitions of enabledness and execution of events and time-steps above define a labelled transition system for a timed DCR graph with decisions and data and marking M. The states of the transition system are markings, with the initial marking being the marking M. There is a transition labelled $e?v$ from M to M' if e is an input event enabled in M and v has the type $T(e)$, and the result of executing e and providing the value v is the marking M'. There is a transition labelled e from M to M' if e is a decision event enabled in M, and the result of executing e is the marking M'. Finally, there is a time step transition labelled n from $M = (Ex, Re, In)$ to M' if the time step n is enabled in M and $M' = (Ex \oplus n, Re \ominus n, In)$ where $(Ex \oplus n)(e) = min\{Ex(e) + n, maxc_G\}$ if $Ex(e)$ is defined and undefined otherwise, and $(Re \ominus n)(e) = max\{Re(e) - n, 0\}$ if $Re(e)$ is defined and undefined otherwise, where $maxc_G = max\{k \mid \exists(e', k, g, e'') \in \rightarrow\!\!\bullet\}$ is the maximum condition appearing at a condition relation. The definition means that for a time step of length n, the time since the last execution of all events is increased by n up-to the maximum delay on any condition in the graph, the deadline on responses is decreased by n or set to 0, if the deadline is less than n. As also explained in Sect. 2, the limits on these operations ensure that there is a finite number of reachable markings for any standard timed DCR graph, meaning that also the transition system has a finite state space. This is however trivially no longer true if the we have unbounded integer types. Indeed, it is easy to prove that DCR graphs with decisions and data, even without time, can encode counter machines and thus Turing complete.

Proposition 1. *DCR graphs with data and decisions are Turing complete.*

However, if one restrict data types to finite domains we regain that the transition systems are finite state.

Proposition 2. *The LTS for any finite timed DCR Graph G with data and decisions has a finite number of reachable states.*

DCR graphs has a notion of acceptance for finite and infinite execution sequences which makes them able to express all regular and ω-regular languages [5].

We define the accepting runs of a DCR graph as follows.

Definition 5 (Accepting runs). *Let G be a timed DCR graph with data and decisions with events E and $\bar{\alpha}$ be a finite or infinite transition sequence $M = M_0 \xrightarrow{\alpha_1} M_1 \xrightarrow{\alpha_2} \cdots$ of length $k \in \infty$ with $M_i = (Ex_i, Re_i, In_i)$. We say that $\bar{\alpha}$ is an* accepting run *and write* $\mathsf{accept_t}(\bar{\alpha})$ *iff*

(i) For all $i \leq k$ and all $e \in E$, if $e \in Re_i$ and $e \in In_i$, then for some $j > i$ we have $\alpha_j = e$ or $\alpha_j = e?v$ or $e \notin Re_j$ or $e \notin In_j$
(ii) If $\bar{\alpha}$ is infinite, it contains infinitely many time steps.

Here (i) captures that if an event ever is pending and included it will eventually be executed or no longer pending and included. (ii) captures that an accepting run is non-zeno, i.e. only finitely many event transitions can happen between each time step transition.

Definition 6 (Trace). *A trace t of a graph G with initial marking M is a sequence $\alpha_1, \alpha_2, \alpha_3, \cdots$ such that there exists an accepting run $M \xrightarrow{\alpha_1} M_1 \xrightarrow{\alpha_2} M_2 \xrightarrow{\alpha_3} \cdots$. We write Traces$(G)$ for the set of all possible traces in the graph G.*

Example traces of our running example are:

1. Low cash expense: Amount?100,Type?1,Submit,Decision,8,Payout
2. Medium cash: Type?1,Amount?800,Submit,Decision,6,Approve?1,1,Payout
3. High cash: Type?1,Amount?6000,Submit,Decision
4. High cash changed to bill: Type?1,Amount?6000,Submit,Decision,Type?2, Submit,Decision,7,Approve?1,Payout

3 Related Work

The first, basic version of Dynamic Condition Response (DCR) graphs were introduced shortly after the similar declarative modelling language DECLARE [19]. Both languages provides a graphical notation for declarative constraints between events inspired by the property specification patterns of [6], with the shared aim to provide a notation for processes requiring more flexibility than the, also often referred to as Knowledge intensive Processes (KiPs) [7]. A key difference between DCR and DECLARE is that the semantics of DECLARE is based on constraints representing constraints in (finitary) LTL [20], which can then be translated to finite automata. DCR can be described as a temporal logic parametrized over a dynamically changing (through inclusion and exclusion) set of events, but the standard semantics of DCR is given directly as in the present paper as a transitions between finitary markings of events. It has been proven that standard DCR graphs can express the union of all omega-regular languages and regular languages [5,17] and is thus formally more expressive than DECLARE by allowing the specification of all omega-regular liveness properties. As stated in the introduction, DCR as been extended in several ways. Some extensions (time and single-instance sub processes) preserve finite state property of models, while combinations of other extensions (fresh events and multi-instance sub processes) makes the model Turing complete [5]. As shown in the present paper, DCR graphs with time and finite data domains is finite state, while unbounded integers again makes the model Turing complete. Related to the present work is the language ReSeDa [21] combining semi-structured data with reactive computations defined using DCR-like constraints.

There have also been several extensions of DECLARE since its inception. In [16], semantics of data-aware constraints is DECLARE is given. The semantics is given in the Event Calculus and thus not expressed directly on DECLARE. DECLARE is extended with support for resources and DMN decisions in [15] and exemplified by the modelling of a flexible process at a emergency department of a hospital. The work in [15] relates to the approach in the present paper in that it also allows constraints to be guarded by decision models. However, the approach does not allow to specify when the event of making the decision happens, as in the present paper. Finally, none of the other works are supported by commercial process design and workflow management systems used in industry.

4 Conclusions and Future Work

We provided a formal semantics for timed DCR graphs with decision events and data, and showed how the model supports merging declarative decision modelling and process modelling in one notation, exemplified by an expense report process. The formal model is fully implemented in commercial process design and workflow management systems provided by DCRSolutions and also freely available for academic use at dcrsolutions.net. As future work we aim to explore extensions of DCR process mining [1] to time, decisions and data.

References

1. Back, C.O., Slaats, T., Hildebrandt, T., Marquard, M.: Discover: accurate & efficient discovery of declarative process models. arXiv, abs/2005.10085 (2020)
2. Bazhenova, E., Zerbato, F., Oliboni, B., Weske, M.: From BPMN process models to DMN decision models. Inf. Syst. **83**, 69–88 (2019)
3. Debois, S., Hildebrandt, T.: The DCR workbench: declarative choreographies for collaborative processes. In: Gay, S., Ravara, A. (eds.) Behavioural Types: From Theory to Tools, pp. 99–124. River Publishers, June 2017
4. Debois, S., Slaats, T.: The analysis of a real life declarative process. In: SSCI 2015, pp. 1374–1382. IEEE (2015)
5. Debois, S., Hildebrandt, T.T., Slaats, T.: Replication, refinement & reachability: complexity in dynamic condition-response graphs. Acta Informatica **55**(6), 489–520 (2017). https://doi.org/10.1007/s00236-017-0303-8
6. Dwyer, M.B., Avrunin, G.S., Corbett, J.C.: Property specification patterns for finite-state verification. In: Proceedings of the Second Workshop on Formal Methods in Software Practice, pp. 7–15 (1998)
7. Goedertier, S., Vanthienen, J., Caron, F.: Declarative business process modelling: principles and modelling languages. Enterp. Inf. Syst. **9**, 161–185 (2013)
8. Object Management Group. Business Process Model and Notation V 2.0.2 (2014)
9. Object Management Group: Decision Model and Notation V 1.3 (2020)
10. Hildebrandt, T., Mukkamala, R.R., Slaats, T.: Nested dynamic condition response graphs. In: Arbab, F., Sirjani, M. (eds.) FSEN 2011. LNCS, vol. 7141, pp. 343–350. Springer, Heidelberg (2012). https://doi.org/10.1007/978-3-642-29320-7_23
11. Hildebrandt, T.T., Mukkamala, R.R.: Declarative event-based workflow as distributed dynamic condition response graphs. In: Honda, K., Mycroft, A. (eds.) Proceedings of the Third Workshop on Programming Language Approaches to Concurrency and communication-cEntric Software, PLACES 2010, Paphos, Cyprus, 21st March 2010. EPTCS, vol. 69, pp. 59–73 (2010)
12. Hildebrandt, T.T., Mukkamala, R.R., Slaats, T., Zanitti, F.: Contracts for cross-organizational workflows as timed dynamic condition response graphs. J. Log. Algebr. Program. **82**(5–7), 164–185 (2013)
13. Marquard, M., Debois, S., Slaats, T., Hildebrandt, T.T.: Forms are declarative processes! In: BPM 2016 Industry Track (2016, to appear)
14. Marquard, M., Shahzad, M., Slaats, T.: Web-based modelling and collaborative simulation of declarative processes. In: Motahari-Nezhad, H.R., Recker, J., Weidlich, M. (eds.) BPM 2015. LNCS, vol. 9253, pp. 209–225. Springer, Cham (2015). https://doi.org/10.1007/978-3-319-23063-4_15
15. Mertens, S., Gailly, F., Poels, G.: Enhancing declarative process models with DMN decision logic. In: Gaaloul, K., Schmidt, R., Nurcan, S., Guerreiro, S., Ma, Q. (eds.) CAISE 2015. LNBIP, vol. 214, pp. 151–165. Springer, Cham (2015). https://doi.org/10.1007/978-3-319-19237-6_10
16. Montali, M., Chesani, F., Mello, P., Maggi, F.M.: Towards data-aware constraints in declare. In: Proceedings of the 28th Annual ACM Symposium on Applied Computing, SAC 2013, pp. 1391–1396. Association for Computing Machinery, New York (2013)
17. Mukkamala, R.R.: A formal model for declarative workflows: dynamic condition response graphs. Ph.D. thesis, IT University of Copenhagen, June 2012

18. Normann, H., Debois, S., Slaats, T., Hildebrandt, T.T.: Zoom and enhance: action refinement via subprocesses in timed declarative processes. In: Polyvyanyy, A., Wynn, M.T., Van Looy, A., Reichert, M. (eds.) BPM 2021. LNCS, vol. 12875, pp. 161–178. Springer, Cham (2021). https://doi.org/10.1007/978-3-030-85469-0_12
19. Pesic, M., Schonenberg, H., Aalst, W.: Declare: full support for loosely-structured processes. In: Proceedings - IEEE International Enterprise Distributed Object Computing Workshop, EDOC, pp. 287–287, November 2007
20. Pnueli, A.: The temporal logic of programs. In: 18th Annual Symposium on Foundations of Computer Science (SFCS 1977), pp. 46–57 (1977)
21. Seco, J.C., Debois, S., Hildebrandt, T.T., Slaats, T.: RESEDA: declaring live event-driven computations as REactive SEmi-Structured DAta. In: 22nd IEEE International Enterprise Distributed Object Computing Conference, EDOC 2018, pp. 75–84 (2018)
22. Slaats, T.: Flexible process notations for cross-organizational case management systems. Ph.D. thesis, IT University of Copenhagen, January 2015

5th International Workshop on Business Processes Meet the Internet-of-Things (BP-Meet-IoT 2021)

5th International Workshop on Business Processes Meet the Internet-of-Things (BP-Meet-IoT 2021)

The arrival of the Internet of Things (IoT) has put into play a huge amount of interconnected and embedded computing devices with sensing and actuating capabilities that are revolutionizing our way of living and doing business. The incorporation of this technology into the business process management field has the potential to create business processes with higher levels of flexibility, efficiency, and responsiveness, providing as a result a better support to the continuous evolution of business requirements. In addition, the proper combination of these two fields can foster the development of innovative solutions, not only in the business domain where BPM emerged but also in the many different application areas in which IoT is applied (e.g., smart cities, smart agriculture, or e-health).

To ensure that BPM and IoT can mutually benefit from each other, several challenges must still be tackled. Specifically, it must be understood how processes can improve IoT and how to exploit IoT for BPM. For further discussion on how to combine these two paradigms we refer to [1].

The objective of this workshop was to attract novel research at the intersection of these two areas by bringing together practitioners and researchers from both communities that are interested in making IoT-enhanced business processes a reality.

The 5th edition of this workshop was co-organized with the 4th International Workshop on Security and Privacy-enhanced Business Process Management (SPBP 2021). Both workshops attracted nine international submissions (three for SPBP and six for BP-Meet-IoT), which were reviewed by three members of the corresponding Program Committees. Finally, five submissions were accepted and selected for presentation in the co-organized event. Specifically, the following three submissions were accepted for inclusion in the BP-Meet-IoT 2021 workshop program: "A Context and Augmented Reality BPMN and BPMS Extension for Industrial Internet of Things Processes" authored by Gregor Grambow, Daniel Hieber, Roy Oberhauser, and Camil Pogolski; "Supporting Users in the Continuous Evolution of Automated Routines in their Smart Spaces" authored by Estefanía Serral, Daniel Schuster, and Yannis Bertrand; and "The Benefits of Sensor-Measurement Aggregation in Discovering IoT Process Models: A Smart-House Case Study" authored by Massimiliano de Leoni and Lucia Pellattiero.

The first paper is focused on the integration of process automation, Industrial Internet of Things (IIoT) context, and AR. To this end it presents BPMN-CARX, a Context and Augmented Reality eXtension (CARX) for Business Process Model and Notation (BPMN), and the CARX Framework, which enable AR and IIoT context integration with existing Business Process Management Systems (BPMSs). The second paper is focused on supporting the evolution of users' daily routines to address changes in users' behavior. To this end, it proposes a solution that combines two state-of-the-art approaches: MAtE, an end-user model-driven approach, and Cortado, an incremental process mining approach. Finally, the third paper focuses on the data collected and exchanged in IoT systems. To this end, it reports on the experience of using real-life data to discover the human habits in a smart house. In particular, how to aggregate event data to the right granularity to further apply process mining discovery techniques.

The co-organized event took place on September 6, 2021, and around 40 attendees were present during the workshop presentations, talk, and discussion.

As with previous editions of this workshop, we hope that the reader finds this selection of papers and talks interesting and useful to get a better insight into the integration of these two fields from a theoretical and practical point of view.

Agnes Koschmider
Francesco Leotta
Estefanía Serral
Victoria Torres

Reference

1. Janiesch, C., et al.: The Internet of things meets business process management: a manifesto. IEEE Syst. Man Cybern. Mag. **6**(4), 34–44 (2020)

Organization

Workshop Chairs

Agnes Koschmider	Kiel University, Germany
Francesco Leotta	Sapienza Università di Roma, Italy
Estefanía Serral	KU Leuven, Belgium
Victoria Torres	Universitat Politècnica de València, Spain

Program Commitee

Andrea Marrella	Sapienza Università di Roma, Italy
Zakaria Maamar	Zayed University, Dubai, UAE
Vicente Pelechano	Universitat Politècnica de València, Spain
Manfred Reichert	University of Ulm, Germany
Andreas Oberweis	Karlsruhe Institute of Technology, Germany
Matthias Weidlich	Humboldt-Universität zu Berlin, Germany
Pnina Soffer	University of Haifa, Israel
Felix Mannhardt	SINTEF Digital, Norway
Udo Kannengießer	Compunity GmbH, Germany
Mathias Weske	Hasso-Plattner-Institute, University of Potsdam, Germany
Sylvain Cherrier	University Marne-la-Vallée, France
Jianwen Su	University of California, Santa Barbara, USA
Christian Janiesch	University of Wurzburg, Germany

A Context and Augmented Reality BPMN and BPMS Extension for Industrial Internet of Things Processes

Gregor Grambow, Daniel Hieber, Roy Oberhauser(✉), and Camil Pogolski

Aalen University, Aalen, Germany
{gregor.grambow,daniel.hieber,roy.oberhauser,camil.pogolski}@hs-aalen.de

Abstract. In the context of Industry 4.0, smart factories enable a new level of highly individualized and very efficient production, driven by highly automated processes and connected Industrial Internet of Things (IIoT) devices. Yet the IIoT process context, crucial for operational process enactment, cannot be readily represented in processes as currently modeled. Despite automation progress, manual tasks performed by humans (such as maintenance) remain, and while complicated tasks can be supported by Augmented Reality (AR) devices, they remain insufficiently integrated into global production processes. To seamlessly integrate process automation, IIoT context, and AR, this paper contributes BPMN-CARX, a Context and Augmented Reality eXtension (CARX) for BPMN (Business Process Model and Notation) and the CARX Framework, which enables AR and IIoT context integration with existing Business Process Management Systems (BPMSs). An Industry 4.0 case study demonstrates its feasibility and applicability.

Keywords: Business process modeling · Industrial Internet of Things · Business Process Model and Notation · Augmented Reality · Context-awareness

1 Introduction

The rise of smart factories with their highly automated processes and networked IIoT devices provides not only advantages, but also new challenges for workers and process engineers alike. First, while machines are becoming more sophisticated and digitized, the IIoT evolution provides many additional capabilities such as automatic sensor measurements and connected predictive maintenance solutions while improving and minimizing maintenance by providing *contextually-relevant data*. Second, *process automation* via Business Process Management (BPM) enables the creation and enactment of workflows that integrate business and factory processes; tasks can be decomposed into self-contained subtasks and a clearly structured model can represent each process, e.g., via BPMN 2.0 [4]. Third, the spread of AR devices enables completely new industrial (training or maintenance) scenarios for supporting humans - primarily via *visual augmentation*,

A. Marrella and B. Weber (Eds.): BPM 2021 Workshops, LNBIP 436, pp. 379–390, 2022.
https://doi.org/10.1007/978-3-030-94343-1_29

where complex, dangerous, or expensive operations can be augmented or simulated for safety, effectiveness, and efficiency. While each of these three factors provides a focused solution for their area, none integrates or includes the other two factors comprehensively. Therefore, holistic solutions for smart factories remain scarce and are often limited to specific use cases. Consequently, further investigation is needed to improve this situation.

In prior work, we developed a holistic concept for contextual process management in the software engineering domain [8], while with VR-BPMN [11] we demonstrated BPMN model visualization in Virtual Reality (VR). This paper contributes BPMN-CARX, the Context and Augmented Reality eXtension for BPMN, a generic approach towards amalgamating IIoT context, process automation, and visual AR. A prototype extends a prevalent BPMN modeling tool and integrates with two commercial BPMSs. Its feasibility and applicability to address the aforementioned issues is demonstrated with a case study.

The paper is structured as follows: Sect. 2 elaborates related work, while Sect. 3 describes our solution concept. In Sect. 4 realization details are provided. Thereafter we introduce a case study conducted with CARX in Sect. 5. Finally, Sect. 6 provides a conclusion and outlook on further upcoming work.

2 Related Work

As to the integrating AR/VR with BPM, besides our prior work with VR-BPMN [11], Poppe et al. [12] implemented a collaborative process modeling approach utilizing AR, but it was limited to modeling and lacked an implementation for process enactment and IoT device integration. BPM to Go combined AR and sensor data with BPM and enactment on a mobile device [14], but does not address BPMN aspects for AR and IIoT context, nor shows how to generically integrate and extend BPMSs to consider AR and IIoT in a unified context.

A combination of AR, IoT, and BPM, can be found in the HoloFlows project [18] which provides process modeling software in mixed reality for the HoloLens, supporting end users during IoT device setup and interaction modeling. While HoloFlows guides users through processes and includes real-time sensor data, no tool to create or model these processes is mentioned. BPMN modeling is not supported. An IIoT solution with AR-supported worker tasks is as yet unrealized.

Regarding IoT with BPM, Schönig et al. [17] showed the feasibility of an IoT BPM framework utilizing BPMN. They designed and implemented an approach for IoT-aware business processes integrating IoT sensor data into predefined processes in a modern BPMS. The solution integrates a smart watch during process execution, providing workers with guidance and real-time sensor data. However, the approach lacks an easy way to model those processes, requiring knowledge in the IoT and BPMN domain. Further, it lacks AR support to augment the environment - only a smart watch interface was provided.

For extending BPMN processes with context, BPMN4SGA [19] extends BPMN with smart glasses, but lacks AR actions via a BPMS and is limited primarily to documentation. BPMN4CPS [7] describes an approach combining BPMN with cyber-physical systems, but lacks the capability to integrate AR directly.

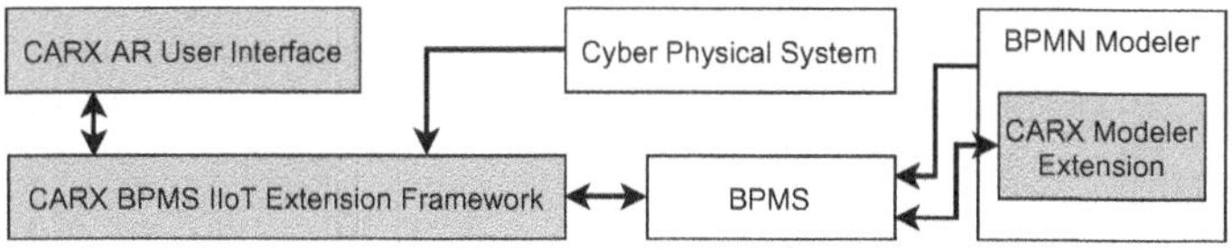

Fig. 1. High-level logical architecture

As to context modeling, domain-specific modeling (DSM) languages like SenSoMod [5] add context to mobile applications, but only cover conventional aspects like email, calendar, etc. Our approach works with predefined templates that are readily extensible and consider industrial process context.

3 Solution Concept

Our solution concept needs to address the following problems when combining IoT and BPMN, as elaborated by Hasić and Serral Asensio [9]: (P1) IoT resource-awareness in tasks; (P2) IoT resource binding at runtime; (P3) IoT resource malfunction event handling; (P4) IoT data retrieval: push vs. pull; (P5) IoT tasks and human resources communication. The CARX DSM extension for BPMN addresses the following requirements:

- R1 Natural modeling support for context-aware and AR business processes (i.e., conformance to current business process modeling norms and notations)
- R2 Easy integration with existing prevalent BPMSs
- R3 Ability to integrate IIoT context information during process modeling
- R4 Real-time sensor data for system and workers during task execution
- R5 Integrated AR support during business process enactment (no need to switch between software solutions or platforms)
- R6 Just-in-time rule execution utilizing the latest IIoT sensor data values
- R7 Worker task assignment in accord with the extended (comprehensive) context by utilizing available data in an intelligent way

The intent with the CARX architecture is to avoid a highly-tailored solution limited to a specific use case, or one that only addresses AR support or IIoT data in processes, or one that only enriches BPMN with additional context. Thus, we introduce our tripartite solution concept consisting of: AR user interface support, a BPMS IIoT extension framework, and our BPMN modeling extension CARX that links these three components. The high-level logical architecture can be seen Fig. 1, where the connections and information flow between the different components is originating from the BPMS extended with CARX.

CARX AR User Interface. The user interface must provide AR support to workers during execution and act as a device to control the BPM task without the need to switch clients (R5), therefore hand free devices with input/output (IO) options like smart glasses should be preferred, as they can be used without limiting the users action range. To also provide a solution which can be integrated with less effort and cost (R2), other AR devices like smartphones and tablets should also be supported. The interface shall be capable of providing current IIoT sensor data to the user (R4, P5).

CARX BPMS IIoT Extension Framework. Rather than extending a single BPMS, we want to provide a generic solution applicable for most BPMS and therefore reduce the integration effort (R2). To achieve this, we introduce a supporting framework providing necessary functionality via generic interfaces to the BPMS. This framework supports IIoT machine data utilization and IIoT context-aware rule execution (R6, P1, P2), which can also be used to resolve resource issues regarding IIoT machines (P3) and intelligent assignment (R7).

BPMN-CARX Realization and Modeler Extension. BPMN-CARX links the other components and enables easy modeling of new AR-supported, IIoT-aware processes. Rather than providing a new Domain Specific Language (DSL), we extend BPMN 2.0, easing the integration with existing process models and their associated tools (R2), providing a natural environment for process engineers (R1), and reusing BPMN elements. In order to handle a task assignment, we utilize a script task that requests the assignment calculation from the CARX Framework. The framework then calculates the optimal assignment in a synchronous way and sends the response to the script task, or, in asynchronous mode - directly handles the assignment via the REST API of the BPMS. To support the creation of new context-aware AR-supported process models, we also provide a CARX modeling tool extension. It guides process engineers for creating service tasks for assignments and adding associated data objects as in- and output data objects, while also displaying all associated context information. The reuse of existing elements further minimizes the integration effort for existing models, as only some generic script tasks have to be added.

4 Realization

The CARX framework was implemented with the Python Django framework [10] for fast prototyping and scalability. Our prototype demonstrates our generic BPMS integration concept via both Camunda [2] and the AristaFlow BPM Suite [15]. While the AristaFlow BPM Suite already provides data stores for users, resources, and machines, the Camunda BPMS only provides a minimal user store and no further context. It was therefore extended with data stores for users, resources, and machines containing all required information. Rather than implementing rule processing ourselves within the supporting framework, we integrated the prevalent Drools [13] rule engine to support easily customizable and accessible context-based logic. The communication between the different applications is handled via REST for the application logic which is supported by most modern BPMS, and MQTT utilizing a Mosquitto broker for the transmission of sensor data in compliance with the OPC-UA industry standard [6]. By utilizing a Publish/Subscribe architecture sensor values from the IIoT domain are available to interested parties, and the BPMS, support framework, or worker can choose when to consume these (solving P4).

4.1 CARX Context Realization

The CARX context model consists of five major blocks (cf. Fig. 2A). The Global System (red) handles all organization-wide rules and context information. The BPMN Process context (violet) is responsible for the process-specific rules and data. The BPMN Task (blue) handles the BPM task specific rules, data, and AR configuration. The AR Context (BPMN-AR green and Template AR grey) consists of two parts and is only applicable for tasks utilizing AR support (R5). BPMN-AR Context is defined in the BPMN modeler, addressing how what to display in the AR-Component. The Template AR Context is defined in the specific AR template (e.g., Unity) and must be modeled in advance. In the following, these are described in more detail.

Global System (Global Context). This is the top-level object for all processes in CARX environment. It consists of a Global Context and the BPMN processes. The Global Context itself contains URLs to all required data stores, containing meta-data for workers, resources, machines, and an optionally connected rule engine. It includes the framework URL, where the intelligent assignment for tasks according to real-time sensor data and worker information can be requested. As a final component, the Global Context contains a rule set. This set defines default organizational rules which are executed before each task. A rule in this rule set consists of the rules name, a boolean indicating if it should be looped until the rule is successfully fulfilled, and an optional key/value list, providing information about which values have to be submitted to the rule engine with the request, e.g., which machine sensors.

BPMN Process (Process Context). The BPMN Process is a default BPMN 2.0 process extended with a Process Context. It further contains all BPMN tasks of the process. The Process Context itself contains a process rule set which contains rules to be executed by default before each task of the process. It is further possible to overwrite the Global Context on the process level, e.g., removing a global rule or changing the URL to a data store.

BPMN Task (Task Context). The BPMN Task consists of a default BPMN 2.0 task extended with the Task Context. Like the other contexts, it contains a rule set, allowing the addition of new rules and the option to remove rules from the Global Context or Process Context. Further, the Task Context must contain all information to provide a foundation for an intelligent task assignment (R7), enable AR support (R5), and provide real time sensor data from connected IIoT devices (R4). To provide real time sensor data during task execution, the context contains information about all connected machines in the machine variable. Rather than selecting a specific machine during creation, the type of machine is selected at design-time and then assigned during runtime by the supporting framework's assignment logic (P2). Smart assignments are enabled by the extension, with further context information consisting of the position where the task must be executed, dangers connected to the execution (e.g., noise or heat hazards), the worker qualification needed to successfully complete a task, and resources required during the task, which could also require special qualifications

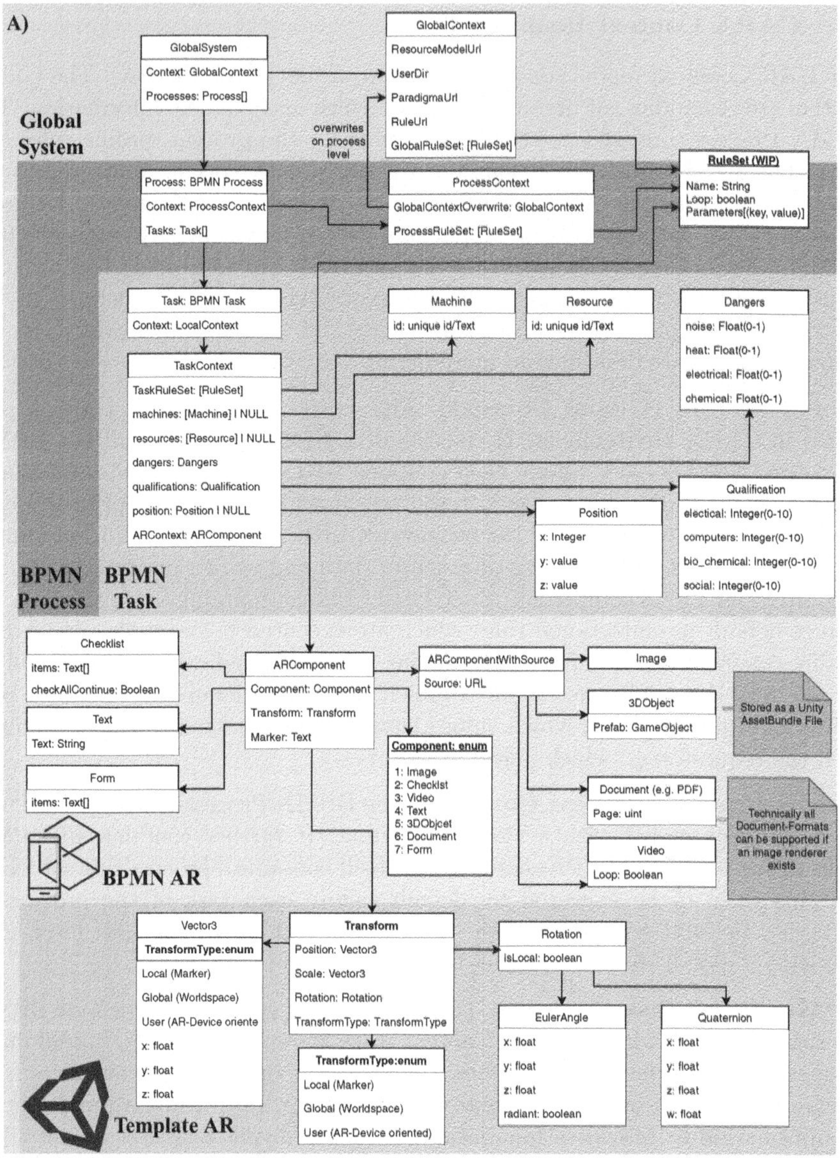

B)

```
...
<camunda:inputParameter name="machineUrl">https://paradigma.hs-aalen.de/api/types/machines</camunda:inputParameter>
<camunda:inputParameter name="machine">Roller</camunda:inputParameter>
<camunda:inputParameter name="rulesUrl">
    https://paradigma.hs-aalen.de:8180/kie-server/services/rest/server/containers/instances/
</camunda:inputParameter>
...
```

Fig. 2. A) CARX context data model B) Camunda CARX XML snippet (Color figure online)

or hazard clearance. The information required for AR support is defined in the AR Context of the Task Context.

AR Context. The AR Context provides all information to display AR relevant elements to the workers during task execution, providing them with assistance and guidance (R5) while supporting workflow control from within the AR Client. The context itself is split in two parts, the BPMN-AR Context and Template AR Context (Fig. 2A). BPMN AR displays the configuration defined during process creation, while Template AR is predefined in the AR templates.

4.2 CARX Framework Realization

To connect the sensor data to the system, all IIoT devices are connected to the Mosquitto broker and provide their data regarding system-wide topics. If the information has to be shown to a worker during a task execution, the AR client can subscribe to the topic and display the latest sensor values. The information about which sensor value is required and to which topic it is published is contained in the machine's data store. To retrieve this metadata, the CARX machine information contains an ID that can be used by the framework to send a request to the machine data store and receive all required data. If sensor values are required during validation of rules defined in a rule set, the framework retrieves the latest sensor data from the Mosquitto broker and sends it, together with the rule requiring validation to the rule engine. In this case, the information as to which sensor values are required, is provided by the parameters values of the specific rule and the machine's properties provided in the Task Context.

To support modeling IIoT context-aware AR-supported processes with our model (Fig. 2A), we extended the Camunda Modeler. This enables context and CARX processes to be modeled in a guided manner without additional software. The required extended context information is then saved as Data Objects inside the XML Camunda Process template (cf. Fig. 2B). The integration of the context in the modeler is shown in Fig. 3. The Global Context (Fig. 3A(1) and Fig. 3B)) can be added via the new tab in the top toolbar, clearly separating it from process-specific configuration. The Task Context and Process Context can be edited by selecting either a task or process and selecting the Context tab (Fig. 3(2)) of the respective element. The Task Context is displayed in Fig. 3C while the Process Context is identical to the Global Context in Fig. 3B. This mirrors the default Camunda Modeler control and eases the use of the extension.

Various facets of the realization address context integration. The URLs defined in the Global/Process Context define where to retrieve specific context information. In the Task Context, the required meta data is requested from the data stores defined at their URLs via REST and directly displayed in the modeler (as seen in Fig. 3 for the Machine or Group field). Both consist of dropdown values, displaying all possible types of Groups this task could be assigned to or Machines that could be used during task execution.

The AR client is realized with the Unity Engine and the AR Foundation framework [3], utilizing its support for the major device-specific AR frameworks ARCore, ARKit, Magic Leap, and HoloLens. According to the CARX concept

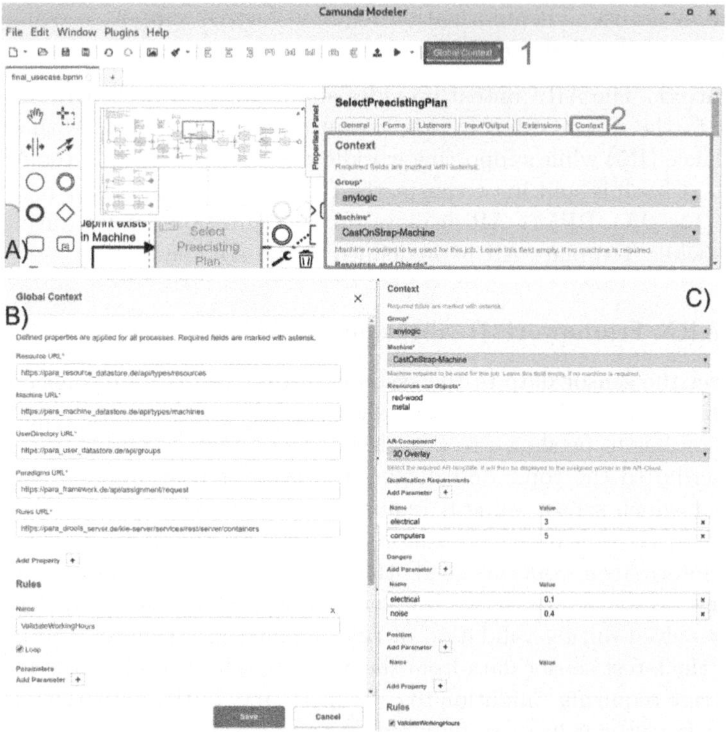

Fig. 3. The context-aware Camunda modeler

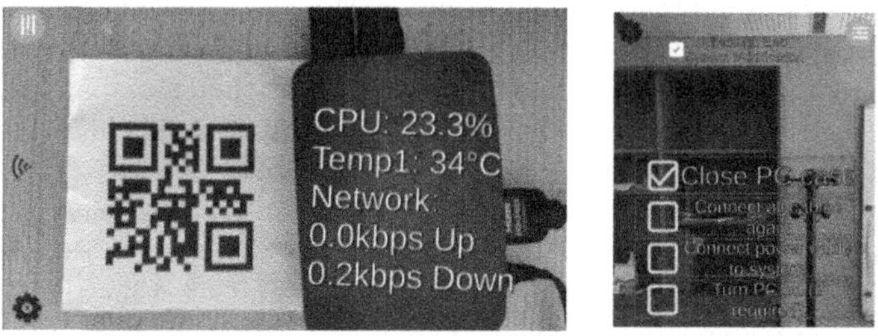

Fig. 4. AR image of live sensor data (left) amd task checklist (right)

in Fig. 2A) we implemented eight possible templates ranging from the display of simple images or documents to checklists (cf. Fig. 4), handling business logic, and 3D overlays, showing complicated procedures. Further, the AR client can display relevant real-time sensor values as shown in Fig. 4. Enabling the worker to react to changes in relevant machine values without the need to constantly check the machines directly or use different software interfaces provide direct IIoT context-aware AR support (R5). Adding further templates to the AR Context is currently supported by extending our Unity AR build.

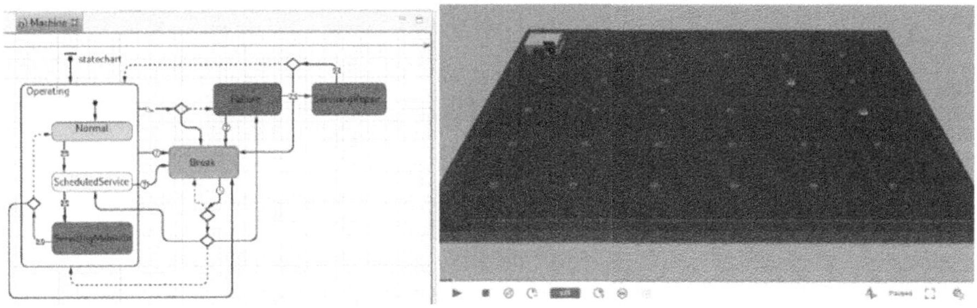

Fig. 5. AnyLogic simulation showing machine model and factory with machine state

5 Case Study

A case study using a simulation with AnyLogic [1] is used for the evaluation, due to COVID-restricted industrial access. A partial overview of the simulation structure can be seen in Fig. 5. All required systems for the case study (including the Camunda engine) were started with a single docker-compose file on a Linux server, minimizing deployment effort. This simulation consists of the maintenance and repair of 28 cast-on-strap machines in a factory hall using permanent employee and external specialist agents. The machines require a scheduled maintenance every 12 and require some kind of repair every 24 operational hours. If such an event is triggered, a worker is assigned via Staff Assignment Rules in a BPMS, the agent prints required information, navigates to the machine, checks if it is safe to perform the task (a machine has an oven - a potential hazard), uses the printed checklist, and walks back to a desktop client, completing the task via the BPMS interface. With CARX we utilize an intelligent assignment service that evaluates safety and other constraints, and further provide guidance to workers via AR-support and IIoT context awareness. OPC-UA was used for transmitting machine data to the system and workers. The simulation showed that by utilizing context, CARX reduced operational costs via better utilization of permanent employees versus external specialists; machine repair downtime was reduced, but at slightly higher maintenance delays.

The agents and machines were added to their respective data store. The maintenance and repair CARX processes are shown in Fig. 6 for AristaFlow and Camunda. The worker is assigned by the CARX BPMS IIoT Extension Framework, which pre-checks for any safety hazards, and is then guided by an AR navigation overlay to the machine while the temperature of the oven is displayed in the AR goggles as in Fig. 4. Figure 7 shows AR preview in the Modeler (AR content was scaled/cropped to better recognize the checklist), where AR content such as AR checklists (e.g., for maintenance, cf. Fig. 4) or AR-based overlay and instruction documents such as multimedia video, audio, or PDF (e.g., for repair tasks). The worker can directly interact with the process in the AR client, so no additional software client is necessary to inform the BPMS.

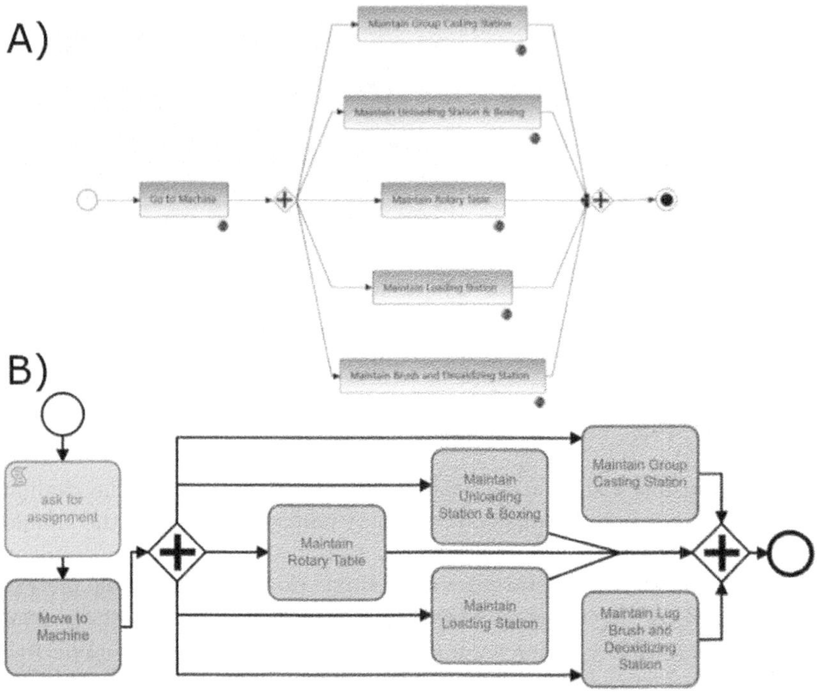

Fig. 6. CARX process maintenance in A) AristaFlow and B) Camunda

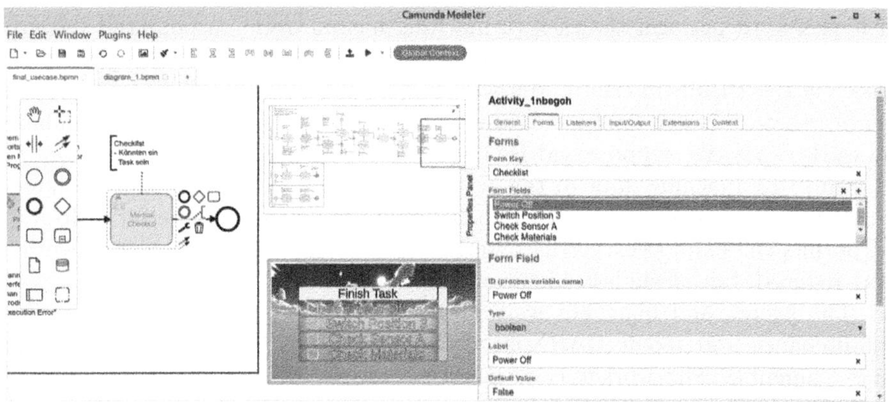

Fig. 7. AR preview in the BPMN-CARX extension in Camunda modeler

By directly including AR, IIoT context, and business processes via the BPMN-CARX modeling extension, the case study shows the feasibility and practicality of integrating AR and IIoT context within conventional process modeling. The CARX process enactment via our CARX BPMS IIoT Framework integration with two BPMSs shows the feasibility of our generic BPMS extension concept towards more holistic and comprehensive integration of AR and IIoT context in BPM automation process enactment. While the case study is directly applicable to real world factories, the navigation for each factory must be addressed individually.

6 Conclusion

This paper described CARX, our holistic solution concept for seamlessly integrating context-awareness for business processes in IIoT settings, visual AR support within business processes, and additional process modeling capabilities in BPMN. Our generic integration approach enables existing BPMS to be easily extended with AR and IIoT capabilities. Natural modeling support for context-aware and AR business processes is supported with the CARX-BPMN extension, conforming to current BPM norms without a totally new notation. AR and IoT domain context information can be integrated during process modeling, real-time sensor data provisioning to the system and workers during task execution is supported, and rule execution is incorporated to utilize real-time IIoT sensor data values. Worker tasks and resources can be assigned in an intelligent way by utilizing the extended comprehensive context. Support for AR is integrated during business process enactment, without the need for AR users to switch between software solutions or platforms. Context-aware AR support to workers provides data, assistance, and guidance during their tasks and next assignments and updates them on the latest data from relevant IIoT devices. BPMN-CARX links IIoT context, process enactment, and BPMN to support implementation of new IIoT context-aware processes or upgrading preexisting IIoT processes. Our realization demonstrated its feasibility, integrating with and extending two commercially available BPMS. The case study showed its applicability in a simulated smart factory environment utilizing OPC-UA, IIoT, and AR.

Future work includes: the optimization of our context-integrated process editor to improve its appearance and usability; integration of our framework and CARX with further BPMSs; further improvements to the CARX BPMN 2.0 extension; integration of human-robot collaboration concepts (e.g., [16]); and a comprehensive empirical evaluation in an industrial environment.

Acknowledgements. The authors thank: Felix Gräber for his advice and assistance with the models and AnyLogic for their simulation suite. Partial funding for this work: PARADIGMS project by "ZIM", a funding program of the Federal Ministry for Economic Affairs and Energy of the Federal Republic of Germany.

References

1. Anylogic. https://www.anylogic.de. Accessed June 2021
2. Camunda. https://camunda.com. Accessed May 2021
3. Unity AR Foundation. https://unity.com/unity/features/arfoundation. Accessed May 2021

4. Business Process Model and Notation. Object Management Group (2013). https://www.omg.org/spec/BPMN/2.0.2/PDF
5. Dörndorfer, J., Hopfensperger, F., Seel, C.: The SenSoMod-modeler - a model-driven architecture approach for mobile context-aware business applications. In: Cappiello, C., Ruiz, M. (eds.) CAiSE 2019. LNBIP, vol. 350, pp. 75–86. Springer, Cham (2019). https://doi.org/10.1007/978-3-030-21297-1_7
6. Enste, U., Mahnke, W.: OPC unified architecture. Automatisierungstechnik **59**(7), 397–404 (2011). https://doi.org/10.1524/auto.2011.0934
7. Graja, I., Kallel, S., Guermouche, N., Kacem, A.H.: BPMN4CPS: a BPMN extension for modeling cyber-physical systems. In: 2016 IEEE 25th International Conference on Enabling Technologies: Infrastructure for Collaborative Enterprises (WETICE), pp. 152–157 (2016). https://doi.org/10.1109/WETICE.2016.41
8. Grambow, G., Oberhauser, R., Reichert, M.: Enabling automatic process-aware collaboration support in software engineering projects. In: Escalona, M.J., Cordeiro, J., Shishkov, B. (eds.) ICSOFT 2011. CCIS, vol. 303, pp. 73–88. Springer, Heidelberg (2013). https://doi.org/10.1007/978-3-642-36177-7_5
9. Hasić, F., Asensio, E.S.: Executing IoT processes in BPMN 2.0: current support and remaining challenges. In: International Conference on Research Challenges in Information Science (RCIS), vol. 13, pp. 1–6 (2019). https://doi.org/10.1109/RCIS.2019.8876998
10. Kaplan-Moss, J., Holovaty, A.: The Definitive Guide to Django, 2 edn. Apress (2009). https://doi.org/10.1007/978-1-4302-0331-5
11. Oberhauser, R., Pogolski, C., Matic, A.: VR-BPMN: visualizing BPMN models in virtual reality. In: Shishkov, B. (ed.) BMSD 2018. LNBIP, vol. 319, pp. 83–97. Springer, Cham (2018). https://doi.org/10.1007/978-3-319-94214-8_6
12. Poppe, E., Brown, R., Johnson, D., Recker, J.: A prototype augmented reality collaborative process modelling tool. In: Proceedings of the Demo Track of the Nineth Conference on BPM, vol. 820. CEUR-WS.org (2011)
13. Proctor, M.: Drools: a rule engine for complex event processing. In: Schürr, A., Varró, D., Varró, G. (eds.) AGTIVE 2011. LNCS, vol. 7233, p. 2. Springer, Heidelberg (2012). https://doi.org/10.1007/978-3-642-34176-2_2
14. Pryss, R., Reichert, M., Bachmeier, A., Albach, J.: BPM to go: supporting business processes in a mobile and sensing world. In: BPM Everywhere, pp. 167–182. Future Strategies Inc. (2015). http://dbis.eprints.uni-ulm.de/1153/
15. Reichert, M.: Enabling flexible and robust business process automation for the agile enterprise. In: The Essence of Software Engineering, pp. 203–220. Springer, Cham (2018). https://doi.org/10.1007/978-3-319-73897-0_12
16. Schmidbauer, C., Schlund, S., Ionescu, T.B., Hader, B.: Adaptive task sharing in human-robot interaction in assembly. In: 2020 IEEE International Conference on Industrial Engineering and Engineering Management, pp. 546–550 (2020). https://doi.org/10.1109/IEEM45057.2020.9309971
17. Schönig, S., Ackermann, L., Jablonski, S., Ermer, A.: IoT meets BPM: a bidirectional communication architecture for IoT-aware process execution. Softw. Syst. Model. **19**(6), 1443–1459 (2020). https://doi.org/10.1007/s10270-020-00785-7
18. Seiger, R., Kühn, R., Korzetz, M., Aßmann, U.: HoloFlows: modelling of processes for the Internet of Things in mixed reality. Softw. Syst. Model. **20**(5), 1465–1489 (2021). https://doi.org/10.1007/s10270-020-00859-6
19. Vogel, J., Zobel, B., Jannaber, S., Thomas, O.: BPMN4SGA: a BPMN extension for smart glasses applications to enable process visualisations. In: Workshops der INFORMATIK 2018 - Architekturen, Prozesse, Sicherheit und Nachhaltigkeit, pp. 259–273. Köllen Druck+Verlag GmbH, Bonn (2018)

Supporting Users in the Continuous Evolution of Automated Routines in Their Smart Spaces

Estefanía Serral[1(✉)], Daniel Schuster[2], and Yannis Bertrand[1]

[1] LIRIS, KU Leuven, Warmoesberg 26, 1000 Brussels, Belgium
{estefania.serralasensio,yannis.bertrand}@kuleuven.be
[2] Fraunhofer Institute for Applied Information Technology FIT, Sankt Augustin, Germany
daniel.schuster@fit.fraunhofer.de

Abstract. Smart spaces' systems help users in their daily routines by automating various tasks. It can however be frustrating for users if the system does not evolve to support changes in their routines. To address this issue, we present an approach that combines two state-of-the-art approaches: *MAtE*, an end-user model-driven approach, and *Cortado*, an incremental process mining approach. *CortadoMAtE* can automatically detect changes in routines and allows the users to easily include the changes in the system step by step, or to even adapt them further before integrating them. In this way, continuous system evolution is addressed enabling the system to stay up to date with user needs.

Keywords: Process mining · Incremental process discovery · Task models · Smart spaces · Routine evolution

1 Introduction

Software evolution is necessary to ensure that software systems stay up to date in supporting changing user needs. This is essential for smart spaces, where the system continuously interacts with the home inhabitants.

We can mainly find two strategies for supporting system evolution in the literature: 1) end-user approaches, also called user-in-the-loop, provide easy-to use interfaces for the end-users to describe the necessary changes in the system; while 2) user behaviour monitoring combined with data mining techniques can automatically identify changes in user behaviour (via, e.g., process checking), changes that can then be used to make the system evolve.

End-user approaches have the advantages that they allow the user to take part in the design of the system; hence, the system typically fits them more closely. Also, the users decide themselves when a change is needed. However, it may be tedious for the users to have to identify and describe these changes

Supported by KU Leuven internal funding.

A. Marrella and B. Weber (Eds.): BPM 2021 Workshops, LNBIP 436, pp. 391–402, 2022.
https://doi.org/10.1007/978-3-030-94343-1_30

themselves [13]. Monitoring methods, on the other hand, have the advantages that they can automatically detect changes and provide them in a formalism ready to be used to update the system. However, they usually need quite some historical data on previous executed tasks to be able to identify changes, which may sometimes do not really fit the user needs.

In [11,13], a user-in-the-loop approach, called MAtE, was presented to address runtime requirements evolution. MAtE uses state-of-the-art executable task models to support 1) the automation of user routines (i.e., set of tasks that are routinely executed in the same way, also known as behaviour patterns) and 2) the evolution of the automated routines by allowing users to describe the necessary changes using an end-user tool. Task models are used to describe routines at a high level of abstraction. At runtime, these models are interpreted to execute the routines the users desire to have automated by their smart homes. At any time during the execution of the system, the user can change the task models and, thus, the system's behavior, using the end-user tool.

In this paper, we propose to combine MAtE with Cortado [9], a state-of-the-art incremental process discovery approach that automatically detects newly observed behavior and allows to incrementally incorporate it. Using Cortado-MAtE, the user has the option to step by step decide which new behaviour should be incorporated into a task model. Once the changes are approved by the user, the task models are modified accordingly to support the approved changes.

The rest of the paper is organized as follows. Section 2 presents MAtE and Cortado. Section 3 describes how these approaches are integrated in order to improve the support for user routine evolution. Section 4 presents a case study of CortadoMAtE. Section 5 describes the related work and finally Sect. 6 concludes the paper and introduces some lines for future work.

2 Background

This section describes MAtE and Cortado, the two approaches were this paper is based on, as well as the corresponding modelling formalisms that the use.

2.1 MAtE: Executable Task Models

Executable task models [11,13] specify how a smart environment system can support the routines of its users. Task models are based on Hierarchical Task Analysis (HTA) [15], which are tree structures that refine high-level tasks into executable ones. Using executable task models, every user routine is described as a hierarchy of tasks. The root task represents the routine as a whole and is broken down into subtasks, which can be composite (they are further broken down) or executable (they are leaf tasks). Two types of refinements can be used to break down a composite or root task: exclusive refinement and temporal refinement. Exclusive refinement (represented by a solid line) decomposes a task into a set of subtasks in such a way that exactly one subtask will be executed; while using a Temporal refinement (represented by a dashed line) all the subtasks shall be

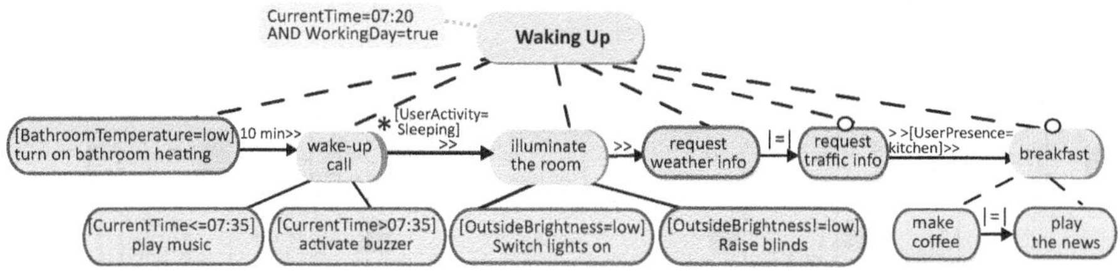

Fig. 1. Example task model for a wake up routine

performed following the order that is graphically depicted by the arrows between sibling tasks. Temporal constraints use Concurrent Task Trees (CTT) operators [8]. For example, in Fig. 1, we use:

- (T1 $\gg$ T2): task T2 is triggered when task T1 finishes.
- Task Independence (T1 $|=|$ T2): T1 and T2 can be performed in any order.

The task refinement process terminates when every leaf task in the tree is associated with a pervasive service (controlled by the smart environment system), which is capable of executing the task. The execution of a routine is context adaptive. Context can be described in a task model in the following constructs:

- Activation condition: it is associated with the root task of each routine, and indicates when the routine gets activated.
- Task precondition: it can be associated with a task to indicate that its execution depends on whether a situation holds. This condition is not expressed graphically in order to keep the diagrams easy to read.
- Iterative task: it is executed repeatedly while the situation associated with the task holds. Iterative tasks are graphically marked with an asterisk.
- Temporal constraints:
  - T1 $\gg$ [s] $\gg$ T2: after the completion of T1, T2 is started as soon as situation s holds.
  - T1 t $\gg$ T2: after the completion of T1, T2 is started as soon as the time period t has elapsed.

Once the routines are specified by a system analyst, they can directly be executed by MAtE. MAtE is an automation engine that builds on the services provided for the smart environment to control the needed devices. We consider a service as a mechanism that provides a coherent set of functionalities described in terms of atomic operations (or methods). These operations allow the system to control the devices of the environment in order to change it and/or sense it. MAtE uses these services to perform the tasks of the routines specified in the task models and to sense context changes. MAtE monitors the context and executes the routines specified in the task model when their corresponding activation condition holds. The execution of the routines is performed in a context-adaptive way according to their specification and the current context. An end-user tool

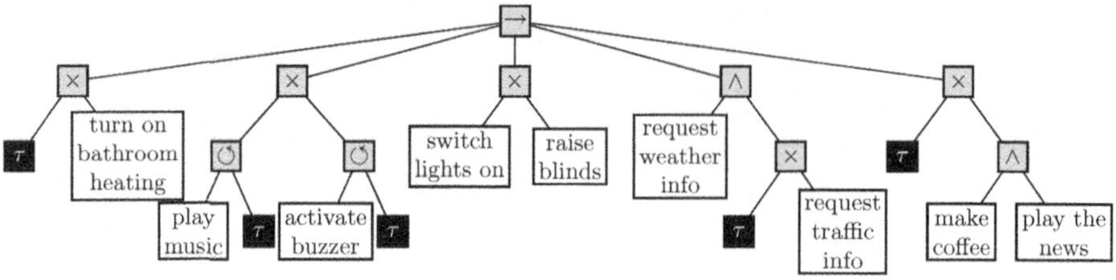

Fig. 2. Process tree modeling the control flow of the task model shown in Fig. 1

is also provided by MAtE to allow end-users to manually evolve the automated routines as they please using the available services [13]. Using this tool, any change that respects the task metamodel syntaxes can be performed to evolve the supported routines. Thus, new behaviour can be created, or those that are already specified can be modified or deleted. Finally, users can also execute the smart environment services using a web User Interface (UI) [13].

2.2 Process Trees and Incremental Process Tree Discovery

Process models allow us to model the control-flow of activities performed within the execution of a process. Process trees [9] are an important process model formalism within the area of *process mining* [1]. Figure 2 shows an example of process tree that represents the control-flow of the task model shown in Fig. 1. Leaf nodes of a process tree represent activities, which can either be visible, e.g., "turn on bathroom heating"; or invisible, τ leaf nodes used to model certain control-flow patterns. Inner nodes of a tree represent operators that specify the control-flow relation among its subtrees. We distinguish four different operators:

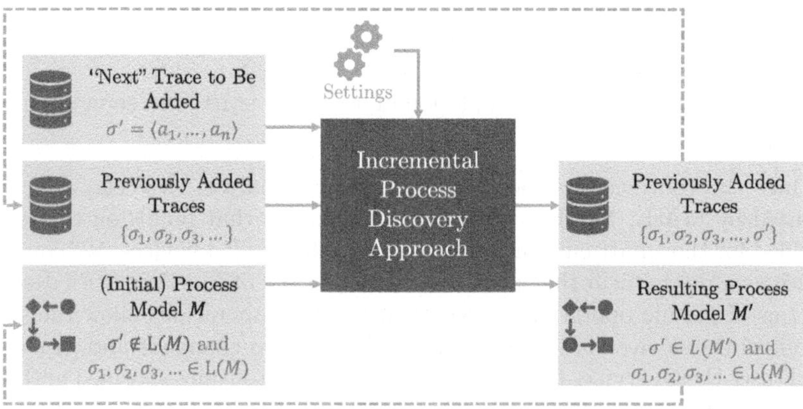

Fig. 3. Overview of the incremental process discovery algorithm [9]. Figure from [10]

1. Sequence ($\rightarrow$): subtrees must be executed in the given order
2. Exclusive-choice ($\times$): exactly one subtree must be executed
3. Loop ($\circlearrowleft$): always consists of two subtrees. The first subtree (do-part) must be executed, afterwards the second subtree (redo-part) is optionally executed. If the second subtree is executed, the first subtree must be executed again.
4. Parallel ($\wedge$): subtrees can be executed in any order and also interleaved

Compared to task models, as most process modelling formalisms used in process mining, process trees purely focus on the control-flow of activities, i.e., no contextual information is present in process trees. In Fig. 1, for instance, the task "turn on bathroom heating" is only executed if the temperature is low. However, in the corresponding process tree (Fig. 2), the activity "turn on bathroom heating" is modelled as an optional activity, i.e., either "turn on bathroom heating" or τ is executed.

Process discovery is an important sub-discipline in process mining [1] and covers techniques to learn a process model from observed process behaviour. Recently, an incremental process tree discovery approach, has been introduced [9] and implemented in a software tool [10] called *Cortado*. A conceptual overview of the approach is given in Fig. 3. Starting from an initial process tree M, Cortado allows a user to incrementally extend/learn a process tree from observed process behaviour. Cortado identifies, in an event log, the executed traces that are not yet represented in the initial process tree. These traces are then shown ordered by frequency to the user, who can select the traces that s/he wants the process tree to also support. By incrementally selecting a trace σ', which represents a single execution of the process, the approach modifies the tree M into M' such that the selected trace and the previously added traces are accepted by the

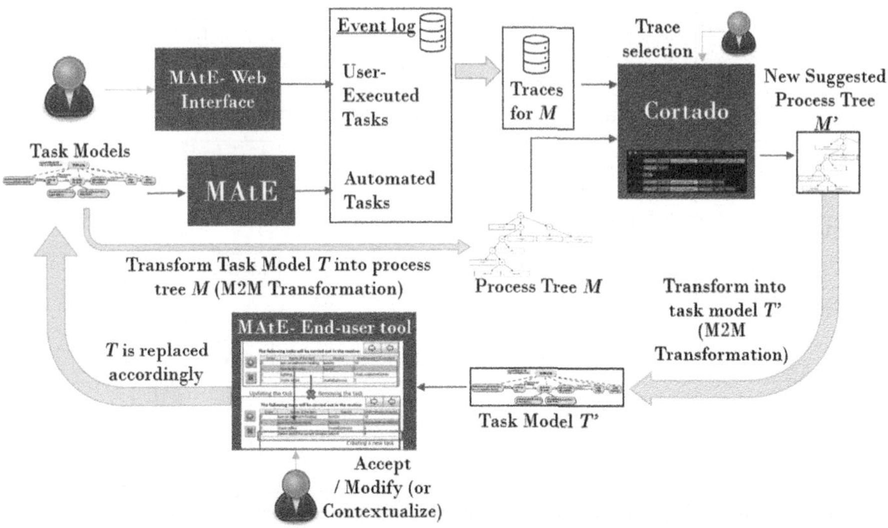

Fig. 4. Approach for continuous evolution of user automated routines

altered process tree, i.e., visualised by $\sigma' \in L(M')$ and $\sigma_1, \sigma_2, \sigma_3, \ldots \in L(M')$ in Fig. 3. The tree M' is then used as an input in the next incremental execution.

In short, the incremental approach allows a user to gradually discover/learn a process model from observed process behaviour. Moreover, a user can extend an existing process model any time to represent new observed behaviour.

3 CortadoMAtE: Addressing Evolution

To support end-users keeping the automation of their routines updated according to their needs, we integrate MAtE with Cortado (see Sect. 2). Figure 4 shows the overall approach of CortadoMAtE. As explained, Cortado needs as input: a) an (initial) process tree, and b) the new traces that are not yet supported by the process tree (*Traces for M* in Fig. 4). In order to provide the necessary inputs, MAtE and the web interface are extended to record the execution of routines (i.e., traces) in the required format. The tasks executed for a routine by MAtE are recorded with an associated *case ID*, which identifies the routine execution to which the executed task belongs. The web interface additionally records the tasks that the user executes via the web interface; these tasks do not have a predefined case ID as they do not belong to any routine yet.

In addition, we need to transform the initial task models into process trees. For this purpose, we define model to model (M2M) mappings as shown in Fig. 5.

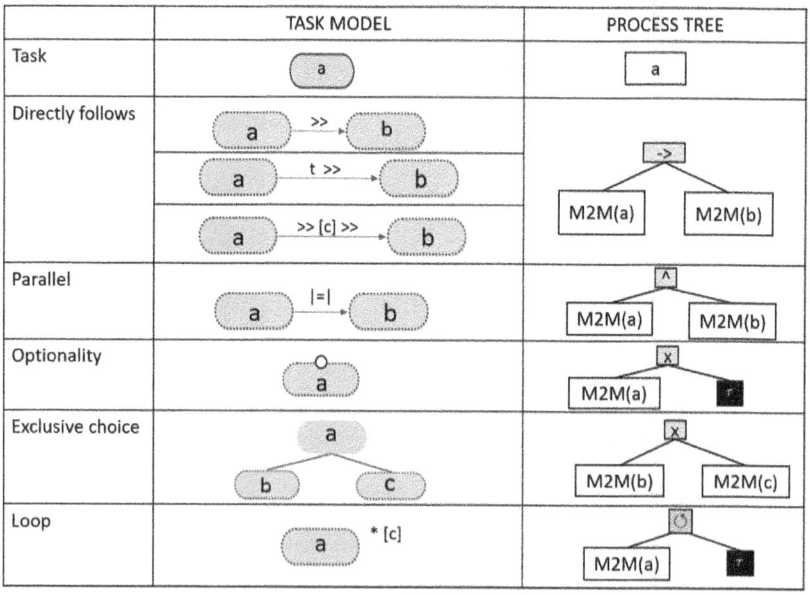

Fig. 5. Main mappings between task models and process trees. Note that the dotted tasks in the table can represent either a leaf or a composite task; also, M2M represents the recursive application of mappings for that particular task.

E.g., in the second and third row of the table, we see that the temporal constraints $\gg$ and $|=|$ in task models are mapped to $\rightarrow$ and $\land$ respectively. As can be seen, composite tasks (of task models) are directly translated into operators in the process tree, i.e., the translation of a composite task corresponds to the translation of its subtasks. As such, mappings are applied in a recursive manner. Also note that context information is omitted in the mapping, as Cortado does not yet support it (see Sect. 6). To enable the reverse mapping (from process tree to task model), context information as well as names of composite tasks are stored as labels in the corresponding process tree tasks or operators.

Taking into account these two inputs, CortadoMAtE detects changes on the current routines using the incremental process tree discovery algorithm. For each task model, we first extract the traces that belong to the execution of that model. Since usually routines are time-bounded, we consider that tasks without case ID belong to the trace that is closer in time. As explained in Sect. 2.2, the identified changes are shown to the user in the form of traces. The user can then select the traces that should also be supported. Once this is done, a new process tree (*New Suggested Process Tree M'* in Fig. 4) is obtained.

This process tree is then transformed back into a task model using the M2M mappings. The user can then select to 1) accept the new routine or 2) modify the suggested routine using the user-interface, which also allows the user to contextualize it if s/he so prefers. The final task model will then replace the old one. Since task models are interpreted at runtime, the system reflects the changes from the next execution of the routine [11]. In short, CortadoMAtE is a semi-automated evolution approach in which the user receives suggestions for changes, but still has the control over the evolution supported by the underlying incremental process discovery approach.

4 Case Study

This section presents a case study of the proposed approach. First, we introduce the experimental setup. Subsequently, we discuss the results.

4.1 Experimental Setup

Figure 6 shows an overview of the experimental setup. The overall idea is to demonstrate that our approach is able to incrementally discover a task model that represents the new observed behaviour. Therefore, we start from a target

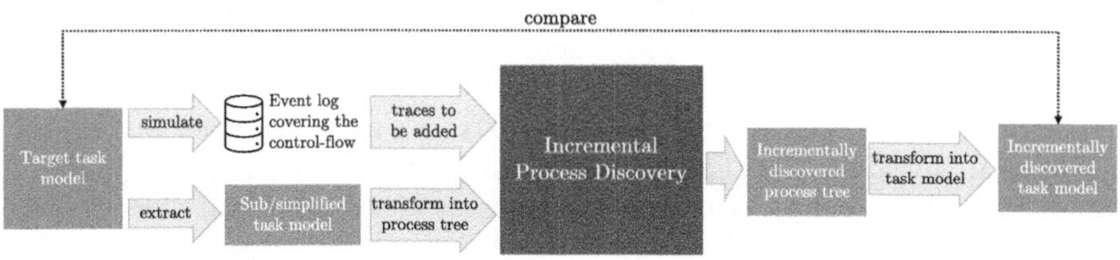

Fig. 6. Overview of the experimental setup

Table 1. Excerpt of the simulated event data from the target task model (Fig. 2). Table shows one execution of the waking up routine (i.e., a trace), and the start of a next execution of the same routine.

Control-flow information			*Context information*				
Case-ID	**Activity/ Task**	**Timestamp**	**Bathroom temperature**	**User presence**	**Outside brightness**	**User activity**	**Actor**
100	turn on bathroom heating	01-01-2000 07:20	low	bedroom	low	awake	system
100	illuminate the room	01-01-2000 07:30	medium	bedroom	low	awake	system
100	request weather info	01-01-2000 07:31	medium	bedroom	low	awake	system
100	-	01-01-2000 07:33	medium	bathroom	low	awake	user 1
100	make coffee	01-01-2000 07:52	medium	kitchen	low	awake	system
100	play the news	01-01-2000 07:52	medium	kitchen	low	awake	system
101	turn on bathroom heating	01-02-2000 07:20	low	bedroom	low	sleeping	system
...	...	...	...	...	...	...	...

task model that we would like to incrementally discover with our approach. First, the target task model is simulated to obtain event data capturing the different possible executions of the task model. Next, from the target task model, we extract a simplified task model and transform it into a process tree (see Sect. 3). This process tree serves as an initial model for the incremental process discovery approach. Now, we apply incremental process tree discovery. Thereby, we incrementally extend the simplified task model, which was transformed into a process tree, by new process behaviour from the simulated event log. Finally, we obtain an incrementally extended process tree. This process tree is converted back into a task model and compared to the target task model.

As a target task model we use the one introduced in Fig. 1. First, we simulate this target task model as stated in Fig. 6. Table 1 shows an excerpt of the event data generated. Each row represents an event. Events with the same case-ID belong to the same execution of the task model, e.g., events with case-id 100 represent the execution of the task model on 01-01-2000. Table 1 is divided into control-flow information and context information. Next, we extract the traces out of the event data, e.g., the case 100 describes the trace ⟨*turn on bathroom heating, illuminate the room, request weather information, make coffee, play the news*⟩. These traces are then incrementally fed into the incremental process tree discovery approach. In total, we simulated the target task model 1,000 times, i.e., we obtained 1,000 traces and a total amount of 7,028 events. Moreover, the simulated event data contains 345 different trace-variants. Note that multiple traces can describe the same sequence of executed tasks.

From the target task model (Fig. 1), we extract a simplified task model which serves as an initial model. Figure 8a shows the simplified task model, and Fig. 7b shows the corresponding process tree representation. Note that, for instance, the activity *illuminate the room* that we observe in the generated event data (Table 1) is not part of the initial task model (Fig. 8a).

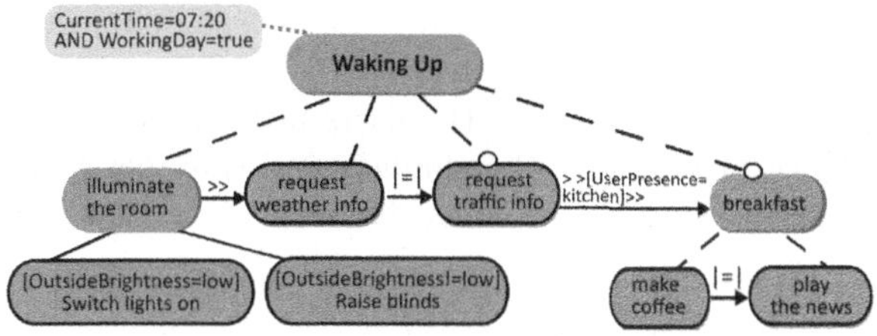

(a) Initial/simplified task model

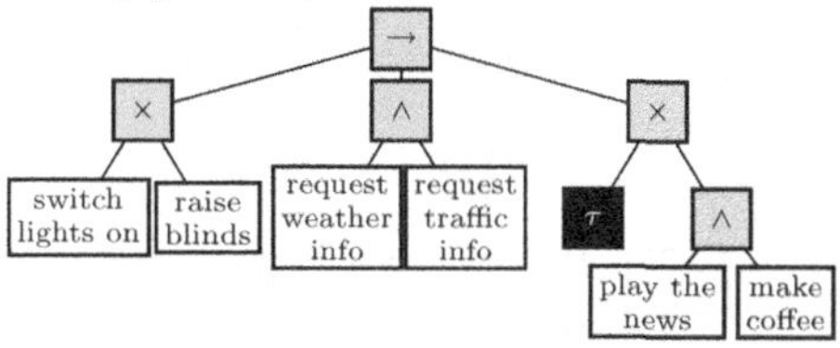

(b) Initial/simplified process tree

Fig. 7. Initial/simplified task model extracted from the target task model (Fig. 1) and corresponding process tree representation

4.2 Results

This section presents the results of the conducted experiment. We show the different obtained process trees and added traces after each incremental execution of the incremental discovery approach. Finally, we compare the resulting task model with the target task model as visualized in Fig. 6.

Figure 8 shows the intermediate trees during the incremental execution. Starting from the initial task model (Fig. 8a) that is converted into a process tree (Fig. 7b), we add the trace ⟨*switch light on*, *request weather info*, *play the news*, *make coffee*⟩, consider Fig. 8a, which is obtained from the simulated event data, as partly shown in Table 1. After adding six traces to our initial model, we obtain the process tree shown in Fig. 8d. This process tree describes the entire event data that we obtained through simulation of the target task model. The experimental results show that it is possible to extend an initially given task model by new behavior such that the resulting model represents the intended behavior.

5 State-of-the-Art in Smart Home System Evolution

There exist three types of approaches to smart space systems' evolution: User-in-the-loop, monitoring with data driven, and combined techniques.

User-in-the-loop Approaches. User-in-the-loop approaches provide easy-to-use interfaces to allow less technically-fluent users to modify their habits [6]. For instance, Kolb et al. propose techniques for end-users to define and modify process models using CTTs [5]. Similarly to MAtE, Koussaifi et al. [6] developed a framework and tool based on model-driven engineering (MDE) to allow users to modify their smart home systems. They can accept or reject models of applications, automatically generated for accepted application proposals. The main drawback of these approaches is that the users need to do the evolution process themselves: they need to identify the changes and they need to introduce these changes in the system using the provided tools.

Monitoring and Data-Driven. Monitoring and data-driven approaches use data generated by the use of the system to detect changes in user routines. Most monitoring-based models follow Machine-learning approaches, based on

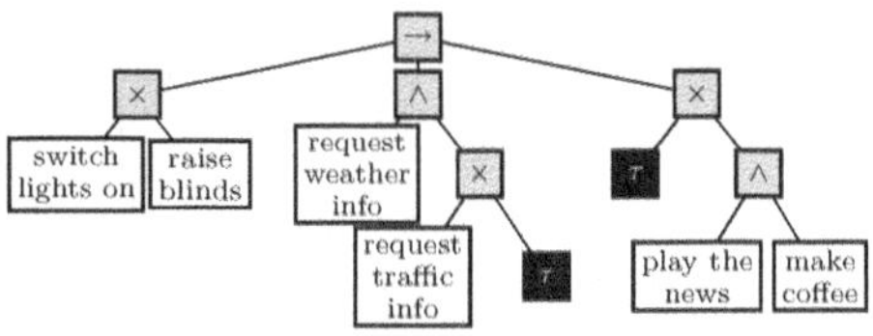

(a) Added trace ⟨*switch light on, request weather info, play the news, make coffee* ⟩

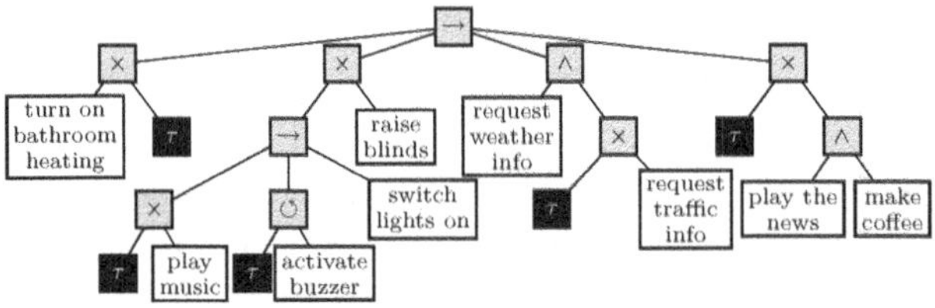

(b) Added traces: ⟨*turn on bathroom heating, raise blinds, request weather info, make coffee, play the news*⟩ and ⟨*turn on bathroom heating, play music, activate buzzer, activate buzzer, switch light on, request weather info, request traffic info, play the news, make coffee*⟩

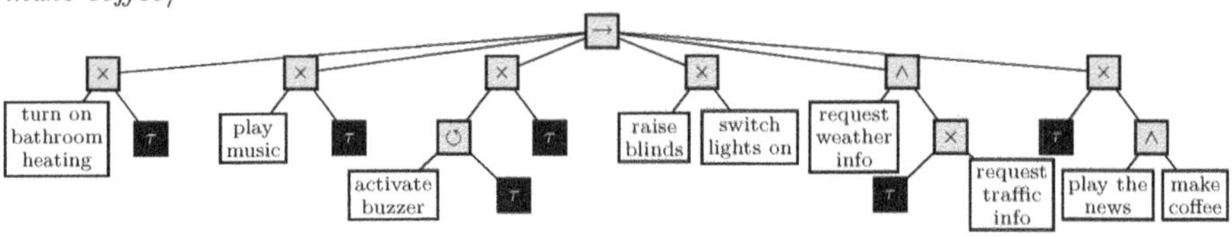

(c) Added trace ⟨*turn on bathroom heating, play music, activate buzzer, activate buzzer, switch light on, request weather info, request traffic info, play the news, make coffee*⟩

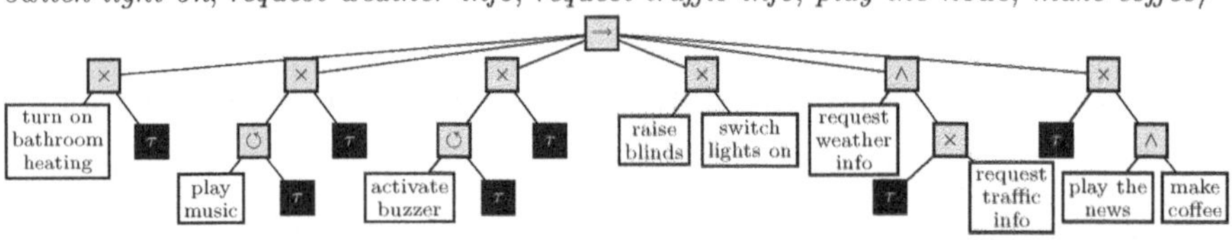

(d) Added trace ⟨*play music, play music, play music, raise blinds, request weather info, request traffic info, play the news, make coffee*⟩

Fig. 8. Incrementally discovered trees starting from the initial one (Fig. 7b)

mathematical and statistical formalisms to model the routines by detecting patterns in the collected data. These approaches have been extensively analysed by Leotta et al. [7]. E.g., MavHome and CASAS [3] use Hidden Markov Models to represent user behaviour as patterns of series of states linked by transitions with given probabilities. The models can be extended with new states to support new user behaviour. Dabrowski et al. [4] use process discovery techniques to detect deviations from the designed requirements. Main drawbacks of these techniques include: 1) they require a great amount of training data (the cold-start problem), hence it may take some time for the system to recognise a habit change, delaying the support of the user's habit by the system [14]; 2) The absence of knowledge on users' actual desires may lead to automating tasks that the user does not wish to automate [13]; and 3) Only tasks that have already been performed, and only in the way they have already been performed, can be learned [13].

Combined Approaches. Some authors propose to combine monitoring approaches with so-called knowledge-driven approaches [2,16]. In these papers, a knowledge base is first constituted to broadly describe the routines performed by the user. A data-driven model is then learned, which better fits the actual routines of the user. The resulting model is often better at recognizing activities [2] Sukor et al. [16] notes that the "cold start" effect is mitigated. However, these techniques are designed for activity recognition, not for smart home evolution.

6 Conclusion and Future Work

In this paper, we have presented CortadoMAtE, an approach that does not only support the automation of routines in smart spaces, but also facilitates the continuous evolution of the automated routines. CortadoMAtE tackles the problems of previous approaches, such as the cold-start problem. Moreover, although CortadoMAtE suggests possible changes to the users, they always have control over the routines that are finally automated; therefore, both users' desires as well as time and energy concerns are taken into account [13]. Task models can be stored, so users can easily revert to previous versions of the system if they are not satisfied with a newly implemented automation. In addition, using the end-user interface provided by MAtE, it is possible to automate tasks regardless of whether or not the users have performed them in the past, which is a big advantage of CortadoMAtE over classical monitoring approaches. Finally, CortadoMAtE also requires less input from the user than typical user-in-the-loop approaches, making the CortadoMAtE more user-friendly.

As future works, the automated discovery of the context of the model should be addressed. Currently, the user manually contextualizes the models generated by Cortado, but the incremental process mining algorithm does not integrate context information yet. However, automating this step would 1) contribute to further reducing the input required from the user and 2) potentially extract unknown context patterns from the data. Context data are already collected by the sensors present in the smart home and could be utilized. In addition, when users are proposed a new task model, it could help them to be able to simulate its execution. This would help in their decision to accept or reject the new model, and would also help them visualize the impact of manual modifications and contextualization. This simulation could be supported by using CPN Tools [12].

References

1. van der Aalst, W.M.P.: Process Mining - Data Science in Action. Springer, Heidelberg (2016). https://doi.org/10.1007/978-3-662-49851-4
2. Azkune, G., Almeida, A., López-de Ipiña, D., Chen, L.: Extending knowledge-driven activity models through data-driven learning techniques. Expert Syst. Appl. **42**(6), 3115–3128 (2015)
3. Cook, D., et al.: MavHome: an agent-based smart home. In: PerCom 2003, pp. 521–524. IEEE Computer Society (2003)
4. Dabrowski, J., Kifetew, F.M., Munante, D., Letier, E., Siena, A., Susi, A.: Discovering requirements through goal-driven process mining. In: REW, pp. 199–203. IEEE (2017)
5. Kolb, J., Reichert, M., Weber, B.: Using concurrent task trees for stakeholder-centered modeling and visualization of business processes. In: Oppl, S., Fleischmann, A. (eds.) S-BPM ONE 2012. CCIS, vol. 284, pp. 237–251. Springer, Heidelberg (2012). https://doi.org/10.1007/978-3-642-29294-1_19
6. Koussaifi, M., Arcangeli, J.P., Trouilhet, S., Bruel, J.-M.: Putting the end-user in the loop in smart ambient systems: an approach based on model-driven engineering. Technical report (2020)
7. Leotta, F., Mecella, M., Sora, D., Catarci, T.: Surveying human habit modeling and mining techniques in smart spaces. Future Internet **11**(1), 23 (2019)
8. Paternó, F.: ConcurTaskTrees: An Engineered Notation for Task Models. Lawrence Erlbaum, Mahwah (2004)
9. Schuster, D., van Zelst, S.J., van der Aalst, W.M.P.: Incremental discovery of hierarchical process models. In: Dalpiaz, F., Zdravkovic, J., Loucopoulos, P. (eds.) RCIS 2020. LNBIP, vol. 385, pp. 417–433. Springer, Cham (2020). https://doi.org/10.1007/978-3-030-50316-1_25
10. Schuster, D., van Zelst, S.J., van der Aalst, W.M.P.: Cortado—an interactive tool for data-driven process discovery and modeling. In: Buchs, D., Carmona, J. (eds.) PETRI NETS 2021. LNCS, vol. 12734, pp. 465–475. Springer, Cham (2021). https://doi.org/10.1007/978-3-030-76983-3_23
11. Serral, E., Valderas, P., Pelechano, V.: Context-adaptive coordination of pervasive services by interpreting models during runtime. Comput. J. **56**(1), 87–114 (2013)
12. Serral, E., De Smedt, J., Snoeck, M., Vanthienen, J.: Context-adaptive petri nets: supporting adaptation for the execution context. Expert Syst. Appl. **42**(23), 9307–9317 (2015)
13. Serral, E., Valderas, P., Pelechano, V.: Addressing the evolution of automated user behaviour patterns by runtime model interpretation. Softw. Syst. Model. **14**(4), 1387–1420 (2013). https://doi.org/10.1007/s10270-013-0371-3
14. Serral, E., Valderas, P., Pelechano, V.: Improving the cold-start problem in user task automation by using models at runtime. In: Pokorny, J., et al. (eds.) Information Systems Development, pp. 671–683. Springer, New York (2011). https://doi.org/10.1007/978-1-4419-9790-6_54
15. Shepherd, A.: Hierarchical Task Analysis. Taylor & Francis, Milton Park (2001)
16. Sukor, A.S.A., Zakaria, A., Rahim, N.A., Kamarudin, L.M., Setchi, R., Nishizaki, H.: A hybrid approach of knowledge-driven and data-driven reasoning for activity recognition in smart homes. J. Intell. Fuzzy Syst. **36**(5), 4177–4188 (2019)

The Benefits of Sensor-Measurement Aggregation in Discovering IoT Process Models: A Smart-House Case Study

Massimiliano de Leoni(✉) and Lucia Pellattiero

University of Padua, Padua, Italy
deleoni@math.unipd.it

Abstract. IoT systems collect and exchange data whose analysis opens up incredible opportunities to improve the human satisfaction with IoT systems. The IoT data can be indeed used to discover human habits and interaction patterns, useful to both improve human experience and further automatize the system. Process Mining can be leveraged on for this purpose, but a gap needs to be bridged between IoT-device event data and logs by aggregating events to take to the right granularity for Process Mining. This papers reports on the experience on real-life data to discover the human habits in a smart house. In particular, the benefits are reported on how to aggregate event data to the right granularity to further apply process-mining discovery techniques. The results illustrate that, when applied on the case study, the proposed technique is able to discover human-habit models that are more readable and accurate, thus providing actionable insights for a subsequent optimization of the human experience with the IoT system.

Keywords: Sensor data · Event-log abstraction · Clustering · Human habits model discovery · Smart house

1 Introduction

The Internet of Things (IoT) is the inter-networking of physical objects, which can range from sensors and actuators to software systems and devices [1]. The interest for IoT systems is nowadays gaining momentum. In general, IoT systems collect and exchange data via local networks or the Internet. Analysing these data opens up incredible opportunities to improve IoT systems from many viewpoint, e.g. considering the human satisfaction. The data harvested from the sensors can be used to discover the personal habits of the subjects who interact with the IoT system. The discovery can potentially provide valuable input to guide an improvement of the interaction of the IoT-system devices. Furthermore, discovering the main interaction patterns can also be used to reduce the needs of human intervention, thus increasing the level of automation.

Personal habits can be seen as instances of a process, namely a sequence of steps/activities/actions aiming at a certain goal (e.g., cooking food, dressing or washing). With this view in mind, Process Mining can provide the techniques to discover personal habits from sensor data. However, "*a challenge is to bridge the gap between*

A. Marrella and B. Weber (Eds.): BPM 2021 Workshops, LNBIP 436, pp. 403–415, 2022.
https://doi.org/10.1007/978-3-030-94343-1_31

clouds of sensor data and event logs for process mining." [1]. As a matter of fact, "*sensor data must be aggregated and interpreted to detect activities that can be used as input for process mining algorithms that support decision making*" [1]. Otherwise, the risk is that the results are given in terms of sensor measurements, which are too low-level to provide interpretable insights. As a matter of fact, existing approaches that apply Process Mining in IoT scenarios overlook the practical need of aggregation (see, e.g., [2–6]). Typical, they tackle the intrinsic variability by focusing on the main paths, thus potentially filtering out plenty of relevant behavior. In fact, they "sweep the actual problem under the carpet", as well as they generate low-level models that provide little insights for analysts.

This paper reports on an experience to employ process-mining techniques to discover the habit of a human subject who was living in a fully-automated house, which was equipped with sensors to monitor the subject's presence in the different rooms. The sensors produced a stream of events, each of which referred to a measurement: the presence of the subject in a room, or environmental information (e.g., the presence/absence of ventilation in room and its temperature, or power and water usage. Process Mining required a set of event traces, each of which is the recording of one instance of human behavior. The stream of events were transformed in traces, each of which containing the events of a different day. As a result, the mined model encodes the typical subject's behavior in one day.

This case study shares the typical IoT challenge indicated above: the events were sensor measurements instead of actual instances of human actions. This triggered the need to tune the granularity of the events and to abstract these low-level measurements to a higher-level event concepts. To this aim, we employed an *event-log abstraction technique* developed in the realm of process mining [7] and specialized here to the IoT case. In a nutshell, the idea is that the IoT system events are split per day, and each day corresponds to different trace. Events of the same trace can be clustered into sessions such that the time distance between the last event of a session and the first event of the subsequent session is larger than a user-defined threshold. Each trace is thus seen as a sequence of sessions of events. These sessions are encoded into data points, which are later clustered. Each cluster is given a name. The abstract event log is created such that the entire session is replaced by a sequence of two high-level events with the name of the cluster, indicating the moments when the session started and concluded.

The abstraction effort has ultimately made possible to mine a model of the human habit from an event log at the right granularity. The resulting *abstract model* is clearly more informative and accurate than the *flat model* obtained without abstracting. The abstraction has in fact enabled to only focus on the salient aspects.

The abstract model is actually the highest level of a *hierarchical model*. Each transition/activity represents an entire sub-process, whose activities are referred to by the low-level events. Recall that each abstract-model activity is associated to a cluster, which contains a multiset of session traces, i.e., a sub event-log. This enables discovering the sub-process associated to each abstract-model activity by using so-constructed sub event-logs. The hierarchical model has now the same granularity as the flat model. The experiments show the hierarchical model to be more structured and readable, while better balancing fitness (i.e., recall) and precision.

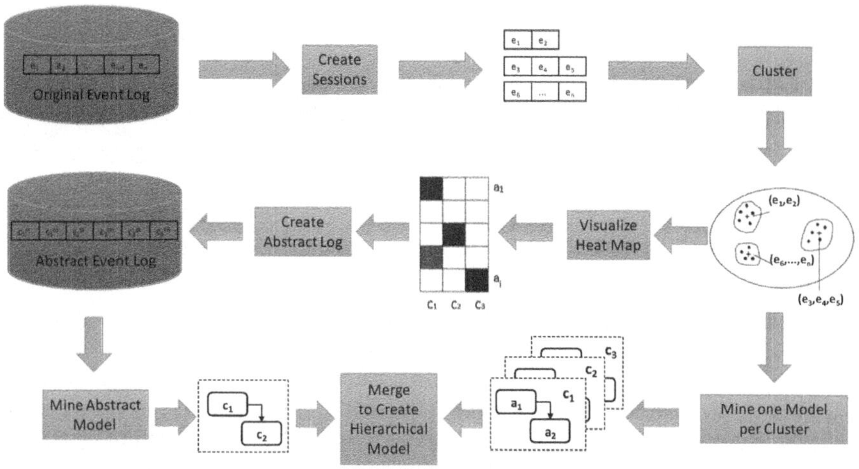

Fig. 1. The abstraction-technique framework for process discovery, extended from [7]. For compactness, we assume that each cluster c_i is given name c_i.

2 The Abstraction Technique

The starting point is an **event log**, which is composed by a set $\mathcal{E}$ of events. Events can be arranged in traces $\sigma = \langle e_1, \ldots, e_n \rangle \in \mathcal{E}^*$ that group events that refer to executions of activities within the same instance of process (e.g., every event that refers to human activities in a domotics house within a certain day). Each event e_i carries information. For the scope of this paper, any e_i is at least associated with the activity $\lambda_A(e_i)$ to which the event refers, and the timestamp $\lambda_T(e_i)$ when the event occurred. Figure 1 summarizes the idea behind the technique: the framework is composed by seven steps, which are discussed below in detail. Note that the first four steps coincide with those in [7], whereas the others are discussed in this paper as new.

Create Sessions. All the traces of the event log are split into sessions; each session of a log trace is a the smallest sub-sequences of the trace such that, for each session, the difference between the timestamp of the last event of the session and that of the first event of the session that follows is larger than a user-defined threshold. For instance, let us consider a trace $\sigma = \langle a_1, b_3, c_4, a_{10}, d_{13} \rangle$. The letter indicates the activity name associated with the event, and the subscript is the timestamp of the event's occurrence (e.g. d occurred at time 13). Assume that the time interval $\Delta = 5$. One can easily see that the time difference between the second occurrence of a and the first of e is greater than Δ: $\lambda_T(a_{10}) - \lambda_T(c_4) = 6 > \Delta = 5$). This yields two sessions: $Ⓢ_{\Delta}(\sigma) = \langle s_1, s_2 \rangle$ where $s_1 = \langle a_1, b_3, c_4 \rangle$ and $s_2 = \langle a_{10}, d_{13} \rangle$. Note that their concatenation results in σ.

Cluster. Each session s is converted into a vector via an encoding function $\text{ENCODE}(s)$ that abstract the behavior observed in the session. All these vectors (e.g. the sessions) are clustered. The encoding can be obtained in many ways. Here we consider $\text{ENCODE}(s) = (v_{x_1}, \ldots, v_{x_n})$, where $x_1, \ldots, x_n$ are the activities observed in the log and any v_{x_i} is obtained in one of the two following alternatives:

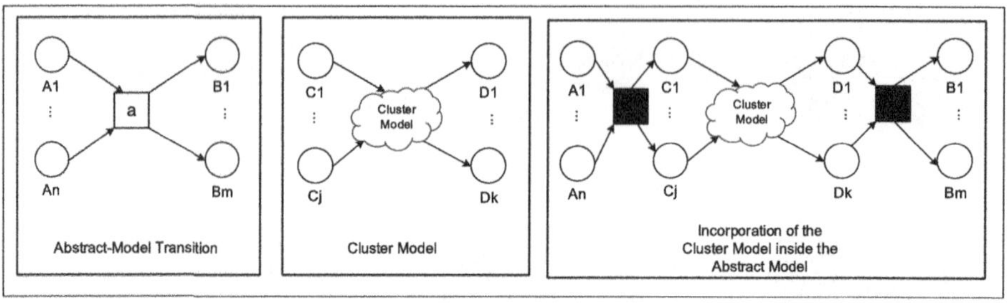

Fig. 2. Merging one cluster model into the abstract model to create the hierarchical model.

Frequency-based encoding. The value of dimension v_{x_k} is the number of events for activity x_k in s. If both starting and completion events are observed for an activity x_k, the number of events is ultimately divided by two, to prevent double counting.

Duration-based encoding. The value of dimension v_{x_k} is the average duration of instances of activity x_k in the session s (a zero value is given if x_k does not occur in s). To further clarify, let us again consider session s_1 introduced above in paragraph *Create Sessions.* We can estimated that the duration of activity a is 2, because a occurred at time 1 and b at time 3. Similarly, b lasted 1 time unit because b was at time 3 and the subsequent event, for c, was at time 4; in the example, the duration of c cannot be estimated because it is the last of the session. In [7], the average duration of c was considered among those executions when the duration was estimated.

Note that the output of the cluster phase can be abstracted as a set $C \in 2^{\mathcal{E}^*}$ of clusters where each cluster is a set of sessions (i.e. sequences). Given a cluster $c \in C$ and an encoding function, the centroid can be easily computed by encoding every session in c and compute the center.

Visualize Heat Map. Clusters are visualized on heatmap to provide analysts with some insights on how sessions were clustered. See example in Fig. 1: each column is a cluster (e.g. a human activities) and each row is a low-level activity (e.g. a sensor measured in a domotics house). The cell related to a cluster c a low-level activity a corresponds to the value of the dimension a for cluster c for the centroid of c. Note that, for comparability of the dimension values at centroid, the values are normalized. The normalization of a given centroid $(y_1, \ldots, y_n)$ is achieved by dividing by the sum of the centroid's values: $(\frac{y_1}{sum}, \ldots, \frac{y_n}{sum})$ where $sum = y_1 + \ldots + y_n$.

Create Abstract Event Log. The third step focuses on visualizing the centroids of the clusters on a heatmap, thus providing information about the most predominant activities for low-level events. This information is used by process analysts to assign cluster names $\text{NAME}(c)$ to each cluster $c \in C$. The fourth step creates the abstract event log. For each log trace σ, we compute the sessions' sequence $\circledS_{\Delta}(\sigma) = \langle s_{\sigma_1}, \ldots, s_{\sigma_m} \rangle$ and then the abstract trace $\sigma_{\text{ABST}} = \langle f^{st}_{\sigma_1}, f^{co}_{\sigma_1} \ldots, f^{st}_{\sigma_n}, f^{co}_{\sigma_n} \rangle$ where, for every $1 \leq i \leq m$, $f^{st}_{\sigma_i} and f^{co}_{\sigma_i}$ are the abstract events for session s_{σ_i}: the activity names of $\lambda_A(f^{st}_{\sigma_i})$ and $\lambda_A(f^{co}_{\sigma_i})$ are equal to the name $\text{NAME}(c_i)$ of the cluster c_i to which s_{σ_i}.

Mine Abstract Model and Mine one Model per Cluster. Here, we can use employ any discovery algorithms available in literature [8]. The step to mine one Model per Cluster is only necessary when one wants to create a hierarchical model. In particular, it aims to

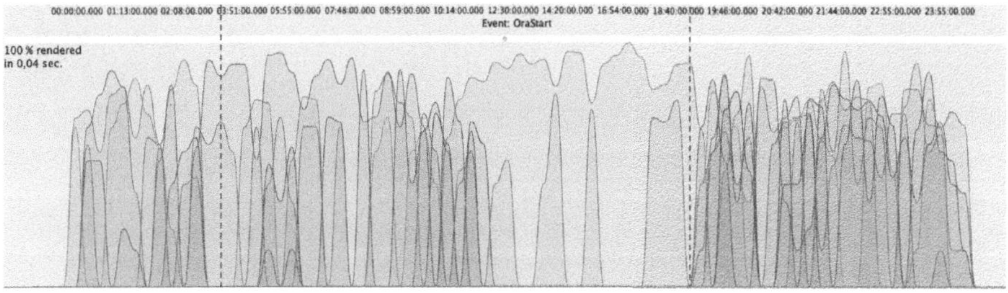

Fig. 3. Curves indicating the distribution of event occurrence over time. Each curve is associated with a different weekday and colored differently. Pink: Sunday, Yellow: Monday, Green: Tuesday, Aqua: Wednesday, Purple: Thursday, Orange: Friday, Light Blue: Saturday. (Color figure online)

mine the sub-process related to each activity of the abstract model. Each cluster $c \in C$ consists of a set of sessions, which is in fact an event log to be used as input to mine a model of the behaviour of the sessions within the cluster. Here we also assume to employ a miner that returns strongly connected Petri nets.

Merge to Create Hierarchical Model. Figure 2 illustrates how one cluster model can be incorporate into the abstract model to create the hierarchical model: the leftmost part shows the portion related to the high-level activity a where the center part depicts the structure of the model of the cluster referring to high-level activity a. The incorporation of the cluster model into the abstract model is illustrated in the rightmost part of the figure, where two invisible transitions are introduced (the black squares) in the abstract model along with the cluster model, which is connected as shown in figure. This procedure is repeated for each cluster, thus finally producing the hierarchical model. Note that often those invisible transitions are not necessary (see, e.g., the case study): in that case, the invisible transitions can finally be removed.

3 A Case Study in Domotics

This case study is based on the smart-home event data that were employed as challenge for the 4th International Workshop on BP-Meet-IoT in 2020.[1] The complete dataset includes three different scenarios: morning routine of one person, daily routine of one person and daily routine of two people. For our experiments we used the second one which includes recordings for all the activities of one user within 21 days. All the datasets provide an event stream of all the activities performed by the user inside the smart home.

3.1 Preprocessing of the Event Dataset

In order to apply our abstraction method we had to perform a preprocessing of the event data. The aims of this phase were two: to remove the activities that represent noise and to split the event stream into traces.

In the original log the noisy activities were conveniently named as "Noise", so they were removed using a filter on the activity name. Secondly, to split the stream into traces

[1] BP-Meets-Iot Challenge page: http://pros.webs.upv.es/sites/bp-meet-iot2020/#six.

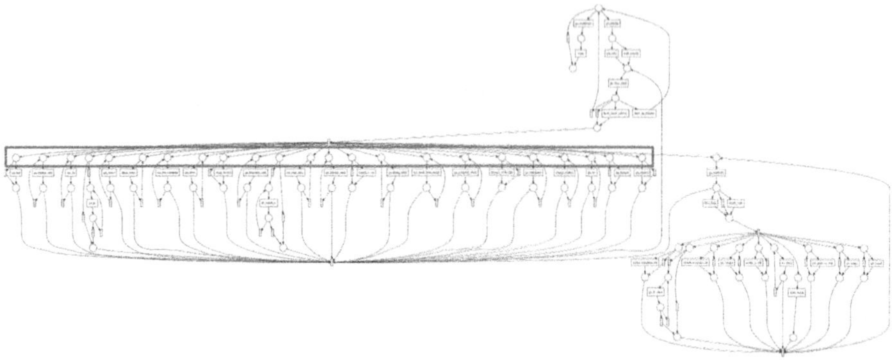

Fig. 4. *Flat model* of human habits, discovered on the smart-home transaction data via Inductive Miner. Due to the structure in the area highlighted in red, the model is very imprecise, allowing for too much behavior. (Color figure online)

we analysed the time distribution of the activities within a day. To do that we extract the hour from the timestamp and we used a dotted chart visualization. We arranged all the activities in an interval from midnight to 11:59pm. Figure 3 shows the time distribution of low-level events over day time. Each curve refers to a different weekday and is associated with a different color. Note that, e.g., between around 12:30pm and 7pm, events only occurred on Saturdays and Sundays, since the house inhabitant was likely at work on the other days. The identification of potential breaking points is based on the analysis of the areas with the lowest density of activities. We can identify two possible breaking points, highlighted by the red dashed lines in the figure: one around 3am and one around 7pm. Based on this analysis, and on the domain knowledge we had about the user habits, we have chosen to split the traces at 7pm, because this is the hour at which the user ends its working day. The examination of the log and the underlying process shows the alternative at 3am to not represent a meaningful splitting: no recognizable pattern was observed around that time.

The result event log was composed by 22 traces, with overall 2918 events for 66 low-level activities.[2]

3.2 Application of the Event-Log Abstraction Technique and Results

The resulting event log was used to discover a *flat model*. The Inductive Miner [8] was employed, using the default configuration: a noise threshold equal to 0.2. The model is depicted in Fig. 4. The model was discovered on 70% of the traces of the post-processed event log, while 30% were used for fitness computation. Conversely, precision was computed using every trace in the post-processed log. Here, we have employed a token-based computation of fitness and precision [8] because the alignment-based computation never completed on the flat model due to the its intrinsic complexity. The fitness value is 0.94, but the precision was extremely poor: 0.034. This clearly

[2] The post-processed event log is available at https://github.com/Ciaaa95/TechniquesForClustering_LowLevelEvents.

indicates that the model allows for an excessive amount of behavior, compared with what recorded in the event log. This is ultimately caused by an excessive level of parallelism: several activities are possible in any order (see in particular the area shown in red). The application of the event-abstraction method requires to provide a session threshold. In this case study, we computed the elapsing time between all couples of consecutive events in the log. Figure 5 shows the box plot of the elapsed-time distribution, where the upper whisker is equal to 1.5 times the interquartile range, namely 21 min. Note that the lower quartile is equal to 1 min, which is also the minimum value. Around 17% of values are outliers, namely larger than the upper box-plot whisker, with values till 847 min. This shows that most of measures are small, but a portion of values are very large. This motivates the reason to use the median as session threshold, which is 3 min. This enabled the application of the methods presented in Sect. 2. We employed *KMeans* and *DBScan* [9] as clustering algorithms, which are the most widespread.

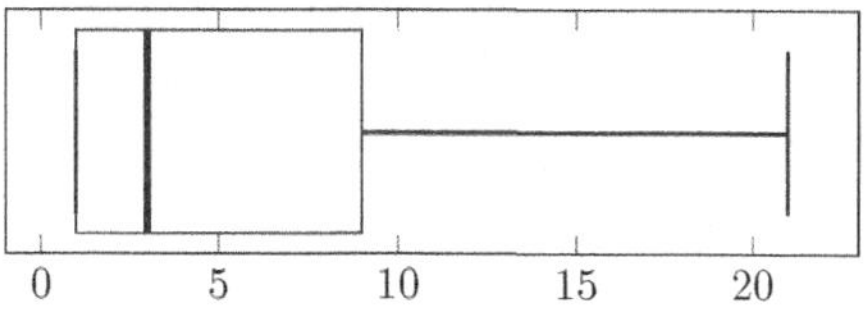

Fig. 5. Box plot of the distribution of elapsed times between consecutive events in minute. Outliers are removed for readability.

Table 1. Results of the abstraction

Encoding	Clustering	Fitness	Precision	Harmonic mean
Frequency	KMeans	0.51	0.69	0.59
Frequency	DBScan	0.56	0.64	0.60
Duration	KMeans	0.20	0.80	0.32
Duration	DBScan	0.15	0.70	0.25

Parameters Tuning for KMeans and DBScan. KMeans requires determining the number of clusters to be obtained. Here, we employed a well-known technique called *Elbow Method*, in which we compute the error curve in a given interval of number of clusters. By displaying this curve in a graphic we can then easily identify the value that offers the best tradeoff between too much generality and excessive specialization. DBScan relies on two parameters: *epsilon*, which determines the cluster dimension of the region to examine at each iteration, and *minPts*, which is the minimum number of samples that identifies an high density region. Parameter *epsilon* was estimated by computing the distance of the nearest neighbour for all the samples, then by ordering the points by distance we chose the value that ensure us to include most of the samples. Then, we computed the number of neighbour inside a region of radius *epsilon* for each sample and we ordered the results by number of neighbours. The resulting distribution presents a initial peak which means that most of the samples have very few neighbours within *epsilon* distance meaning that their surroundings have low density of points. After the peak the number decreases until it reach a local minimum. This minimum identifies

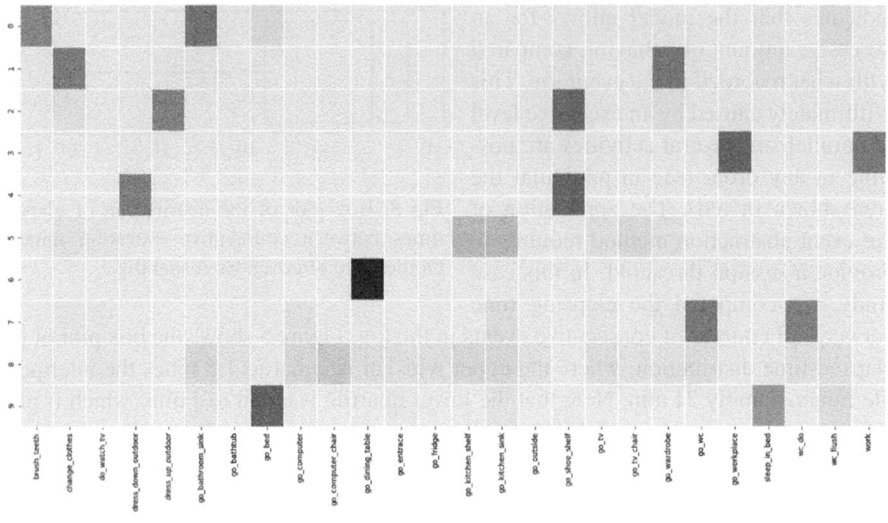

Fig. 6. Heatmap for the abstraction obtained with DBScan clustering and frequency encoding

Table 2. Comparison between domain knowledge based abstraction and our abstraction method

Cluster	Most predominant activity (ies)	Automatic cluster name	Manual cluster name	Distance
0	brush_teeth, go_bathroom_sink	Go bathroom sink0	WC	0.29
1	go_wardrobe, change_clothes	Go wardrobe1	Go work, Walk	0.50
2	dress_up_outdoor, go_shoe_shelf	Go shoe shelf2	Go work, Walk	0.50
3	go_workplace, work	Go workplace3	Go work	0.25
4	dress_down_outdoor, go_shoe_shelf	Go shoe shelf4	Go work, Walk	0.50
5	go_kitchen_sink	Go kitchen sink5	Drink	1.00
6	go_dining_table	Go dining table6	Eat cold, Eat warm	0.40
7	go_wc, wc_do	Go WC7	WC	0.57
8	go_computer_chair	Go computer chair8	Use computer	0.33
9	go_bed, sleep_in_bed	Go bed9	Prepared to sleep	0.44

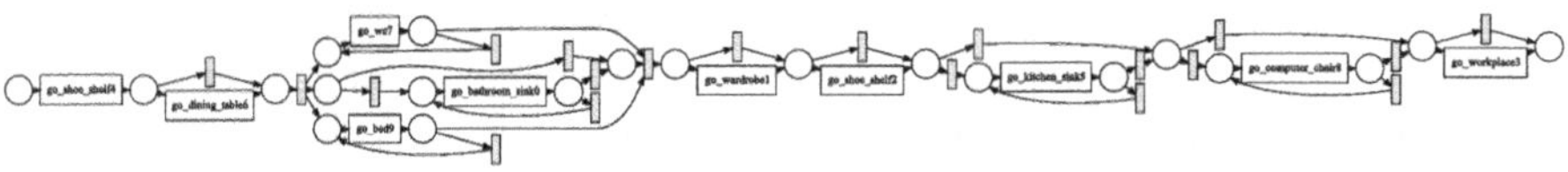

Fig. 7. Abstract model obtained using the frequency encoding and DBScan clustering, using Inductive Miner with noise threshold of 0.2

a good estimation for *minPts* because it represents the breaking point between low-density and high-density regions.

Determination of the Best Abstract Model. We evaluate the quality of the abstract models by (1) abstracting the log trace using DBScan and KMeans and via both a frequency and time encoding of the sessions, and (2) discovering a model via inductive miner, using the same configuration as for the flat model (cf. the beginning of this

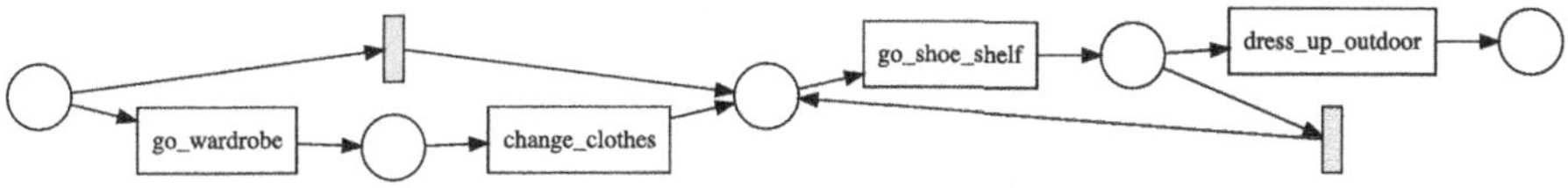

Fig. 8. Model for cluster number 2 named go_shoe_shelf

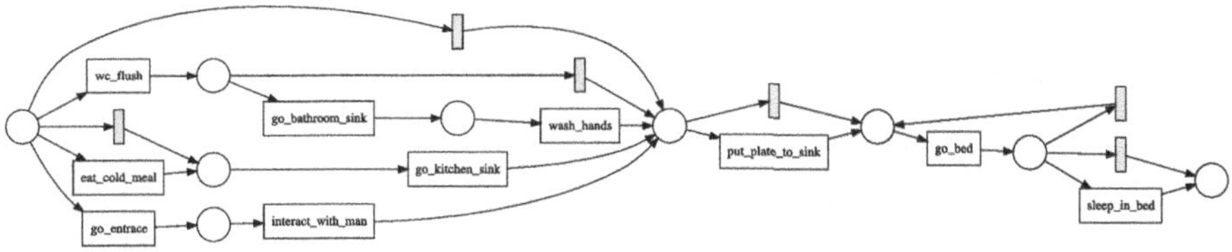

Fig. 9. Model for cluster number 9 named go_bed

Section). To evaluate the models we have calculated fitness and precision by computing alignments, which has become a "de facto" standard [8]. We used 70% of the traces as training set to mine the model and 30% to calculate the fitness, whereas for the precision we use the entire log. Table 1 summarizes the results that we have obtained for fitness and precision, highlighting their harmonic mean. We observe that frequency encoding generates better abstract models in terms of harmonic means of fitness and precision, both via KMeans and DBScan. Therefore, the used clustering algorithm is largely irrelevant. We finally opted for DBScan with the highest mean of 0.6. Figure 6 shows the heatmap obtained for this abstraction, which highlights how each cluster has one or two evident dominant low-level activities (the cells of the darkest colors). Table 2 reports on the associations between cluster number, name, and dominant activities in the three leftmost columns: the name of each cluster coincides with that of the most dominant activity in the cluster, followed by the cluster number. Figure 7 shows the abstract model.

Automatic vs Manual Cluster Creation. The dataset used in this case study is accompanied by a document that provides a manual creation of the clusters of low-level events. The manual abstraction was provided by the challenge as a separate log, in which each trace identifies a high-level activity.[3] It is hence worthwhile comparing our clusters - hereafter *automatic* - with those *manual* created via domain knowledge. For each automatic cluster, we compute the closest manual cluster. The distance between an automatic A and a manual cluster M is $2 \cdot |Att_A \cap Att_M| / |Att_A| + |Att_M|$ where Att_A and Att_M are respectively the set of activities in A and M with a relative frequency above 0.01 in the centroid (cf. definition of frequency in paragraph *Cluster* in Sect. 2).

Note that the manual clusters were already given a name by the experts, using domain knowledge. We have made the automatic cluster names correspond to the dominant activity name. The results are reported in Table 2: for each automatic cluster, the table reports the automatic cluster name (i.e. via our technique), the manual cluster

[3] The dataset description is available at http://pros.webs.upv.es/sites/bp-meet-iot2020/challenge/BP_Meets_Iot2020_Challenge_Dataset.pdf (Accessed June 2th, 2021).

Table 3. Comparison between flat model and hierarchical model based.

	Fitness	Precision	Harmonic mean
Flat model	0.91	0.034	0.065
Hierarchical model	0.66	0.35	0.45

names (i.e. via the domain knowledge) and the distance. Sometimes, two manual clusters were equally distant from the corresponding automatic (see cluster 1, 2 e 4). A qualitative comparison of the automatic and manual cluster names illustrates that our technique seems to discover high-level activities that reflect similar concepts as the manual clusters. See, e.g., cluster 0 (*Go bathroom sink* vs *WC*), cluster 3 (*Go workplace* vs *Go work*, or cluster 5 (*Go kitchen sink* vs *Drink*). Furthermore, clusters 2 e 4 have the same most prominent activity (i.e., *go_shoe_shelf*), but the second prominent changes: *dress_up_outdoor* and *dress_down_outdoor*, respectively. This means that, very likely, automatic cluster 2 should correspond to manual cluster *Go work*, and automatic cluster 5 should corresponds to *Walk*. Unfortunately, there were manual clusters, such as *Wash dishes* and *Doing exercises*, that were not matched by any automatic cluster. This is due to the fact that those manual clusters refer to low-level events that occur more rarely in the log. As a consequence, our technique has failed to group them in a standalone cluster, clustering them along with other events that were more frequent and thus predominant.

Hierarchical Model. The case-study last step was that to mine the inner workflow of the high level activities. Figures 8 and 9 shows two example models, respectively for cluster 2 (go_shoe_shelf) and cluster 9 (go_bed). These models were mined using the Inductive Miner with the same noise threshold as when we mined the abstract model.

In accordance with the technique discussed in Sect. 2, we replaced each high-level activity of the model in Fig. 7 with a whole process model that was mined for the correspond cluster. By doing that we obtain the hierarchical model shown in Fig. 10. For a quantitative evaluation, we have compared fitness and precision of this hierarchical model with the flat model in Fig. 4. To ensure a failr comparison, fitness and precision is again computed based on token replaying [8] because the alignment-based computation never completed on the flat model due to the its intrinsic complexity. The precision and fitness values are summarized in Table 3 and compared with those of the flat model: although the fitness of the hierarchical model is lower than that of the flat model, the precision is ten-times higher. This large difference is mainly due to the fact that the hierarchical model is much structured to capture the order relations between the activities. In sum, it follows that the harmonic mean of fitness and precision shows that the hierarchical is significantly better model. More qualitatively, if we compared the lasagna-like structure of the hierarchical model with a more spaghetti-like structure of the flat model, we can certainly conclude that the hierarchical model is more readable and suitable to convey insights to process analysts.

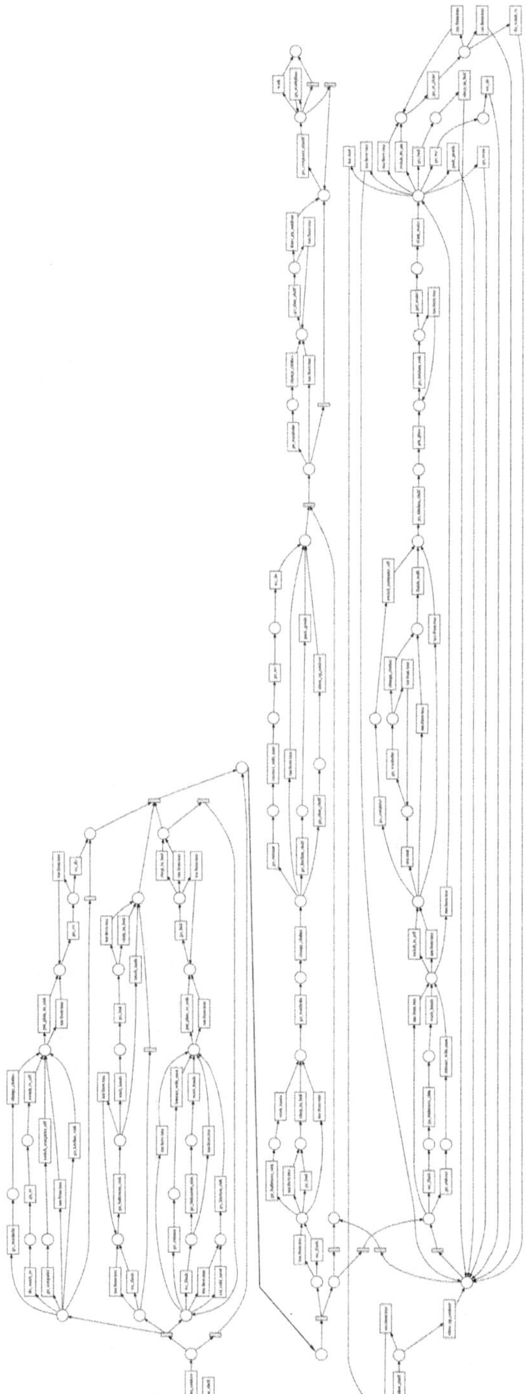

Fig. 10. Hierarchical model of the daily human habits.

4 Discussion and Conclusion

This paper has reported on the successful experience to employ event-log abstraction techniques to learn the daily human habits. Data were harvested from sensors and devices in a domotics house. Event-log abstraction is likely necessary in many IoT settings that generate low-level logging data. If these data were directly used to discover some model of the IoT system usage or human habits, the model would possibly be of little use because it would talk in terms of technical details (e.g., from sensor), instead of higher-level concepts. The abstraction enables tuning the granularity to a level that is meaningful from a conceptual/business perspective.

Even when one does not need to change the event granularity, the event-log abstraction technique leads to hierarchical models where concepts are represented at different levels. The human-habit case study has illustrated that it is possible to create hierarchical models that *(i)* can better balance fitness and precision and *(ii)* are significantly more insightful than flat models.

We acknowledge the low precision of the abstract and hierarchical model for the case study declinated in this paper. We expect similar values for those of other IoT systems. This is not caused by the abstraction techniques: in fact, the flat model scores almost zero in precision. This is likely due to the use of Petri nets, which is less suitable for processes that show high behavioral variability, in comparison with more structured processes. Future research directions ought to investigate declarative modelling approaches [10] or hybrid. The latter refers to a modelling notation that can alternate parts with a declarative notation with others using a Petri-net-like notation [11].

Every model for the case study was discovered via Inductive Miner [8], with the same configuration of parameters. However, a more thorough assessment requires the employment of different process-discovery techniques and the analysis of the quality of results, when abstraction techniques are used. We do not expect though to see significant differences.

Furthermore, we also tried to employ contextual information to provide additional information to improve the clustering accuracy. The dataset of the human-habit case study indeed contained information related to the room temperature and humidity or the level of ventilation (e.g. because of open windows). As an example, the human presence in a room might have a different meaning depending on the room temperature and whether the window is open or close. However, this additional information showed to not impact the clustering, which makes us believe that information is not relevant, at least for the case study in question.

The case study certainly hints at the relevance of the abstraction approach employed here. However, for a conclusive, definitive assessment, we plan to conduct further case studies in different IoT system settings.

References

1. Janiesch, C., et al.: The internet of things meets business process management: a manifesto. IEEE Sys. Man Cybern. Mag. **6**(4), 34–44 (2020)
2. Hemmer, A., Badonnel, R., Chrisment, I.: A process mining approach for supporting IoT predictive security. In: Proceedings of OMS IEEE/IFIP Network Operations and Management Symposium (OMS 2020) (2020)
3. Coltellese, S., Maria Maggi, F., Marrella, A., Massarelli, L., Querzoni, L.: Triage of IoT attacks through process mining. In: Panetto, H., Debruyne, C., Hepp, M., Lewis, D., Ardagna, C.A., Meersman, R. (eds.) OTM 2019. LNCS, vol. 11877, pp. 326–344. Springer, Cham (2019). https://doi.org/10.1007/978-3-030-33246-4_22
4. Dimaggio, M., Leotta, F., Mecella, M., Sora, D.: Process-based habit mining: experiments and techniques. In: 2016 IEEE 13th International Conference on Ubiquitous Intelligence & Computing, pp. 145–152 (2016)
5. Tax, N., Sidorova, N., van der Aalst, W.M.P.: Discovering more precise process models from event logs by filtering out chaotic activities. J. Intell. Inf. Syst. **52**(1), 107–139 (2018). https://doi.org/10.1007/s10844-018-0507-6
6. Cameranesi, M., Diamantini, C., Potena, D.: Discovering process models of activities of daily living from sensors. In: Teniente, E., Weidlich, M. (eds.) BPM 2017. LNBIP, vol. 308, pp. 285–297. Springer, Cham (2018). https://doi.org/10.1007/978-3-319-74030-0_21
7. de Leoni, M., Dündar, S.: Event-log abstraction using batch session identification and clustering. In: Proceedings of the 35th Annual ACM Symposium on Applied Computing, SAC '20, 36–44. Association for Computing Machinery, New York (2020)
8. van der Aalst, W.M.P.: Process Mining - Data Science in Action. Springer, Heidelberg (2016). https://doi.org/10.1007/978-3-662-49851-4
9. Mitchell, T.M.: Machine Learning. McGraw-Hill, New York (1997)
10. Pesic, M., Schonenberg, H., van der Aalst, W.M.P.: DECLARE: full support for loosely-structured processes. In: EDOC 2007, p. 287. IEEE (2007)
11. van der Aalst, W.M.P., De Masellis, R., Di Francescomarino, C., Ghidini, C.: Learning hybrid process models from events. In: Carmona, J., Engels, G., Kumar, A. (eds.) BPM 2017. LNCS, vol. 10445, pp. 59–76. Springer, Cham (2017). https://doi.org/10.1007/978-3-319-65000-5_4

The Fourth Workshop on Security and Privacy-enhanced Business Process Management (SPBP 2021)

The Fourth Workshop on Security and Privacy-enhanced Business Process Management (SPBP 2021)

Despite the growing demand for business processes that comply with security and privacy policies, security and privacy incidents caused by erroneous workflow specifications are regrettably common. This is because business process management and security and privacy are seldom addressed together, thereby hindering the development of trustworthy, privacy and security-compliant business processes. The fourth edition of the International Workshop on Security and Privacy-Enhanced Business Process Management (SPBP 2021) aimed to establish a venue to discuss business process privacy and integrity management.

The workshop was co-organized with the 5th International Workshop on Business Processes Meet the Internet-of-Things (BP-Meet-IoT 2021). Together, the workshops received nine submissions. Five of the papers were accepted, and of these, two were selected for the SPBP 2021 workshop. The paper by Schardong et al. discusses how to match metadata on blockchain for self-sovereign identities. In another paper, Brucker and Yalman present an assessment of confidentiality enhanced lifecycles. The workshop program also included an invited talk given by Agnes Koschmider. In her talk, the speaker highlighted motivation, methods, and challenges for privacy-aware process mining.

We wish to thank all those who contributed to making SPBP 2021 a success: the authors who submitted papers, the members of the Program Committee who carefully reviewed the submissions, and the speakers who presented their work at the workshop. We also express our gratitude to the BPM 2021 Workshop Chairs for their support in preparing the workshop.

November 2021

Raimundas Matulevičius
Nicolas Mayer

Organization

Workshop Chairs

Raimundas Matulevičius	University of Tartu, Estonia
Nicolas Mayer	Luxembourg Institute of Science and Technology, Luxembourg

Publicity Chair

Mubashar Iqbal	University of Tartu, Estonia

Program Committee

Abdelmalek Benzekri	Université Toulouse III - Paul Sabatier, France
Achim D. Brucker	University of Exeter, UK
Vladislav Fomin	Vilnius University, Lithuania
Virginia Franqueira	University of Kent, UK
Michael Henke	TU Dortmund and Fraunhofer IML, Germany
Julius Holderer	University of Freiburg, Germany
Mubashar Iqbal	University of Tartu, Estonia
Jan Jürjens	University of Koblenz-Landau, Germany
Christos Kalloniatis	University of the Aegean, Greece
Dimka Karastoyanova	University of Groningen, The Netherlands
Stephan Kuehnel	Martin Luther University Halle-Wittenberg, Germany
Lukas Malina	Brno University of Technology, Czech Republic
Giovanni Meroni	Politecnico di Milano, Italy
Alex Norta	Tallinn University of Technology, Estonia
Oscar Pastor	Universitat Politècnica de València, Spain
Constantinos Patsakis	University of Piraeus, Greece
Jens Myrup Pedersen	Aalborg University, Denmark
Günther Pernul	Universität Regensburg, Germany
Gerald Quirchmayr	University of Vienna, Austria
Uwe Roth	Luxembourg Institute of Science and Technology, Luxembourg
Thomas Schaberreiter	CS-Aware Corporation, Austria

Stefan Schauer	AIT Austrian Institute of Technology GmbH, Austria
Mark Strembeck	Vienna University of Economics and Business, Austria
Xiwei Xu	CSIRO, Australia

Matching Metadata on Blockchain for Self-Sovereign Identity

Frederico Schardong[1,2(✉)], Ricardo Custódio[1], Láercio Pioli[1], and João Meyer[1]

[1] Universidade Federal de Santa Catarina, Florianópolis, Brazil
ricardo.custodio@ufsc.br, {laercio.pioli,joao.meyer}@posgrad.ufsc.br
[2] Instituto Federal do Rio Grande do Sul, Campus Rolante, Rolante, Brazil
frederico.schardong@rolante.ifrs.edu.br

Abstract. Self-Sovereign Identity (SSI) is a privacy-preserving identity paradigm where users own and manage their digital identities. SSI is also referred to as blockchain identity, as it is commonly implemented using distributed ledger technologies. In this work, we describe the problem of schema matching on blockchain-based SSI implementations, systematically review the literature for tools that attack this problem, introduce a novel solution, and empirically compare it with the works reported from our literature review. Our solution uses pre-trained word vectors to find semantic similarities between user queries and schemas on the blockchain. Experimental evaluation shows that it outperforms existing solutions regarding queries that approach natural language.

Keywords: Self-Sovereign Identity · SSI · Blockchain · Schema matching · Metadata matching

1 Introduction

Privacy is perhaps the biggest concern that people have when using electronic services on the Web. Ensuring that information is securely transferred, stored, and processed by Service Providers (SPs) are complex problems that have been studied for decades by computer scientists. In this context, manipulating personal data is critical and requires even more care, as it can cause irreparable damage to their owners if unduly revealed or altered. To this end, a new identity management model named Self-Sovereign Identity (SSI) [2] was introduced based on the idea of reducing the presence of Identity Providers (IdPs) by having users storing and managing their data. This new identity model is also called blockchain identity as it is commonly implemented using distributed ledger technologies [10]. The use of blockchain provides high availability, immutability, and no centralized authoritarian control for identity.

This study was supported by the Federal Institute of Education, Science and Technology of Rio Grande do Sul (IFRS).

A. Marrella and B. Weber (Eds.): BPM 2021 Workshops, LNBIP 436, pp. 421–433, 2022.
https://doi.org/10.1007/978-3-030-94343-1_32

The objective of this work is to study the problem of matching metadata on blockchain-based SSI systems. Generally, a blockchain stores transactions describing new users joining the network, metadata[1] describing how users should organize their data to be exchanged with others, among other operations.

As a study case, we adopt Sovrin [18], one of the first SSI offerings made available to the general public and perhaps the most studied [10,11,14]. However, for the use of Sovrin, there are some challenges. The main one is related to the monetary cost. It happens that most transactions are free, but some have a monetary cost to be handled. Registering a new block describing a new schema is currently the most expensive, costing 50 USD [19]. Therefore, reusing existing schemas reduces the capital expenditure of users. Figure 1 presents the sequence of actions in a toy example where a Human Resources (HR) company joins the ledger and publicizes that it wants to receive Curriculum Vitae (CV) in a specific format. Due to the blockchain's structure, where one block is placed after another, searching if similar metadata was published in the ledger is not a trivial task. One would have to go through all the blocks and inspect their contents. Manually matching schemas is not only time-consuming but also error-prone.

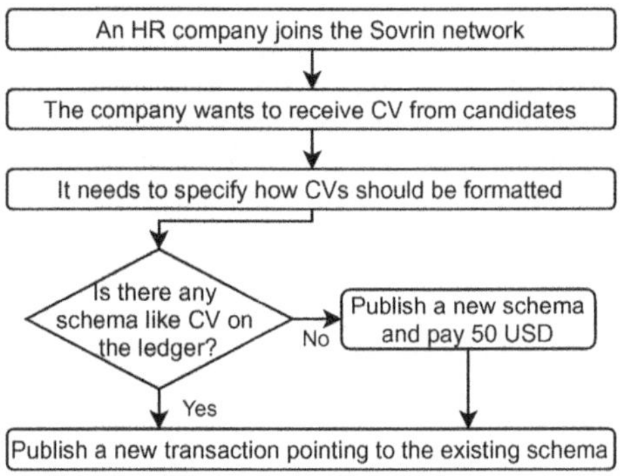

Fig. 1. Action flowchart of an HR company from joining the Sovrin network to publishing a schema that describes how candidates should format their CV.

Our contribution in this work is two-fold: (i) systematically review the blockchain-based SSI scientific literature to identify and investigate which research materials have been published that present any proposal, technique, or tool for retrieving metadata from blockchain-based identity solutions; and (ii) based on the analysis of these existing proposals, we introduced a novel tool to perform metadata matching on Sovrin that outperforms existing solutions by finding semantic similarities between user queries and schemas on the blockchain. Additionally, it can be easily expanded to other blockchain-based SSI systems. The rest of this paper is organized as follows. In Sect. 2 we provide the fundamentals of SSI needed to follow this study. Next, in Sect. 3 we detail the

[1] The terms metadata and schemas are used interchangeably in this paper.

systematic literature review, while in Sect. 4 we describe our proposed solution, and Sect. 5 presents some experiments carried out with our proposal showing that it outperforms current solutions. Finally, in Sect. 6 our final observations are made.

2 Self-Sovereign Identity

Perhaps the most naive and straightforward way to authenticate users of electronic services on the Web is to register and establish access credentials for each service. In this model, each SP has its own private identity system. Therefore, users have the arduous task of accessing a new service, registering, and defining their access credentials. In this model, the user is responsible for memorizing their access credentials for each of the services. This model is suitable when users use a few services. However, if the number of services is high, users tend to reuse their credentials, posing security challenges [7]. Likewise, the greater the number of services, the greater the user's inability to keep track of their sensitive data. In this scenario, users tend to forget which services keep their data. If data leaks, it is challenging to know which SP was responsible for the problem.

One way to minimize these difficulties, inherent to the one identity per service model, is using a specialized service, where identity management tasks, such as registration and authentication, are performed by an IdP [8]. In this model, users are no longer registered with SPs. Thus, whenever a user needs to authenticate to a service, the SP redirects to their identity provider. Once the user's IdP authenticates it, the SP is communicated that the user is whom they claim to be [16]. From a user perspective, having just one username and password for all SPs dramatically reduces the risk of incorrectly managing access credentials and improves security. However, the IdP-based model creates silos of personal information, which are high-value assets for attackers.

In practice, however, it is common to have many identity providers. Users register with one or more of these providers, those they trust. When the number of identity providers increases, it is reasonable to organize them in a federation [4]. This federation, using some "where are you from" scheme, allows the user to be associated with an IdP, which the SP accepts. So, instead of having multiple options for IdPs in service, there would be only one link to the federation.

The manipulation of private and sensitive user data by IdPs can raise privacy issues. Recently, the possibility of users keeping their identity with themselves has been discussed [10]. In this model, the user holds credentials with its attributes issued by the IdPs it is associated with. Although the user can be its own IdP, SPs might not trust self-issued credentials. Nonetheless, credentials are said to be anonymous [3], as they are not directly shared with SPs. Instead, the user presents mathematical proofs regarding the data contained in credentials, such as being older than 18 years old, along with proof that the IdP did not revoke this credential. This model is known as Self-Sovereign Identity (SSI) [2].

The advantages of SSI are manifold. We list a few: (i) users are not tracked by IdPs, because SPs do not contact IdPs to get users' data; (ii) users have total control of their data, revealing only the minimum necessary to those who wish, thus helping to avoid that their sensitive information is misused or sold; and (iii) multiple self-sovereign digital identities can be created, allowing the use of one identity for each transaction, anonymizing the owner of these identities.

Most SSI systems are implemented using distributed ledger technology (*i.e.* blockchains) because of their immutability and fault tolerance [10,14]. As data stored in a blockchain cannot be changed, credentials or other critical information are usually not saved on-chain. What goes on the ledger are revocation registries (or pointers to the registries), end-points for communication, and schemas describing what information each type of credential has [18]. The latter is the subject of this research. Effectively searching and finding registered schemas to reuse them in new credential formats results in not needing to register new schemas, which is the most expensive kind of operation on Sovrin [19].

3 Systematic Literature Review

The main idea of secondary studies is to follow a systematic method to gather evidence and answer specific research questions [9]. In this paper, we conduct a Systematic Literature Review (SLR) to discover the state of the art regarding searching schemas on blockchain-based SSI.

3.1 Research Method

The review protocol we followed has five steps [9]: (i) research questions definition; (ii) search strategy for primary studies; (iii) definition of inclusion criteria; (iv) classification of the papers; and (v) data extraction.

To accomplish the first step, we follow the suggestion of Petticrew and Roberts [15] and adopt the PICOC (Population, Intervention, Comparison, Outcome, Context) guideline to construct the research question. *Population* refers to a set of elements that we are investigating, blockchain. *Intervention* refers to the element that addresses the study, schema matching. *Comparison* seeks to compare the intervention with some element, which is not used in this study. *Outcome* describes the obtained results, including a practice point of view of them. Our study uses the terms technique, technology, strategy, method, approach, and solution. Finally, *context* is the context in which the comparison takes place, self-sovereign identity. As a consequence of the PICOC guideline, we define the research question guiding this SLR as follows.

Which techniques and technologies were used to search or match schema in blockchain-based self-sovereign identity solutions?

Next, our search strategy consists of using the terms defined in the PICOC guideline in addition to synonyms to create the search string. It is important to note that when running the search string, we noticed that the terms referring to the *outcomes* reduced considerably the number of papers returned by the search libraries. Therefore, we omitted the *outcome* terms to increase the number of papers. The final search string is presented below.

(blockchain **OR** ledger)
AND
(ontology **OR** retrieve **OR** matching **OR** similarity **OR** crosswalk **OR** mapping)
AND
(self-sovereign **OR** self-sovereignty **OR** self sovereign **OR** self-sovereignty **OR** decentralized identity **OR** decentralised identity **OR** distributed identity)

The third step is to define a set of inclusion criteria. These criteria aim to select research studies that fit the research question. In this SLR, we defined three inclusion criteria: (i) IC1: articles written in English; (ii) IC2: articles that necessarily have a title and abstract; and (iii) IC3: articles that propose a technique to search or match schemas in blockchain-based identity management solutions. Papers should meet all IC to be selected for data extraction.

The fourth step, namely the classification of papers, was not performed as any article that meets all IC has equal value in this SLR. Lastly, the data extraction step consisted of reading the selected papers and identifying: (i) what algorithm/technique/technology was used to carry out the schema matching; and (ii) if the proposed solution is an algorithm, tool, framework, or other.

3.2 Execution

We conducted our search string on ACM Digital Library, Engineering Village, IEEE Xplore, ScienceDirect, Scopus, and SpringerLink on March 28, 2021, and retrieved a total of 34 primary studies. All the libraries were configured to execute the search string considering only the metadata fields (*i.e.* title, keywords, and abstract). However, in the ACM library, we had to run the search string only on the abstract field, and in Science Direct, we searched on the entire paper. In the second stage, we identified and removed 11 duplicated articles. Next, our IC were applied. As a result, IC1 removed no paper, IC2 removed four books of proceedings. Finally, we read all the 19 papers' metadata (*i.e.* keyword, title, and abstract) and applied IC3, which discarded 18 articles.

The remaining paper was read and analysed [12]. Then, to increase the scope of our SLR, a snowballing process was carried. We reviewed its references (backward snowballing) and other articles that cited it (forward snowballing) to identify other potential primary studies. Fifty-five references were reviewed in the backward and forward snowballing. Two new works [17,21] were included in our set of final results. Figure 2 depicts the execution of our SLR protocol.

3.3 Report

With regards to the research question driving this SLR, our findings are summarized in Table 1. It lists the tools systematically discovered and what technologies

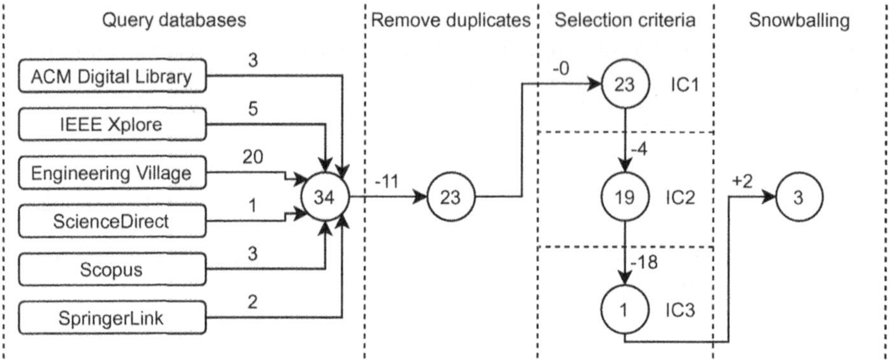

Fig. 2. Number of articles in each stage of the execution of our protocol.

were used to search for information in the blockchain. Furthermore, we included our tool in the last row for comparison and detailed it in the next section. Next, we report the three tools found in this SLR.

Table 1. Comparison of the works found in our SLR and this work.

Work	Technique	Proposal	Search technology
[12]	Full-text Search	Tool	Apache Solr
[17]	Full-text Search	Tool	Elasticsearch
[21]	Basic Search [12]	Tool	Apache Solr
This work	Semantic Search	Tool	spaCy

The authors of [12] proposed a tool to perform a full-text search on Hyperledger Indy, which is the blockchain technology underlying Sovrin. This tool performs schema matching based on textual input. It stores a local copy of the ledger on a database and integrates it with Apache Solr[2] to do the matching.

Next, the authors of [17] proposed a tool, named Indyscan, that stores the transactions on a local database and uses Elasticsearch[3] to search for information on the Sovrin ledger based on user-inputted search terms. According to [12], Indyscan uses MongoDB for storing the blockchain's transactions. Moreover, an HTTP API wrapper was developed to display the ledger transactions as HTML pages and allow users to execute the searches.

Finally, the last work found in our SLR [21] also implemented an HTTP API to display and search for transactions on the Sovrin ledger. According to [12], this application performs a "basic search". For instance, when searching for a transaction with a person's name, which is the case of the first transaction of

[2] https://solr.apache.org/.
[3] https://www.elastic.co/.

the ledger, any changed letter failed to find it. This tool also uses Apache Solr to perform the searches, and the transactions are stored in a PostgreSQL database.

4 Semantic-Based Schema Matching

In this section, we describe our proposed solution to the problem of matching schemas in blockchain-based SSI. We use as a study case the Sovrin ledger, which is the first publicly available SSI system [18,20]. The Sovrin ledger is a public-permissioned blockchain where only registered members can write to it, *i.e.*, create digital identities, and declare credential formats, but anyone can download and read them [20].

Following the design principles introduced by the authors of [12], which pointed out that sending requests to an online ledger for every system query could impact the performance of the system, we maintain a local copy of the ledger to perform schema matching. As of April of 2021, there are almost 60 thousand transactions registered in the Sovrin ledger. However, only 149 of these are transactions that register new schemas. Therefore, our prototype's persistent storage has less than 100 kilobytes, as only transactions registering schema are persisted. To ensure that the local storage is up-to-date, a pooling service was created and is executed at a constant interval.

Our approach to attack the schema matching problem consists of using a natural language processing tool named spaCy [6]. Through the usage of pre-trained word-vectors [13], semantic similarities between user-inputted terms and existing schemas in the blockchain can be calculated. For instance, a pre-trained word-vector of the English language is likely to find a high similarity between the terms *surname* and *last name*. This sort of matching enables our solution to find similar schemas with high accuracy. We used spaCy because it outperforms other offerings [1] and has detailed documentation and active development [6].

Figure 3 shows an overview of our solution. Users can query existing schemas by inputting either natural language queries or presenting schemas formatted in JSON, as represented in Fig. 3 by the queries of Alice and Bob to the Command Line Interface (CLI). Both the pooling service and CLI are built using Python 3 and embedded into Docker containers. The transactions are stored in an instance of RocksDB, a persistent key-value store optimized for low latency queries [5]. Our implementation is open-source and publicly available[4].

For every query made to the CLI, the three most similar schemas are returned by default along with their score of similarity (calculated by spaCy) and transaction number as referenced in the Sovrin ledger. Two examples are provided: (i) the output of the query "money" is shown in Table 2; and (ii) the results for the JSON-formatted query `["student", "university", "degree"]` is shown in Table 3. These two concrete examples help illustrate the usefulness of semantic search. On the results of the former query, the term `money` is only present in the first result, which is the only schema in the ledger at this point to have

[4] https://github.com/fredericoschardong/sovrin-schema-matching/.

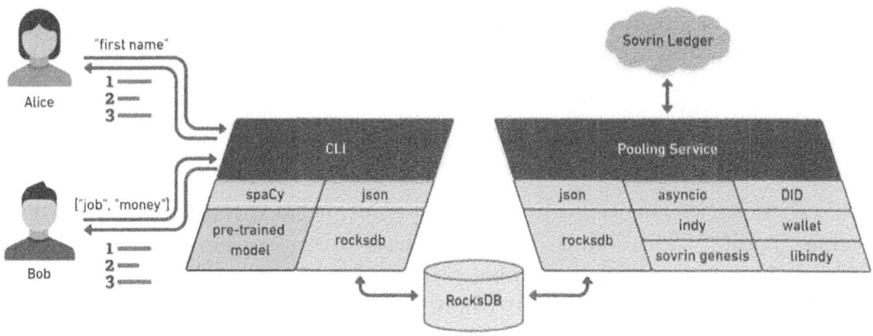

Fig. 3. An overview of the proposed solution. The CLI and the pooling service are written in Python 3, where blue parallelograms represent the modules. On the CLI, a green parallelogram shows the pre-trained model of the English language made available by spaCy. The purple parallelograms of the pooling service (DID, wallet, etc.) represent Sovrin-related components.

this term. Nonetheless, the other two schemas have related terms such as `loan amount`, `interest rate`, `credit score`, `total deposits`, and others. Regarding the results of the latter query, the first and second schemas have the same score as they are composed of the same terms but arranged in a different order. None of the returned schemas have the terms `student` or `university`, but all of them have `degree` and other fields related to education such as `institution` and `GPA`.

Table 2. Results for the query "money"

Score	Trans. #	Schema values
0.595	59556	`State, Listing Agent, Lot Number, Buyer First Name, Contract Signed Date, Purchase Price, Postal code, Buyer Last Name, Subidivision, City, Street Address, Buyer Agent, Estimated Completion date, Model Name, Earnest Money`
0.590	59555	`Application Status, Loan Number, Loan Amount, Date of Approval, First Name, Subdivision, Lot Number, Last Name, Earliest Closing Date, Loan Term in Months, APR, Interest Rate, Lender Name`
0.569	59551	`Credit Score, Account Type, Institution Name, DOB, Total Deposits, SWIFT BIC, IBAN, Last Name, First Name, Statement Period, Average Montly Balance Last 12 months, Total Withrawwals, Account Number`

Table 3. Results for the query `["student","university","degree"]`

Score	Trans. #	Schema values
0.660	54788	`degree, last_name, axuall_proof_id, institution, status, year, first_name`
0.660	54802	`first_name, institution, axuall_proof_id, last_name, degree, year, status`
0.620	33627	`DEMO-GPA, DEMO-Major, DEMO-Degree, DEMO-College Name, DEMO-Student Name`

5 Empirical Evaluation

In this section, we describe the method developed to conduct experiments and compare our tool with the solutions found on the SLR, then report the results of its execution and conduct a performance analysis.

5.1 Experiment Plan

We have defined three schema matching queries to evaluate our proposed solution and how it compares to the tools in the literature. For each query, a set of correct schemas, *i.e.* schemas that the evaluated technique should return, were manually selected by the fourth author and revised by the first author.

The first query is "address", and the 32 schemas that contain the terms `address`, `country`, `state`, `street`, `zip`, or `city` were manually selected. Next, the second query is "first name". The correct results are the 67 schemas that include `first name`, `given name`, `member name`, `full name`, or `student name`. The reasoning behind these two queries is to evaluate how our solution compares to the existing literature with regards to straightforward queries that might not gain significant advantage using a semantically aware search algorithm. On the other hand, our third query is "company job", and aims to evaluate the gains of performing semantic-aware searches. Correct results included the 23 schemas with any of those two terms or `organization`, `employer`, `department`, `role`, `contract`, or `payment`. All manually selected schemas were not chosen a priori, but rather during the selection of correct schemas on May 16, 2021.

To measure and compare the performance of our tool with the related work regarding those three queries, we have used the f-score, which is the harmonic mean of precision and recall. A confusion matrix for each query and tool was created to calculate those measurements, where the gold standard is the correct schema matching following our manual selection. Regarding our tool, only the results with a score greater than 0.6 were considered schema predictions.

5.2 Experimental Evaluation

The experimental evaluation happened on May 16, 2021, when the Sovrin ledger had 149 schemas. We ran the three queries on [17] and [21] through their publicly accessible websites and in our tool. It was not possible to empirically evaluate the work of Lux *et al.* [12] as their solution is not publicly available.

Having the correct and incorrect classification of each tool, we used `sklearn`'s `classification_report`[5] to assemble the confusion matrices, which are omitted for brevity, and to calculate the f-score. The measurements are made for the two prediction classes: correct and incorrect prediction of schemas, respectively abbreviated to `True` and `False`. Table 4 presents the f-score for each class and their average, along with the support, which is the absolute value of schemas in the `True` and `False` classes.

Table 4. The number of correct and incorrect schemas based on our manual selection (support) and the f-score of predictions.

Query	Technique	Support		f-score		
		True	False	True	False	Avg.
"address"	[12]	36	112	–	–	–
	[21]			0.65	0.92	0.79
	[17]			0.46	0.90	0.68
	This work			0.29	0.84	0.56
"first name"	[12]	67	81	–	–	–
	[21]			0.66	0.83	0.74
	[17]			0.48	0.78	0.63
	This work			0.76	0.84	0.80
"company job"	[12]	23	125	–	–	–
	[21]			0.00	0.92	0.46
	[17]			0.00	0.92	0.46
	This work			0.55	0.94	0.74

The existing tools produced better results (*i.e.* higher f-score) for the single-word query "address". Our solution returned few and incorrect predictions, which resulted in a low f-score of 0.29 for the `True` class. However, with regards to the query "first name", in which many of the correct schemas do not have the terms `first` and `name`, our tool had a precision of 0.88 and recall of 0.67, thus resulting in an f-score of 0.76 for the `True` class, surpassing the other offerings. Moreover, the advantage of our solution is evidenced in the last query, "company job", in which the other solutions were unable to predict a single schema correctly. That happened because no schema has the two terms `company` and `job`.

[5] https://scikit-learn.org/stable/modules/generated/sklearn.metrics.classification_report.html.

5.3 Performance

In addition to the production-ready ledger called *live*, which currently has 149 schemas, Sovrin also offers two ledgers for development, *builder* and *sandbox*, which currently have, respectively, 875 and 34543 schemas. We evaluated the performance of our solution on the three ledgers running different amounts of queries and measuring how long they took on commodity hardware. Figure 4 shows how long, on average, each test took. In total, 11 tests composed of $[2^0, 2^1, 2^2, ..., 2^{10}]$ unique two-term queries were performed for each ledger. Although the running time of our tool has scaled linearly with the number of schemas on a ledger, it executed 10 queries per second on *live* and 4 on *builder* but took 20 s to run one query on *sandbox*. Therefore, we conclude that it has adequate performance for the short and middle-term future. Many improvements can be made, for instance, to use multi-threading on a CPU level or to use a Graphics Processing Unit (GPU).

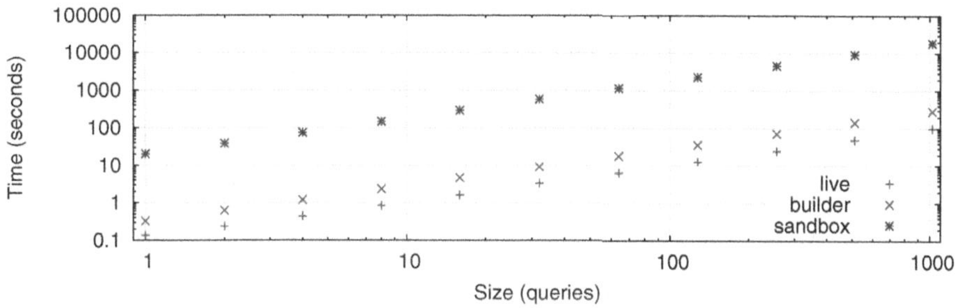

Fig. 4. The performance of our semantically-aware schema matching tool.

6 Conclusion and Future Works

In this study, we investigated the problem of searching metadata in blockchain-based SSI. An SLR was conducted to identify the research materials that have been published considering this issue. Moreover, we proposed a novel tool to find similar schemas in this context. As a study case, our tool performs searches on the Sovrin [18] ledger, a publicly available blockchain-based SSI system. Nonetheless, it can be easily expanded to other SSI systems. We have conducted three experiments to measure and compare the proposed solution with the works found in the literature and a performance analysis regarding its scalability.

The experimental results show that it outperforms the existing works in scenarios where the query approaches natural language, which benefits non-specialists, thus popularizing SSI. Both the research method adopted in our SLR and experiments can be reused in future studies. In this sense, we intend to continue this research and: (i) experiment with a more extensive set of queries; (ii) evaluate how different pre-trained word-vectors impact the result; and (iii) add pre-trained word vectors of different languages to capture schemas in languages other than English.

References

1. Al Omran, F.N.A., Treude, C.: Choosing an NLP library for analyzing software documentation: a systematic literature review and a series of experiments. In: 2017 IEEE/ACM 14th International Conference on Mining Software Repositories (MSR), pp. 187–197. IEEE (2017). https://doi.org/10.1109/MSR.2017.42
2. Allen, C.: The path to self-sovereign identity (2016). http://www.lifewithalacrity.com/2016/04/the-path-to-self-soverereign-identity.html. Accessed 24 Feb 2021
3. Camenisch, J., Lysyanskaya, A.: An efficient system for non-transferable anonymous credentials with optional anonymity revocation. In: Pfitzmann, B. (ed.) EUROCRYPT 2001. LNCS, vol. 2045, pp. 93–118. Springer, Heidelberg (2001). https://doi.org/10.1007/3-540-44987-6_7
4. Chadwick, D.W.: Federated identity management. In: Aldini, A., Barthe, G., Gorrieri, R. (eds.) FOSAD 2007-2009. LNCS, vol. 5705, pp. 96–120. Springer, Heidelberg (2009). https://doi.org/10.1007/978-3-642-03829-7_3
5. Facebook Database Engineering Team: Rocksdb (2013). https://rocksdb.org/. Accessed 06 Apr 2021
6. Honnibal, M., Montani, I., Van Landeghem, S., Boyd, A.: spaCy: industrial-strength natural language processing in python (2020). https://doi.org/10.5281/zenodo.1212303
7. Ives, B., Walsh, K.R., Schneider, H.: The domino effect of password reuse. Commun. ACM **47**(4), 75–78 (2004)
8. Jones, M.B., McIntosh, M.: Identity metasystem interoperability version 1.0. OASIS Standard (2009)
9. Keele, S., et al.: Guidelines for performing systematic literature reviews in software engineering. Technical report (2007)
10. Kuperberg, M.: Blockchain-based identity management: a survey from the enterprise and ecosystem perspective. IEEE Trans. Eng. Manag. **67**, 1–20 (2019). https://doi.org/10.1109/TEM.2019.2926471
11. Lim, S.Y., et al.: Blockchain technology the identity management and authentication service disruptor: a survey. Int. J. Adv. Sci. Eng. Inf. Technol. **8**(4–2), 1735–1745 (2018). https://doi.org/10.18517/ijaseit.8.4-2.6838
12. Lux, Z.A., Beierle, F., Zickau, S., Göndör, S.: Full-text search for verifiable credential metadata on distributed ledgers. In: 2019 Sixth International Conference on Internet of Things: Systems, Management and Security (IOTSMS), pp. 519–528. IEEE (2019). https://doi.org/10.1109/IOTSMS48152.2019.8939249
13. Mikolov, T., Chen, K., Corrado, G., Dean, J.: Efficient estimation of word representations in vector space. arXiv preprint arXiv:1301.3781 (2013)
14. Mühle, A., Grüner, A., Gayvoronskaya, T., Meinel, C.: A survey on essential components of a self-sovereign identity. Comput. Sci. Rev. **30**, 80–86 (2018). https://doi.org/10.1016/j.cosrev.2018.10.002
15. Petticrew, M., Roberts, H.: Systematic Reviews in the Social Sciences: A Practical Guide. Wiley, Hoboken (2008)
16. Sakimura, N., Bradley, J., Jones, M., De Medeiros, B., Mortimore, C.: Openid connect core 1.0. The OpenID Foundation, p. S3 (2014)

17. Staš, P.: Hyperledger indy transaction explorer (2019). https://indyscan.io/. Accessed 21 Mar 2021
18. The Sovrin Foundation: Sovrin (2016). https://sovrin.org. Accessed 24 Feb 2021
19. The Sovrin Foundation: Write To The Sovrin Public Ledger (2019). https://sovrin.org/issue-credentials/. Accessed 24 Feb 2021
20. Tobin, A.: WriteSovrin: what goes on the ledger? (2018). https://sovrin.org/wp-content/uploads/2017/04/What-Goes-On-The-Ledger.pdf. Accessed 24 Feb 2021
21. Whitehead, A.: Sovrin main net (2019). https://sovrin-mainnet-browser.vonx.io/. Accessed 21 Apr 2021

Confidentiality Enhanced Life-Cycle Assessment

Achim D. Brucker and Sakine Yalman(✉)

Department of Computer Science, University of Exeter, Exeter, UK
{a.brucker,sy359}@exeter.ac.uk
http://www.brucker.ch/
https://emps.exeter.ac.uk/computer-science/staff/sy359

Abstract. The environmental impact of products is an important factor in buying decisions of customers and it is also an increasing concern of law makers. Hence, companies are interested in determining the ecological footprint of their products. Life-cycle assessment (LCA) is a standardized method for computing the ecological footprint of a product.

Today, LCA is usually not computed in real-time and neither is LCA using actual sensor data: in contrast it is computed "offline" using "historic" values based on exemplary measurements. With the rise of the Internet of Things (IoT), LCA computations can be based on actual production processes. While an LCA based on actual sensor data is desirable from an ecological perspective, it also can reveal trade secrets, e.g., details about production processes or business relationships.

In this paper, we present an approach, using secure multi-party computation, to enable the confidential data sharing required for an LCA computation using sensor data.

Keywords: Life-cycle assessment · LCA · Confidential computation · Secure multi-party computation · SMPC

1 Introduction

It is well-known that sharing information within collaborative business processes, e.g., supply chains can, on the one hand, be very beneficial [6,12]. On the other hand, sharing data can reveal trade secrets the attack surface for cyber attacks [31]. Overall, security and confidentiality concerns are one of the main factors that prevent close collaborations within supply chains [18]. At the same time, the environmental impact of products is becoming an important factor in buying decisions of customers and is also an increasing concern of law makers [11]. Ultimately, a low ecological footprint of a product is a competitive advantage [25]. Determining the environmental impact of a product, including the impact of producing or delivering materials required for its creation, requires the sharing of data between partners of the product's supply chain.

The most common for computing the ecological footprint of a product is called life-cycle assessment (LCA) [14,19]. Today, LCA is usually not computed

A. Marrella and B. Weber (Eds.): BPM 2021 Workshops, LNBIP 436, pp. 434–446, 2022.
https://doi.org/10.1007/978-3-030-94343-1_33

in real-time and neither is LCA using actual sensor data: in contrast, it is computed "offline" using "historic" values (read from generic databases rather than individual, item specific production data) based on exemplary measurements. LCA based on actual sensor data collected during the production and transport of a specific product is desirable from an ecological perspective, as the result will be more precise and product specific (e.g., allowing for a better comparison of the ecological footprint between products). Still, due to the security risks of sharing data within a supply chain (see [18,31] for a discussion of the risk of data sharing in supply chains), companies are not willing to share the data necessary for LCA based on sensor data [5,26].

Sharing confidential data securely among partners of a supply chain is not a new challenge (see, e.g., [16,21,22,24]). Many of these solutions are based on privacy-enhancing technologies such as homomorphic encryption [2] or secure multi-party computation (SMPC) [13], which come, when applied naïvely, with a severe performance penalty and need to be adapted to each use case.

In this paper, we present an approach that allows an LCA that minimizes the amount of confidential data that needs to be shared. The approach can easily be integrated into business process engines, improving the confidentiality of data process within such systems. In more detail, our contributions are two-fold:

1. a method for computing LCA that ensures the confidentiality of the data of the participants of the supply chain that is based on a combination of decomposing the LCA computation and the use of secure multi-party computation.
2. an evaluation of our LCA method using a family of artificial examples, showing that our approach performs, on average, better than a naïve application of secure multi-party computation to LCA.

2 Life-Cycle Assessment in a Nutshell

Life-cycle assessment (LCA) [5] is an approach for assessing the ecological impact of all phases of the life-cycle of a product, starting from obtaining the raw materials to the disposal of the product ("cradle-to-grave").

Figure 1 illustrates an LCA example of an aluminum production supply chain (inspired by [1]) from mining bauxite to producing the actual aluminum ingots. The suppliers in this process include companies producing anodes (potentially using different processes, having a different ecological footprint). For an LCA, all suppliers send their *unit process data* to the main company *A*. The unit process data includes information about both the economic flows and the environmental flows within a supply chain. The economic flows describe the detailed supplier-consumer-relationship (i.e., how much Alumina Company B is ordering from Company E) and the environmental flows describe in detail flows from and into the global environment (i.e., how much SO_2 is emitted by Company B).

After receiving all unit process data: bauxite mining, alumina production, anode production (produced by two companies), aluminum electrolysis (produced by two companies), and ingot casting respectively, company A does the

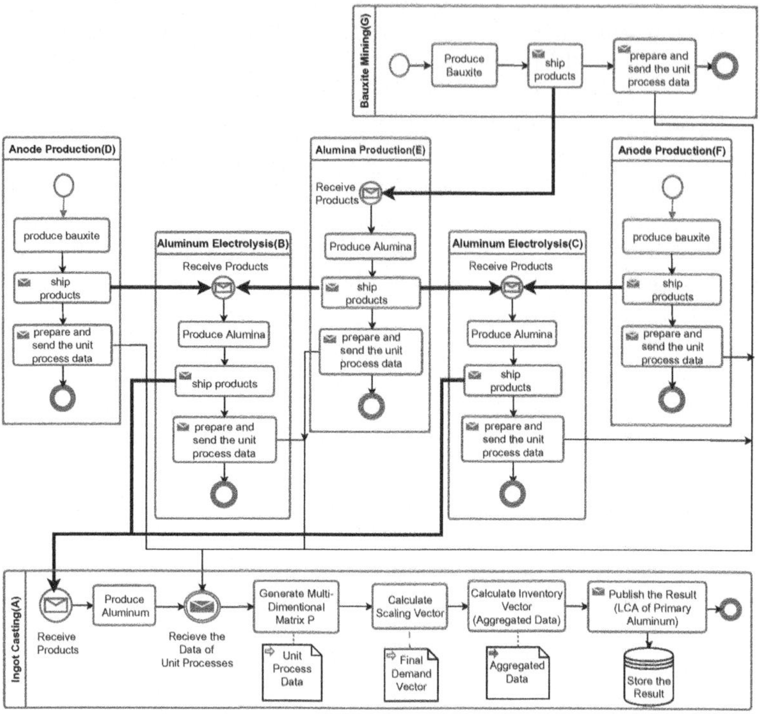

Fig. 1. The BPMN diagram of LCA of aluminum production

actual LCA computation to get the inventory vector g (the aggregated environmental flows matrix):

$$s = \mathcal{A}^{-1} \cdot f \qquad\qquad g = \mathcal{B} \cdot s \tag{1}$$

Where s is a $p \times 1$ matrix (i.e., a vector) representing scaling factors for unit processes, $\mathcal{A}$ is a $m \times p$ matrix representing the economic flows, $\mathcal{B}$ is a $n \times p$ matrix representing the environmental flows; p is the number of companies (unit processes) within the supply chain, m (n) is the number of economic (environmental) flows. Finally, f is the final demand vector, representing the quantity of produced material. To summarize, the computation effort required for an LCA is mainly determined by the number of companies and the number of both economic and environmental flows. Due to space reasons, we omit further details that are not required for understanding the rest of this paper. We refer readers interested in these details to [14,15].

3 Confidentiality Requirements in LCA

LCA can, in particular if precise sensor data from the actual production processes is used, reveal sensitive information about the production processes. Moreover,

LCA can also reveal sensitive information about the business relationships of the partners of a supply chain. In this section, we discuss exemplary attacks that we identified while threat modeling (following an approach inspired by [29]) LCA. Due to space limitations, we cannot present the threat modeling results in full detail.

3.1 Attacks on Business Relationships

Exposing business relationships between companies in a supply chain can be considered a breach of confidentiality, it can reveal secret agreements between companies. This could, e.g., have an impact on future collaborations. Also, supplier-consumer relationships can be considered confidential, as they might, e.g., reveal information about pricing or brand manufactures supplying also supermarkets with "white label" products that the supermarkets then offer as their own brand.

Attacks from members of the supply chain, i.e., an *insider attacker*, can be mounted by both suppliers and consumers. A dishonest supplier (e.g., company B in Fig. 1), might want to learn whether there are any other competitive suppliers like C that supply the same material (here: Aluminum Electrolysis) to company A. If there are other competitive companies, the dishonest supplier might also be interested in knowing the sub-suppliers of the competitive companies in the system. This information can, e.g., be used to directly negotiate with these sub-suppliers to out the direct supplier and increase profits.

Moreover, a consumer might want to learn who the sub-suppliers of its suppliers in order to switch them or make a special contract with them. In our example, consumer company A can want to know the competitive suppliers (suppliers D and F) of supplier companies B and C. Similarly, knowing that E is supplying to both B and C is a fact that E might want to keep confidential.

3.2 Attacks on Data and Process Confidentiality

LCA requires companies to provide information, e.g., the economic flows, that can give insights into their profit margins. Moreover, the knowledge of precise ecological flows might reveal information about the efficiency of production processes, which is often considered a trade secret. Therefore, most companies are reluctant to share such information [5,26]. Consequently, this information is confidential. Again, a threat model focusing on confidentiality if the exchanged information needs to consider both, dishonest suppliers and dishonest consumers.

For example, detailed environmental footprints of a material (e.g., Aluminum Electrolysis) might reveal details about the efficiency of a production unit. Therefore, this information should not be revealed to a competitor. Thus, while, in our example, company C might be willing to share this information with its customer (A), they do not want to share it with a competitor (e.g., company B). Actually, they might only want to share this information in aggregated form with A.

3.3 Deriving Security and Privacy Requirements

From our threat model that we briefly sketched in the last section, we derived security and privacy requirements for LCA. Firstly, suppliers of a consumer company should not be aware of each other. They may know how many participants and the list of environmental flows in the computation. However, they should not know who the other suppliers are, what they supply to the consumer company, how much/many materials they supply, and the value of environmental flows of other suppliers or the consumer company. Consumer companies should not learn the individual data sets of their suppliers. Also, a consumer company should not know how many materials its suppliers get from their sub-suppliers and what the materials are. Not only about productions processes but also business relationships need to be kept secret. Therefore, in a supply chain, a consumer company should not know whether its suppliers have any supplier company or who/how many they are. As an ideal system, all companies in a supply chain should be aware of the companies that they have direct relationships with, and they should just learn the information (the total result) that they would use for analyzing the market, evaluating their own profits and producing more sustainable product.

4 Improving the Confidentiality in LCA

As we have seen in the last section, LCA can reveal confidential data within a supply chain. Consequently, many companies are reluctant to share data that is required for a close supply chain collaboration (see, e.g., [16]) in general, or, in particular, LCA (see, e.g., [22]). To overcome this challenge, we are presenting a novel approach for LCA that uses secure multi-party computation (SMPC) [13] to provide increased protection of data required completing an LCA. Compared to naïvely applying SMPC to computed Eq. 1 jointly, our approach, provides additional security properties (e.g., a supplier does not need to reveal their sub-suppliers) and, in most cases, also performance improvements.

4.1 Integrating SMPC into LCA

Our approach for improving confidentiality in LCA is based on three observations: 1. SMPC requires a quadratic number of messages being sent between participants, hence, the performance of LCA using SMPC naïvely scales at best quadratic with the numbers of participants within a supply chain, 2. usually, suppliers consider their suppliers confidential, and 3. while the number of partners within a supply chain can be large, usually each company only has a relatively small number of direct suppliers. These observations led to the idea of a recursive LCA.

Recall our example from Sect. 2 (respectively, Fig. 1): seven companies (called A-G) form the supply chain for aluminum ingots. They provide the following unit processes: Bauxite Mining (G), Anode Production (D), Aluminum Electrolysis (B), Alumina Production (E), Aluminum Electrolysis (C), Anode Production (F), and

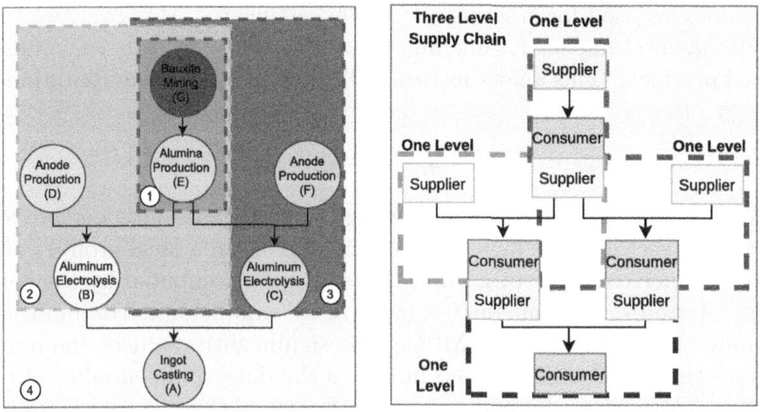

(a) Recursive Computation (b) Supply Chain Participants

Fig. 2. Hierarchical grouping of our example supply chain

Ingot Casting (A). This supply chain can naturally be divided into a hierarchy along the "direct supplier relationship" (see Fig. 2a). Each group of companies (e.g., group 2 in Fig. 2a) has one consumer (company B for group 2) and several suppliers (D and E for group 2). Within a large supply chain, an individual company can be both a supplier and a consumer. For example, company B is participating as a supplier in group 4 and as a consumer in group 2 (see Fig. 2b). The core idea of our approach is to do local LCAs for each group using SMPC. We start by establishing a secure communication infrastructure.

Creating a Secure Communication Infrastructure. For each level, we establish a public-key infrastructure (PKI) using X.509 certificates. The consumer company takes over the role of the certificate authority (CA). Suppliers can use pseudonyms when joining a PKI and, by default, we recommend using different pseudonyms when joining different supply chains or if a company participates in different levels of the same supply chain. This ensures that only the consumer company knows its lat suppliers, but a supplier cannot learn the real identity of the other suppliers within the same group. While the final setup looks like a PKI with hierarchical CAs, there is an important difference: the CAs are independent of each other, i.e., there is no common root CA for the complete supply chain.

LCA Using SMPC. To protect the confidential information that companies provide as part of an LCA, we use secure multi-party computation (SMPC). The fundamental security property of SMPC is that all participants only learn their own input into the joint computation and, if published, the final output. As SMPC requires that all partners of the joint communication exchange messages with each other, each member of a group can learn the size of the group and the pseudonyms used for creating the X.509 certificates. Moreover, we compute the environmental impact for one unit of production (e.g., 1 piece or 1 kg of the produced product). This allows us to simplify the joint LCA computation within one group i to

$$g_i = \mathcal{B}_i \cdot s_i^n \tag{2}$$

Where $\mathcal{B}_i$ is the $n \times p_i$ matrix representing the local flow and s_i^n is a $p_i \times 1$ normalized scale vector (scale factors for each company in a local group) computed by the consumer company of group i. As s_i^n can be computed by the consumer company of group i without further input of its suppliers, its computation does not require the application of SMPC. This significantly reduces the number of operations that require SMPC and improves the data confidentiality of the economic flows. In most real-world scenarios, the size of these groups will be rather small compared to the overall number of participants of the supply chain (i.e., $p_i \ll p$). Therefore, we expect the computation for one group to be significantly faster than an SMPC-based LCA for the whole supply chain. Moreover, independent groups (in our example group 2 consisting out of the companies B, D, and E and group 3 consisting out of the companies C, F, and E) can do the LCA in parallel, resulting in a further speedup.

Putting Everything Together. We assume that LCA is initiated by the main consumer party, e.g., the company producing the final product or the company recycling the final product. In our example, the main company is company A, producing aluminum ingots. The consumer company contacts "down-stream" its direct suppliers (e.g., B) and, if necessary invites them to join its local PKI. Suppliers (e.g., B) that themselves have suppliers (e.g., D and E are suppliers to B), initiate recursively an LCA for the product they deliver to their consumer (e.g., A). After B obtained the results of its local LCA, it provides this as input "up-stream" to the LCA initiated by A.

Our implementation collects data from "cradle-to-grave" (up-stream collection). Recall Fig. 2a, we first run a LCA for the group 1 (companies E and G). In this case, we cannot use SMPC, because company E has just one supplier (G). We ensure SMPC for the rest of the supply chain. Company E uses the result in up-stream computations as a supplier. We can execute the next two LCA computations (group 2 and group 3) in parallel, as they are independent of each other. The computation for group 2 is between the companies B, D, and E (using the result from the previous computation of group 1). We continue in this way until we reach the resulting group 4 with the final consumer company A.

4.2 Discussion

We now show how we mitigate the attacks from Sect. 3.1 and Sect. 3.2.

Attacking Business Relationships. Mitigation against attacks revealing business relationships within a supply chain, are mostly built on two pillars:

1. The hierarchical setup ensures that consumers can only talk to their direct suppliers, preventing them to learn (down-stream) anything about indirect suppliers. Similarly, a supplier cannot learn anything about the customers of their customers (up-stream). 2. Allowing companies to use pseudonymous handles for their X.509 certificates allows minimizes the risk that suppliers to the same consumer learn about each other. A company could further obfuscate their participation (this requires cooperation by the consumer company acting as CA), e.g., by taking over the role of multiple suppliers. By adding artificial suppliers using pseudonyms, the supplier company can, moreover, minimize the risk of revealing the number of its suppliers.

Attacks on Data and Process Confidentiality. To mitigate attacks on the confidentiality of data or internal production details, we mainly rely on the application of SMPC. In general, this gives us the guarantee that a company only learns their own input into the LCA computation and potentially the final output. Clearly, if only two parties are involved in the computation or, similarly, only a few environmental flows are considered, knowing one's own inputs and the final result might allow one to compute the input of the other party. To minimize this risk, our implementation checks that each group executing an LCA computation has at least three participants. Similar checks could be added to ensure a minimum number of economical and/or environmental flows.

5 Runtime Evaluation

To evaluate the performance of our approach, we developed a prototype using SCALE-MAMBA [4] as SMPC implementation. The remaining infrastructure, e.g., for creating the PKIs, distributing the SMPC-programs to the suppliers, and for executing the local computation of each participant is implemented in Python.

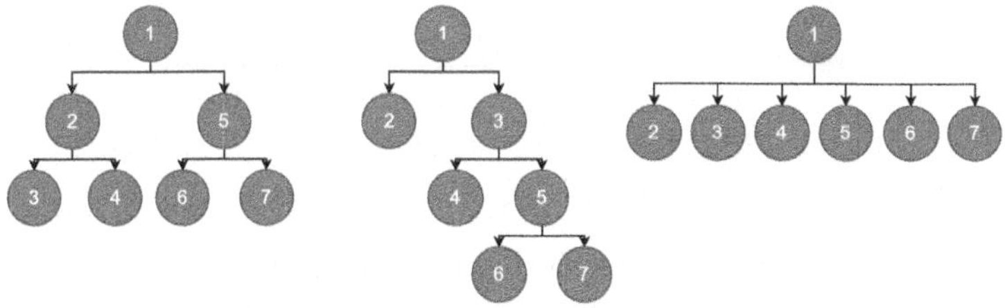

Fig. 3. Structure of the test scenarios 1–3 (From Left to Right).

We evaluated our prototype using three different scenarios (see Fig. 3):

- *Scenario 1* is a supply chain resembling a balanced binary tree, i.e., each company has two direct suppliers (i.e., each local group has the size three).

Only companies on the last level have no suppliers. The total depth of the supply chain grows logarithmic with the total number of companies. All LCAs for groups on the same level can be executed in parallel.
- *Scenario 2* simulates a supply chain resembling a linear list, i.e., each company has one supplier with no further suppliers and one company that, again, has two suppliers.
- *Scenario 3* resembles a "flat" supply chain with one consumer company and $n-1$ suppliers.

We have chosen these three scenarios, as they represent edge cases: the first scenario is the best case for our approach, as it allows for the maximum degree of parallelism for a given number of companies within a supply chain. The second scenario shows, in a certain way, the opposite behavior: all local LCA computations need to be executed sequentially. The third scenario is used to study the impact of a different number of direct suppliers on the runtime of an individual SMPC-based LCA. Overall, we expect the structure of a real-world supply chain to be a mixture of scenario 1 and 2. Of course, in a real-world supply chain, not all companies will have exactly two scenarios. Finally, for the LCA computation (recall Eq. 2) we generate the values for all required flows randomly for each computation.

We used a server with a Xeon E4-2680v4 CPU (with 28 CPU threads, i.e., 14 physical cores) and 256 GB RAM. This system powerful enough to run all our experimental scenarios in parallel, without resource conflicts. By using the local network interface for communication, we excluded any network latency impacts from our analysis.

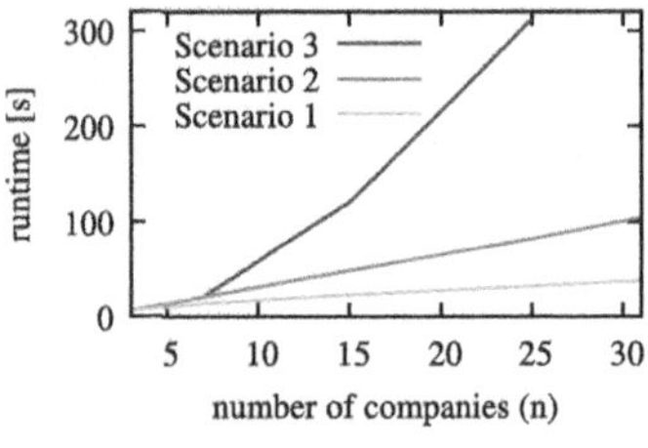

n	t_1[s]	$\frac{\delta_1}{t_1}$	t_2[s]	$\frac{\delta_2}{t_2}$	t_3[s]	$\frac{\delta_3}{t_3}$
3	7	0.06	7	0.07	7	0.06
7	14	0.03	21	0.04	21	0.02
15	23	0.03	49	0.02	121	0.04
25	-	-	83	0.01	312	0.03
31	38	0.02	105	0.01	-	-

Fig. 4. Runtimes of test scenarios

Figure 4 shows the result of executing our test scenarios for various sizes of the supply chain (n is the total number of companies in a supply chain). As scenario 1 requires $2^m - 1$ companies, we excluded the setup for 25 companies, and as running scenario 3 with 31 companies requires 31 threads being executed in parallel, we excluded this as well from our experiment. The reported mean runtimes (t_i) are the average taken over 16 executions of the same scenario. For all scenarios the quotient of the standard deviation and the mean ($\frac{\delta_i}{t_i}$) is at most 0.06, which confirms our assumptions that the values used for the actual computation do not have a significant influence on the overall runtime. In other words, the runtime for an LCA depends mostly on the structure of the supply chain:

- The runtimes for *scenario 1* grow rough logarithmic with the number of companies. This is strong evidence that executing local LCA computations in parallel works well.
- The runtimes for *scenario 2* grow roughly linearly, as one would expect for a supply chain whose lengths grow linearly.
- The runtimes for *scenario 3* grow quadratic. As this scenario models a "flat" supply chain, its runtime is mostly determined by one large SMPC over all inputs from all partners of the supply chain.

From scenario 3, we can conclude that our approach always performs at least as good as the direct application of SMPC for the LCA of a supply chain. As such a flat supply chain is a rare exception, we expect that our approach results in most cases in a significant performance gain and, at the same time, provides additional security properties.

6 Related Work

From the business domain, there are several related works, e.g., showing that concerns about the confidentiality and lack of trust are one of the main reasons preventing close collaborations in supply chains [18,31] in general and for LCA in particular [5,26]. Moreover, different approaches to LCA (e.g., [14,19]) differ in the way how the ecological impact is computed. Still, they share the same principle and need to collect the same data. As our runtime evaluation shows that the actual computation has only little impact on the runtime of our approach, we expect our approach to work equally well with the various LCA variants.

Outside of the LCA domain, there are works integrating privacy enhancing technologies (such as SMPC) in the execution of business processes (e.g., [16,21, 22,24,28]). Similar to our work, they present domain specific adaptions of these technologies to solve a particular information-sharing problem. And, there are several frameworks (e.g., [3,23,30]) for assessing security and privacy properties of business processes. For example, Anica [3] allows assessing systems with respect to their security levels while focus on privacy aspects: [23] present a privacy-aware business process modeling platform for inferring and enforcing privacy constraints and PLEAK [30] is a tool for analyzing privacy-enhanced BPMN models.

Moreover, there are plenty of works for integrating security requirements into BPMN modeling (e.g., [8–10,20,27] and some of them (e.g., [9,10]) also support the analysis of security properties, e.g., to ensure that both on the level of BPMN and on the level of the implementation, the business process-driven system complies to its security requirements. Finally, there are approaches (e.g., [7,17]) for monitoring business process executions. While these approaches work well when information should not be exchanged at all, they do not work well with privacy requirements where information is allowed to be exchanged, but only under certain conditions or certain abstractions of data are allowed to be shared.

7 Conclusion and Future Work

We presented an approach for LCA of supply chains that minimizes the data being shared between participants of the supply chain while, at the same time, improving the performance. We believe that this combination can enable closer collaborations within a supply chain in general and, in particular, enable precise and real-time LCAs for products that are necessary to determine the environmental impact of products resulting from agile supply chains.

As future work, we plan to develop a security and privacy analysis that goes beyond the rather abstract security guarantees provided by SMPC. SMPC guarantees that participants of a computation only learn their own inputs and the result of the computation ignores the information that participants of a supply chain can infer from information learned within one or several LCAs. We plan to extend our threat analysis to include such inferred information, supporting companies in their decision to join (or not join) a supply chain.

Acknowledgements. This work was partially funded by the Turkish Ministry of National Education.

Availability. Our prototype is available at https://git.logicalhacking.com/PrivacyPreservingLCA/ConfidentialLCA under an Apache license (SPDX-License-Identifier: Apache-2.0).

References

1. Life cycle assessment of aluminium: Inventory data for the primary aluminium industry (2007). http://www.world-aluminium.org/media/filer_public/2013/01/15/fl0000166.pdf
2. Acar, A., Aksu, H., Uluagac, A.S., Conti, M.: A survey on homomorphic encryption schemes. ACM Comput. Surv. **51**(4) (2018). https://doi.org/10.1145/3214303
3. Accorsi, R., Lehmann, A.: Automatic information flow analysis of business process models. In: Barros, A., Gal, A., Kindler, E. (eds.) BPM 2012. LNCS, vol. 7481, pp. 172–187. Springer, Heidelberg (2012). https://doi.org/10.1007/978-3-642-32885-5_13
4. Aly, A., et al.: SCALE-MAMBA v1.13 (2021)
5. Arbuckle, P., Kahn, E., Kriesberg, A.: Challenges to sharing data and models for life cycle assessment. ACM J. Data Inf. Qual. **9**(1) (2017). https://doi.org/10.1145/3106236
6. Asgari, N., Nikbakhsh, E., Hill, A., Farahani, R.Z.: Supply chain management 1982–2015. J. Manag. Math. **27**(3) (2016). https://doi.org/10.1093/imaman/dpw004

7. Asim, M., Yautsiukhin, A., Brucker, A.D., Baker, T., Shi, Q., Lempereur, B.: Security policy monitoring of BPMN-based service compositions. J. Softw. Evol. Proc. **30**(1) (2018). https://doi.org/10.1002/smr.1944
8. Belluccini, S., Nicola, R.D., Dumas, M., Pullonen, P., Re, B., Tiezzi, F.: Verification of privacy-enhanced collaborations. In: FormaliSE. ACM (2020). https://doi.org/10.1145/3372020.3391553
9. Brucker, A.D., Hang, I.: Secure and compliant implementation of business process-driven systems. In: La Rosa, M., Soffer, P. (eds.) BPM 2012. LNBIP, vol. 132, pp. 662–674. Springer, Heidelberg (2013). https://doi.org/10.1007/978-3-642-36285-9_66
10. Brucker, A.D., Hang, I., Lückemeyer, G., Ruparel, R.: SecureBPMN: modeling and enforcing access control requirements in business processes. In: Atluri, V., Vaidya, J., Kern, A., Kantarcioglu, M. (eds.) SACMAT. ACM (2012). https://doi.org/10.1145/2295136.2295160
11. Camilleri, M.A.: Corporate sustainability and responsibility: creating value for business, society and the environment. Asian J. Sustain. Soc. Responsib. **2**(1), 59–74 (2017). https://doi.org/10.1186/s41180-017-0016-5
12. Ciancimino, E., Cannella, S., Bruccoleri, M., Framinan, J.M.: On the bullwhip avoidance phase: the synchronised supply chain. Eur. J. Oper. Res. **221**(1) (2012). https://doi.org/10.1016/j.ejor.2012.02.039
13. Evans, D., Kolesnikov, V., Rosulek, M.: A pragmatic introduction to secure multi-party computation. Found. Trends Priv. Secur. **2**(2–3) (2018). https://doi.org/10.1561/3300000019
14. Hauschild, M.Z., Rosenbaum, R.K., Olsen, S.I. (eds.): Life Cycle Assessment. Springer, Cham (2018). https://doi.org/10.1007/978-3-319-56475-3
15. Heijungs, R., Suh, S.: The Computational Structure of Life Cycle Assessment. Eco-Efficiency in Industry and Science. Springer, Heidelberg (2002)
16. Hong, Y., Vaidya, J., Wang, S.: A survey of privacy-aware supply chain collaboration. Inf. Syst. **28**(1) (2014). https://doi.org/10.2308/isys-50692
17. Hotz, L., von Riegen, S., Pokahr, A., Braubach, L., Schwinghammer, T.: Monitoring BPMN-processes with rules in a distributed environment. In: Aït-Kaci, H., Hu, Y., Nalepa, G.J., Palmirani, M., Roman, D. (eds.) RuleML2012, CEUR, vol. 874. CEUR-WS.org (2012)
18. Huong Tran, T.T., Childerhouse, P., Deakins, E.: Supply chain information sharing: challenges and risk mitigation strategies. J. Manuf. Technol. Manag. **27**(8) (2016). https://doi.org/10.1108/JMTM-03-2016-0033
19. Ibn-Mohammed, T., et al.: Integrated hybrid life cycle assessment and supply chain environmental profile evaluations of lead-based (lead zirconate titanate) versus lead-free (potassium sodium niobate) piezoelectric ceramics. Energy Environ. Sci. **9** (2016). https://doi.org/10.1039/c6ee02429g
20. Irshad, H., Shafiq, B., Vaidya, J., Bashir, M.A., Shamail, S., Adam, N.R.: Preserving privacy in collaborative business process composition. In: Obaidat, M.S., Lorenz, P., Samarati, P. (eds.) SECRYPT. SciTePress (2015). https://doi.org/10.5220/0005567801120123
21. Kerschbaum, F.: Secure and sustainable benchmarking in clouds - a multi-party cloud application with an untrusted service provider. Bus. Inf. Syst. Eng. **3**(3) (2011). https://doi.org/10.1007/s12599-011-0153-9
22. Kuczenski, B., Sahin, C., El Abbadi, A.: Privacy-preserving aggregation in life cycle assessment. Environ. Syst. Decis. **37**(1), 13–21 (2016). https://doi.org/10.1007/s10669-016-9620-7

23. Labda, W., Mehandjiev, N., Sampaio, P.: Modeling of privacy-aware business processes in BPMN to protect personal data. In: Cho, Y., Shin, S.Y., Kim, S., Hung, C., Hong, J. (eds.) SAC. ACM (2014). https://doi.org/10.1145/2554850.2555014
24. Li, L., Zhang, H.: Confidentiality and information sharing in supply chain coordination. Manag. Sci. **54**(8) (2008). https://doi.org/10.1287/mnsc.1070.0851
25. Parida, V., Wincent, J.: Why and how to compete through sustainability: a review and outline of trends influencing firm and network-level transformation. Int. Entrep. Manag. J. **15**(1), 1–19 (2019). https://doi.org/10.1007/s11365-019-00558-9
26. Sahin, C., Kuczenski, B., Egecioglu, Ö., Abbadi, A.E.: Towards practical privacy-preserving life cycle assessment computations. In: Ahn, G., Pretschner, A., Ghinita, G. (eds.) CODAS. ACM (2017). https://doi.org/10.1145/3029806.3029845
27. Salnitri, M., Dalpiaz, F., Giorgini, P.: Designing secure business processes with SecBPMN. Softw. Syst. Model. **16**(3), 737–757 (2015). https://doi.org/10.1007/s10270-015-0499-4
28. Schwarzbach, B., Glöckner, M., Makarov, S., Franczyk, B., Ludwig, A.: Privacy preserving BPMS for collaborative BPaaS. In: Ganzha, M., Maciaszek, L.A., Paprzycki, M. (eds.) FedCSIS, ACSIS, vol. 11 (2017). https://doi.org/10.15439/2017F330
29. Shostack, A.: Threat Modeling: Designing for Security. Wiley, Hoboken (2014)
30. Toots, A., et al.: Business process privacy analysis in pleak. In: Hähnle, R., van der Aalst, W. (eds.) FASE 2019. LNCS, vol. 11424, pp. 306–312. Springer, Cham (2019). https://doi.org/10.1007/978-3-030-16722-6_18
31. Tran, T.T.H., Childerhouse, P., Deakins, E.: Supply chain information sharing: challenges and risk mitigation strategies. J. Manuf. Technol. Manag. **27**(8) (2016). https://doi.org/10.1108/JMTM-03-2016-0033

Author Index

The manufacturer's authorised representative in the EU is Springer Nature Customer Service Centre GmbH, Europaplatz 3, 69115 Heidelberg, Germany. If you have any concerns regarding our products, please contact ProductSafety@springernature.com

Printed and bound by CPI Group (UK) Ltd, Croydon, CR0 4YY
30/09/2026
02225312-0007